Prevention and Management of Soil Erosion and Torrential Floods

Slobodan Milutinović
University of Niš, Serbia

Snežana Živković
Faculty of Occupational Safety, University of Niš, Serbia

IGI Global
PUBLISHER of TIMELY KNOWLEDGE

A volume in the Advances in Environmental Engineering and Green Technologies (AEEGT) Book Series

Published in the United States of America by
IGI Global
Engineering Science Reference (an imprint of IGI Global)
701 E. Chocolate Avenue
Hershey PA, USA 17033
Tel: 717-533-8845
Fax: 717-533-8661
E-mail: cust@igi-global.com
Web site: http://www.igi-global.com

Library of Congress Cataloging-in-Publication Data

Names: Milutinović, Slobodan, 1960-editor. | Živković, Snežana, 1971- editor.
Title: Prevention and management of soil erosion and torrential floods / Slobodan Milutinović and Snežana Živković, editor.
Description: Hershey, PA : Engineering Science Reference, an imprint of IGI Global, [2022] | Includes bibliographical references and index. | Summary: "Chapter contributors to this publication discuss about the importance of the prevention and management practices of soil erosion and torrential floods, with respect to the exchange of knowledge and best practices with the aim to add value in sharing knowledge among different stakeholders to provide answers to the ongoing concerns in the changed climate conditions about the soil erosion and torrential floods prevention"-- Provided by publisher.
Identifiers: LCCN 2021037991 (print) | LCCN 2021037992 (ebook) | ISBN 9781799884590 (h/c) | ISBN 9781799884606 (s/c) | ISBN 9781799884613 (ebook)
Subjects: LCSH: Flood control. | Soil conservation. | Soil erosion.
Classification: LCC TC530 .P73 2022 (print) | LCC TC530 (ebook) | DDC 627/.4--dc23/eng/20211014
LC record available at https://lccn.loc.gov/2021037991
LC ebook record available at https://lccn.loc.gov/2021037992

This book is published in the IGI Global book series Advances in Environmental Engineering and Green Technologies (AEEGT) (ISSN: 2326-9162; eISSN: 2326-9170)

British Cataloguing in Publication Data
A Cataloguing in Publication record for this book is available from the British Library.

For electronic access to this publication, please contact: eresources@igi-global.com.

Advances in Environmental Engineering and Green Technologies (AEEGT) Book Series

Sang-Bing Tsai
Zhongshan Institute, University of Electronic Science and Technology of China, China & Wuyi University, China
Ming-Lang Tseng
Lunghwa University of Science and Technology, Taiwan
Yuchi Wang
University of Electronic Science and Technology of China Zhongshan Institute, China

ISSN:2326-9162
EISSN:2326-9170

Mission

Growing awareness and an increased focus on environmental issues such as climate change, energy use, and loss of non-renewable resources have brought about a greater need for research that provides potential solutions to these problems. Research in environmental science and engineering continues to play a vital role in uncovering new opportunities for a "green" future.

The **Advances in Environmental Engineering and Green Technologies (AEEGT)** book series is a mouthpiece for research in all aspects of environmental science, earth science, and green initiatives. This series supports the ongoing research in this field through publishing books that discuss topics within environmental engineering or that deal with the interdisciplinary field of green technologies.

Coverage

- Biofilters and Biofiltration
- Green Technology
- Waste Management
- Contaminated Site Remediation
- Renewable Energy
- Radioactive Waste Treatment
- Electric Vehicles
- Alternative Power Sources
- Air Quality
- Green Transportation

IGI Global is currently accepting manuscripts for publication within this series. To submit a proposal for a volume in this series, please contact our Acquisition Editors at Acquisitions@igi-global.com or visit: http://www.igi-global.com/publish/.

The Advances in Environmental Engineering and Green Technologies (AEEGT) Book Series (ISSN 2326-9162) is published by IGI Global, 701 E. Chocolate Avenue, Hershey, PA 17033-1240, USA, www.igi-global.com. This series is composed of titles available for purchase individually; each title is edited to be contextually exclusive from any other title within the series. For pricing and ordering information please visit http://www.igi-global.com/book-series/advances-environmental-engineering-green-technologies/73679. Postmaster: Send all address changes to above address.

Titles in this Series

For a list of additional titles in this series, please visit: www.igi-global.com/book-series/advances-environmental-engineering-green-technologies/73679

Modeling and Control of Static Converters for Hybrid Storage Systems
Arezki Fekik (Akli Mohand Oulhadj University, Bouira, Algeria) and Nacereddine Benamrouche (University of Tizi Ouzou, Algeria)
Engineering Science Reference • © 2022 • 355pp • H/C (ISBN: 9781799874478) • US $225.00

Disruptive Technologies and Eco-Innovation for Sustainable Development
Ulas Akkucuk (Boğaziçi University, Turkey)
Engineering Science Reference • © 2022 • 340pp • H/C (ISBN: 9781799889007) • US $195.00

Handbook of Research on Microbial Remediation and Microbial Biotechnology for Sustainable Soil
Junaid Ahmad Malik (Government Degree College, Bijbehara, India)
Engineering Science Reference • © 2021 • 806pp • H/C (ISBN: 9781799870623) • US $295.00

Handbook of Research on Strategic Management for Current Energy Investments
Serhat Yüksel (Istanbul Medipol University, Turkey) and Hasan Dinçer (Istanbul Medipol University, Turkey)
Engineering Science Reference • © 2021 • 422pp • H/C (ISBN: 9781799883357) • US $295.00

Forest Fire Danger Prediction Using Deterministic-Probabilistic Approach
Nikolay Viktorovich Baranovskiy (National Research Tomsk Polytechnic University, Russia)
Engineering Science Reference • © 2021 • 297pp • H/C (ISBN: 9781799872504) • US $195.00

Role of IoT in Green Energy Systems
Vasaki Ponnusamy (Universiti Tunku Abdul Rahman, Malaysia) Noor Zaman (Taylor's University, Malaysia) Low Tang Jung (Universiti Teknologi PETRONAS, Malaysia) and Anang Hudaya Muhamad Amin (Higher Colleges of Technology, UAE)
Engineering Science Reference • © 2021 • 405pp • H/C (ISBN: 9781799867098) • US $195.00

Combating and Controlling Nagana and Tick-Borne Diseases in Livestock
Caleb Oburu Orenge (Department of Veterinary Anatomy and Physiology, Faculty of Veterinary Medicine and Surgery, Egerton University, Kenya)
Engineering Science Reference • © 2021 • 440pp • H/C (ISBN: 9781799864332) • US $185.00

Artificial Intelligence and IoT-Based Technologies for Sustainable Farming and Smart Agriculture
Pradeep Tomar (Gautam Buddha University, India) and Gurjit Kaur (Delhi Technological University, India)

IGI Global
PUBLISHER of TIMELY KNOWLEDGE

701 East Chocolate Avenue, Hershey, PA 17033, USA
Tel: 717-533-8845 x100 • Fax: 717-533-8661
E-Mail: cust@igi-global.com • www.igi-global.com

Table of Contents

Detailed Table of Contents

The EU countries are obliged to harmonize their legislation in the field of flood protection, and thus torrential floods, in accordance with the Water Framework Directive (WFD) which was adopted in 2000. Two EU countries, Austria and Italy, and three Western Balkan countries were selected for the strategic and legal framework of torrential flood control: Serbia, North Macedonia, and Bosnia and Herzegovina. In addition to the legal framework of torrential flood control in EU countries, policies and strategies related to this area were studied for comparative analysis with non-EU countries. The strategic framework for the protection of water resources, and in particular torrential flood protection, is lacking in all Western Balkan countries. The aim of this chapter is to determine the directions of future strategic directions and torrential flood control policies in the Western Balkans based on the experiences of EU countries, advantages and disadvantages of the existing strategic, and legal frameworks.

Watersheds in the Lesser Himalayan region are highly susceptible to natural hazards, particularly those instigated by action and movement of water, such as soil erosion, flood, and mass movements of lands. Hilly watersheds with diversified land use and fragile ecosystems are responsible for accelerating soil erosion. Soil erosion is one of the most implicit hazards as it degrades water and soil quality in a watershed. The study prioritizes the soil erosion-susceptible zones in the Tons river watershed (India) in the Lesser Himalayan region. The interrelationships and role of morphometry, soil quality, slope, and land use together as four components in soil erosion are studied. Remote sensing data and multi-criteria decision method (MCDM) framework has been used to estimate soil erosion susceptibility of sub-watersheds. Results showed that morphometric parameters like elongation ratio and slope of sub-watersheds play a major role in determining the state of erosion.

stressors from the environment, regardless of origin. Torrential floods represent a hydrological hazard whose frequency and intensity have increased in recent years, mainly due to climate changes. In order to effectively manage the risks of torrents, it is necessary to apply early warning systems, since torrential floods are formed very quickly, especially on the watercourses of a small catchment area. The early warning system is part of a comprehensive torrential flood risk management system, seen as a technical entity for the collection, transformation, and rapid distribution of data. Modern early warning systems are the successors of rudimentary methods used in the past, and they are based on ICT and mobile applications developed in relation to the requirements of end users. The chapter presents an analysis of characteristic examples of the use. The main conclusion of the chapter indicates the need to implement early warning systems in national emergency management structures.

This chapter aimed to investigate the record of climatic and environmental change in the sedimentary filling of sebkha Mhabeul and their effect on hydric and eolian erosion within the wetland and its watershed. Along a 37 cm core, the sedimentary, geochemical, and geophysical signals at the Holocene-Anthropocene transition were followed. Sampling was carried out each 1 cm to obtain 37 samples. All studied parameters and clustering techniques indicate that the first 7 cm represent the Anthropocene strata. According to the age model, this upper part of the core records the last 300 yrs. The sedimentary record of the Anthropocene is marked by an increasing rate of sedimentation, grain size fining, heavy metals (Pb, Cu, Ni, Mn, and Fe) enrichment, which is related to increased erosion. Other intrinsic parameters such as CE, pH, Na, K, and CaCO3 enhance sediment erodibility. The measurement of the magnetic susceptibility along a 37 cm core collected from the sebkha Mhabeul shows an obvious upward increase related to a high content of heavy metals for the first 7 cm.

The Sidi El Hani Wetland is located in Eastern Tunisia. It represents the natural outlet of an endorheic system, Mechertate-Chrita-Sidi El Hani, and it collects all the eroded sediment from this watershed. In this chapter, the visual core description focused on three reference sandy bands and on the concept of grey scale variability in order to infer the clay pan response to the climatic variability and erosion during the last two millennia. First, in the uppermost part, the stage Warming Present (WP) stretches from (1954-80= 1874) to 1993, i.e. ≈120yrs; the establishment of modern conditions is characterized by stable conditions with high grey scale. Added to a small salt crust, this period is dominated by a clayey sedimentation. Second, the stage C4 is called the Late Little Ice Age (Late LIA); it stretches between the 80yrBP and 400yrBP, i.e., 320yrs. It is characterized by intermediate GS values; the clayey sedimentation makes up the twofold and threefold laminates. Based on laser granulometer, the genetic approach shows the interplay of eolian and hydraulic erosion.

Preface

Heavy rainfall in the summer of 2021 caused catastrophic floods in Western and Central Europe, especially in the German states of North Rhine-Westphalia and Rhineland-Palatinate, in Luxembourg, as well as along the rivers Ahr, Erft, and, in particular, the rivers Meuse and its tributaries in Belgium and the Netherlands. The floods killed at least 184 people in Germany and 38 in Belgium, with significant damage to infrastructure - housing, highways, railways, and bridges, as well as facilities vital to the local economy. Many villages remained inaccessible for days due to road closures, making evacuation and emergency response difficult. In addition to the fact that the land was already partially saturated before the rainfall, other objective circumstances affected the scale of this catastrophic event. This primarily refers to the characteristics of the funnel-like relief, with steep slopes, which affected the appearance of torrents. These factors have also been locally modified by differences in land cover, infrastructure, and water management, mitigating or increasing the devastation of extreme floods.

These, of course, are neither the first nor the last torrential floods to hit the world. Indeed, floods and storms were the most prevalent extreme weather events in the last two decades, counting for 44% of the overall occurrence of disasters, and affecting more than 1.65 million people (UNDRR, 2020). The biggest impact of the floods was in Asia, where 1.5 billion people were affected (93% of the total number of those affected in the world), Africa, and the Americas. Europe has also been hit by a large number of flood events caused by heavy rainfall. In the last 20 years alone, more than 400 major floods have been recorded (many of which have been catastrophic), affecting more than 8.7 million people, killing more than 2,000, and causing more than € 72 billion in monetary losses (WHO, 2017). It is evident that the trend of increasing damage of floods caused by extreme rainfall at the global level is increasing. Deforestation in general, and in inundation areas in particular, increasing economic activity in endangered areas, and global warming are the most significant causes (UNISDR, 2015).

Moreover, there are clear indications that these factors will contribute to a further increase in the devastating frequency and strength of floods (UNISDR, 2015). Today, there is scientific agreement that the amount and frequency of extreme rainfall will increase in the future due to climate change, which will inevitably cause more frequent and extreme floods. If climate scenarios predicting an increase in global mean temperature by 4°C by the end of the 21st century were to accomplish, direct flood damage during the 21st century could be tripled and flood-related damage could be almost sixfold if no adjustment measures are taken, whereby 73% of the world's population would face an increased risk (Alfieri at al, 2017). Even if an increase in the mean global temperature would be 2°C, which is considered easily achievable in the conditions that exist on Earth today, it could cause a 170% increase in flood-related damage compared to current levels (Alfieri et al, 2017).

Extreme rainfall often triggers flash floods, torrential floods, and mudslides. Torrential flooding is a natural process of increased intensity, which is an outpouring of water with a high concentration of sediment from the riverbed. Unlike the river floods, which are characteristic of large plain rivers and which are realized in the lower parts of the basin, torrential floods are associated with smaller, occasional, periodic, and permanent watercourses whose basins are affected by erosion processes. Also, concerning river floods, appear suddenly after intense rains and the sudden melting of snow and last for a shorter time. All above mentioned water-related natural hazards occur as a result of intensive rainfall showers causing the sudden occurrence of maximal water discharges and a high concentration of solid phase in torrential watersheds of small areas, and steep slopes, and are accompanied by high soil erosion intensity (Norbiato et al, 2008). Some estimates are that on average, almost 1 cm of the topsoil is lost annually, or 12 to 15 tons per hectare (FAO, 2015; Ashiagbor et al., 2013). Soil loss due to water-induced erosion is considered a major threat to soil and can have unforeseeable consequences and negative impacts on ecosystem services, crop production, water resources, carbon stocks, and biodiversity (Olson et al, 2016; Orgiazzi & Panagos, 2018).

Water erosion leads to physical degradation of the soil due to the reduction of the porosity of the surface layer and the reduction of the infiltration capacity of water into the soil. All this leads to an increase in surface runoff and sediment transport. In addition, it can cause other impacts, such as deposition, floods, infrastructure damage, and landslides (Mello et al., 2020).

Water erosion is one of the most important causes of reduced yields due to reduced productivity of agricultural land, caused by the removal of the most fertile topsoil (Panagos et al., 2018), and, consequently, land degradation and nutrient losses (Elnashar et al., 2021; Pena et al., 2020). It causes a significant accumulation of sediment downstream, reducing the life of dams, the production of electricity in hydropower plants, and the available amount of water in the reservoirs. In addition, water erosion affects surface water quality. All these impacts make water erosion a significant factor in ecosystem damage and therefore it is important to study the degree and impact factors of water erosion, including their quantification, qualitative assessment, and spatial aspect, to reduce the degree of soil degradation, create conditions for its restoration and reduce the risks of the impact of eroded soil on the ecosystem.

Since torrential floods endanger all elements of space (population, its activities, material goods, and nature), they can have the characteristics of a natural disaster. Consequently, their research and risk assessment requires an interdisciplinary approach, based on the careful assessment of the natural and social components of space (UNISDR, 2004).

Prevention and Management of Soil Erosion and Torrential Floods investigates the problems of erosion and torrential floods and opportunities for the prevention, management, and control of these destructive processes. It highlights the importance of the prevention and management practices of soil erosion and torrential floods concerning the exchange of knowledge and best practices. Covering topics such as the impacts of climate change and natural disasters, dam maintenance, wind erosion, agriculture, it is ideal for environmentalists, environmental engineers, crisis response specialists, policymakers, government officials, academicians, students, experts, practitioners, and researchers in the fields of soil erosion, torrential flood, environmental protection, sustainable development, engineering, and management.

Therefore, this publication is exploring soil protection and torrential floods as an issue that is gaining considerable importance throughout the world. It is another attempt to contribute to the study of the above-mentioned factors, without pretending to give definitive answers to some questions related to the phenomena of soil erosion and future impacts of climate change on these phenomena. Rather, the chapters that follow will offer different angles of observation, related to the world regions they analyze and in which these phenomena occur.

ORGANIZATION OF THE BOOK

The chapters in this book explore the many issues in which the field of soil protection and torrential floods issues is appearing. It is a result of the valuable participation of several researchers, students, academics, and policy makers.

Chapter 1 discusses the quality-of-life concept in the context of the relation with the environment and environmental quality. Quality of life here is perceived as a complex and multi-dimensional construct that warrants multiple approaches from different theoretical perspectives, arguing that the evaluation of the quality of life determined by the environment can be facilitated using objective and subjective measurement methodologies, discussed in the chapter.

Chapter 2 deals with the theoretical understandings of natural risk management and crisis management and their role in the development of an effective societal system by its transformation. After the presenting of the overview of the characteristics of natural disaster risk management and crisis management, including their phases and main steps, the authors discuss the different roles of the main actors, by analyzing a case of a catastrophic flood that happened in Serbia in 2015.

Chapter 3 analyses the strategic and legal framework of flood protection, and more specifically torrential floods in two EU countries - Austria and Italy, and three Western Balkan countries - Serbia, North Macedonia, and Bosnia and Herzegovina. Based on the comparative analysis, the chapter points at the lack of a contemporary strategic framework and adequate policies in non-EU countries and provides recommendations for future directions in the alignment of torrential flooding policies on the Western Balkans countries with EU experiences.

Chapter 4 is devoted to the methodology for prioritization of soil erosion susceptible sub-watersheds. To assess the vulnerability of the watershed towards soil erosion the authors employed the Multi-Criteria Decision Method (MCDM) framework, arguing that the prioritization by using the quantitative morphometric analysis used alone as vulnerability analysis method should be complemented by additional criteria – soil quality, land use patterns, slope characteristics, etc. With the help of satellite-based remote sensing, data to study watersheds can be accurately extracted and their analysis for prioritizing the sub-watersheds may further help provide useful input in formulating conservation strategies of soil and water in watersheds.

Chapters 5, 6, and 7 are devoted to and deal with the effects of soil erosion on agriculture. Chapter 5 emphasizes the different effects of soil erosion, including nutrient loss, land-use changes, reduced productivity, siltation of water bodies among other effects like affecting the livelihood of marginal communities dependent on agriculture globally and public health. The Chapter is focused on the data review aimed at a better understanding of soil erosion prevention methods. Similarly, the level of damage associated with soil erosion has been analyzed in Chapter 6, with emphasis on the impacts they may have on the global carbon cycle, phosphorus loss, dust emissions, eutrophication, and soil biodiversity. It points out causes and types of soil degradation with emphasis on soil erosion and offers insight into guidelines for soil degradation prevention. Chapter 7 analyzes existing agriculture practices, starting from the assumption that sustainability in agriculture is an integral prerequisite for sustainable socio-economic development. The chapter discusses the possible impact of polyculture farming on some determinants of sustainable agricultural development, especially from the point of view of economic viability, biodiversity, and land degradation, arguing that the increase of the area under polyculture is one of the main solutions to the present environmental problems.

Chapter 8 is devoted to wind erosion and shelterbelts as one of the most effective measures to influence wind erosion-related impacts. Additionally, the chapter discusses the potential of shelterbelts to help fights climate change due to their function of carbon sequestration.

Chapter 9 deals with the legacy of dams, particularly their maintenance and sedimentation, concerning current and future climate change issues. The author argued for proper maintenance policies and, more specifically, nature-based solutions in climate change adaptation.

Chapter 10 discusses early warning systems as a part of a comprehensive torrential flood risk management system, aimed to collect; transform, and rapidly distribute the data. The Chapter provides an analysis of characteristic examples in Western Balkan countries, concluding with the guidelines for the improvement of early warning systems in national emergency management structures.

Chapters 11 and 12 deal with climatic and environmental change in the sedimentary filling and their effect on water and eolian erosion, studying the sedimentary record of salt lakes and wetlands in Tunisia as a case. Chapter 11, using a multi-proxies study, described environmental and climatic changes during the Anthropocene strata within the sedimentary filling of sebkha Mhabeul in Tunisia, pointing at the possible effects on eolian and hydric erosion and deposition within this wetland and its hydrological catchment. Chapter 12 presents the results of visual core description research on eroded land in eastern Tunisia, establishing a clear connection between the long-term climate variability prediction and visual characteristics of sedimentary layers. Both chapters pointed at the increasing rate of sedimentation during the Anthropocene, with high values of metals in sedimentary flux.

Chapter 13 offers a review of historical torrent floods on the territory of Serbia, as a territory naturally destined for torrents and torrential floods. Given the distribution of hilly and mountainous areas in Serbia and the developed hydro-graphic network, torrential floods occur very often, almost every year.

The chapters presented in this book can contribute to a better understanding of torrential floods and erosion-caused soil degradation processes, both for academics and practitioners, thus helping in the development of soil erosion mitigation recommendations and policies at different scales. It is expected that the output of this book will contribute strengthen the preparedness for forecasting, assessing, and monitoring the occurrence of torrential flooding and its consequences on soil degradation.

Slobodan Milutinović
University of Niš, Serbia

Snežana Živković
University of Niš, Serbia

REFERENCES

Alfieri, L., Bisselink, B., Dottori, F., Naumann, G., de Roo, A., Salamon, P., Wyser, K., & Feyen, L. (2017). Global projections of river flood risk in a warmer world. *Earth's Future*, *5*(2), 171–182. doi:10.1002/2016EF000485

Ashiagbor, G., Forkuo, E. K., Laari, P., & Aabeyir, R. (2013). Modeling soil erosion using RUSLE and GIS tools. *International Journal of Remote Sensing and Geosciences*, *2*(4), 1–17.

de Mello, C. R., Alves, G. J., Beskow, S., & Norton, L. D. (2020). Daily rainfall erosivity as an indicator for natural disasters: Assessment in mountainous regions of southeastern Brazil. *Natural Hazards*, *103*(1), 947–966. doi:10.100711069-020-04020-w

Elnashar, A., Zeng, H., Wu, B., Fenta, A. A., Nabil, M., & Duerler, R. (2021). Soil erosion assessment in the Blue Nile Basin driven by a novel RUSLE-GEE framework. *The Science of the Total Environment*, *793*, 148–466. doi:10.1016/j.scitotenv.2021.148466 PMID:34175609

FAO. (2015). *Status of the world's soil resources (SWSR)–main report*. Food and agriculture organization of the United Nations and intergovernmental technical panel on soils, Rome, Italy, 650. Retrieved from https://www.fao.org/3/i5199e/I5199E.pdf

Norbiato, D., Borga, M., Esposti, S. D., Gaume, E., & Anquetin, S. (2008). Flash flood warning based on rainfall thresholds and soil moisture conditions: An assessment for gauged and ungauged basins. *Journal of Hydrology (Amsterdam)*, *362*(3-4), 274–290. doi:10.1016/j.jhydrol.2008.08.023

Olson, K. R., Al-Kaisi, M., Lal, R., & Cihacek, L. (2016). Impact of soil erosion on soil organic carbon stocks. *Journal of Soil and Water Conservation*, *71*(3), 61A–67A. doi:10.2489/jswc.71.3.61A

Orgiazzi, A., & Panagos, P. (2018). Soil biodiversity and soil erosion: It is time to get married: Adding an earthworm factor to soil erosion modelling. *Global Ecology and Biogeography*, *27*(10), 1155–1167. doi:10.1111/geb.12782

Panagos, P., Borrelli, P., Poesen, J., Ballabio, C., Lugato, E., Meusburger, K., Montanarella, L., & Alewell, C. (2015). The new assessment of soil loss by water erosion in Europe Environ. *Environmental Science & Policy*, *54*, 438–447. doi:10.1016/j.envsci.2015.08.012

Pena, S. B., Abreu, M. M., Magalhães, M. R., & Cortez, N. (2020). Water erosion aspects of land degradation neutrality to landscape planning tools at national scale. *Geoderma*, *363*, 114093. doi:10.1016/j.geoderma.2019.114093

UNDRR. (2020). *The Human Cost of Disasters 2000-2019*. Retrieved from https://www.preventionweb.net/files/74124_humancostofdisasters20002019reportu.pdf

UNISDR. (2004). *Living with risk - A global review of disaster reduction initiatives. Volume II*. United Nations Office for Disaster Risk Reduction. Retrieved from https://www.undrr.org/publication/living-risk-global-review-disaster-reduction-initiatives

UNISDR. (2015). *The Human Cost of Weather-Related Disasters 1995–2015*. CRED & UNISDR. Available online: https://reliefweb.int/sites/reliefweb.int/files/resources/COP21_WeatherDisastersReport_2015_FINAL.pdf

WHO. (2017). *Flooding: Managing Health Risks in the WHO European Region*. Copenhagen, Denmark: World Health Organisation Regional Office for Europe. Retrieved from https://apps.who.int/iris/handle/10665/329518

Acknowledgment

Chapters 1, 3, 6, 8, and 10 of this publication was generated in the context of the Project **"Soil Erosion and Torrential Flood Prevention: Curriculum Development at the Universities of Western Balkan Countries" (SETOF)**, co-financed by EU funds within the framework of the Erasmus+ Programme.

The editors should like to express their sincere gratitude to all reviewers for their effort to improve the chapter quality. We are very grateful to Dr. Tamara Rađenović for her selfless help in communicating with the authors and technical editing, and Mrs. Aleksandra Petković for language editing and proofreading. The continued assistance of the Editorial Department of IGI global is also highly appreciated.

Finally, we would like to take this opportunity to thank all the authors for their hard work to develop and revise the chapters.

Slobodan Milutinović

Snežana Živković

Chapter 1
Environmental Protection and Quality of Life

Snežana Živković
https://orcid.org/0000-0002-7931-9334
University of Niš, Serbia

Slobodan Milutinović
https://orcid.org/0000-0002-4384-3687
University of Niš, Serbia

ABSTRACT

Development of environmental protection together with economic and social development can be considered sustainable only if they support individual quality of life. Conceptually, quality of life is closely related to sustainable development, since sustainability implies a balance between environmental, social, and economic qualities. Environmental quality is reflected in its ability to meet the basic human needs. Quality of life is a complex and multi-dimensional construct that warrants multiple approaches from different theoretical perspectives. Evaluation of the quality of life determined by the environment can be facilitated using objective and subjective measurements. Regardless of how these two indicators are classified, both are considered equally beneficial and valuable for research. Considering all the above mentioned, the aim of this chapter is to shed light on the importance of environmental protection for the quality of life, as well as the necessity to measure quality of life determined by environmental factors in order to adequately manage them.

INTRODUCTION

To properly analyze the connection between quality of life and environmental conditions, it is necessary to consider all human activities leading to environmental pollution. Rapid urbanization and population growth due to industrialization and technological development, higher expectations caused by a higher standard of living, differences between expectation levels and actual consumption patterns, a lack of public environmental awareness of the negative environmental impacts, as well as excessive consump-

DOI: 10.4018/978-1-7998-8459-0.ch001

tion of resources make the goals of protecting the environment and improving the quality of life all the more difficult to achieve.

Nowadays, there is a growing awareness of how the environment can impact individual lives. People have long been concerned about environmental degradation and pollution. Environmental protection can ensure the full preservation of environmental quality, biodiversity, and geodiversity, and the rational use of natural resources, which is a basic condition for sustainable development, a healthy life, and good quality of life.

A large part of individual quality of life depends on the quality of the living surroundings. Such quality of life, which relies on the environment, is a good criterion of sustainability since environmental, economic, and social development can be considered sustainable only if they support individual quality of life. The environment in which one lives plays a major role in influencing one's behavior; hence it affects experiences, decisions and wellbeing, ultimately affecting one's quality of life (Rastogi et al., 2021).

The relationship that comes to the forefront is that between the individual and the environment, as the environment determines how different socially-conditioned needs are to be met. The objective living conditions, which determine how the needs are being expressed and met, are a common topic in sociological studies. Such an approach to quality of life allows the different environmental effects on the individuals to be studied, including the effects from the physical as well as social and economic environments (De Groot & Steg, 2006). Moreover, in addition to diagnosing the effects of the state of the environment, the quality-of-life principle can also be used to analyze the impact of environmental changes on individuals. This is of paramount importance, considering that continuous environmental change is an integral aspect of sustainable development. With their mere existence, humans act on and alter the environment, adapting it to themselves and their own needs. Unlike other living organisms, humans create and connect cultural, social, economic, and political events.

Conceptually, quality of life is closely related to sustainable development, because sustainability refers to a balance between environmental, social, and economic qualities, which is why the concept should be introduced as an evaluation measure of sustainability and the environment at an individual level. Quality of life is defined as a range within which individual needs are met, which depends on the physical, social, and economic environments (Diener, 2000). It could be argued that quality of life reflects the social dimension of sustainable development. The idea of quality of life of healthy individuals was first brought into focus by WHO in 1993 (WHO, 1993). Due to the increasing significance of environmental conditions for the concept of quality of life, the Parliamentary Assembly of the Council of Europe drafted an additional protocol to the European Convention on Human Rights concerning the right to a healthy environment (Council of Europe, Parliamentary Assembly, 2009).

ENVIRONMENTAL QUALITY

Environmental quality is reflected in its ability to meet the basic human needs, and the more human needs the perceived space can meet, the bigger the subjective feeling of happiness and satisfaction will be within it (Pickett et al, 2001). In this context, the field of quality of life is fairly broad, as it pertains to the relationships between economy, education, employment, social security, and life satisfaction. Environmental quality is no less essential because most people appreciate the beauty and health of the place where they live and care about the depletion of natural resources (Brajša-Žganec, 2011). Preservation of the environment and natural resources is also one of the crucial factors in preserving well-being over time (Liere & Dunlap, 1980). Environmental policies play a key role in dealing with global health priorities and improving environmentally responsible behavior (Felix & Garcia-Vega, 2012).

Environmental quality is a key factor for people's well-being, given that the quality of the local living environment directly influences people's health and well-being (Coan & Holman, 2008). Undisturbed environments act as sources of satisfaction, enhancing mental well-being and allowing people to recover from the stress of everyday life and engage in physical activities.

Access to resources such as green spaces, forests, and rivers, is an important aspect of quality of life. In addition, a healthy economy relies not only on healthy and productive workers but also on natural resources such as water, trees, animals, and crops (Zheng, 2010). On the other hand, extreme environmental calamities, such as natural disasters (earthquakes, cyclones, floods, droughts, and volcanic eruptions) and epidemics, can also increase the number of fatalities, injuries, and diseases. Unfortunately, in times of the current Covid-19 pandemic, we are all witnesses to how an epidemic can affect their incidence. In the long term, drastic environmental change can also undermine health via climate change (Ahmad & Yamano, 2011). In addition to its impact on human health, the environment is also important because most people ascribe some significance to the beauty and health of the place in which they live and because they care about preventing the degradation of our planet and the depletion of natural resources (Balestra & Dottori, 2012). People directly benefit from environmentally friendly goods and services, such as water, clean air, land, forests, and access to green spaces, as these allow them to meet their basic needs and enjoy their free time and the company of other people (Balestra & Sultan, 2013). Reducing flood risk may improve citizens' quality of life (Daniel & Michaela, 2021).

The creation of various income-increasing opportunities is an important part of successful flood risk management. In developed countries, water management companies should consider new projects related to flood control. In this way, they will hire new employees and cooperate with other companies to provide equipment. On the other hand, newcomers to the project will also improve their quality of life because of their income. Public works projects have specific content that can be used to help people who cannot make money. They will be engaged in public works, and the main goal is to establish a flood control system. Activities related to health and education can also be used to reduce flood risk while improving quality of life. These activities may include a special introduction to flood issues in school materials, with a special emphasis on health, and brief training on flood preparedness and rapid response. This may help raise awareness of flooding in households and communities. Activities should establish close links between the Ministry of Health, the Ministry of Education and the national emergency agency. Volunteer activities should be a good tool to reduce flood risk, and at the same time, it should be a good tool for Eurostat to improve social interaction and leisure as a dimension of quality of life. Social development and quality of life. The quality of a society and the quality of individual lives are directly connected. Both qualities pertain to the overall societal well-being and are aimed at allowing each member of the society to realize their goals and aspirations. This means that the lives of individuals are connected and intertwined with the personal system of values they adopted while growing up, with the system of values of their community, with their culture, lore, and traditions, and, finally, with their education and the healthcare system. The concepts of well-being and quality of life refer to evaluative judgements about selected aspects or the entirety of a life-situation or life-path, for an individual, group or society (Gasper, 2010).

Social development is defined as the change of the main social subsystem, which implies that the change is 'positive', i.e., leading toward progress. Yet, many parts of the world are affected by economic, political, cultural, and environmental changes, whose quality is questionable. Accordingly, there is a standard issue concerning the criteria to be used to evaluate the expectations or the 'positivity' of such changes. The potential antagonism of certain events further complicates the issue. Finally, the multidimensional nature of the development process encumbers any comparison and assessment of the development of different countries, the assessment of different development strategies, and the role various factors play within them.

The comparison issue can be overcome by selecting a single unique dimension of development, but this can also produce some difficulties. For instance, economic development can be expressed by the value of produced goods per capita, but the material standard of the population does not depend only on the system's economic efficiency but also on the way in which the produced goods are distributed – and this depends not on the economy but on social policy. Furthermore, there is the question of how to measure and assess the social costs of economic development, which comprise natural and cultural resources destruction, unemployment, rising crime and traffic accident rates in large cities, and so forth. Thus, the prerequisite for the measurement and comparison of different types of development is the existence of a universal value and scale. However, many development theoreticians fail to see such a unique criterion of evaluation, hence their skepticism over whether it is even possible to measure development and, ultimately, whether it is possible to formulate a general theory of development (Wackernagel et al. 2002).

HIERARCHY OF NEEDS

A rational basis of development, which involves the satisfaction of human needs, must not be in conflict with economic, political, or any other interpretations of development, the only difference being that only specific needs that are prioritized and that do not cover the entire spectrum of universal needs would be considered. It is now well known that universal human needs were divided into categories within the hierarchical structure first developed by Maslow, who proposed five categories, and later modified by Alderfer, who reduced Maslow's categories to three levels of need (Alderfer, 1969): existence, relatedness, and growth (self-actualization).

The fundamental law of the hierarchy of needs is that a higher-level need cannot be met until a lower-level one has been met, because lower-level needs are more pronounced and are required for reaching the higher-level needs. It is therefore understandable that quality of life is reduced to the economic dimension in countries struggling with low economic development and prioritizing the basic needs for the survival of their population majority. On the other hand, post-industrial development, which occurs at a high level of economic development, enables the articulation of higher social and self-actualizing needs, rendering people more sensitive to limitations to their satisfaction. They are characterized by a noticeable dematerialization of the system of life values, so quality of life is dependent on the realization of social and self-actualization values. Quality of life concerns people in both developed and developing countries.

Sirgy (1986) developed a comprehensive quality-of-life theory by following the rules of the hierarchy of needs, or Maslow's perspective of human development. According to Sirgy, developed countries have people who are mostly preoccupied with satisfying higher-order needs (social, esteem, and self-actualization needs), while less-developed countries are mostly preoccupied with satisfying lower-order needs (biological and safety-related needs). The higher the level of need satisfaction in a society, the better the quality of life. Accordingly, social institutions have to be established and developed in such a way as to enable the satisfaction of various human needs, all for the purpose of improving the quality of life in a society.

If quality of life is the main parameter of development, the development strategy has to be calculated according to the laws governing human satisfaction. Considering that the structure of these factors is not the same on different levels of economic, political, and environmental development, it needs to be adapted to the situation in these social subsystems and to the state of the natural environment.

ASPECTS OF SUSTAINABLE DEVELOPMENT

This chapter discusses the limits of sustainable development from the perspective of environmental psychology. Sustainability is perceived as a balanced relationship between people and the environment. Sustainability implies finding the balance between the sustainability of environmental, social, and economic aspects (World Commission on Environment and Development, 1987). For instance, let us consider an initiative to increase the fuel process in order to reduce carbon dioxide emissions. The sustainability of such an initiative cannot be assessed without considering the social and economic effects. A certain number of people, especially those with lower incomes, would have to give up on driving a car, which would on the other hand significantly increase the degree of social injustice. The population's purchasing power would then decrease, accompanied by life satisfaction. To achieve sustainable development, the sustainability of any of the three aspects (environmental, social, and economic) should by no means significantly undermine the sustainability of the other two.

The Vital Aspects of Quality of Life and Sustainable Development

Environmentally focused sustainability programs are not sustainable if they undermine important individual needs and values. Let us imagine that there is a rigorous pro-environment program prescribing minimal consumption of meat and fish, considerably limiting household energy consumption, banning people from accessing areas with high biodiversity, and so on. On the one hand, the implementation of such an environmental development program would be considered sustainable from the environmental perspective, as it would help the natural environment remain undisturbed and endure for a long time. On the other hand, from the human psychology standpoint, such a rigorous initiative would endanger human well-being due to drastic changes, reduced comfort, and limited freedom. Therefore, people's quality of life would be diminished, whereas the public resistance toward the program would only increase. Naturally, this raises the issue of whether such a program could be characterized as sustainable at all.

Experts may sometimes wrongly interpret people's preferences, as they tend to overestimate certain parameters, such as the appearance of materials, while underestimating the relevance of other parameters, such as freedom of choice and privacy. Quality-of-life measurement requires that an individual evaluate the relevance of every individual aspect. The more relevant aspects will then be prioritized in terms of needs and values, while the others will be deemed less important for a person's quality of life (Fawcett et al., 2008).

Sustainability programs and interventions usually transform the environment, but it is crucial to understand how these transformations affect the personal quality of life. Steg et al. (2002) investigated people's satisfaction with specific quality-of-life indicators before and after one such intervention, namely reduced household energy use. After a month of reduced energy use, the group of respondents reported a change in their quality of life, which was reflected in their satisfaction with environmental quality and nature/biodiversity. In contrast, their satisfaction with other quality-of-life aspects remained unchanged. These were the respondents' exact expectations prior to the intervention.

THE CONCEPT OF QUALITY OF LIFE

The understanding of the term *quality of life* has actually gone through several stages of development. In the beginning, during the 1950s, quality of life was considered a synonym for economic well-being or material standard of living. That is why studies of quality of life were commonly conducted within the economic sciences. Thus, quality of life was understood as a broad term referring mostly to the general well-being of society. According to one definition, well-being does not only refer to life itself, but also refers to the ways people react, feel, and experience life with all its domains (Fahey et al. 2005).

As the notion of social progress developed and expanded from the category of economic growth and development to other development dimensions, quality of life became increasingly regarded as consumption and equal distribution of goods across various aspects of social life. In the early 1980s, quality-of-life studies shifted their focus toward examining subjective assessments, as it had been observed that quality of life is not solely an objective category measured with material and social conditions for the satisfaction of different needs – it is also subjectively determined.

The subjective aspects refer to different individuals and groups of people being unequally satisfied with the same or equally objective conditions. It turned out that there are numerous individual differences regarding life satisfaction as a quality-of-life indicator, which are more due to someone's personality traits than due to their objective situation. System of values, expectation level, and frustration tolerance are psychological concepts that may explain the weak connection between objective living conditions and general life satisfaction.

It should primarily be emphasized that quality of life is indeed a subjective and mental phenomenon, dependent on the satisfaction of all universal needs structured as a hierarchy of needs, which is why it can be associated with the concept of life satisfaction. Today, most researchers of the topic acknowledge the experiential aspect of quality of life. However, economists and social scientists believe it primarily depends on the objective circumstances of one's life, i.e., on the socio-economic context, thus eventually equating quality of life with economic and social well-being. It follows that quality of life is unevenly distributed among social groups living in different conditions or not having equal access to the same conditions. The groups that enjoy a better overall social status also have a better quality of life (Seferagić & Popovski, 1989). Allardt (1977) attempted to explain such a dual understanding of quality of life by referring to objective predicaments as the level of living or welfare and subjective predicaments as quality of life or happiness.

Seferagić and Popovski (1989) concluded that the problem lies at the individual level and that individual quality of life depends on the individual's membership in a social group, but that everything ultimately leads to the individualization of provided objective conditions. Therefore, the above considerations recognize the importance of individual psychological structure for quality of life, but fail to quantify the relationships between objective and subjective factors underlying quality of life. This, in turn, allows further arbitrary weighing of their importance (Allardt, 1977).

Pastuović et al. (1995) define the individual quality of life as a complex and combined experience of a person's life (dis)satisfaction stemming from continuous evaluation and re-evaluation of how successful they are at meeting their own needs. Such an understanding of quality of life allows the objective circumstances and the psychological structure of individuals to be considered as its determinants, which interact, thus overcoming the deficiencies of partial conceptualizations and studies that do not consider them at all. In fact, how individuals assess the effects they achieve in their various life roles (professional, social, parental, political, and so on) depends on the values toward which they are striving, specific wants, expectation levels, and different personality traits. At the same time, better living conditions favor the improvement of quality of life, whereas a low living standard, political oppression, poor healthcare, and a degraded environment will typically diminish the quality of life. However, whether and to what extent such circumstances will be created depends on the individuals' mental structure, which in turn determines their reaction to the objective circumstances and their ability to adapt to the changes in their natural and social environment. Thus, even in the objectively best and the objectively worst conditions, there will always be individuals who will regard their life as either very satisfactory or worthless.

Quality of life is a complex and multidimensional construct, which requires multiple approaches from a variety of theoretical perspectives. Social scientists have had some difficulty developing the instruments for quantitative as well as qualitative measurement and assessment of quality of life. It is impossible to precisely determine the values and attitudes regarding natural beauty, fresh air, noise, smoke, and large crowds due to people's individual differences, i.e. due to different categories affecting individuals differently, whether positively or negatively.

The concept of quality of life has three main characteristics. First, it refers to the individual perception of life situations, not to the general quality of life of a specific country's population. Second, it is a multidimensional concept encompassing multiple life domains, such as housing, education, employment, work and private life balance, access to institutions and public services, and the interaction between all of these. Third, it integrates objective information about the living conditions and the subjective views and attitudes to present a picture of the overall well-being in a society.

The main steps toward achieving quality of life are to develop a framework for the assessment of environmental indicators that are relevant for quality of life and to select indicators for environmental quality assessment, environmentally responsible behavior, and use of environmental services, according to Eurostat (2015). Eurostat lists the following indicators:

- **Material living conditions** refer to economic, financial, and material conditions required to establish, implement, maintain, and continuously improve the environmental management system.
- **Productive or main activity** is an important indicator for quality of life and the possibility of employment for people with the knowledge of and increased understanding of environmental policy, environmental aspects, actual or potential impact, and their own contributions to the efficiency of the environmental management system.
- **Health**, which refers to the healthcare system, indicates the importance of gaining trust and meeting the requirements of environmental standards and health, safety, and environmental protection policies. The emphasis is on preventive rather than corrective measures.
- **Education** is an important quality-of-life indicator in terms of educational possibilities serving to improve the quality of life.
- **Social relations and leisure** have a strong influence on life satisfaction and happiness. Social relations can be considered as social capital, both for the individual and for society as a whole, and they can influence people's quality of life in many of its aspects.
- **Economic and physical safety** indicate the need to provide each member of a society with equal possibilities so as to guarantee economic security, which influences quality of life.

- **Governance and basic rights** refer to the respect of human rights, as the rule of law, a competent government, and civil society form the foundation of modern democracy and affect the quality of life of all people.
- **Natural and living environment** refers to an organization's intention to achieve its goals by implementing an environmental management system, including improving the environment by fulfilling obligations regarding coordination and accomplishment of environmental goals.
- **Overall experience of life.**

Each of the above indicators has a fairly clear influence on quality-of-life assessment and fulfilment. The influence is particularly pronounced when observing all indicators at the same time, because each of them grows in importance if implemented with other indicators. For instance, it would be fair to question the meaning of economic, financial, and material security if indicators such as education, health, or environmental quality are not taken into consideration.

ENVIRONMENTAL PROTECTION INDICATORS BASED ON QUALITY OF LIFE

Quality of Life Measurement Methods

Objective and subjective measurements are used to evaluate quality of life determined by the environment. Objective measurements describe to what extent environmental characteristics meet the criteria that are considered relevant to good quality of life. The criteria involve technological measurements and expert opinions on the state of the environment (e.g. drinking water quality or SO2 in the air), which can be evaluated at the social or individual level. Objective measurements describe environmental quality and the assumed effects on quality of life, although they are not without their limitations (Diener et al., 2009). In order to set objective criteria, it is necessary to determine which part of life people find the most important and what the satisfaction criteria levels are in terms of quality of life. Even though needs and values are mostly universal, this does not apply to all indicators involved in the quality-of-life assessment. Likewise, the significance people attach to certain elements and their satisfaction differ in different environments. Specifically, people's quality of life depends on personal expectations, values, attitudes, and previous knowledge (Diener & Biswas-Diener, 2011). Consequently, it is impossible to assess quality of life based solely on objective methods. Accordingly, there is the question of how to investigate and evaluate quality of life. The prevalent answer found in literature is to use social indicators, which include objective indicators, such as economic indicators (GDP, GDP per capita, household income, purchasing power, etc.), as well as complex social indicators, which comprise data on the school system, healthcare system, and crime rates Diener & Suh, 1997).

The concept of quality of life can be used in the context of the economy, which emphasizes material values such as income and wealth in the measurement of well-being (Shucksmith et al., 2009). GDP and similar economic measures were long considered as the main indicators for the well-being of a country. However, such measures, based on money and national income, provide insight only into the financial state, without offering any information about other dimensions, such as education, health, natural environmental conditions, human rights, and so forth. In addition, the increased material well-being of a country does not necessarily lead to an improved living standard or a higher quality of life of individuals. All of the above suggests that life does not depend only on financial variables but also includes many others.

This led to the publication of numerous studies clearly demonstrating that a higher living standard in the material sense does not necessarily lead to improved subjective well-being (Slavuj, 2012). An example can be found in the 'depression of the wealthy' in Switzerland. According to studies, Switzerland has long been among the leading countries regarding suicide rates, and it is generally well known that the country's living standard in the material sense is considerably high. This example only confirms that a better material standard of living does not always improve subjective well-being.

After reaching a certain level of basic needs satisfaction, personal income as well as other objective indicators become only weakly associated with the subjective assessment of one's well-being (Inglehart & Klingemann, 2000). When the basic material needs have been met, other life domains become considerably more important for a subjective experience of quality of life. The economic indicators do not paint a full picture of quality of life but are nonetheless an unavoidable foundation for the emergence of certain economic presumptions. In addition, in terms of quality of life, they offer the possibility for simpler and faster economic comparisons of specific areas and hence provide the guidelines for the development of that area.

It is also necessary to mention Hazel Henderson, who, without disregarding the decisive role and function of economy in quality-of-life improvement, claimed that income per capita is a rather weak indicator of human well-being, which is why she proposed that it be replaced by less tangible measurements, such as political participation, literacy, education, and health (Henderson, 2002). Another noteworthy contribution is Maslow's theory on human needs classification developed in the 1960s, which is a common starting point in the majority of extensive studies. As has been correctly observed in one publication on quality-of-life indicators, the concept can be approached from several angles – philosophical, ethical, and psychological, but also in terms of health, human rights, and care for the environment.

The modern concept of quality of life is based on economic as well as various subjective indicators. In the studies on quality of life, subjective indicators are defined via different concepts only to ultimately arrive at the factors such as happiness, satisfaction, and positive influence (Galinha & Pais-Ribeiro, 2008. Protection of the environment and natural resources is a crucial factor for ensuring long-term well-being. However, measuring environmental indicators is a complicated task. First of all, it is complicated because the degree of the environmental impact of current factors is impossible to determine for future well-being. Second of all, it is complicated because there are a variety of incomparable indicators that can, in fact, meet the set standards.

The biggest issue underlying quality-of-life research is the operative definition of quality of life, which is more implicit than explicit, as it fails at being sufficiently different from the factors influencing it or connected to it (Ferriss, 2006). Quality of life indicators or dimensions are quite often treated equally, so such models may result in inaccurate explanations or in the impossibility to improve quality of life. Therefore, it is important to draw a clear line between quality of life and the factors influencing it. Subjective measurements based on self-assessment are used to supplement objective methods. Individual measurements allow the studying of the subjective assessment of environmental characteristics and the fulfilment of specific needs.

Regardless of how indicators are classified, both types of indicators are considered equally good and valuable to any research, whereby each type has its own advantages and disadvantages. This is where the researcher's role really comes into focus, specifically whether they will be more prone to rely on the objective indicators and statistical data or on the subjective indicators. To obtain the fullest possible picture of quality of life, both objective and subjective indicators should ideally be used (Costanza et al., 2008). The two indicator types are commonly divided into three basic categories: the environment, economy, and social life.

Quality of Life Measurement Methods

Studying the relationship between environmental and quality-of-life factors requires complex and multidimensional methods. The multidimensional methods integrate multiple aspects of quality of life, and they can provide an estimate of how and to what extent these are dependent on environmental factors. For example, one may measure the satisfaction with specific quality-of-life elements in various surroundings and examine how the differences among environmental factors can lead to different levels of satisfaction with life factors (Diener & Biswas-Diener, 2011). This can be done through a comparison between the actual circumstances as well as hypothetical situations.

According to Štreimikienė (2015), environmental indicators can be divided into the following categories according to their relevance to quality of life: **environmental quality, environmentally responsible behavior,** and **consumption of environmental services**.

Environmental quality indicators

Indicators of environmental quality comprise a number of environmental factors, such as land, water, and air. However, the majority of relevant data pertain to the impact of air pollution on human health, which means that the indicators of air pollution garner the most attention. In this section, objective measurements of air quality concern the concentrations of PM10 particles, ground-level ozone, and CO2 emissions from transport, considering that the transport sector poses a great problem for sustainable development in the EU. CO2 emissions are the chief cause of climate change.

The problems caused by ever-increasing pollution from transport are particularly important. With the raised standard of living, the people's purchasing power also increases, including the purchase of new cars, which in turn increases the CO2 emissions. Yet, the use of more energy-efficient cars can help reduce the emissions of greenhouse gases. This is precisely why the Serbian Ministry of Environmental Protection introduced a ban on the import of used cars with Euro 3 engines as of 1 January 2021, while the citizens who already own older cars will have to undergo stricter exhaust emission controls to be allowed to drive them. Such a decision by the Ministry resulted from the fact that numerous Serbian cities have been among the most heavily polluted in the world for many months back, as well as from the report by the Serbian Environmental Protection Agency stating that transport contributes to the overall PM air pollution with six percent and to the total nitrogen oxide pollution with 20 percent. Nevertheless, despite their negative impact, the positive aspects of cars also need to be considered, including their usefulness to individuals in terms of mobility. Car use increases individual freedom but also affects different needs and values, such as the need for independence, for enjoying one's life, and for improving one's status (De Groot & Steg, 2006). Similarly, freedom is determined not only by the degree of car use but also by other environmental factors, such as access to nature and various institutions.

EU citizens have become wealthier and the improvement of their living standard led to increased consumption, thus generating more waste. In this region only, around 3 billion tons of waste are deposited annually, of which 90 million tons represent hazardous waste. According to Eurostat, this amounts to approximately 6 tons of solid waste per every man, woman, and child.

In addition, over a half of the population live in urban areas, where the ozone levels exceed the maximum allowed limits. The fraction of PM_{10} particles considered the most contaminating is smaller than 2.5 micrometers and referred to as $PM_{2.5}$ particles. $PM_{2.5}$ always accompany PM_{10} particles; they reach the upper respiratory tract even easier due to being smaller and they are the ones carrying heavy metals (Arruti et al., 2010). Epidemiological studies conducted over the last 20 years have identified a significant correlation between short- and long-term exposure to increased environmental PM concentrations and increased morbidity (mostly cardiovascular and respiratory diseases) and mortality rates (Goldberg et al. 2001).

Another crucial aspect for human well-being is access to clean drinking water. Water management is a major and growing challenge in many parts of the world. Many people do not have access to sufficient amounts of water, or at least to water that is safe for drinking, in their surroundings. Even though considerable progress has been made in the EU countries, whose wastewater treatment processes greatly reduced the water pollution in industrial and public utility facilities, diffuse pollution caused by agricultural and urban effluents still remains a challenge. Improved freshwater quality is not always easy to identify. Biochemical oxygen demand is the main water quality indicator in rivers. Specifically, the amount of oxygen used by aquatic microorganisms during biochemical oxidation of organic substances is called the biological oxygen demand. Organic pollution of river water often leads to rapid water deoxygenation, increased ammonia concentrations, and the consequent death of fish and aquatic invertebrates. The most important sources of organic waste are household and industrial waste waters.

Indicators of Environmentally Responsible Behavior

Environmentally responsible behavior manifests itself through activities such as energy conservation, the use of renewable resources, and sustainable consumption (Osbaldiston & Sheldon, 2003).

The chief indicators of environmentally responsible behavior were chosen based on Eurostat data. They include energy productivity, resource productivity, the share of renewables in final energy consumption, packaging waste recycling rate, and wastewater sludge production and disposal per capita. These indicators directly affect quality of life as they are the main triggers for the impact on environmental quality.

Increased use of renewables is a priority in the EU energy and environmental policies, because such use reduces greenhouse gas emissions and enables more efficient energy consumption. Over the last 20 years, Europe has seen a significant improvement in drinking water quality, but continuous efforts are still necessary to further improve the quality of water resources. For example, the bioaccumulation of mercury and other persistent organic pollutants can be high enough to cause health issues among vulnerable populations, such as pregnant women. Between 1990 and 1995, the amount of generated waste in Europe increased by ten percent. Most of the waste thrown away is either burned or deposited in landfills (67%), but both methods damage the environment. Recycling is the principal measure of the policy to reduce the negative environmental impact of accumulated waste.

Therefore, it should be borne in mind that environmentally responsible behavior is closely connected with efficient and economical use of energy and resources, use of renewables instead of fossil fuels, waste recycling, and proper management and disposal of waste water (Thogersen, 2006)

Consumption of Environmental Services

Almost 75 percent of European citizens live in urban environments. The Sixth Environment Action Programme of the EU, specifically its thematic strategy on the protection of the urban environment emphasizes the effects on the health and quality of life of city dwellers. The aim is to improve the urban environment in order to make it more attractive and healthier to live, work, and invest in, while attempting to reduce the harmful effects on the wider environment. The quality of life and health of urban dwellers greatly depends on the quality of the urban environment itself, with a highly complex interaction between the urban environment and the social, economic, and cultural factors.

The main indicators for the consumption of environmental services and amenities provided are selected based on Eurostat data. They comprise an index of sufficiency, total freshwater abstraction per capita, fish production per capita, and forested and other wooded land per capita. The increase in these indicators indicates an upsurge in the use of services provided by the environment, which has a direct positive impact on quality of life.

The index of sufficiency shows to what extent the specific Sites of Community Importance are appropriately covered by the plant and animal species as well as habitats listed in Habitats Directive Annexes I and II. This is an important quality-of-life indicator connected to the measures for biodiversity protection.

With regard to water, as a life-essential amenity, it is also a vital economic resource with a key role in the regulation cycle of climatic factors. There are significant differences between the amounts of abstracted water per capita for each EU member country. This partly reflects the availability of resources, as the country's climate, as well as industrial and agricultural structure, are also key factors.

The fish stock in water is considered a natural resource provided by the environment to fulfil human needs. Fish are a natural, biological, and renewable resource. With the exception of fish farming, fish found in the wild are usually not owned until caught, although some lakes and rivers may be privately owned.

In Europe, forests have been traditionally associated with their primary function, which is to provide wood for further processing. However, their function as recreation or tourist sites has become increasingly important in many countries in the region – particularly their benefits for economic growth, health and well-being, and quality of life.

Other important indicators include access to green spaces and satisfaction with the quality of local environment; unfortunately, these are not included in Eurostat databases.

Besides, it should be noted that, in addition to objective data and indicators, subjective data on environmental quality also provide some key information about environmental conditions.

One of the subjective indicators regarding access to green spaces refers to the complaints about the lack of access to recreational or green zones. Access to green spaces is crucial to quality of life, because undisturbed environments provide a source of satisfaction (Milligan et al., 2004), improve mental well-being (Brown & Grant, 2007), and allow people to recover from everyday stress (Mace et al., 1999) and to pursue physical activities. The plant composition is a fundamental element in public green spaces, improving the environment and people's quality of life (Santos et al., 2021).

This leads to a conclusion that the three aforementioned groups of indicators positively impact environmental quality, which in turn results in higher consumption of environmental services (Štreimikienė, 2015). Perhaps the finest example of the claim that environmentally responsible behavior directly improves environmental quality can be found in the footage and images testifying that nature has noticeably regenerated itself during the isolation of worldwide populations due to the Covid-19 pandemic, following an extended period of pollution. With people spending less time in nature and having fewer opportunities to behave environmentally irresponsible, nature has been given a chance to restore itself.

Keles' Categorization of Environmental Factors

Keles (2012) offered a categorization of environmental aspects that can either improve or deteriorate quality of life. According to him, the first and the most crucial factor is access to environmental infrastructure and services, such as water and sanitation systems, solid waste management, drainage, and transportation. When people do not have adequate access to the said amenities or when their quality is inferior, a series of health issues immediately arise.

The second factor is the pollution due to urban waste and activities, which significantly affects every day urban life, causing air and water pollution and land degradation. It has been estimated that 300 to 700 premature deaths might be avoided every year if a minimum of WHO's clean air standards is followed in practice. According to WHO statistical data, environmental conditions are responsible for one quarter of the deaths caused by respiratory or other infectious diseases (Keles, 2010).

The third factor affecting quality of life is resource degradation. For instance, urban development can damage neighboring ecosystems through construction on sensitive and fertile land, as well as through inadequate disposal of urban and industrial waste. Cultural and historical heritage is another resource prone to loss due to neglect and ignorance (Oktay et al., 2009).

The fourth factor involves environmental hazards from natural sources, such as earthquakes, floods, etc., as well as from anthropogenic sources, such as industrial, transport, or fire-related accidents. Finally, global environmental issues, such as greenhouse gases, rise in sea levels, climate change, and pollution of international waters, generate considerable risk to the environment (Leitmann, 1999).

Among the aforementioned factors, it should be noted that the decisions regarding the use of urban land are crucial determinants for environmental protection, and thus urban and rural life. Fluctuations in the real estate markets combined with inefficient land management policies and practices have resulted in the degradation of environmental capacities.

From a different standpoint, a certain level of quality that guarantees sustainable welfare for the people could also have socio-cultural, economic, and spatial characteristics. The fact that people have wishes that they intend to fulfil is what constitutes the socio-cultural dimension of the issue. On the other hand, if they wish to consume specific material and non-material goods on a specific level, there is a case of an essentially economic dimension of the issue – limited resources. To fulfil their wishes, people use the space around them for living, working, and recreation. Considering that they create new physical conditions, i.e. a new environment for living, working, and recreation, spatial factors will necessarily play a special part in raising the level of quality of life. There are certain indicators that these changes the equilibrium between processes in the biological and non-biological spheres is disturbed by these changes. This is why it is of paramount importance for people to behave environmentally responsibly (Van der Heuvel, 1992).

Environment and Quality of Life in Serbia

In Serbia, there is a very pronounced difference between its regions. The southern and eastern regions are characterized by their less developed economy. Therefore, the quality of life of people in Serbia differs from region to region. In less developed parts of the country, there are untapped natural potentials and resources, which could be used in a sustainable way. By exploiting these potentials, the quality of life in the area would improve. Replacing non-renewable energy sources with renewable sources is a form of sustainable use of natural resources (Furlan & Mortarino, 2018), and it can be said that Serbia is rich in renewable energy sources. It is certainly important to use these sources, but only their use largely depends on the degree of environmental awareness (Arundati et al., 2020). Sustainable development focuses on meeting the needs of people and overcoming the conflict between the economy and the environment. By realizing this model in practice, it is possible to create conditions so that the economy and ecology can go hand in hand. In order for this development to proceed without contradictions, it is necessary to simultaneously respect social issues, starting from the lowest (local) to the highest (global) level of their manifestation (Strbac et al, 2012). An aesthetically uncultivated environment has an adverse effect on humans. A clean environment, aesthetically refined, is a precondition of modern ecological culture (Koković, 2010). In Serbia, a relatively low level of environmental awareness is expressed, and thus the average citizen does not have a positive attitude towards the need to act in a way to reduce pollution, use energy rationally and change the attitude towards non-renewable resources. According to the reports included in the National Development Strategy of Serbia, almost a quarter of the citizens do not have enough self-confidence, and do not trust other people's views, while in some absurd way, they still believe in their superiority and omniscience. Due to the weak economic situation, the youth in Serbia have little chance to build better lifestyles. Considering the fact that the quality of life refers to both the individual and the region, i.e. the local community, the mentioned aspects include the psychosocial and economic aspects of quality. The economic aspect is related to investments and the quality of the economy. The psychosocial aspect is associated with physical and mental health, social security, social institutions (health, education, court), political stability and the environment (Arsovski, 2005). It is necessary to find ways to use the natural potentials of Serbia, and thus to accelerate the economic development of the region, through social welfare and more pronounced care for the environment. However, no major progress can be achieved without adequate education, development policy, and without raising social awareness. In the end, it is concluded that sustainable development is related to the quality of all processes which man undertakes and in which he is the main actor. Raising awareness of a different, more careful treatment of nature, and thus indirectly of society, leads to raising the quality of life.

CONCLUSION

Environmental management directly or indirectly influences the overall quality of life and life satisfaction. When environmental management is reliant upon the principles of quality management, it can be expected to have an effective impact on all quality-of-life elements listed by Eurostat. The principles refer to environmental management, which is achieved through several standard requirements. Consequently, the principles of quality management can be adjusted to apply to environmental management. Concerning the principles of quality management adjusted to the environment, there is the principle of customer focus, which emphasises the environment and the stakeholders. The principle of leadership focuses on teamwork, environmental policy development, and continuous learning about the ways to protect the environment. The principle of people involvement states that the environment can be conserved only if all members of a social community work together toward this goal. The principle of the process approach means that the environment is either preserved or destroyed in various processes. The principle of continual improvement is the core principle for managing the demand to improve environmental protection every year. The principle of a factual approach to decision-making refers to the re-examination and application of scientific methods planned at all levels of decision-making regarding the environment. Finally, the principle of mutually beneficial supplier relations refers to the development of relations with all the relevant environmental actors for the purpose of coordinating environmental protection. The practical significance of this paper is that we need to look at all human activities that lead to environmental pollution in order to be able to adequately analyze the relationship between quality of life and the ecological state of the environment.

FUTURE RESEARCH DIRECTIONS

Future research should address some of the curtail quality of life issues at the forefront of local as well as global political and environmental agendas at the beginning of the 21st century. Also, future research should examine how and in what way satisfying and endangering the environmental aspects of quality of life affects society and how it affects the individual. Future research should examine the relationship between the individual and collective responses to environmental problems, and examine the extent to which the subjective perception of quality of life coincides with objective measures of environmental assessment. Because quality of life depends not only on the physical and social 'quality' of the environment, but is also a result of the way people interact with their environment, future research should examine the possible ways of human interaction with their environment. Also, they should examine what kind of environment we need in order to have a sense of subjective well-being and better quality of life.

ACKNOWLEDGMENT

This research was supported by the European Commission under the Project "Soil Erosion and Torrential Flood Prevention: Curriculum Development at the Universities of Western Balkan Countries (SETOF)" [grant number: 598403-EPP-1-2018-1-RS-EPPKA2-CBHE-JP] and by the Ministry of Education, Science and Technological Development of the Republic of Serbia (Agreement No. 451-03-9/2021-14/200148).

REFERENCES

Ahmad, N., & Yamano, N. (2011). *Carbon dioxide emissions embodied in goods and services: Domestic consumption versus production*. OECD Science, Technology and Industry Working Papers, No. 2003/15. Paris: OECD Publishing.

Alderfer, C. P. (1969). An empirical test of a new theory of human needs. *Organizational Behavior and Human Performance*, *4*(2), 142–175. doi:10.1016/0030-5073(69)90004-X

Allardt, E. (1977). On the relationship between objective and subjective predicaments (No. 16). Helsinki: Research Group for Comparative Sociology, University of Helsinki.

Arruti, A., Fernández-Olmo, I., & Irabien, Á. (2010). Evaluation of the contribution of local sources to trace metals levels in urban PM2. 5 and PM10 in the Cantabria region (Northern Spain). *Journal of Environmental Monitoring*, *12*(7), 1451–1458. doi:10.1039/b926740a PMID:20517581

Arsovski, S. (2005). Put od kvaliteta do kvaliteta života. *Zbornik radova Festival kvaliteta, 32*, 3-16.

Arundati, R., Sutiono, H. T., & Suryono, I. A. (2020). Effect of ecological awareness, personal norms and ecological attitude to conservation behavior. *Proceedings on Engineering*, *2*(2), 187–196. doi:10.24874/PES02.02.009

Balestra, C., & Dottori, D. (2012). Aging society, health and the environment. *Journal of Population Economics*, *25*(3), 1045–1076. doi:10.100700148-011-0380-x

Balestra, C., & Sultan, J. (2013). *Home sweet home: The determinants of residential satisfaction and its relation with well-being*. OECD Statistics Working Papers 2013/05. Paris: OECD Publishing.

Brajša-Žganec, A., Merkaš, M., & Šverko, I. (2011). Quality of life and leisure activities: How do leisure activities contribute to subjective well-being? *Social Indicators Research*, *102*(1), 81–91. doi:10.100711205-010-9724-2

Brown, C., & Grant, M. (2007). Natural medicine for planners. *Town and Country Planning-Special Issue: Planning for Healthy Communities*, *76*(2), 67–69.

Coan, T. G., & Holman, M. R. (2008). Voting green. *Social Science Quarterly*, *89*(5), 1121–1135. doi:10.1111/j.1540-6237.2008.00564.x

Costanza, R., Fisher, B., Ali, S., Beer, C., Bond, L., Boumans, R., ... Gayer, D. E. (2008). An integrative approach to quality of life measurement, research, and policy. *Sapiens*, *1*(1), 17–21. doi:10.5194apiens-1-11-2008

Council of Europe Parliamentary Assembly. (2009). *Drafting an additional protocol to the European Convention on Human Rights concerning the right to a healthy environment*. Recommendation 1885. Strasbourg Cedex, France: Council of Europe.

Daniel, A., & Michaela, C. (2021). Mental health and health-related quality of life in victims of the 2013 flood disaster in Germany–A longitudinal study of health-related flood consequences and evaluation of institutionalized low-threshold psycho-social support. *International Journal of Disaster Risk Reduction*, *58*, 102179. doi:10.1016/j.ijdrr.2021.102179

De Groot, J., & Steg, L. (2006). Impact of transport pricing on quality of life, acceptability, and intentions to reduce car use: An exploratory study in five European countries. *Journal of Transport Geography*, *14*(6), 463–470. doi:10.1016/j.jtrangeo.2006.02.011

Diener, E. (2000). Subjective well-being: The science of happiness and a proposal for a national index. *The American Psychologist*, *55*(1), 34–43. doi:10.1037/0003-066X.55.1.34 PMID:11392863

Diener, E., & Biswas-Diener, R. (2011). *Happiness: Unlocking the mysteries of psychological wealth*. John Wiley & Sons.

Diener, E., Lucas, R., Helliwell, J. F., Schimmack, U., & Helliwell, J. (2009). *Well-being for public policy*. Oxford University Press. doi:10.1093/acprof:oso/9780195334074.001.0001

Diener, E., & Suh, E. (1997). Measuring quality of life: Economic, social, and subjective indicators. *Social Indicators Research*, *40*(1-2), 189–216. doi:10.1023/A:1006859511756

Eurostat. (2015). *Quality of life: Facts and views* (2015 edition). Luxembourg: Publications Office of the European Union.

Fahey, T., Whelan, C. T., & Maître, B. (2005). *First European quality of life survey: Income inequalities and deprivation*. Office for Official Publications in the European Communities.

Fawcett, W., Ellingham, I., & Platt, S. (2008). Reconciling the architectural preferences of architects and the public: The ordered preference model. *Environment and Behavior*, *40*(5), 599–618. doi:10.1177/0013916507304695

Felix, R., & Garcia-Vega, J. (2012). Quality of life in Mexico: A formative measurement approach. *Applied Research in Quality of Life*, *7*(3), 223–238. doi:10.100711482-011-9164-4

Ferriss, A. L. (2006). A theory of social structure and the quality of life. *Applied Research in Quality of Life*, *1*(1), 117–123. doi:10.100711482-006-9003-1

Furlan, C., & Mortarino, C. (2018). Forecasting the impact of renewable energies in competition with non-renewable sources. *Renewable & Sustainable Energy Reviews*, *81*, 1879–1886. doi:10.1016/j.rser.2017.05.284

Galinha, I. C. C., & Pais-Ribeiro, J. L. (2008). The structure and stability of subjective well-being: A structure equation modelling analysis. *Applied Research in Quality of Life*, *3*(4), 293–314. doi:10.100711482-009-9063-0

Gasper, D. (2010). Understanding the diversity of conceptions of well-being and quality of life. *Journal of Socio-Economics*, *39*(3), 351–360. doi:10.1016/j.socec.2009.11.006

Goldberg, M. S., Burnett, R. T., Bailar, J. C. III, Brook, J., Bonvalot, Y., Tamblyn, R., Singh, R., & Valois, M. F. (2001). The association between daily mortality and ambient air particle pollution in Montreal, Quebec: 1. Nonaccidental mortality. *Environmental Research*, *86*(1), 12–25. doi:10.1006/enrs.2001.4242 PMID:11386737

Henderson, H. (2002). Quality of life: Issues and questions. In D. Lamberton (Ed.), *Managing the global: Globalization, employment and quality of life (Humanity and security and global governance)* (pp. 51–95). I. B. Tauris.

Inglehart, R., & Klingemann, H. D. (2000). Genes, culture, democracy, and happiness. Culture and subjective well-being. In E. Diener & E. M. Suh (Eds.), *Culture and subjective well-being* (pp. 165–183). MIT Press.

Keles, R. (2010). Globalization and Ecological Changes, *20th Congress of Spiritual Health and Diseases*, Bodrum.

Keles, R. (2012). The quality of life and the environment. *Procedia: Social and Behavioral Sciences, 35*, 23–32. doi:10.1016/j.sbspro.2012.02.059

Koković, D. (2010). Ekologija kao način života. *Svarog: Časopis za društvene i prirodne nauke, 1*(1), 75-84.

Leitmann, J. (1999). *Sustaining cities: Environmental planning and management in urban design.* McGraw-Hill.

Liere, K. D. V., & Dunlap, R. E. (1980). The social bases of environmental concern: A review of hypotheses, explanations and empirical evidence. *Public Opinion Quarterly, 44*(2), 181–197. doi:10.1086/268583

Mace, B. L., Bell, P. A., & Loomis, R. J. (1999). Aesthetic, affective, and cognitive effects of noise on natural landscape assessment. *Society & Natural Resources, 12*(3), 225–242. doi:10.1080/089419299279713

Milligan, C., Gatrell, A., & Bingley, A. (2004). Cultivating health. Therapeutic landscapes and older people in Northern England. *Social Science & Medicine, 58*(9), 1781–1793. doi:10.1016/S0277-9536(03)00397-6 PMID:14990378

Oktay, D., Rüstemli, A., & Marans, R. W. (2009). Neighborhood satisfaction, sense of community, and attachment: Initial findings from Famagusta quality of urban life study. *ITU A/Z Journal, 6*(1), 6-20.

Osbaldiston, R., & Sheldon, K. M. (2003). Promoting internalized motivation for environmentally responsible behavior: A prospective study of environmental goals. *Journal of Environmental Psychology, 23*(4), 349–357. doi:10.1016/S0272-4944(03)00035-5

Pastuović, N., Kolesarić, V., & Križmanić, M. (1995). Psychological variables as predictors of quality of life. *Review of Psychology, 2*(1-2), 49–62.

Pickett, S. T., Cadenasso, M. L., Grove, J. M., Nilon, C. H., Pouyat, R. V., Zipperer, W. C., & Costanza, R. (2001). Urban ecological systems: Linking terrestrial ecological, physical, and socioeconomic components of metropolitan areas. *Annual Review of Ecology and Systematics, 32*(1), 127–157. doi:10.1146/annurev.ecolsys.32.081501.114012

Rastogi, K., Surekha, C., Nayar, S., & Gupta, T. (2021). Impact of Urban Environmental Quality, Residential Satisfaction, Personality on Quality of Life among Delhi Residents. *SPAST Abstracts, 1*(1).

Santos, E. B. D., Nogueira, F. M., & Talgatti, D. M. (2021). Plant Species Composition and the Perception of the Afforestation in Urban Public Green Spaces in a Municipality in Eastern Brazilian Amazon. *Sustainability, 13*(18), 1–15. doi:10.3390u131810332

Seferagić, D., & Popovski, V. (1989). Kvaliteta života kao cilj društvenog razvoja. *Revija za Sociologiju, 20*(1-2), 73–87.

Shucksmith, M., Cameron, S., Merridew, T., & Pichler, F. (2009). Urban–rural differences in quality of life across the European Union. *Regional Studies, 43*(10), 1275–1289. doi:10.1080/00343400802378750

Sirgy, M. J. (1986). A quality-of-life theory derived from Maslow's developmental perspective: 'Quality' is related to progressive satisfaction of a hierarchy of needs, lower order and higher. *American Journal of Economics and Sociology, 45*(3), 329–342. doi:10.1111/j.1536-7150.1986.tb02394.x

Slavuj, L. (2012). Objective and subjective indicators in the research of the quality of life concept. *Geoadria, 17*(1), 73–92. doi:10.15291/geoadria.238

Steg, L., Abrahamse, W., Vlek, C., Wiersma, G., Noorman, K. J., Oostra, M., Van Kampen, P., Smeets, P., Graas, S., Grimberg, H., & Roos, J. (2002). *Energiebesparing door gedragsverandering. De invloed van informatie, feedback en commitment op energiegebruik, psychologische achtergronden en kwaliteit van leven (Reducing energy use via behavioural change)*. University of Groningen, Department of Psychology.

Štrbac, N., Vuković, M., Voza, D., & Sokić, M. (2012). Održivi razvoj i zaštita životne sredine. *Zbornik radova: Reciklaža i održivi razvoj*, 18 – 29.

Štreimikienė, D. (2015). Environmental indicators for the assessment of quality of life. *Intelektinė Ekonomika, 9*(1), 67–79. doi:10.1016/j.intele.2015.10.001

Thogersen, J. (2006). Norms for environmentally responsible behaviour: An extended taxonomy. *Journal of Environmental Psychology, 26*(4), 247–261. doi:10.1016/j.jenvp.2006.09.004

Van der Heuvel, H. M. (1992). The quality of the daily environment: Ways of cooperation. *Urban Ecology: Seventh Conference on Urban and Regional Research. Ankara: General Directorate of Technical Research and Implementation.*

Wackernagel, M., Schulz, N. B., Deumling, D., Linares, A. C., Jenkins, M., Kapos, V., Monfreda, C., Loh, J., Myers, N., Norgaard, R., & Randers, J. (2002). Tracking the ecological overshoot of the human economy. *Proceedings of the National Academy of Sciences of the United States of America, 99*(14), 9266–9271. doi:10.1073/pnas.142033699 PMID:12089326

WHO. (1993). *Report of WHOQOL Focus Group Work*. World Health Organization.

World Commission on Environment and Development. (1987). *Our common future*. Oxford University Press.

Zheng, Y. (2010). Association analysis on pro-environmental behaviors and environmental consciousness in main cities of East Asia. *Behaviormetrika, 37*(1), 55–69. doi:10.2333/bhmk.37.55

ADDITIONAL READING

Baumol, W. J., & Oates, W. E. (1993). *Economics, environmental policy, and the quality of life*. Gregg Revivals.

Conant, J. K., & Balint, P. J. (2016). *The Life Cycles of the Council on Environmental Quality and the Environmental Protection Agency: 1970-2035*. Oxford University Press. doi:10.1093/oso/9780190203702.001.0001

Gelissen, J. (2007). Explaining popular support for environmental protection: A multilevel analysis of 50 nations. *Environment and Behavior, 39*(3), 392–415. doi:10.1177/0013916506292014

Levett, R. (1998). Sustainability indicators—Integrating quality of life and environmental protection. *Journal of the Royal Statistical Society. Series A, (Statistics in Society)*, *161*(3), 291–302. doi:10.1111/1467-985X.00109

Li, Y. (2018). Study of the effect of environmental education on environmental awareness and environmental attitude based on environmental protection law of the People's Republic of China. *Eurasia Journal of Mathematics, Science and Technology Education*, *14*(6), 2277–2285. doi:10.29333/ejmste/86214

Martinez, J., Mikkelsen, C. A., & Phillips, R. (Eds.). (2021). *Handbook of quality of life and sustainability*. Springer. doi:10.1007/978-3-030-50540-0

Muralidharan, K. (2021). *Sustainable Development and Quality of Life: Through Lean, Green and Clean Concepts*. Springer Nature. doi:10.1007/978-981-16-1835-2

KEY TERMS AND DEFINITIONS

Ecological Awareness: Is to be informed and know the impact of our actions and activities on our environment and the ecosystems around us: individually or collectively, locally and globally.

Environmental Protection: Considers the full preservation of environmental quality, biodiversity, and geodiversity, and the rational use of natural resources, which is a basic condition for sustainable development, a healthy life, and a good quality of life.

Environmental Quality: Reflects the ability of environment to meet the basic human needs.

Quality of Life: Reflects the social dimension of sustainable development and represents the range within which individual needs are met.

Sustainability: Reflects the balanced relationship between people and the environment. It refers to a balance between environmental, social, and economic qualities.

Chapter 2
Natural Disasters and Risk Management: A Theoretical Overview

Milica Jovanović Vujatović
Innovation Center, University of Niš, Serbia

Sandra Milanović
https://orcid.org/0000-0002-0582-045X
Innovation Center, University of Niš, Serbia

Ivana Janjić
Innovation Center, University of Niš, Serbia

ABSTRACT

In recent years, natural disasters have compelled public authorities, organizations, and citizens to increase their efforts in properly planning and implementing effective risk management procedures. Accordingly, in literature contemporary concepts such as natural disaster risk management and crisis management emerged. Therefore, the chapter aims to shed light on the significance of natural disaster risk management and crisis management in the development of an effective societal system by its transformation and to point out the positive and negative factors influencing these management activities. The authors will firstly give an overview of these two concepts, their elements, and development phases, and afterward, the investigation of possible positive and negative factors of natural disaster risk management will be introduced. The chapter will make a significant contribution to filling the gap in the literature on mitigating the influence of natural disasters and risk management.

INTRODUCTION

In the contemporary and turbulent business environment, increased competition and accelerated dynamics of social, economic, and political processes (Janjić & Rađenović, 2019) as well as the process of glo-

DOI: 10.4018/978-1-7998-8459-0.ch002

balization, are the cause of various economic, social, environmental, and political crises. Globalization can be observed as a well-known and recent phenomenon in human history, which creates new ideas, identities, and competition (Janjić et al., 2020), where competitiveness presents the basic assumption of the economic prosperity of each country (Krstić & Janjić, 2019). Crises occur and evolve in parallel with the development of human societies. The epochs of time, which were marked by various crises, were imbued with numerous socio-cultural changes. Despite the tendencies of the modern unpredictable and complex environment (Jovanović et al., 2020) and efforts of organizations to reduce the factors that can lead to accidents, crises are inevitable due to the complexity of modern technology, human factor, and the very interdependent nature of the industry (Ray, 1999).

Crisis management and risk management are relatively new concepts, both theoretically as teaching and scientific disciplines, research fields, and practice in corporate business, government administration, public companies, and non-profit organizations. The belated and slow entry of these concepts into the academic sphere was influenced by terminological confusion related to different types of proactive and reactive management of dangers and uncertainties that threaten society, individual organizations, and groups. The crisis contains the seeds of success, as well as the root of failure. Finding potential for success is the essence of crisis management (Kyhn, 2008).

Climate changes are becoming more serious, and the consequences are being observed on all continents through numerous natural disasters. Natural disasters such as earthquakes, floods, droughts, landslides, erosions pose the greatest threats to sustainable development globally. Natural disasters are a natural syndrome and part of human existence, but they are among the most common types of crises, which produce significant material damage, causing the greatest danger to the economy and life conditions. Predicting, preventing, and reducing the impact of disasters is possible through the systematic efforts of the state and companies, which are reflected in the adequate management of crisis and risk, adoption of appropriate regulations, engagement and training of human resources, and using appropriate equipment and resources.

The aim of this chapter is to point out the significance of natural risk management and crisis management in the development of an effective societal system by its transformation. The purpose of the chapter is to fill the gap in the literature and merge theoretical and practical solutions in the field of natural disaster risks management.

The chapter is structured as follows: Firstly, it is presented the theoretical background of natural risk management. Secondly, the authors give an overview of the characteristics of natural disaster risk management and crisis management. The third part of the chapter points out the solutions and recommendations for natural risk management in the case of societal transformation with the special emphasis on the role of the state in the case of floods in the Republic of Serbia. Lastly, future research directions and concluding remarks are given.

BACKGROUND

The process of globalization in the last decades of the twentieth century has led to political, economic, and social integration, reducing the importance of national boundaries (Jovanović & Đokić, 2020). In globalization and the knowledge economy, knowledge and ideas at the corporate, national, and global levels play a key role in solving various problems (Jovanović, 2018). The economic reality of modern companies and national economies is faced with the processes of internationalization and globalization

(Jovanović & Đokić, 2019). Their interdependence emphasizes the importance of managing certain problems at the global or regional level. It is also the case with natural disaster risk management. Namely, it is necessary to look at global trends that encourage the risk of natural disasters (earthquakes, windstorms, and floods) and strive to mitigate them through policies and practices at the global and national levels (Gencer, 2013). In addition, companies are expected to continuously analyze the macro environment in terms of various exogenous factors, including environmental factors (Đorđević et al., 2020).

The drastic growth of losses due to natural disasters began in the 1950s, as a result of the concentration of the population in urban areas, the development of coastal areas, as well as the way of life of people, and the development of technologies, but also due to global warming (consequential changes in the natural environment). Without planned measures at the global level, such trends will deepen further (Smolka, 2006). At the global level, natural disasters kill on average 60,000 people per year. It was responsible for 0.1% of deaths over the past decade. In addition, droughts and floods were the most fatal disaster events (Ritchie & Roser, 2014). According to the Emergency Events Database (EM-DAT, 2017), that provides The Centre for Research on the Epidemiology of Disasters (CRED), the total number of people affected by disasters in 2016 (569.4 million) was the highest since 2006, far above its 2006-2015 annual average (224.1 million). Taking into account the distribution of disaster occurrences across continents for the 2006-2016 period, the data are as follows: Asia was most often hit (46.7%), followed by America (24.3%), Africa (16.9%), Europe (8.2%), and Oceania (3.8%).

In the period from 2010 to 2015, the biggest natural disasters in Serbia included: extreme temperatures, floods and fires. The floods affected 1,615,820 people, with 57 deaths during this period. When it comes to extreme temperatures, they affected 118,264 people and the damage amounted to 132,260,000 US dollars. Also, in the same period, two people died as a result of the earthquake (Petrović & Grujović, 2015).

Disaster is defined as a "situation or event which overwhelms local capacity, necessitating a request to the national or international level for external assistance, or is recognized as such by a multilateral agency or by at least two sources, such as national, regional or international assistance groups and the media" (Centre for Research on the Epidemiology of Disasters, 2004). Natural disasters involve natural variations that call into question the socio-economic development and lives of people. It is the result of variations on the earth's surface (Zhang et al., 2006). Although the term natural disaster indicates a process caused by nature, this understanding of the term does not indicate the origin or initiator of that process. It is also called a natural hazard. The following can be distinguished: 1) geo-tectonic hazards - processes originating from the earth's crust such as volcanic eruptions, earthquakes, uplift or subsidence of land, and 2) hydrometeorological hazards - processes originating from the earth's atmosphere that led to floods, droughts, storms and extreme weather conditions (Ranke, 2016). Also, there are other explanations of these phenomena in the literature. Hazard is any phenomenon that can cause the destruction of life and property. It becomes a disaster when the potential to cause destruction is fulfilled. When there is a danger to human life and property, a hazard is called a disaster. This means that not all hazards are disasters (Babu, 2017).

Natural disaster risk can be seen as the exposure or chance of loss (human, physical, economic, natural, social) due to the danger that threatens a certain area in a certain period. It is the probability that the influence of that danger will occur multiplied by the implications of that influence (Asian Disaster Preparedness Centre, 2000). Exposure involves people, property, and other segments that are potentially subject to losses. It refers to people or various assets in a dangerous area. On the other hand, the risk is the possibility of negative consequences taking into account the probability that a risky event will occur. Disaster is the inability of the affected community to solve the problem of negative consequences due to human, material, economic and environmental losses with its resources (United Nations International Strategy for Disaster Reduction, 2009).

About 90% of natural disasters in Europe are directly or indirectly conditioned by weather and climate. About 95% of the economic losses caused by catastrophic events are the result of these weather and climate disasters. The average number of annual disasters caused by weather and climate in Europe has increased by about 65% in the period 1998-2007 compared to an annual average in 1980 (Popović et al., 2009).

The economic consequences of natural disasters can be seen as direct and indirect. Direct losses include losses due to the destruction of the property of private entities and the business sector, but also the property of the state - bridges, roads, schools, hospitals, communication infrastructure. Indirect effects can be observed through a slowdown or interruption of economic activity - reduced demand, interruption in production, lack of input, interruption in distribution channels. Human and economic losses are particularly high among the poorer population (Szlafsztein, 2015).

MAIN FOCUS OF THE CHAPTER – NATURAL DISASTER RISK MANAGEMENT AND CRISIS MANAGEMENT

Theoretical Framework of Crisis Management

Crisis management as a scientific-theoretical discipline and a rationally designed concept of pragmatic action entered the historical scene in the second half of the 20th century. The origin of this term is in the political sphere. It is believed that US President J. F. Kennedy was the first which used the term crisis management during the Cuban Missile Crisis in 1962, when Cold War tensions culminated in the installation of Soviet nuclear warheads in Cuba (Kešetović & Toth, 2012). On the other hand, crisis management is a relatively new concept in theoretical and practical terms. Implementation of crisis management is accompanied by terminological ambiguity related to basic phenomena - emergency, crisis, conflict, and catastrophe, but also different types of management that represent an attempt to manage uncertainties and dangers.

Crisis management has been thoroughly researched in the past few years. There are numerous definitions, and diverse authors have contributed to the basis of this concept (Fearn-Banks, 2002; Boin et al., 2005; Elliot et al., 2005; Glaesser, 2006; Coombs, 2007; Kouzmin, 2008). Comprehensiveness is still being accomplished, and the definition is still a work in progress. The ability of an organization to act quickly, efficiently, and effectively in possible operations aimed at reducing future threats to human health and safety, minimizing damage to public or private corporations, and minimizing the negative impact on the company operations is known as crisis management (Drennan & McConnell, 2007). Other scholars have developed similar conceptual approaches, linking crisis management with dealing with unforeseen occurrences. Crisis management can be defined as a holistic approach that begins with readiness and prevention and continues through the process of reaction, recovering, and learning (Fearn-Banks, 2002).

The task of crisis management should be to demonstrate quick and precise corrective actions, taking into account the needs of stakeholders and society, a clear presentation of the recovery strategy, communication in clear terms, delivery of consistent messages of the recovery plan. The crisis team should gather as many facts about the situation as possible, in an attempt to get a full picture of it in order to take management actions (Aghahowa, 2008).

Figure 1. Crisis management and similar concepts
Source: Adapted to Kešetović, 2008

The Crisis Management and Similar Related Concepts

Apart from crisis and crisis management, there are similar yet separate concepts, linked by a similar philosophical framework. Therefore, it is necessary to distinguish between the following terms: issue management, risk management, emergency management, and disaster (catastrophe) management (Figure 1).

Issue management is a management process aimed at preserving markets, reducing risk, creating opportunities, and managing an organization's image as an asset for the benefit of both the firm and its shareholders (Gaunt & Ollenburger, 1995). There are few differences between issue and crisis management. Issue management means primarily identifying trends and events that may have an impact on the company, whereas crisis management is a more reactive discipline that mainly focuses on the situation after a crisis has occurred (Monstad, 2003). In comparison with crisis management, issue management focuses less on the action and more on anticipation. Issue management is proactive in determining the possibility for change and influencing the implementation of appropriate decisions before the changes have a detrimental impact on the company.

The changing corporate environment, which is defined by risks from political, economic, natural, and technical resources, indicates the importance of risk management (Wu & Olson, 2010). *Risk management* includes the management activities of organizing, coordinating, planning, controlling, and directing. It is a systematic way to measuring, evaluating, and managing the risk that an organization faces by aligning strategy, people, technology, processes, and knowledge. In the contemporary world, risk management is the process of assessing risk and implementing risk management methods. It is a matter of determining the extent to which an organization is vulnerable to prospective losses and determining the best course of action for dealing with such risks.

Emergency management is "a complex policy subsystem that involves an intergovernmental, multi-phased effort to mitigate, prepare for, respond to, and recover from disasters" (Donahue & Joyce, 2001). An emergency is a sudden, fast-occurring, usually unexpected event or happening that requires quick

Figure 2. The main stages of crisis management
Source: Mitroff, 1994

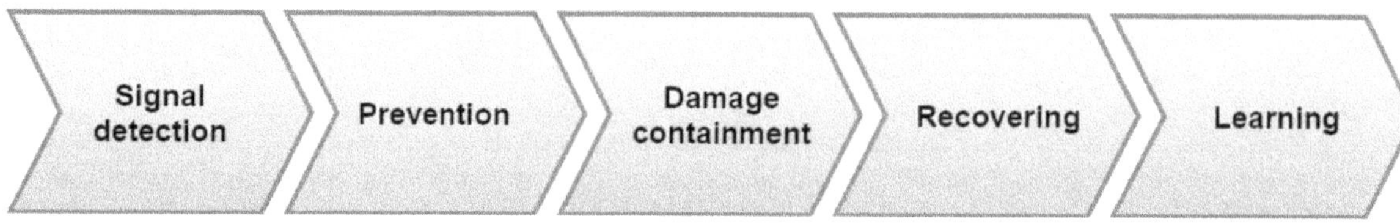

action (Keković et al., 2011), which can be prepared for, but rarely with precision. It is a broader term than crisis, because every crisis is also a certain emergency, while every emergency does not have to be a crisis. (Kešetović, 2008). Emergencies usually evolve during the incubation period, so that they often grow unnoticed and lead to an acute crisis.

Disaster management (catastrophe management) entails taking steps to mitigate or eliminate the consequences of natural and other disasters of catastrophic proportions. It differs from risk management in the sense that catastrophe, as an event that cannot be handled in theory. Disaster is "a situation that overwhelms local capacity, necessitating a request for external assistance at the national and international levels, or is recognized by a multilateral agency or at least two sources, such as national, regional, or international assistance groups and the media" (Moe et al., 2007, p. 787). Disaster management initiatives, strive to prevent or avoid possible losses from hazards, promote early and appropriate aid to disaster victims, and seek to achieve rapid and successful recovery.

The Main Stages of Crisis Management

In conditions of natural disasters, there are different approaches to the development and implementation plans of management crises, intending to find effective ways, methods, and crises management tools for timely identification and effective response to the risk and crisis events. In the field of crisis management theory and practice, different scholars (Mitroff & Pearson, 1993; Pheng et al., 1999; Coombs, 2007), suggested some stages of crisis management (Figure 2).

Stage One: Signal detection provides the proactive measures that a company can take before a problem occurs. Signal detection is the process of detecting warning indicators of impending problems within and out of an organization. Crisis signals can be any type of qualitative or quantitative data of "organizational imperfections" that indicates a departure from normality. There have also been various attempts to develop early warning systems, with an emphasis on natural disasters (Zschau & Küppers, 2003), geophysics and financial markets (Sornette, 2004), and financial crisis (Berg & Pattillo, 1999).

Stage Two: Prevention entails attempting to mitigate recognized hazards that could result in a crisis. The activities which are undertaken to avoid a natural phenomenon or prospective hazard from having detrimental impacts on people or economic assets are referred to as prevention. To reduce or eliminate disaster impact on the company, society, and the environment its necessary to define regulatory and physical steps for timely recognition, early warning, and effective intervention at an early stage of crisis development.

Stage Three: Damage containment includes a system of measures to address the negative consequences caused by the crisis. Damage containment is entirely based on reducing the harm from unavoidable events as it occurs after a crisis has happened and averting it from spreading damage to untouched areas of an organization or environment (Canyon, 2013).

Stage Four: The main goal of the recovering stage is to recover from the crisis by performing and adopting short-term and long-term recovery strategies and programs, to help the company resume normal operations. Recovery by natural disaster impacts entails the restoration of normal community activities and stabilization of the disaster conditions, as well as provides an opportunity to enhance organizational capacity to support economic, social, and physical development (Berke et al., 1993).

Stage Five: Learning presents the last phase of crisis management. Learning is achieved by a continuous process of critical examination of previous practices, improving performances, and achieving success. The importance of this stage is represented in recognition of the opportunity to learn from previous catastrophes and to avoid future crises (Kovoor-Misra et al., 2000).

The Main Elements of Natural Disaster Risk Management

The literature provides a distinction between the terms "crisis management" and "risk management" (Zhang et al., 2006). Disaster risk management is defined as a set of planned actions - projects and programs, and tools used to reduce disaster risk or mitigate damage. It is based on three elements: 1) risk assessment, 2) disaster prevention and mitigation, and 3) disaster preparedness (Deutsche Gesellschaft für Technische Zusammenarbeit, 2002). On the one hand, crisis management is based on post-disaster activities, while on the other hand, risk management is focused on the activities in the pre-disaster stage, during the disaster, and in the post-disaster stage. As a result, risk management rather than crisis management is accepted.

Risk management of natural disasters can be divided into four phases (Figure 3). The first step is to identify the risk of a natural disaster and then predict the potential impact of that disaster. Risk identification is based on hazard assessment, vulnerability studies, and risk analysis. Hazard assessment enables the determination of the possible location and probability of the occurrence of a dangerous natural phenomenon in a certain period, based on historical data, scientific information, and teams of experts for the research of this data. Vulnerability studies define the social, physical, and economic impact of that dangerous natural event. Risk analysis combines information from previous assessments and analyses based on which the probability of loss due to a dangerous event is estimated (Freeman et al., 2003).

If the anticipated risk is high, it is necessary to control it by defining certain control measures - for example, the construction of a dam to protect against floods or insurance. Planning documents that include risk reduction measures are an important component of natural disaster risk management. These documents determine the levels of local risk prevention. Furthermore, it is necessary to determine the method for risk assessment of a certain area, and then define clear measures for risk reduction. Ultimately, it is necessary to predict the costs taking into account the level of resistance. Determining whether the measures comply with the characteristics of the area - industrial activities, agricultural sector, environmental issues, cultural heritage, social and ethical values, also plays an important role. In this way, the acceptance of risk reduction measures by people from certain areas is encouraged (Ranke, 2016). Given that resources are limited, it is necessary to consider all risk control measures in order to reap the greatest benefits from them (Mechler, 2003).

The earthquake in Managua (Nicaragua) and the tropical cyclone Tracy in Darwin (Australia) which occurred in the early 1970s, indicated the importance of managing the risk of natural disasters, primarily by identifying this risk. For this purpose, the Geo Risks Research Group was set up in 1974, as well

Figure 3. Four phases of natural disasters risk management
Source: Mechler, 2003

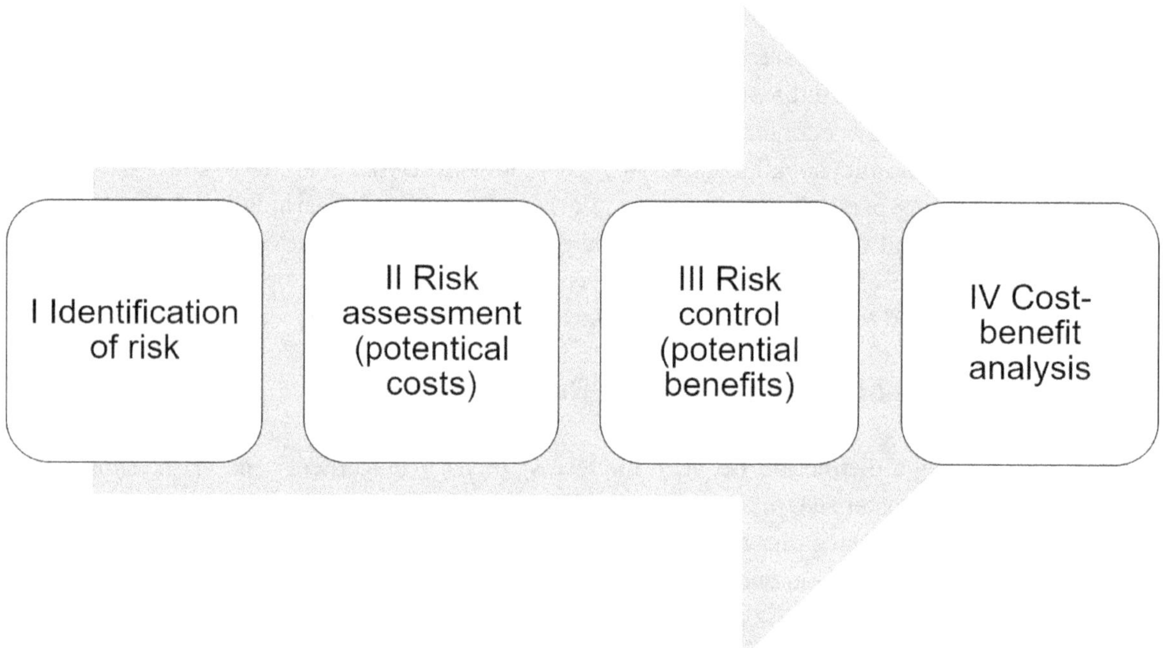

as a database of global natural disasters that facilitates the risk identification phase. The first model for assessing natural disasters' risk was developed in 1987. This model enables the determination of average annual and potential maximum losses, which is the goal of the next phase of natural disaster risk management. When it comes to the third phase, the best way to control risk is to insure against natural disasters but also through loss prevention and mitigation - 1) land-use planning, 2) construction assets and 3) contingency strategies. Finally, it is necessary to consider the possibilities for financing losses, which requires the cooperation of different entities – affected private individuals and business entities, financial institutions, and governments/public authorities (Smolka, 2006).

Over the past few years, the literature has emphasized the importance of integrated natural disaster risk management. It is the concept that includes all types of natural disasters and a consistent strategy and management plan. In addition, it suggests risk management practices in the pre-disaster period, during and after the disaster. This concept is characterized by holistic risk management with multilevel, multidimensional, and multidisciplinary collaboration. It is also the management based on performance (Mileti, 1999). This concept is built around the following elements (Zhang et al., 2006):

- a multidimensional and multidisciplinary approach is required to ensure effective disaster risk reduction;
- the development of holistic sustainable policies at the state level, while the implementation of these policies takes place from the national level to the local level;
- creating support tools for local levels;
- integrated disaster risk management becomes part of regional plans and projects;

- decisions are taken based on information obtained through hazard mapping and vulnerability assessment;
- consolidation of knowledge at all levels and its use in policy implementation and decision-making;
- promotion of the identification of responsibilities at the local and regional levels.

SOLUTIONS AND RECOMMENDATIONS FOR THE NATURAL DISASTER RISK MANAGEMENT

Natural disaster risks are differently managed in every country. For example, New Zealand has a government structure with shared responsibilities for natural risk management among all levels of jurisdiction. Management is based on the concept of 4Rs, meaning that reduction, readiness, response, and recovery are the main steps to be followed in the case of a natural disaster emergency (Crawford et al., 2018). Thaler et al. (2019) proposed the model of societal transformation as a solution for natural risk management that is grounded on the data from Austria, France, and Ireland. It is very intriguing to analyze how society coexists with natural risks in its environment (Bertoldo, 2021) and to overlook social circumstances during materialized consequences of natural disasters (Ivčević et al., 2019). This model could be an outcome of the risk analysis and the strategy for its prevention. In environmental, financial, or social crisis, all factors should be analyzed in order to take action. The transformation of a social system may bring the system into balance. However, it is important to point out that most of these changes are triggered by specific hazards and conditions in some areas. Therefore, previous risk analysis will contribute to the goals' setting, effective planning activities, and corresponding decision making (Smith et al., 2015).

According to Thaler et al. (2019), there is an equal distribution of barriers and drivers for the transformation of societies in the case of natural disaster risk management. They have studied data from several cases. Their conclusions are unprecedented that the transformation of society was i initiated mostly by individuals or other interested parties at the local level who were faced with natural disasters and their negative consequences. Except for the local citizens as drivers of societal changes who are most often at the bottom of the transformation process, there are positive examples of mutual efforts of local government and citizens to come up with a solution for natural risks. Additionally, even though the initiatives were driven by individuals, local governments (i.e., municipalities) were legal representatives of the transformation process because only they could be legally responsible for every negative consequence of the natural risk management. Nevertheless, Zuccaro et al. (2020) assert that local organizations (such as NGOs) are the ones whose aim is to disseminate all information about natural risks and to galvanize cultural changes. Examples of transformation frequently appeared after hazardous situations, such as floods, when everybody understood the importance of changes. Understanding phase as a driving force for changes was accompanied by phases in which planning and managing natural risks were transformation initiators. A phenomenon of natural disasters could be previously planned and managed with not necessary for the risk to manifest itself. Therefore, some local entities demonstrated entrepreneurship behavior and social engagement in the situation of transformation and managed to propose and implement social changes from the bottom to the top political levels. It is vital to estimate how policy-makers perceive natural risks and how their resource management, especially human, transmits on local citizens (Milanović, 2017). Some preventive measures are not only in the domain of government. Inhabitants of the area that are constantly under the pressure of natural disasters risk may eventually decide to migrate in order to mitigate its negative effects (Banerjee et al., 2019) and to change residence for the one with labor market efficiency (Milanović & Marjanović, 2019).

According to Thaler et al. (2019), barriers to societal transformation and natural disaster risk management, are drivers but with the opposite direction. So, if local citizens or local government are not interested in transformation, if there are not enough resources, most often financial (Milanović & Marjanović, 2020), or if the group cohesion is very low, it will have a negative effect on the transformation process. In some cases, institutional barriers were stronger than citizens' local activism to transform the society, and authorities were using their power-distance in order to make inhabitants obey their decisions. Furthermore, the mismatch between demand and supply of resources (such as information, capital, competence) is one of the obstacles for planning and implementation activities for both risk management and societal transformation. For example, information on the natural risks is not publicly available as much as it is preferably (Zuccaro et al., 2020). Organizations that have dedicated their work to promoting the importance of natural risks reduction and management face reduced availability of information or require a high level of education to be understood (Thaler et al., 2019). A conflict of interests between local government authorities and adequate solutions for natural disaster risk management may occur, which could cause obstacles in the process of promotion risk management proposals on the government level (Bertoldo, 2021). Lastly, de Almeida et al. (2015) advise restraint on the matter of risk management because it is mostly grounded on the data analysis of cases that have changed over time. For example, the human population has changed its occupation vastly, land use has evolved significantly, and climate changes have become a reality.

When drivers are exploited and barriers overcame, practice shows positive examples of natural disaster risk management. Moreover, if the local community finds a way to make the interconnection between natural disaster risk management and some local environmental problem, the outcome will have multiplied effects and a high probability of a permanent solution to the problem. But Thaler et al. (2019) assert that institutional changes are inevitable and will give the local community a wider range of rights to initiate societal transformation.

According to Abramov et al. (2021), natural risk management could be supported by technology. During the era of Industry 4.0, the utilization of digital tools sheds light on the pull of information required for natural risk management and decision making under the pressure of large-scale hazards. On the contrary, Krausmann et al. (2019) stipulate that the advancement in technology has brought new hazards that are following natural disasters such as explosions, nuclear contaminations, degradation, etc. This risk is called the "Natech" risk. Namely, it also requires data analysis, planning, and management activities as its source – natural risk. This type of risk must be included in natural risk management plans and procedures because it derives from it and the scope of the damage is wider if a natural disaster appears close to some technological facility (Nascimento & Alencar, 2016).

On the other hand, Bertoldo (2021) suggests that scientists are one of the factors of natural disaster risk strategies and with their competencies, they can influence public opinion. During the period of exploitation of their knowledge, according to the data from the Mediterranean area, four psychological factors emerged in natural risk management: "knowledge diversity, local history, and trust, shifting risk rationalization and risk objectification" (Bertoldo, 2021, p. 2). Firstly, when it comes to knowledge, mutual exchange must exist in the relation scientists – citizens. Scientists can give innovative solutions and propose natural risks management strategies. Those are underpinned with contemporary scientific analysis of previous experience with natural disasters and supported by technology in predicting future trends (United Nations General Assembly, 2015). Similarly, local citizens have their previous experience and strategies for risk management in natural disasters. The inclusion of business models will contribute to the development of know-how in the risk management field (Tanović et al., 2020). Hence, an exchange of knowledge will mitigate the negative consequences of natural adversity, make the decision-making process more resilient and endorse sustainable development (Crawford et al., 2018; Zuccaro et al., 2020). Secondly, local history may be a significant factor of natural risk management as a common awareness of local inhabitants about previous natural disasters in the living area and how local government has reacted to the manifestation of the risk. As seen in the study of Thaler et al. (2019), the government can be a driver of a change, but also an obstacle on the path. In the same way, Bertoldo (2021) grouped authorities with public confidence and those subject to public criticism. Thirdly, the transition of risk rationalization among people was employed in common denial of the risk or total inactivity to take precautions to manage the natural risk, Depending on the social culture, the same people think that the government has to manage natural risks or shift the responsibility on the individual level. Lastly, Bertoldo (2021) highlights the fourth psychological factor – risk objectivization. The existence of risk objectivization means that risk is anchored in the social culture, accepted by all stakeholders, and that risk management practices could be defined following it. If this is not the case, then the experience of natural disasters will be forgotten.

The Role of the State in the Conditions of Natural Disasters - an Example of Floods in Serbia

In the process of natural disasters risk management, the state has an important role, especially in conditions of very rare losses and risks that exceed the capacity of the private sector. Even more important is its role in prevention: building regulations, providing critical facilities, creating strategies for emergencies, infrastructure. The state can manage these risks directly and indirectly before disasters - through preventive and mitigating activities, but also after disasters - through rehabilitation activities and quick response to emergencies. It must propose plans, policies, and strategies for managing the risks of natural disasters and assess the effectiveness of the proposed actions (Szlafsztein, 2015). Central governments should assist the private sector during the post-disaster period through some national disaster risk management programs, but also cooperation with local governments (Freeman et al., 2003). The function of the state in the process of natural disasters risk management includes the following: 1) to create an effective national disaster management strategy that determines the national resilience to be reached; 2) to determine a land management plan following risk plan; 3) to define disaster prevention and emergency management plan; 4) to allocate resources in order to implement defined plans; 5) to raise awareness and support private sector (businesses and individuals) in connection with the events and risks of natural disasters and in particular to provide community support after disasters; 6) to inform people how to get involved in risk reduction activities; 7) to encourage communities through technical assistance and rehabilitation funds (Ranke, 2016).

Since 2009, the Republic of Serbia has initiated institutional reform of the system of defense against natural disasters. The Sector for Emergency Situations was founded as a key coordinating entity for these purposes. The National Strategy for Protection and Rescue in Emergency Situations and the Law on Emergency Situations were also adopted. In order to understand the extent of damage caused by natural disasters in Serbia, it is necessary to identify economic sectors that are dependent on weather and climate, and at the same time, have a large contribution to gross national income (Petrović & Grujović, 2015). In addition, Republic Directorate for Water, as a body of the Ministry of Agriculture, Forestry, and Water Management is responsible for the development of a legal framework in water management (Water law, Spatial Plan of the Republic of Serbia, Water Management Plan of Serbia - Strategy). An important role in flood defense also has public water enterprise "Srbijavode", public water enterprise "Vode Vojvodine", public water enterprise "Beogradvode" and HIDMET (Ministry of Agriculture, Forestry and Water Management, 2014).

In the past few years, due to frequent floods that hit Serbia (2014, 2016, 2018, 2019), the state has allocated significant funds to repair the consequences and mitigate the damage. Due to the underdevelopment of the flood insurance market in Serbia, the state bears the greatest burden of repairing damage after natural disasters (Ćuzović, 2019).

The state's role in resolving the catastrophic consequences of natural disasters can also be seen in the example of the catastrophic floods that hit Serbia in 2014. Floods can occur in Serbia throughout the year, and the peak is in the spring when there is a high level of precipitation and melting snow from the mountains. The floods of 2014 damaged the infrastructure for flood protection that failed to prevent floods due to underground erosion beneath their foundations. More floods are expected by the end of the 21st century, with an increase (close to 20 percent) in the frequency of floods over 100 years for large rivers, such as the Danube, Sava, and Tisza. However, Serbia needs to cope with not only too much water but also too little. The water flow of the rivers has been reduced in some areas due to changes in precipitation. The average annual flow in Serbia has decreased by about 13% by 2020 and is estimated by 19% by 2100, which can also be accompanied by a decrease in water quality (Simić, 2019). The damage and losses from these floods amount to about 1.5 billion euros. The amount applies to all sectors, including private houses, educational institutions, roads, manufacturing, and agriculture. Two-thirds of that amount goes to production sectors, where damage and losses of 1.07 billion euros are estimated. The damage to the manufacturing sector alone is estimated at half a billion, and the losses at 569.4 million. Most of the arable land in the flooded areas has been destroyed and the damage amounts to about 108 million euros, while the losses are estimated at 120 million euros. Roads and railways, as well as water and electricity transmission infrastructure, were also affected and was consequently affected the business sector. The effects of the floods on the infrastructure sector (transport, communications, and water supply) are estimated at 192.1 million euros. When it comes to issues such as - the environment and management, the estimated damage is 17.2 million, and losses 10.6 million euros (Perović, 2018, p. 114). For the repair of damages due to the floods that occurred in March 2016, state aid was paid for damaged residential buildings 174.1 million dinars, for cleaning agricultural land 52.6 million dinars, and for helping business entities 984 thousand dinars. In 2017, 31.6 million dinars were spent on the repair of damaged buildings in the floods. In 2018, state aid for damaged and destroyed residential buildings in the floods during May and June amounted to 426 million dinars. In the floods that occurred during June 2019, about a thousand residential buildings owned by citizens were damaged, for which the Government of the Republic of Serbia provided non-refundable financial assistance of 120 to 600 thousand dinars, in accordance with the category to which they belong (Ćuzović, 2019).

FUTURE RESEARCH DIRECTIONS

Previous studies have addressed potential positive and negative stimulants that could encourage or discourage natural disaster risk management practices. Therefore, United Nations have recognized the importance of creating both a multinational and multidimensional framework to create resilient societies to disasters (such as natural or caused by humans). The Sendai Framework for Disaster Risk Reduction 2015–2030 introduced by United Nations has four following priorities: an understanding based on knowledge about disaster risks, the establishment of multilevel disaster risk governance, devoting resources for mitigating the disaster risk, and anticipated readiness as proactive behavior of all stakeholders before even disaster occurs (United Nations General Assembly, 2015, p. 14). Following this initiative, every country should develop its framework for the society resilient to natural disasters.

Likewise, natural risk modeling can also contribute to society by delivering models that use historical and vulnerability data as inputs and produce outputs in the form of individual and total losses. Consequently, these models contribute to the public promotion of natural risks and influence policy-makers decisions (Crawford et al., 2018). If the scope of modeling is narrowed down to the modeling of stakeholders' participation in the management process, it would positively affect natural disaster risk management and sustainable development. Hedelin et al. (2017) summarized outcomes of the stakeholders' participation modeling such as the formation of the pull of knowledge based on both science and experience, filling the gap between science and practice, offering multiple-choice solutions by scientists for decision-makers, and bringing prosperity in mutual collaboration between all stakeholders with the free expression of opinion.

Specifically, there is room for improvement in creating a composite index that will measure successful practices in natural risk management in the Republic of Serbia. When it comes to composite index, it is obtained as a mixture of different variables collected to measure some phenomenon such as natural disaster resilience. Most of these indexes and their weights of the sub-indexes are developed on the basis of questionnaire results, pull of indicators, Analytical Hierarchy Process (AHP), or Delphi methods (Kusumastuti et al., 2014). Currently, there are indexes measuring levels of vulnerability and resilience connected to disasters risk management, such as those developed by Kusumastuti and associates (2014). They calculated the resilience index towards natural disasters as a ratio of preparedness and vulnerability composite indexes. These composite indexes were obtained by aggregating indicators of social and community capacity, economics, institution, and infrastructure indicators using the AHP method for both vulnerability and preparedness indexes, while vulnerability had one more group of indicators – hazard. This resilience index indicates which of the composite indexes should be improved. Moreover, Asadzadeh et al. (2017) indicate that the calculation of the community disaster resilience composite index should estimate the strength of the community to be proactive and reactive in case the natural disaster occurs. Marin et al. (2021) developed a disaster risk assessment index composed of hazard, exposure, vulnerability, and resilience indicators at the municipality level that could be used in a proactive and reactive manner while making decisions on natural disaster management activities.

Both modeling and computation of composite indexes depend on reliable databases such as DesInventar Sendai or the Emergency Events Database (EM-DAT). For example, EM-DAT provided reliable data for the study of Tselios and Tompkins (2020) on the drivers of natural and technological disasters in 228 countries over a more than fifty-year period. However, available datasets are also the main shortcomings of many studies (Kusumastuti et al., 2014; Tselios & Tompkins, 2020; Dzator & Dzator, 2021), and scientists and policy-makers are encouraged to bridge the gap in the existing datasets.

In line with the previous research, similar practices and solutions could be implemented in the Republic of Serbia in the context of soil and torrential floods.

CONCLUSION

Natural disaster risk management emerged from the practice of individual management activities aimed at mitigating the negative effects of natural disasters. Most often the decision-making process in the case of natural risks is under the jurisdiction of local authorities because not all areas have the same exposure and vulnerability to the negative environmental hazards. Natural disasters common for Western Balkan countries, such as soil erosion and torrential floods, are not common in Italy, which shares a higher exposure to earthquakes. Therefore, natural disasters impose the need for different societal changes to adapt to emerging living conditions. The aim of this chapter was to shed light on the importance of natural disaster risk management and crisis management and to determine the positive and negative factors of societal changes required for mitigating the influence of natural catastrophes. It is important to bridge the gap between scientific results derived from the literature and the political solutions of policy-makers in line with the interests of all stakeholders.

The authors presented the theoretical background of natural disaster risk management and crisis management. Natural disaster risk management is a set of planned actions - projects and programs, and tools used to reduce disaster risk or mitigate damage, while crisis management is based on post-disaster activities. Risk management is focused on the activities in the pre-disaster stage, during the disaster, and in the post-disaster stage. Both concepts have their elements and phases of development. If society is in the zone of natural disaster risk, it will try to mitigate the risk and enhance its resilience. Therefore, it is significant to analyze factors that lead to societal changes in order to adapt to different circumstances in the environment. Such factors have positive and negative effects and most often are dependent on local authorities and government decisions.

Practical solutions for natural disaster risk management are connected to socioeconomic and political drivers of disaster propensity (Tselios & Tompkins, 2020; Dzator & Dzator, 2021). According to Tselios and Tompkins (2020), socioeconomic factors are more influential on the country's propensity to natural disasters than political factors. This further means that if a country has a high level of poverty, discrimination, overpopulation, low educated population, low level of urbanization, or environmental issues, its exposure to the risk of natural disaster is larger than in countries on the higher level of development. The same rule applies to technological disasters risk. So, if governments strive to decrease country's exposure to disaster risk, it should increase investments in economic growth in order to decrease poverty, encourage educational attainment because it is correlated to economic development, endeavor higher levels of urbanization that is oriented away from risky areas (Tselios & Tompkins, 2020). Besides previous, natural disaster risk management is also dependent on human capital, financial capital, rule of law, and public opinion on the importance of natural disaster risk management (Dzator & Dzator, 2021). The same factors are very important if a natural disaster occurs.

Scientists can also contribute to the decision-making process by providing a scientific approach to the problem of natural disasters, such as developing a comprehensive composite index that will indicate the main points of building resilience to the natural risks and the need for change in existing resilience measures. Therefore, the chapter attempted to raise awareness on natural disaster risk management, crisis management, and drivers and barriers of societal adjustment to natural disasters.

ACKNOWLEDGMENT

This research was supported by the Ministry of Education, Science and Technological Development of the Republic of Serbia - Agreements on the implementation and financing of scientific research in 2021 [number 451-03-9/2021-14/200371].

REFERENCES

Abramov, V., Popov, N., & Shilin, M. (2021). Geo-information Support Tools for Natural Risks Management within Northern Sea Route. *Transportation Research Procedia*, *54*, 144–149. doi:10.1016/j.trpro.2021.02.058

Aghahowa, E. N. (2008). *Developing Key Performance Indicators for Airport Safety and Security: A Study of Three Scottish Airports* (Doctoral thesis). School of the Built Environment, Heriot-Watt University, Edinburgh, UK. Retrieved from https://www.ros-test.hw.ac.uk/handle/10399/2147

Asadzadeh, A., Kötter, T., Salehi, P., & Birkmann, J. (2017). Operationalizing a concept: The systematic review of composite indicator building for measuring community disaster resilience. *International Journal of Disaster Risk Reduction*, *25*, 147–162. doi:10.1016/j.ijdrr.2017.09.015

Asian Disaster Preparedness Centre. (2000, October). *Workshop on Urban Disaster Mitigation*. University of Karlsruhe.

Babu, S. (2017). *Hazard vs Disaster: The principle behind disaster management*. https://eco-intelligent.com/2017/01/21/hazard-vs-disaster-the-principle-behind-disaster-management/

Banerjee, S., Black, R., Mishra, A., & Kniveton, D. (2019). Assessing the vulnerability of remittance-recipient and nonrecipient households in rural communities affected by extreme weather events: Case studies from South-West China and North-East India. *Population Space and Place*, *25*(2), e2157. doi:10.1002/psp.2157

Berg, A., & Pattillo, C. (1999). Predicting currency crises: The indicators approach and an alternative. *Journal of International Money and Finance*, *18*(4), 561–586. doi:10.1016/S0261-5606(99)00024-8

Berke, P. R., Kartez, J., & Wenger, D. (1993). Recovery after disaster: Achieving sustainable development, mitigation and equity. *Disasters*, *17*(2), 93–109. doi:10.1111/j.1467-7717.1993.tb01137.x PMID:20958760

Bertoldo, R. (2021). A comparative frame for a contextualized analysis of local natural risk management. *International Journal of Disaster Risk Reduction*, *52*, 101945. doi:10.1016/j.ijdrr.2020.101945

Boin, A., Hart, P., Stern, E., & Sundelius, B. (2005). *The Politics of Crisis Management: Public Leadership under Pressure*. Cambridge University Press. doi:10.1017/CBO9780511490880

Canyon, D. (2013). Pre-crisis damage containment and leadership policy in health services. *Leadership in Health Services*, *26*(4), 283–293. doi:10.1108/LHS-04-2012-0010

Centre for Research on the Epidemiology of Disasters. (2004). *Thirty years of natural disasters 1974-2003: The numbers*. https://www.unisdr.org/files/1078_8761.pdf

Coombs, W. T. (2007). *Ongoing crisis communication: Planning, managing, and responding*. Sage Publication, Thousand Oaks.

Crawford, M. H., Crowley, K., Potter, S. H., Saunders, W. S. A., & Johnston, D. M. (2018). Risk modelling as a tool to support natural hazard risk management in New Zealand local government. *International Journal of Disaster Risk Reduction*, *28*, 610–619. doi:10.1016/j.ijdrr.2018.01.011

Ćuzović, Đ. (2019). Rizik i ekonomske štete od poplava–osvrt na međunarodna iskustva i prilike u Srbiji. *Revija*, *2*, 22–30.

De Almeida, A. T., Cavalcante, C. A. V., Alencar, M. H., Ferreira, R. J. P., de Almeida-Filho, A. T., & Garcez, T. V. (2015). *Multicriteria and Multiobjective Models for Risk, Reliability and Maintenance Decision Analysis*. International Series in Operations Research & Management Science., doi:10.1007/978-3-319-17969-8

Deutsche Gesellschaft für Technische Zusammenarbeit. (2002). *Disaster risk management*. https://scarp.ubc.ca/sites/scarp.ubc.ca/files/Disaster%20Risk%20Management.pdf

Donahue, A., & Joyce, P. (2001). A Framework for Analysing Emergency Management with an Application to Federal Budgeting. *Public Administration Review*, *61*(6), 728–740. doi:10.1111/0033-3352.00143

Đorđević, A., Jovanović, M., & Rađenović, Ž. (2020). Marketing plan kao deo biznis plana preduzeća. In D. Živković & I. Simić (Eds.), *Inovativno preduzetništvo* (pp. 125–161). Innovation Centre Universty of Niš.

Drennan, L., & McConnell, A. (2007). *Risk and Crisis Management in the Public Sector*. Routledge.

Dzator, J., & Dzator, M. (2021). Natural Disasters: Macroeconomic Implications and Measurement Issues. In Economic Effects of Natural Disasters (pp. 317-333). Academic Press. doi:10.1016/B978-0-12-817465-4.00020-0

Elliot, D., Harris, K., & Baron, S. (2005). Crisis Management and Services Marketing. *Journal of Services Marketing*, *19*(5), 336–345. doi:10.1108/08876040510609943

Emergency Events Database. (2017). *Annual Disaster Statistical Review 2016: The numbers and trends*. https://reliefweb.int/report/world/annual-disaster-statistical-review-2016-numbers-and-trends

Fearn-Banks, K. (2002). *Crisis Communications: A Casebook Approach*. Associates Lawrence Erlbaum, Inc.

Freeman, P. K., Martin, L. A., Linnerooth-Bayer, J., Mechler, R., Pflug, G., & Warner, K. (2003). *National Systems for Comprehensive Disaster Management: Financing Reconstruction*. https://publications.iadb.org/publications/english/document/Disaster-Risk-Management-National-Systems-for-the-Comprehensive-Management-of-Disaster-Risk-and-Financial-Strategies-for-Natural-Disaster-Reconstruction.pdf

Gaunt, P., & Ollenburger, J. (1995). Issue management revisited: A tool that deserves another look. *Public Relations Review*, *21*(3), 199–210. doi:10.1016/0363-8111(95)90021-7

Gencer, E. A. (2013). The Interplay Between Urban Development, Vulnerability, and Risk Management. *Mediterranean Studies (Warrensburg, Mo.)*, *7*, 7–43.

Glaesser, D. (2006). *Crisis Management in the Tourism Industry* (2nd ed.). Butterworth Heinemann. doi:10.4324/9780080464596

Hedelin, B., Evers, M., Alkan-Olsson, J., & Jonsson, A. (2017). Participatory modelling for sustainable development: Key issues derived from five cases of natural resource and disaster risk management. *Environmental Science & Policy, 76*, 185-196. doi:10.1016/j.envsci.2017.07.001

Ivčević, A., Mazurek, H., Siame, L., Moussa, A. B., & Bellier, O. (2019). Indicators in risk management: Are they a user-friendly interface between natural hazards and societal responses? Challenges and opportunities after UN Sendai conference in 2015. *International Journal of Disaster Risk Reduction, 41*, 101301. doi:10.1016/j.ijdrr.2019.101301

Janjić, I., & Rađenović, T. (2019). The importance of managing innovation in modern enterprises. *Ekonomika (Nis), 65*(3), 45–54. doi:10.5937/ekonomika1903045J

Janjić, I., Đorđević, A., & Đokić, M. (2020). Internacionalizacija poslovanja. In D. Živković & I. Simić (Eds.), *Inovativno preduzetništvo* (pp. 333–370). Innovation Centre Universty of Niš.

Jovanović, M. (2018). Innovation of Serbia in relation to neighboring countries as a determinant of competitiveness. *Ekonomika (Nis), 64*(1), 57–78. doi:10.5937/ekonomika1801065J

Jovanović, M., & Đokić, M. (2019). National competitiveness and goods market efficiency in Serbia and neighboring countries. In B. Krstić (Ed.), *Enhancing competitiveness of national economies and enterprises* (pp. 109–142). Faculty of Economics University of Niš.

Jovanović, M., & Đokić, M. (2020). Innovativeness of Serbian Enterprises as a Factor of Competitiveness Improvement at the International Market. In M. Janowicz-Lomott, K. Lyskawa, P. Polychronidou, & A. Karasavvoglou (Eds.), *Economic and Financial Challenges for Balkan and Eastern European Countries* (pp. 101–115). Springer Proceedings in Business and Economics. doi:10.1007/978-3-030-39927-6_7

Jovanović, M., Đokić, M., & Janjić, I. (2020). Plan organizacije. In D. Živković & I. Simić (Eds.), *Inovativno preduzetništvo* (pp. 209–250). Innovation Centre Universty of Niš.

Keković, Z., Marić, P., & Komazec, N. (2011). Republic of Serbia natural and other disaster risk assessment – methodology. *Journal of Criminalistics and Law, 16*(2), 1-17.

Kešetović, Ž. (2008). *Krizni menadžment.* Beograd: Fakultet bezbednosti, Univerzitet u Beogradu.

Kešetović, Ž., & Toth, I. (2012). *Problemi kriznog menadžmenta.* Velika Gorica: Veleučilište Velika Gorica.

Kouzmin, A. (2008). Crisis Management in Crisis? *Administrative Theory & Praxis, 30*(2), 155–183. doi:10.1080/10841806.2008.11029631

Kovoor-Misrca, S., Zammuto, R. F., & Mitroff, I. I. (2000). Crisis preparation in organizations: Prescription versus reality. *Technological Forecasting and Social Change, 63*(1), 43–62. doi:10.1016/S0040-1625(99)00049-9

Krausmann, E., Girgin, S., & Necci, A. (2019). Natural hazard impacts on industry and critical infrastructure: Natech risk drivers and risk management performance indicators. *International Journal of Disaster Risk Reduction*, *40*, 101163. doi:10.1016/j.ijdrr.2019.101163

Krstić, B., & Janjić, I. (2019). Institutions as a factor of competitiveness: The case of South-eastern countries. In B. Krstić (Ed.), *Building national competitiveness* (pp. 1–20). Faculty of Economics, University of Niš.

Kusumastuti, R. D., Husodo, Z. A., Suardi, L., & Danarsari, D. N. (2014). Developing a resilience index towards natural disasters in Indonesia. *International Journal of Disaster Risk Reduction*, *10*, 327–340. doi:10.1016/j.ijdrr.2014.10.007

Kyhn, H. S. (2008). *Situational Crisis Communication Theory: Its Use in a Complex Crisis with Scandinavian Airlines Grounding of Dash 8-Q400 Airplane* (Master thesis). Airbus School of Business, Handelshojskolen. Retrieved from https://dl.icdst.org/pdfs/files/0e8d769c306ae51d9075fa87693ede99.pdf

Marin, G., Modica, M., Paleari, S., & Zoboli, R. (2021). Assessing disaster risk by integrating natural and socio-economic dimensions: A decision-support tool. *Socio-Economic Planning Sciences, 101032*. Advance online publication. doi:10.1016/j.seps.2021.101032

Mechler, R. (2003). Natural Disaster Risk and Cost-Benefit Analysis. In A. Kreimer, M. Arnold, & A. Carlin (Eds.), *Building Safer Cities: The Future of Disaster Risk* (pp. 45–55). The World Bank Disaster Management Facility.

Milanović, S. (2017). Human Resource Management in the Conditions of Globalization. *TEME*, *XLI*(1), 297–299. doi:10.22190/TEME1701297M

Milanović, S., & Marjanović, I. (2019). Labour Market Efficiency - Case of EU Members and Candidate Countries. In B. Krstić (Ed.), *Enhancing Competitiveness of National Economies and Enterprises* (pp. 65–86). Faculty of Economics University of Niš.

Milanović, S., & Marjanović, I. (2020). Obezbeđenje potrebnih resursa. In D. Živković & I. Simić (Eds.), *Inovativno preduzetništvo* (pp. 251–298). Innovation Centre Universty of Niš.

Mileti, D. S. (1999). *Disaster by Design: A Reassessment of Natural Hazards in United States*. Joseph Henry Press.

Ministry of Agriculture, Forestry, and Water Management. (2014). *Workshop on Water-Food-Energy Ecosystems Nexus Assessment in the Sava, River Basin, Zagreb, Croatia*. https://unece.org/fileadmin/DAM/env/Water_Convention_Sava_River_Workshop/8.section_3_merita_borota.pdf

Mitroff, I. I. (1994). Crisis Management and environmentalism: A natural fit. *California Management Review*, *36*(2), 101–113. doi:10.2307/41165747

Mitroff, I., & Pearson, C. M. (1993). *Crisis Management: a Diagnostic Guide for Improving your organizations crisis-preparedness*. Jossey-Bass.

Moe, T. L., Gehbauer, F., Sentz, S., & Mueller, M. (2007). Balanced Scorecard for Natural Disaster Management Projects. *Disaster Prevention and Management*, *16*(5), 785–806. doi:10.1108/09653560710837073

Monstad, T. H. (2003). *Issues and crisis management – same, same but different* (Master thesis). Uppsala University, Sweden. Retrieved from www.dis.uu.se/mkv/education/essays/Helauppsatser/2003/up2003 31.pdf

Nascimento, K. R. D. S., & Alencar, M. H. (2016). Management of risks in natural disasters: A systematic review of the literature on NATECH events. *Journal of Loss Prevention in the Process Industries, 44*, 347–359. doi:10.1016/j.jlp.2016.10.003

Perović, M. (2018). *Model procene ekonomskih i finansijskih efekata uticaja prirodnih katastrofa i mogućnosti njihovog ublažavanja u Srbiji*. Univerzitet Privredna akademija u Novom Sadu, Fakultet za ekonomiju i inženjerski menadžment.

Petrović, G., & Grujović, M. (2015). The economic damage from natural disasters in Serbia and Sumadija district. *Ekonomski signali: poslovni magazin, 10*(2), 99-107.

Pheng, L., Ho, D., & Ann, Y. (1999). Crisis management: A survey of property development firms. *Property Management, 17*(3), 231–251. doi:10.1108/02637479910286916

Popović, T., Đurđević, V., Živković, M., Jović, B., & Jovanović, M. (2009). Promena klime u Srbiji i očekivani uticaji. In Peta regionalna konferencija EnE09-Životna sredina ka Evropi, Ambasadori životne sredine i PKS (pp. 6-11). Agencija za zaštitu životne sredine.

Ranke, U. (2016). *Natural Disaster Risk Management*. Springer. doi:10.1007/978-3-319-20675-2

Ray, S. J. (1999). *Strategic Communications in Crisis Management: Lessons from the Airline Industry*. Quorum Books.

Ritchie, H., & Roser, M. (2014). *Natural Disasters*. https://ourworldindata.org/natural-disasters

Simić, T. (2019). Uticaj klimatskih promena na ljudsku bezbednost u Republici Srbiji. In Ljudska bezbednost - Zbornik studentskih radova (pp. 33-43). Centar za istraživanje javnih politika.

Smith, M., Wallace, K., Lewis, L., & Wagner, C. (2015). A structured elicitation method to identify key direct risk factors for the management of natural resources. *Heliyon, 1*(3), e00043. doi:10.1016/j.heliyon.2015.e00043 PMID:27441228

Smolka, A. (2006). Natural disasters and the challenge of extreme events: Risk management from an insurance perspective. *Philosophical Transactions of the Royal Society A, 364*(1845), 2147–2165. doi:10.1098/rsta.2006.1818 PMID:16844653

Sornette, D. (2004). *Why stock markets crash: Critical events in complex financial systems*. Princeton University Press.

Szlafsztein, C. F. (2015). Management of natural disasters in the Brazilian Amazon region. *Natural Hazards, 76*(3), 1745–1757. doi:10.100711069-014-1567-8

Tanović, M., Milanović, S., & Rađenović, Ž. (2020). Upravljanje preduzećem. In D. Živković & I. Simić (Eds.), *Inovativno preduzetništvo* (pp. 299–332). Innovation Centre Universty of Niš.

Thaler, T., Attems, M. S., Bonnefond, M., Clarke, D., Gatien-Tournat, A., Gralepois, M., Fournier, M., Murphy, C., Rauter, M., Papathoma-Köhle, M., Servain, S., & Fuchs, S. (2019). Drivers and barriers of adaptation initiatives–How societal transformation affects natural hazard management and risk mitigation in Europe. *The Science of the Total Environment*, *650*, 1073–1082. doi:10.1016/j.scitotenv.2018.08.306 PMID:30308796

Tselios, V., & Tompkins, E. L. (2020). Can we prevent disasters using socioeconomic and political policy tools? *International Journal of Disaster Risk Reduction*, *51*, 101764. doi:10.1016/j.ijdrr.2020.101764

United Nations General Assembly. (2015). *The Sendai Framework for Disaster Risk Reduction 2015–2030* (Assembly Resolution A/Res/69/283). Author.

United Nations International Strategy for Disaster Reduction. (2009). *UNISDR terminology on disaster risk reduction*. https://www.unisdr.org/files/7817_UNISDRTerminologyEnglish.pdf

Wu, D., & Olson, D. L. (2010). Enterprise risk management: Coping with model risk in a large bank. *The Journal of the Operational Research Society*, *61*(2), 179–190. doi:10.1057/jors.2008.144

Zhang, J. Q., Okada, N., & Tatano, H. (2006). Integrated natural disaster risk management: Comprehensive and integrated model and Chinese strategy choice. *Ziran Zaihai Xuebao*, *15*(1), 115–128.

Zschau, J., & Küppers, A. N. (2003). *Early warning systems for natural disaster reduction*. Springer-Verlag. doi:10.1007/978-3-642-55903-7

Zuccaro, G., Leone, M. F., & Martucci, C. (2020). Future research and innovation priorities in the field of natural hazards, disaster risk reduction, disaster risk management and climate change adaptation: A shared vision from the ESPREssO project. *International Journal of Disaster Risk Reduction*, *51*, 101783. doi:10.1016/j.ijdrr.2020.101783

ADDITIONAL READING

Batabyal, A. A., & Beladi, H. (2001). Aspects of the theory of financial risk management for natural disasters. *Applied Mathematics Letters*, *14*(7), 875–880. doi:10.1016/S0893-9659(01)00058-1

Chapman, D. (1994). *Natural hazards*. Oxford University Press.

Chen, C.-W., Tseng, C.-P., Hsu, W.-K., & Chiang, W. L. (2012). A novel strategy to determine the insurance and risk control plan for natural disaster risk management. *Natural Hazards*, *64*(2), 1391–1403. doi:10.100711069-012-0305-3

Dias, P., Bertoldo, R., Guignard, S., & Schleyer-Lindenmann, A. (2021). Trust in public risk management and social dynamics in the case of flood risk. *Rev. CES Psico*, *14*(1), 49–63. doi:10.21615/cesp.14.1.5

Dragićević, S., Filipović, D., Kostadinov, S., Nikolić, J., & Stojanović, B. (2009). Zaštita od prirodnih nepogoda i tehnoloških udesa–Strategija prostornog razvoja Republike Srbije. Univerzitet u Beogradu, Geografski fakultet.

Ivčević, A., Mazurek, H., Siame, L., Bertoldo, R., Statzu, V., Agharroud, K., ... & Bellier, O. (2021). The importance of raising risk awareness: lessons learned about risk awareness sessions from the Mediterranean region (North Morocco and West Sardinia, Italy). Natural Hazards and Earth System Sciences Discussions, 1-25. doi:10.21615/cesp.14.1.5

Link, F., Harris, J., Irarrázaval, F., Valenzuela, F., Welz, J., & Barth, K. (2014), "Coping with Natural Disasters and Urban Risk: An Approach to Urban Sustainability from Socio-Environmental Fragmentation and Urban Vulnerability Assessment", From Sustainable to Resilient Cities: Global Concerns and Urban Efforts (Research in Urban Sociology, Vol. 14), Emerald Group Publishing Limited, Bingley, pp. 35-58. doi:10.1108/S1047-004220140000014002

Ranković, N., Nedeljković, J., Zlatić, M., Stanišić, M., & Nonić, D. (2016). Kretanje obima šteta od prirodnih nepogoda u šumama Srbije i uticaj temperature i padavina. Glasnik Šumarskog fakulteta, 114, 201-218.

Sayah Mofazali, A., & Jahangiri, K. (2018). Towards a customized foresight model on "disaster risk management" in developing countries. *Foresight*, *20*(5), 467–487. Advance online publication. doi:10.1108/FS-01-2018-0002

KEY TERMS AND DEFINITIONS

Climate Change: A change in global or regional climate trends, mostly influenced by carbon dioxide emissions.

Crisis Management: A set of planned actions - projects and programs, and tools used for post-disaster activities.

Mitigation: The action of reducing the severity, seriousness, or painfulness of something.

Natural Disaster Risk: Natural disaster risk can be seen as the potential probability of loss (human, physical, economic, natural, social) due to the natural danger that threatens a certain area in a certain period.

Resistance: The ability not to be affected by something, especially adversely such as a natural disaster.

Risk Management: A set of planned actions - projects and programs, and tools used to reduce disaster risk or mitigate damage.

Societal Transformation: A way of changing cultural and social institutions over time with the impact on society.

Chapter 3
Strategic and Legal Framework of Torrential Flood Control in Some Western Balkan and EU Countries

Nada Dragović
Faculty of Forestry, University of Belgrade, Serbia

Tijana Vulević
Faculty of Forestry, University of Belgrade, Serbia

Muhamed Bajrić
Faculty of Forestry, University of Sarajevo, Bosnia and Herzegovina

Johannes Huebl
University of Natural Resources and Life Sciences, Vienna, Austria

Paolo Porto
University Mediterranea of Reggio Calabria, Italy

Ivan Blinkov
"Hans Em" Faculty of Forestry, Ss. Cyril and Methodius University in Skopje, North Macedonia

ABSTRACT

The EU countries are obliged to harmonize their legislation in the field of flood protection, and thus torrential floods, in accordance with the Water Framework Directive (WFD) which was adopted in 2000. Two EU countries, Austria and Italy, and three Western Balkan countries were selected for the strategic and legal framework of torrential flood control: Serbia, North Macedonia, and Bosnia and Herzegovina. In addition to the legal framework of torrential flood control in EU countries, policies and strategies related to this area were studied for comparative analysis with non-EU countries. The strategic framework for the protection of water resources, and in particular torrential flood protection, is lacking in all Western Balkan countries. The aim of this chapter is to determine the directions of future strategic directions and torrential flood control policies in the Western Balkans based on the experiences of EU countries, advantages and disadvantages of the existing strategic, and legal frameworks.

DOI: 10.4018/978-1-7998-8459-0.ch003

INTRODUCTION

As a result of natural and anthropogenic factors, with the increasingly expressive impact of climate change, torrential floods occur in European countries, causing extensive material damage and the loss of human lives. In addition to intensifying flood protection work, the need for legal regulation and their harmonization with the countries of the region and the EU (Nikolić Popadić, 2020).

The most important document of the European Union related to water resources and flood control are Directive 2000/60/EC of the European Parliament and of the Council of 23 October 2000 establishing a framework for Community action in the field of water policy as amended by Decision 455/2001/EC and Directives 2008/32/EC, 2008/105/EC, 2009/31/EC and 2013/39/EU (*Water Framework Directive*, WFD) and *Directive* 2007/60/EC of the European Parliament and of the Council of 23 October 2007 *on the assessment and management of flood risks.*

Water resources are crucial for life, society, economy, and environmental sustainability. Water protection is, therefore, one of the priority areas regulated in EU legislation from the 1970s. An early beginning, the European water legislation had some problems in the efficiency of their implementation, such as the low acceptance rate by the target groups, disparity noted between the level of governance (watersheds and administrative boundaries), and the fact that directives are based on standards that regulate certain segments, but no problems as a whole (Boeuf & Fritsch, 2016).

WFD, adopted in December 2020 to achieve a good water status until 2027, transform the EU water policy from the fragmented regulatory framework to a holistic approach that integrates ecological, economic, and social functions through the entire river basin (Kallis, 2005; Fritsch & Benson, 2020).

Floods Directive (2007/60/EC) has been established a framework for assessing and managing flood risks, represents a new approach to flood management, that is closely coordinated with the WFD. It encourages setting targets and defining measures to flood risk maps. Flood risk maps are tools combining flood hazard formulates country-specific physical interventions after a long process of assessing the flood hazards and risks described in the Directive. The interventions are then defined in a flood management plan, which is the final step of the planning process (Dragović et al., 2016).

BACKGROUND

The field of water legislation was regulated at the level of the European Union 20 years ago (Water Framework Directive-WFD), but the laws of some countries, and especially bylaws, have not yet been harmonized with them. One of the key elements of the WFD is the River Basin Management Plan, which is a binding document, but unfortunately, in some WB countries, it has not yet been adopted for all river basins. Boeuf and Fritsch (2016) analyzed 89 papers published in journals on the application of WFD in European countries. According to their analysis, EU member states have developed policies for the consistent application of the WFD, while Mediterranean countries are lagging behind. They also point to the need to encourage EU candidate countries to implement the directive, as well as to increase research that includes a comparative analysis of the application of WFD in groups of countries. The degree of application of the Flood Directive (FD) and the approach to flood risk management varies across EU and WBC countries. This diversity of approaches has a negative effect, especially on transboundary rivers where incompatibility of approaches can reduce the effectiveness of flood risk management (Priest et al., 2016). Research on the application of FD in some EU countries indicates the problem of greater commitment to establishing a procedure for its implementation than essentially its application in practice (Priest et al., 2016). The EU Soil Thematic Strategy is an important document with which national strategies and laws related to the protection of soil from degradation, i.e., erosion as the most pronounced form of degradation and one of the most significant causes of torrential floods, have been harmonized. For this reason, in some countries (e.g. Italy), flood protection is defined in the legal and strategic framework of land protection. Considering the current situation, there is an evident need for greater harmonization of the WFD and Flood Directive (FD) with the legislation related to climate change, both in the EU countries and in the WBC.

Watercourses in the Western Balkans are classified into two categories (I and II) according to the position of watercourses concerning the state border, size and characteristics of the basin, regime, and characteristics of watercourses in terms of water use, water protection, and protection from harmful effects of water. Torrential flows belong to the second-order flows, and the local community is responsible for their regulation, protection against harmful effects, as well as for the development of the Operational Plan for protection against torrential floods (Dragović et al., 2016). Unfortunately, very few municipalities in Serbia, Bosnia and Herzegovina, and Northern Macedonia have adopted such a plan. It follows from all the above that in addition to the need to harmonize the legal and strategic framework in the Western Balkans countries (WBC) with EU legislation, there is an even greater need for consistent implementation of already adopted acts at the state level.

STRATEGIC AND LEGAL FRAMEWORK OF TORRENTIAL FLOODS CONTROL IN SOME EU COUNTRIES

In recent decades, EU states to experience flooding that led to the devastation and extensive material damages caused mainly by climate change. Flood control is regulated in EU states through multiple laws, such as The Water Framework Directive, the Forest Act, etc. The work outlines the legislative framework for flood control for the EU states: Austria and Italy.

The Strategic and Legal Framework of Torrential Floods Control in Austria[1]

Hazards emerging from torrents, avalanches, and erosion have mainly regional effects, but in the mountainous country of Austria, natural hazards control as such is a task of national importance. Floods and debris flow affected a large area of the Austrian territory (Figure 1). As of 2015, 11,922 torrents have been counted in Austria. About 70% of the federal territory (83,855 km^2) is subject to the care of the Austrian Service for Torrent and Avalanche Control. In the Federal Provinces of Vorarlberg, Tyrol, Carinthia, and Salzburg, this area covers more than 80% of the provincial territories. In Austria, about 120,000 buildings and 1,500 km of transport routes are exposed to the risk of torrents.

There is a long tradition of natural disaster mitigation in Alpine regions. In the Alps, basic techniques of “torrent control” have been known for a long time. The first torrent control works were constructed around the year 1500. Originally, control plans took advantage of the possibilities of using “living” construction material (bioengineering), aimed at preventing the erosion of loose rock, or promoted the protective effect of forests. It was the work of Josef Duile that started the practical implementation of structural mitigation measures (Duile, 1826). The severe floods in Tyrol and Carinthia in the year 1882 initiated, as in France, a rethinking of natural hazards prevention. Julius Graf Falkenhayn (k.u.k. Minister

Figure 1. The area affected by floods and debris floods in Austria
(Source:https://boku.ac.at/fileadmin/data/H03000/H87000/H87100/IAN_Reports/REP0111_Band4.pdf)

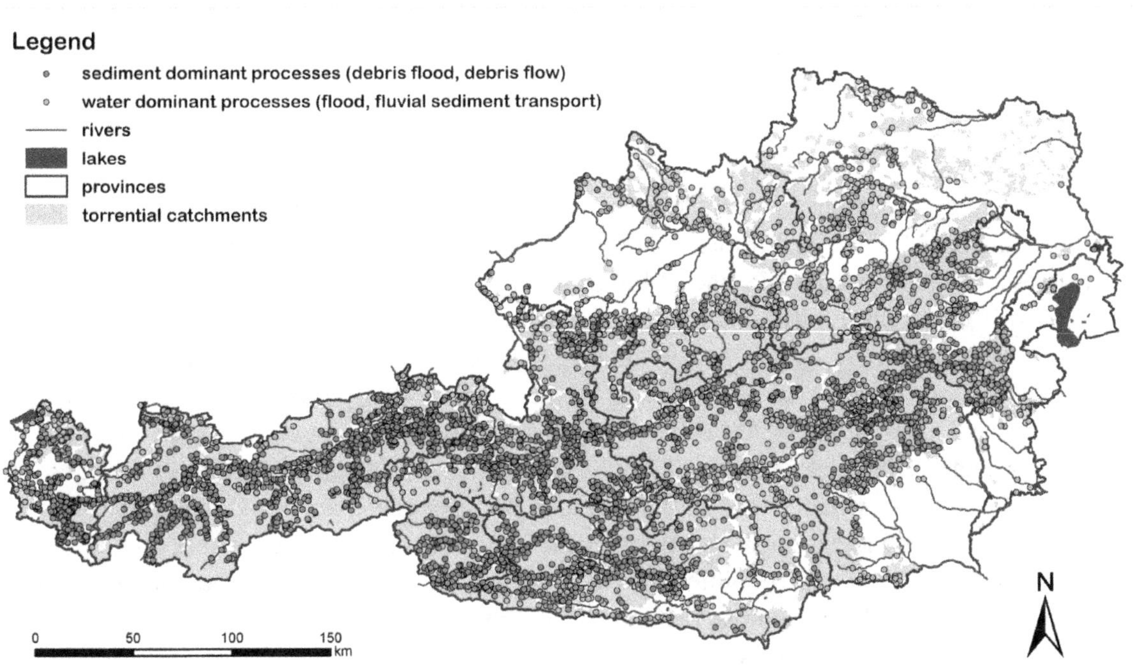

of Agriculture) travelled to France in the year 1883 to learn of the methods used and implement them in Austria (Wang, 1901). In this time, two technical domains characterized the progression of the torrent control: one was hydraulic engineering, and the other was forest engineering. Hydraulic engineering was mainly applied to the lower part of the torrent, while forest engineering addressed the headwaters. In 1884 the emperor of the Austrian-Hungarian monarchy passed the Torrent control act (Imperial Law Gazette concerning provisions on the non-hazardous drainage of mountain waters of 30 June 1884, RGB. No 117/1884), stating that torrent control works have to be executed on a national level, with public funding, and following a systematic approach. The government implemented a legal basis for financing and organizing the torrent control service to ensure further development. The educational course „Afforestation and torrent control in mountainous areas" was established in 1879 at the University of Natural Resources and Life Sciences, Vienna (BOKU), and served as the scientific basis for the staff of the newly established service.

Nowadays, the protection against torrents is laid down in the Austrian constitution as a competence of the Federal Government (Art. 10), both concerning legislation and execution. Based on the Forest Act (Austrian Forest Act of 1975, Federal Law Gazette No 440/1975, as last amended by Federal Law Gazette I No 56/2016), the Federal Government attends to this task via a decentralized agency immediately subordinated to the Ministry of Agriculture, Regions, and Tourism (BMLRT), labelled as the Austrian Service for Torrent and Avalanche Control (WLV or die. wildbach).

In the Forest Act (§ 102), all tasks of its offices are laid down, among them

- the development up of hazard zone maps,

- the planning and implementation of technical and forest-biological control measures,
- the consulting services and expert activities,
- the care for the torrent and avalanche catchment areas,
- the administration of the subsidies allocated, and
- the representation of the public interest concerning the protection against Alpine natural hazards.

The Forest Act also includes provisions on the headquarters and organization of the offices. Presently there are 7 provincial headquarters (Vienna, Lower Austria, and Burgenland in the same one) and 21 regional headquarters (Figure 2). Further, four Centres of Expertise (Natural Hazards Information, Torrential Processes, Geology and Avalanches, Monitoring) and two central Services (Digital Infrastructure, Central Payroll Accounting) fulfill important planning tasks.

The Austrian Service for Torrent and Avalanche Control and its offices (including the Centres of expertise and central services) has about 300 staff members working in the fields of technology and administration and approximately 700 workers employed based on contracts for work and services to fulfill the tasks defined by the Forest Act of 1975. Besides this, the legal foundations of the hazard map for torrential watersheds, which is one of the forestall land use plans, are laid down in the Forest Act (§ 7, 8, 11) and its associated ordinances (1976, 2011, 2021). As far as its legal effect is concerned, the hazard zone map is only an expert opinion, but its content (hazard zones) may become binding due to its being laid down in the local land-use planning. The provincial laws regulating land use and building contain building restrictions for areas exposed to natural hazards. The identification of the hazard zones in the zoning and development plans allows authorities to assess the risk for each parcel and, if necessary, to determine provisions for making a site apt for development.

The legal basis of engineering works along water bodies is laid down in the Water Management Act 1959 ("Wasserrechtsgesetz", Federal Law Gazette No 215/1959, as last amended by Federal Law Gazette I No 54/2014). The steps for implementing the EU Floods Directive (2006/60/EC) are delineated in this act.

Torrent, avalanche, and erosion control measures are financed by the Disaster Fund "Katastrophenfond" of the Federal State (Disaster Fund Act, Federal Law Gazette No 201/1996 as last amended by Federal Law Gazette I No 46/2016). Subsidies are granted subject to the provisions of the Hydraulic Structures Development Act ("Wasserbautenförderungsgesetz", Federal Law Gazette No 148/1985, as last amended by Federal Law Gazette I No 98/2013), which defines the terms and conditions under which subsidization is provided as well as the principles of the planning and implementation of control measures.

Every year the Federal Government dedicates funds (subsidies) of about 85 million Euros from the Disaster Fund to torrent, avalanche, and erosion control. Together with contributions from the Federal Provinces and beneficiaries (municipalities, water corporations, and others), almost € 160 million are thus available annually for investments inactive control measures.

About 78% of the available funds are spent on torrent control, 9% on avalanche control, 8% on rockfall and landslide control measures, and 5% on land-use management measures (protection forests, catchment area management). The share of maintenance activities is growing, that of new constructions is slightly decreasing.

However, comprehensive protection against alpine natural hazards also includes organizational measures (emergency alert, alarm, and evacuation), monitoring systems, and civil disaster control, tasks which are mostly implemented by the Federal Provinces. Modern natural hazard management can best be explained using the principle of the risk cycle, which begins with the event (disaster) and comprises disaster intervention, repair, reconstruction, prevention, and measures for disaster preparedness. The

Figure 2. Organization chart of the Austrian Service for Torrent and Avalanche Control (2016)
(Source: www.bmlrt.gv.at/english/forests/forest-and-natural-hazards/die-wildbach-in-austria.html, 2020, modified)

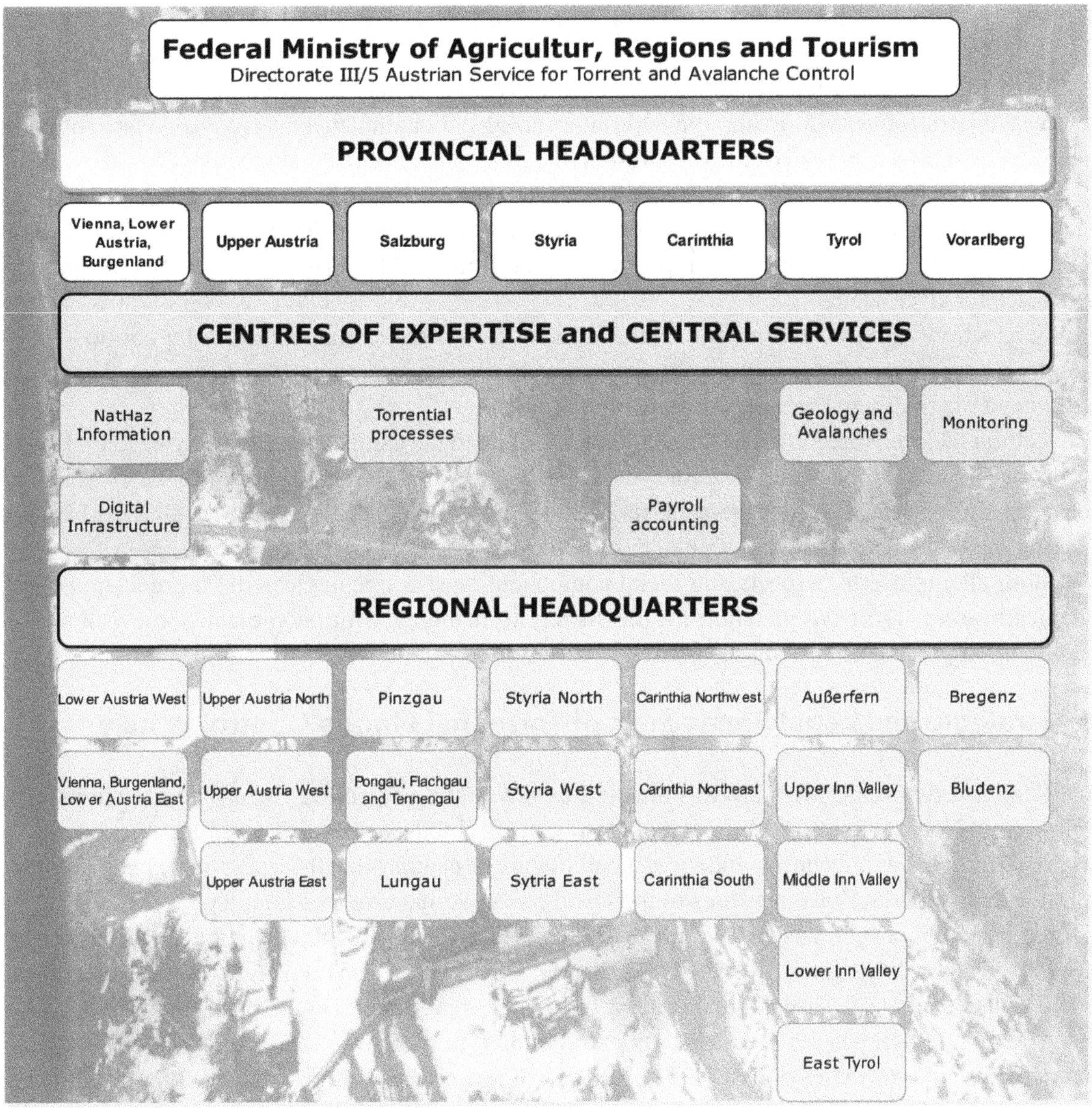

objective is to improve and enhance society's preparedness and resilience to cope with future natural disasters. Provision of these security services requires the cooperation of experts of different technical disciplines and many public and private organizations. A task of natural hazard management is also to harmonize all normative documents and relevant technical plans to serve the goal of protecting against natural hazards.

To coordinate these tasks, the political field of strategy "Protection against natural hazards" has been established at the Federal Ministry of Agriculture, Regions, and Tourism. The main objective is to manage these tasks beyond the specific competencies of the individual ministries, regional units, and technical fields.

Based on European directives (e.g. Water Framework Directive, Directive 2000/60/EC) different aspects in designing control measures have to been taken into account. Especially ecological aspects play a significant role in future planning. Engineers have to consider the biological quality (fish, benthic invertebrates, aquatic flora) and the hydromorphological quality such as riverbank structure, river continuity, or substrate of the river bed. Additionally, the topic „sediment connectivity" is of increasing importance in water-related protection concepts. This also means to include woody debris and sediment transfer into the protection strategy, because both factors may increase the risk of torrential processes. On the other hand, these immanent components during floods and debris flows are of high ecological importance for river habitats.

In connection with forestall measures, the influence of climatic change has to be considered. For example, the spreading of invasive species, dangerous forest pest insects, and the increasing risk of forest fires change the traditional methods of forest management.

Structural measures are permanently adapted according to the gained experience by torrential events. The major challenge concerning structural measures is to extend the durability of these measures and to develop appropriate asset management. In 2022, the liable standards „ÖNORM" (ÖNORM B4800), replacing the ONR Series 248XX, will be published.

Besides all engineering aspects, the social component must not be neglected. Future torrent control strategies have to shift from planning for the society to planning with people and society in a multidisciplinary approach, using tools for proactive Risk Management.

The Strategic and Legal Framework of Torrential Floods Control in Italy

During the last two centuries, Italy was affected by devastating flood events in several regions of the country, from North to South. Historical records document flood disasters from 1850 to date (see Salvati et al., 2010), but it was only during the second half of the 20th century that the tragic consequences related to these events posed serious questions to the flood risk governance of the country.

An important input to focus on the need to revise the national and local legislations in the field of torrential flood control was given after the flood disasters occurred in 1951 during which the area covered by the Po catchment (Polesine, Northern Italy), a smaller area located in Calabria (Southern Italy) and part of Sicily and Sardinia were devasted by extraordinary rainfall events. Calculations, made ex-post, documented economic losses of ca. 206.6 million Euros for the Po catchment and ca. 15.5 million Euros for Calabria, Sardinia, and Sicily (Lastoria et al., 2006).

Losses of similar magnitude occurred 15 years later (November 1966), when the River Arno overflowed in Florence (Central Italy). In that case, it was difficult to calculate the devastating effect of that flood disaster that caused the death of 35 people, damages to homes, buildings, and treasured artworks and the left ca. 70,000 citizens without electricity, gas, or heating for days. However, according to a report provided by the Arno Basin Authority, recent estimates suggest that if a similar flood were to occur in recent years, the damage to Florentine homes and businesses could total ca. 15.5 billion Euros (Galloway et al., 2017).

Before the above events, most of the actions were related to the areas directly hit by the disasters and can be considered only isolated attempts to contrast the problem. Important examples in this direction were the activities carried out in Calabria (Southern Italy) after the big floods in 1951 and 1953. In those cases, the 'Special Law' 1177/55 of 26 November 1955 and the public work projects in the 1950s provided by the 'Cassa per il Mezzogiorno' (established by the national government in 1950 to encourage the development of public works and infrastructure of Southern Italy) are important programs that pre-emptied 'ordinary' planning practices and favored the formation of an approach oriented to implement ad-hoc, 'extraordinary' plans and projects funded from above, rather than manage from below the 'ordinary' government of urban transformations.

At a national scale, the situation improved only after the occurrence of the 1966 flood event in Florence, for which an inter-ministerial commission led by Prof. Giulio De Marchi (1890-1972) was nominated to establish new strategies for flood risk management and to promote actions of priority to help the areas hit by the disasters. However, the suggestions given by the De Marchi Commission took time to reach a political consensus and another crucial event occurred in Valtellina (Lombardia, Northern Italy) in 1987. In that case, 53 people were killed by an incredibly warm summer that caused the glaciers to release a great quantity of water, which summed up to exceptional storm events that occurred in just three days. Economic damages that followed this event were calculated at ca. 1.4 billion US$ (Guzzetti et al., 1994).

The Valtellina event posed an important milestone on the need to focus the attention on the problems generated by flood disasters because, after that tragedy, the national government established the Soil Protection Act (183/1989) with the aim of planning and managing in an integrated way the catchments of the Italian rivers (Boardman, Poesen, 2006).

With the Law 183/89 of 18 May 1989, the relative body of norms and regulation developed rapidly in the 1990s, and the importance of the Civil Protection Mechanism took place. In this respect, following three more tragic events occurred in Piedmont (1994), Sarno (1998), and Soverato (2000), the Department for Civil Protection (DCP), first introduced in the early 1970s, installed a network of advanced early warning and alerting centres, to help the areas at risk in preventing further damages.

The Law 183/89 is a milestone for soil protection in Italy as it attributed a fundamental role to planning and programming within each single river catchment and introduced several key principles in ruling the flood risk management that was incorporated a few years later by the EU Water Framework Directive (WFD, 2000/60/EC) and the EU Floods Directive (FD, 2007/60/EC). The general aim of this aw was the adoption of the 'River Basin Authorities' (RBA) as planning and management units that were identified based on hydrographic criteria. Each of these units was assigned to a single 'River Basin' (RB) based on the establishment of a network of river basins (RBs) of regional, inter-regional and national importance. The map of the RBAs is depicted in Figure 3 and includes 40 RBAs consisting of 6 National RBAs (numbered in Figure 3), 13 Inter-regional RBAs, and the remaining 21 RBAs distributed in Regional, International and 'Pilot' (the latter includes only one unit, the Serchio Catchment, adopted for special studies).

Each RBA was responsible for a 'plan of flood and landslide management' (River Basin Plan, RBP), or PAI (Piano di Assetto Idrogeologico, in Italian), intending to identify areas of different risk within each unit. In this respect, the class R4 (highest risk) includes areas where a significant risk of deaths or injuries could occur together with loss of property and damages to economic activities; the class R3 includes areas prone to risk that could cause more limited harm to people, properties or lifelines; the classes R2 and R1 are related to areas where minor or limited risks with no direct threat to persons and economic activities could occur. However, even if the logical structure of the Soil Protection Act and the related actions were very convincing from a theoretical point of view, the development of the RBPs proved to be more demanding than initially thought and lasted for more than a decade (Mysiak et al., 2013). It must be recognized that the activity of the Civil Protection Service (CPS) managed by the

Figure 3. Location and geographical boundaries of the 40 River Basin Authorities established by the National Law 183/1989

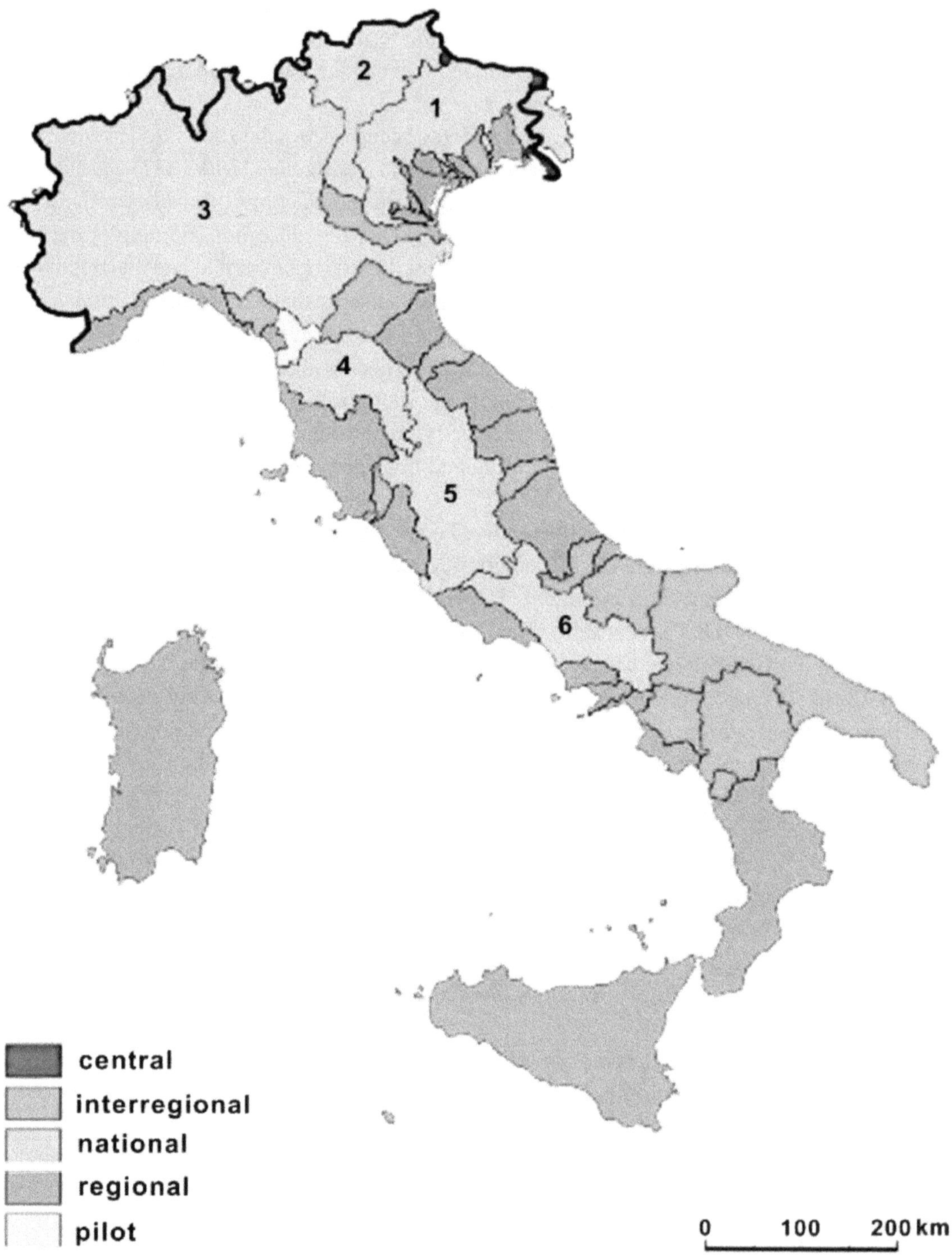

Department for Civil Protection (DCP) was very intense in that period but it was, at the same time, very complicated because it required coordination of multiple governing bodies at the local scale. These units, so-called Functional Multi-hazard Centers (FMCs), had a regional competence and were reunited into a national network together with other independent centers having the tasks of scientific and technical support. The national network was managed by the DCP that was subordinated to the Prime Minister (national government).

In the meantime, the Law 59/1997 of March 1997 transferred the matters of civil protection from the central government to the regions, and this action made the activity of the DCP more complex. Since that political act, each region should have been in charge of regional legislation on civil protection and could decide to change administrative responsibilities and assign additional functions to the provinces and municipalities. Following this structure, each region should have implemented a regional prevention program based on the indications given by the national program; the provinces should have been in charge of collecting data and information for implementing the activity of the CPS, and the mayors should have directed and coordinated the services of assistance to people.

Unfortunately, the tragic events of Piedmont (1994), Sarno (1998), and Soverato (2000) complicated the regular applications of the directives indicated by the Law 183/89 and a new decree-law 179/2002, established by the Ministry for Environment, Land and Sea Protection, introduced the need to consider urgent actions (state of emergency) for the areas hit by flood catastrophes. Based on that law, the national government was empowered to bypass the basic coordination program, for their citizens' safety and protection, which caused a conflict of interests and competencies between state, regions, and local governance.

A solution to this problem was given by the Law 152/2006 that followed the basic indications of the EU Water Framework Directive (WFD, 2000/60/EC) and gave more strength and a new impetus to the activities of the River Basin Authorities that are now consolidated units of planning and management. Based on this law, the so-called 'Hydrographic Districts' were identified. More specifically, 5 River Basin District (Po River, Eastern Alps, Northern Appennines, Central Apennines, and Southern Appennines) and 2 Regional Authorities (Sicily and Sardinia) were established (see Figure 4). The old River Basin Authorities were incorporated into these new geographical units, but they still maintain their main tasks.

Following the EU Floods Directive (FD, 2007/60/EC), and considering the national legislation in force (Legislative Decree 152/2006), the new Legislative Decree 49/2010 (Barbano et al., 2012) focused the attention on the Hydrographic Districts and established the new actions to be taken consisting of the two following steps:

1. preparation of Flood Hazard Maps and Flood Risk maps (art. 6);
2. preparation of the Flood Risk Management Plans, FRMPs (art.7, modified by the Law 116 of August 2014).

The maps covered those areas of the country exposed to significant flood risk and were prepared according to three levels of probability: Low probability (extreme events), medium probability (likely return period $\geq$ 100 years), and high probability. These new maps enhanced and integrated the contents of the PAIs, according to the national guidelines provided by the Environmental Ministry of Land and Sea, with the input of ISPRA, RBDs, and the technical board State-Regions. Examples of Flood Hazard and Flood Risk maps can be found at the following and related links provided by ISPRA https://www.

Figure 4. Location and geographical boundaries of the 7 Hydrographic Districts established by the National Law 152/2006

isprambiente.gov.it/en/publications/reports/landslides-and-floods-in-italy-hazard-and-risk-indicators-2013-summary-report-2018.

The Flood Risk Management Plans are essentially prepared based on the Flood hazard maps and flood risk maps and are periodically reviewed and, if necessary, updated if changes occurred, also taking into account the likely impacts of climate change on the occurrence of floods. Examples of FRMPs can be found at the following link (provided by the EU Commission) for each Hydrographic District https://ec.europa.eu/environment/water/participation/map_mc/countries/italy_en.htm.

STRATEGIC AND LEGAL FRAMEWORK OF TORRENTIAL FLOODS CONTROL IN SOME WESTERN BALKAN COUNTRIES

Floods are one of the main hazards in the Western Balkan countries, increased by topographic and land characteristics, heavy precipitations, removal of forest cover, uncontrolled urbanization, and reduced discharge capacity of regulated river sections (Figure 5). 1,539 torrents are registered in Macedonia, 935 in B&H, and more than 11,500 in Serbia. During catastrophic torrential floods in B&H and Serbia, in May 2014, 76 lives were lost, 2.6 million people were endangered, and about 12,000 km^2 were flooded with material damage higher than 3 billion euros. During a torrential event in Macedonia, near Skopje, in August 2016, 22 people died, and costs of losses and damages were significant.

River basin management in WBC countries is influenced by structural and political organization in certain countries, based on local *Law on Waters*, implementation of EU Directives, as well as the relevant by-laws. A well-established legal, strategic and organizational framework in Western Balkan countries is the foundation for good prevention from floods and preservation of human and material goods.

The Strategic and Legal Framework of Torrential Floods Control in Bosnia and Herzegovina

Presently the Law on Waters in Bosnia and Herzegovina (B&H) has certain exceptions that essentially distinguish B&H from the neighbouring, i.e. Southeast European countries regarding the national laws on water. These specifics arise from the character of constitutional organization of the country, which involves three administrational units of the Federation of Bosnia and Herzegovina (FB&H), the Republika Srpska (RS), and the Brčko District (BD). Further, the legal framework in the water sector is harmonized according to the constitutional organization of B&H, and consists of: Annex IV of the General Framework Agreement for Peace in B&H from the Constitution of B&H, Constitution of FB&H, Constitution of RS, Statute of BD, laws, and bylaws adopted at the levels of state of B&H, entities FB&H, RS, BD, cantons, and municipalities.

Strategic and Legal Framework for the Federation of Bosnia and Herzegovina

Regarding the jurisdiction in the water management sector in the FB&H, the Law on Waters (Law on Waters, 2006) is used as a basic document, as well as several bylaws required for its implementation.

The Law on Waters stipulates that water management includes water protection, water usage, protection against harmful effects of water, and regulation of watercourses and other waters. Among other things, issues related to water resources, water facilities, territorial bases of water management (Figure

Figure 5. Location of floods 2010-2015 in the Western Balkans
Source: A study for the European Commission: Flood Prevention and Management, Gap analysis and needs assessment in the context of implementing the EU Floods Directive, 2015, Executive summary

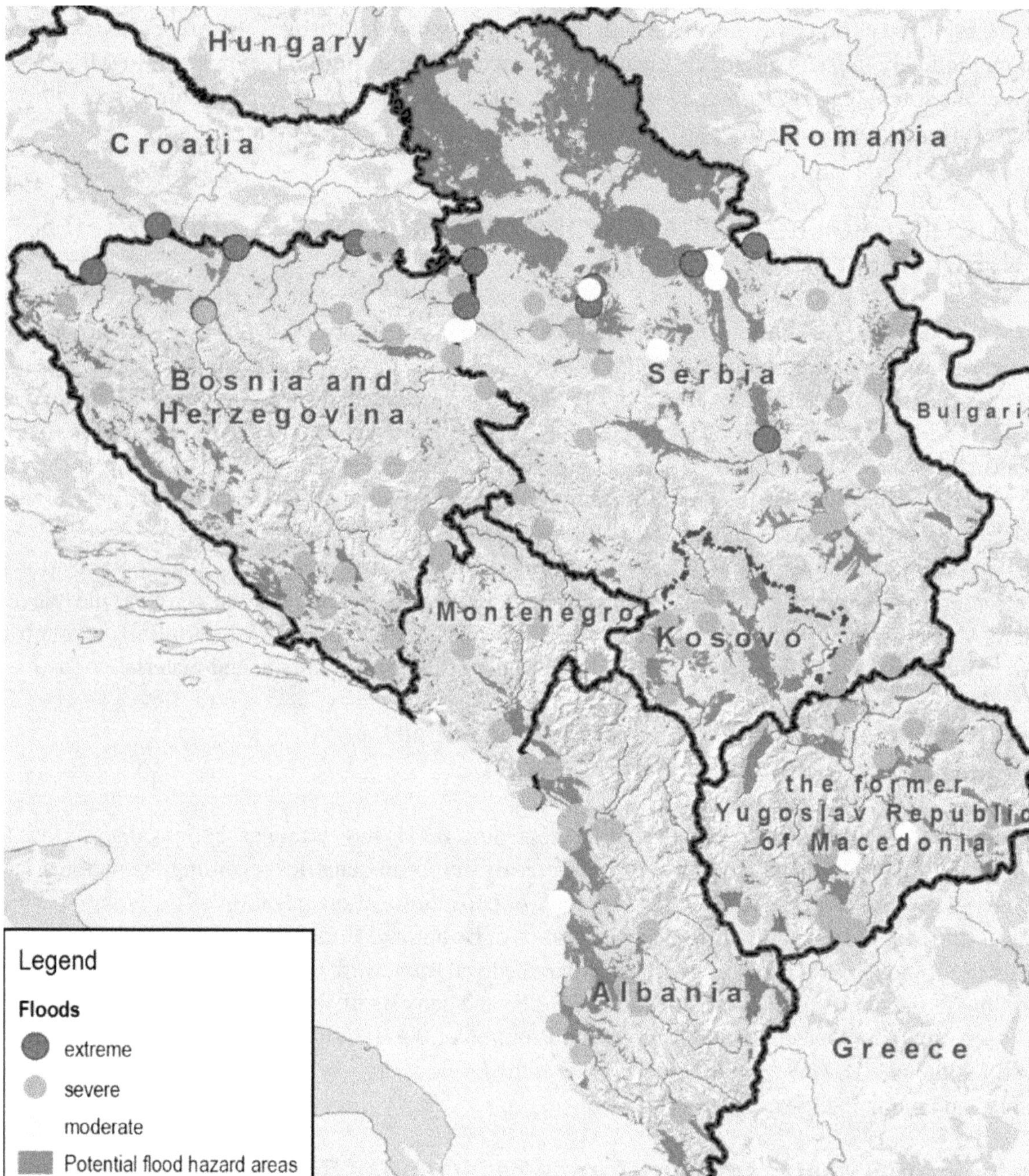

6), water management institutions, water management plans, the role of the public in water management, and financing of water management systems are regulated.

Figure 6. Water catchment area in B&H
(Source: http://www.voders.org/index.php/slivovi)

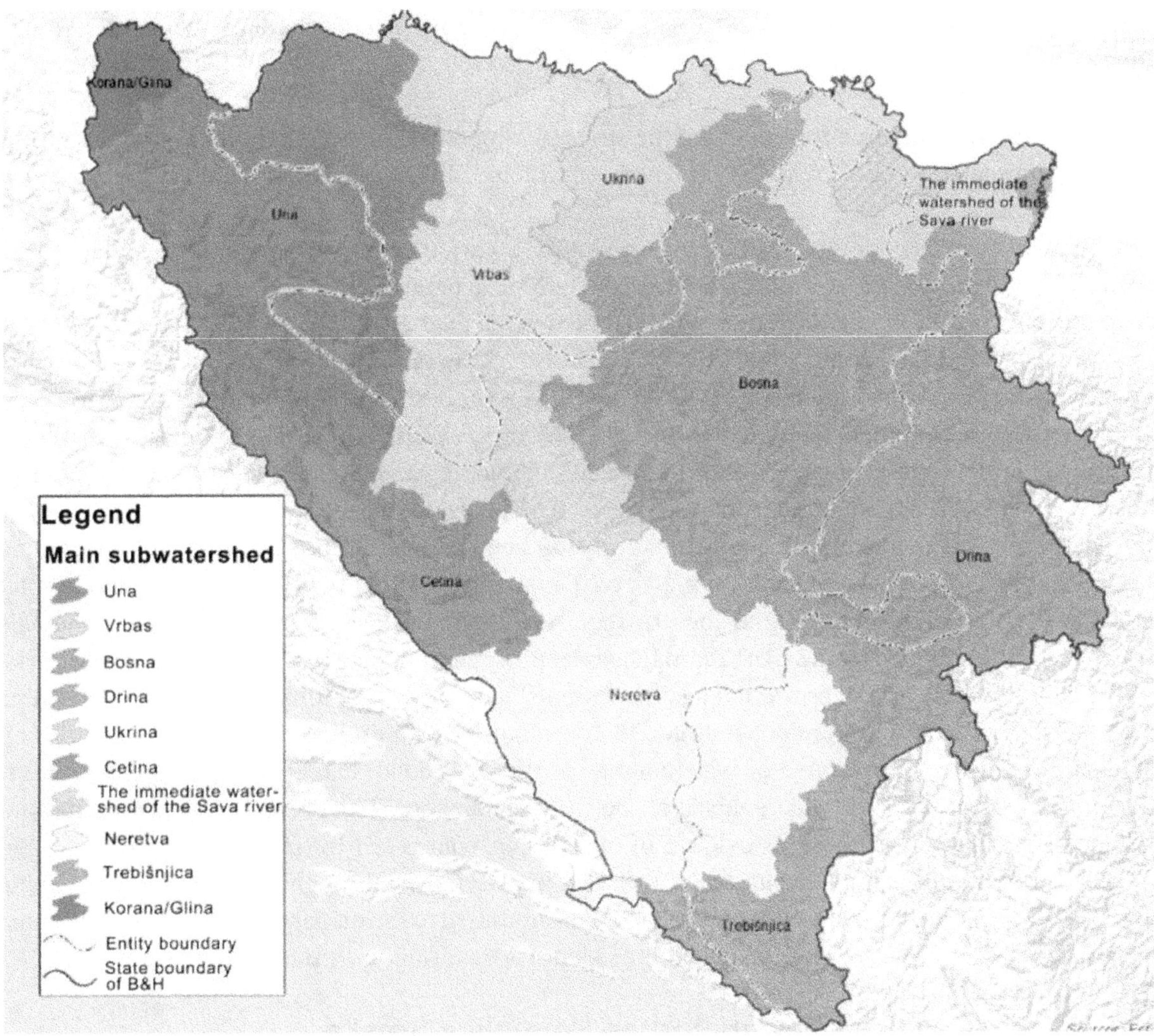

The law (Law on Waters, 2006) also regulates the issue of the institutional framework in the water sector in the FB&H and the financing of this activity, as well as the coordination with the entity of RS and BD in the creation of water management plans, cooperation with the RS and BD authorities on the matters of issuing water, acts, inspections and other issues. Regarding the FB&H, the jurisdictions in water management are also given to cantonal and municipal authorities, stipulated through laws on named administrational levels.

The FB&H Law on Waters regulates the issue of adopting water management plans for the Sava River Basin and the Adriatic Sea Basin, as well as implementing measures and activities aimed at reducing or preventing damage in human lives and material goods caused by contaminated waters and eliminating the consequences of the damage.

The key document that deals with this issue in a more detailed way is the Water Management Strategy of the Federation of Bosnia and Herzegovina 2010 – 2022 (2012). The strategy elaborates the legal framework of water management in B&H and the FB&H, the institutional framework of water management in the FB&H, the economic framework of water management, water use, water protection, water management goals, and measures by water areas. Within the water management goals and measures are the objectives of developing and adopting plans for protection against harmful effects of water and reduction of erosion.

The total funding required for achieving the strategic objectives in the water management sector for 12 years is estimated to be 2,760,695,000 BAM, of which 621,000,000 BAM would be necessary for water protection.

By the Law on Waters of the FB&H (Law on Waters, 2006), there are two water areas in the FB&H: the *Sava River Basin* and the *Adriatic Sea Basin*. Water management plans for these two river basins are prepared to implement the FB&H Water Management Strategy 2010–2022 (2012). The content of the Water Management Plan is prescribed by Article 25 of the Law on Waters of the FB&H.

Water management plans are made exclusively for the river basins (according to the old terminology "basin") because according to the Water Framework Directive (WFD) (2000/60/EC), management must be done with one river basin plan regardless of the administrative boundaries, divisions, and jurisdictions. Cantons in the FB&H are not obliged to make water management plans because they belong to some of the water areas for which management plan has already been prepared. The obligations of the cantonal authorities, arising from the management plans, are developing the Cantonal Operational Flood Protection Plans (COP) and the financing of the projects from the river basin management plan. They also carry out activities defined by the COP, through annual programs adopted by the Cantonal governments.

Cantonal programs aim at protecting against torrents and floods according to priorities determined based on established criteria in cooperation and coordination with civil protection organizations and municipalities. They are in charge of watercourses of the II (second) category while agencies are required to finance projects for watercourses of the I (first) category. The division of watercourses into two categories (I and II) is given in Article 5 of the Law on Waters of FB&H.

The monitoring of the implementation of the strategy is achieved through cantonal authority annual reports on the implemented projects from the program submitted to the management plan holders, which are the Agency for the Sava River Basin and the Agency for Adriatic Sea Basin.

Strategic and Legal Framework for the Republika Srpska

In the entity of RS, the Law on Waters (Law on Waters, 2012) was adopted, as well as several bylaws necessary for the implementation of the law. The Law on Waters in the RS regulates the manner of integrated water management on the territory of the RS, which includes water protection, water use, protection against harmful effects of water, regulation of watercourses and other water bodies. This law also regulates the institutional framework, the manner of financing activities, coordination with the FB&H in water management, and other issues related to integrated water management. The law determines the obligation to adopt management plans for the Sava and Trebišnjica river basins.

The RS Flood Defense Plan (Flood Defense Plan RS, 2019) was adopted by the Government of the RS, and it is a basic document for the coordination and implementation of activities of importance for flood protection and rescue.

The Ministry of Agriculture Forestry, and Water Management adopts the main operational plan for flood defence every year, by which the method of active flood defence on already built water management facilities, at a time of forthcoming danger of large (flood) waters, is determined.

The Strategy of Integrated Water Management of the Republika Srpska has been adopted and is valid until 2024 (Institute for Water Management, 2012). The planning model applied in the strategy is flexible and interactive planning. According to this model, water resources are analyzed, the approximate scope of needs is considered, possible conflicts of interest in the field of use, arrangement, and protection of water, conflicts of interest in land use, and priorities in conflict situations are defined. The outcomes which arise from these analyses are adaptive solutions that are not unambiguous and which can be adapted to inevitable changes in the direction of development. Such planning directs development towards resource possibilities but leaves enough space for the realization of flexible, adaptive solutions.

The main goal of the Water Management Strategy is to achieve a single, managed, and fully harmonized water regime in the territory of RS, in each of its two river basins (by Article 23 of the Law on Waters of Republika Srpska).

Within the Water Management Strategy, as a separate chapter, the issue of Protection against erosion and torrents is addressed, which defines the following: the need for regulation of torrent basins, the strategy of conservation, regulation and protection of basins, measures and works for regulation of torrent basins and erosion areas, also areas on which anti-erosion measures should be performed, the economic importance of anti-erosion land management.

The necessary investments to address alls mentioned issues were assessed in Annex 3 of the framework Plan for Water Management Development of RS (Ministry of Agriculture, Forestry and Water Management RS, 2006). The assessment was made according to the methodology of Gavrilović. Accordingly, for the total of 178 selected catchments in the entity of RS, 123.545.461 EUR is required for water management. For each separate torrential basins, according to the same methodology, annual needs for construction works (m^3 / km^2) and afforestation (ha / km^2) are estimated at 6.000.000 EUR/year. This amount includes about 4.000.000 EUR for new anti-erosion actions, through projects, and 2.000.000 EUR/year is the amount required to repair damaged anti-erosion facilities and their further maintenance. It can be seen that 1/3 of the total annual costs are planned for rehabilitation and maintenance, which if the plan is implemented, will have a positive effect on the optimal functioning of anti-erosion activities and already built facilities against torrential floods.

Strategic and Legal Framework for the Brčko District

The legal framework for the implementation of protection against harmful effects of water consists of Statute of the BD of B&H (Statute of the BD of B&H, 2009), Law on Public Administration of the BD of B&H (Law on Public Administration of the BD of B&H, 2019), the Law on the Government of the BD of B&H (Law on the Government of the BD of B&H, 2019) and the Law on Waters Protection (Law on Waters Protection, 2007).

Water protection is implemented based on the Main Operational Plan for Flood Defense in the Brcko District (MOP), which is adopted every year by the Government of the BD. This plan defines the organizations, manner, and procedure of performing flood defence in the area of Brčko District where protective water management facilities have been built, as well as where they have not been built. Floodplains with built and unbuilt water management facilities for flood prevention are divided into sectors, and further into sections. For the area with constructed protective water management facilities in the MOP, a regional head of flood defence is appointed, as well as his assistant.

The Strategic and Legal Framework of Torrential Floods Control in Serbia

Works on protecting against erosion and torrential floods in Serbia have been carried out since 1907. In a long tradition, this area was regulated by different legal and other regulations, and the works were funded mostly within the forestry sector, and after 1967 strictly within the water sector. The first Law on Regulation of Torrent in Serbia was adopted in 1930, which regulates the work on protection of torrential watersheds as well as their financing. After World War II, in 1954, the Law on Land Protection and Regulation of Torrents was enacted, which was slightly amended in the new Law on Protection of Land from Erosion and Regulation of Torrents in 1960. The law was repealed in 1965 and since then, this area has been regulated by the Law on Waters (Dragović et al., 2007).

The basic legal document in the field of water management is the Law on Water, which regulates "the legal status of water, integrated water management, management of water facilities and water land, sources and methods of financing water activities as well as and other issues of importance to water management" (Law on Water, 2018). The main watersheds in Serbia are depicted in Figure 7.

In the current Law on Water, there is no special chapter regulating the area of soil erosion and prevention of torrential flooding, but it is regulated only by individual articles. The chapter Water facilities list, among others, facilities for "flood protection, erosion, and torrents", with precise definition. The chapter Management of water facilities defines that the Public Water Management Company manages "…by water facilities for protection against erosion and torrents in the waterway area accumulations", as well as "by water facilities for watercourse regulation and flood protection on II order waters and water facilities for protection against erosion and torrents are managed by the local self-government unit on whose territory the facilities are located". Chapter named Programme of Measures also contains a special subchapter "Protection against the harmful effects of erosion and torrents" in which several articles of the Law defined: the manner of determining the erosion area; works and measures for preventing the harmful effects of erosion and torrents (Figure 8) and obligations of performing works and measures. The Law on Water also found that the jobs of general interest that are financed: construction, reconstruction, rehabilitation, maintenance, and management of water facilities for protection against erosion and torrents of public property and the execution of works and measures to protect against erosion and torrents.

In recent decades, little has been invested in water protection, building new and maintaining existing facilities, and this is the reason why the floods that occurred in 2014 have catastrophic consequences (Dragovic & Ristic, 2013).

The necessary preconditions for the implementation of the Law on Water are the adoption of bylaws, taking into account the relevant EU directives. According to the Law on Water, 30 bylaws were adopted, of which 13 ordinances, 8 regulations, several decisions, and ordinances, as well as the Operational plan for flood control for the current year. The legal acts of particular importance are the Regulation on the annual program for monitoring over the water status, and the Regulation on the program for water management, which are adopted each year.

The Law on Disaster Risk Reduction and Emergency Management regulates "disaster risk reduction, prevention, and strengthening of the resilience and readiness of individuals and communities to respond to the consequences of disasters, protection and rescue of people, material, cultural and other goods...early warning, notification, and alarm..." (Law on Disaster Risk Reduction and Emergency Management, 2018).

The Law on Meteorological and Hydrometry Activities defines the system of early warning of meteorological and hydrological disasters. The Republic Hydrometeorological Institute of Serbia (RHMZ), among other activities, provides meteorological and hydrological support to flood defence and protection.

The Institute produces and periodically novels endangered maps and maps of the risk of meteorological natural disasters and participates in the production of flood risk maps based on the prescribed methodology and within its scope assesses the endangerment of the Republic of Serbia (Law on Meteorological and Hydrological Activities, 2010).

In this study, we cited only laws that greatly affect the area of water and flood protection.

Strategic planning and normative activities that are the basis for water management in the territory of the Republic of Serbia are defined by the Law on Water. The basic strategic document in the field of water is the Water Management Strategy on the territory of the Republic of Serbia until 2034. The sub-chapter "Flood Protection with Outer Waters" gave a view of the existing state of flood protection with a special focus on torrential flood protection measures and facilities built for this purpose (Water Management Strategy, 2017).

The current state of flood protection in Serbia is not satisfactory despite many built facilities and undertaken measures. It is estimated that 18% of Serbia's territory is potentially threatened by flooding. The Republic Water Directorate has drafted a Preliminary Flood Risk Assessment (2012) that identified flood areas on Serbian territory. Endangered maps and flood risk maps were done only for areas along the Danube and in the Velika Morava river basin, while for other areas of Serbia's territory that are largely threatened by torrential flooding, those maps have not yet been produced.

The strategic goal is to reduce the risk of harmful effects of water, which is achievable by adequate risk management involving multiple subjects and includes the following phases: response to a flood event, recovery, and preparedness for the next event. According to strategy, to sustain the current condition in the area of erosion and torrential flood protection, it is necessary to build technical works in the amount of 350,000 m^3 (about 4 m^3/km^2), and biological works In the amount of around 34,000 ha (0,4 ha/km^2). To improve the situation in this segment of water protection in the coming period, technical works in the amount of about 1,000,000 m3 (12 m^3/km^2) and biological works covering the area of about 100,000 ha should be conducted. To achieve this goal, it is necessary to invest around €230 million until 2034, while the preservation of the projected functions of the constructed facilities and performed works requires annual investments of around €7,5 million. The negative in this Strategy is that the planned level of financing of flood protection until 2024 is the lowest, compared to the other water sectors.

A significant contribution to the adoption of planning documents in the field of water to harmonize with the WFD is the proposal of the Work Program and the dynamics of the water management plan for the territory of the Republic of Serbia 2021-2027, which is awaiting adoption. The Water Management Plan is the basic instrument that implements the principles of the WFD and represents a strategic framework for water management.

Although several legal regulations on the water are currently in effect, the transposition of the EU legislation has not been fully achieved. The relevant pieces of legislation fully complying with the Water Framework Directive and the Flood Directive are under preparation after adaptation by-laws and decrees follow. As important requirements of the Flood Directive, preliminary assessments have been carried out in the country. Detailed flood risk and flood hazard maps were prepared under the International Commission for the Protection of the Danube River (ICPDR) initiative for the Morava and Danube River Basin and other projects. Maps were produced for 27 flood basins out of 99 identified as areas of potential significant flood risk (Dragović et al., 2018).

Figure 7. Main watersheds in Serbia
(Source: Water management strategy of the territory of the Republic of Serbia, 2017)

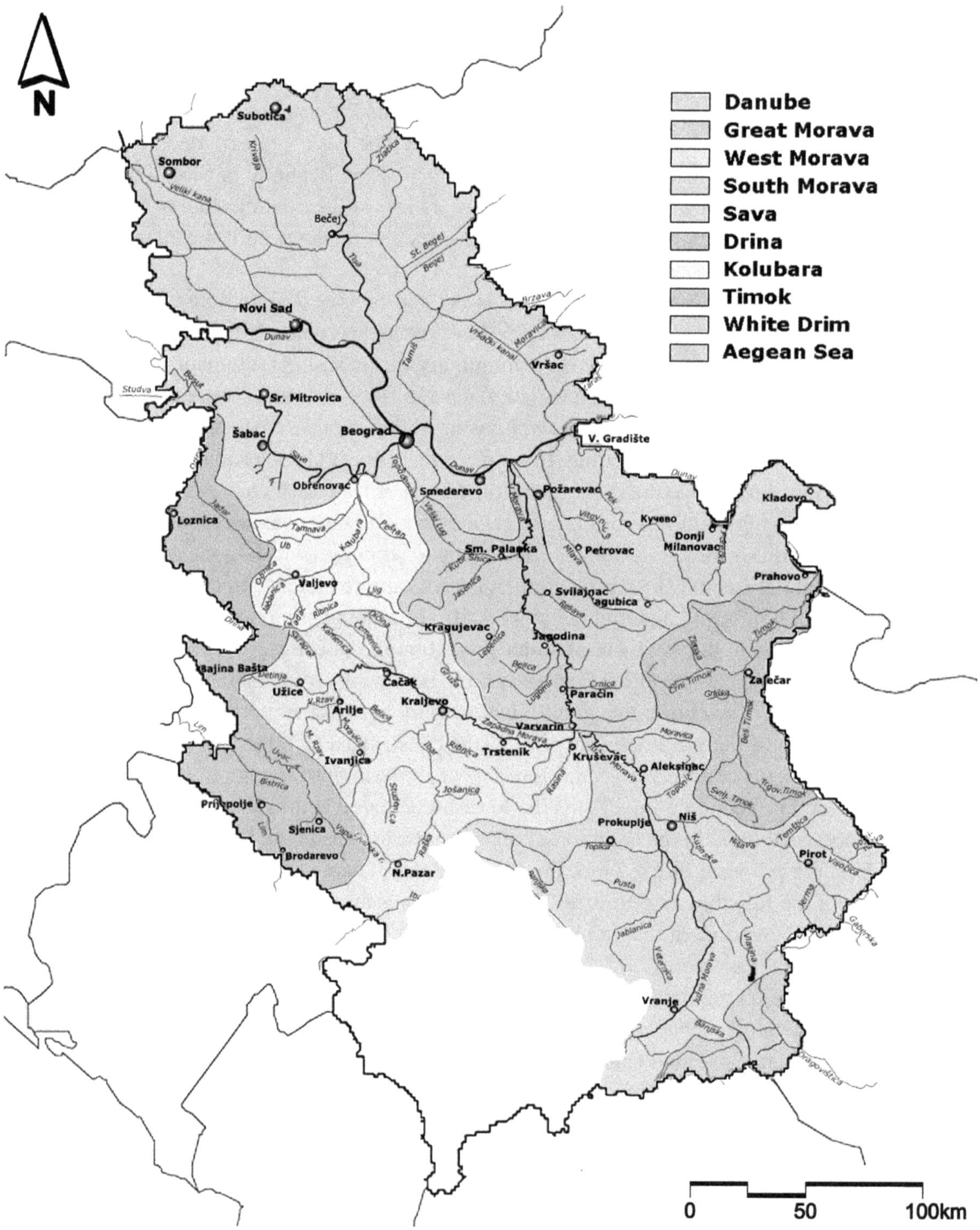

Figure 8. Technical and biotechnical works in the territory of Serbia
(Source: Water management strategy of the territory of the Republic of Serbia, 2017)

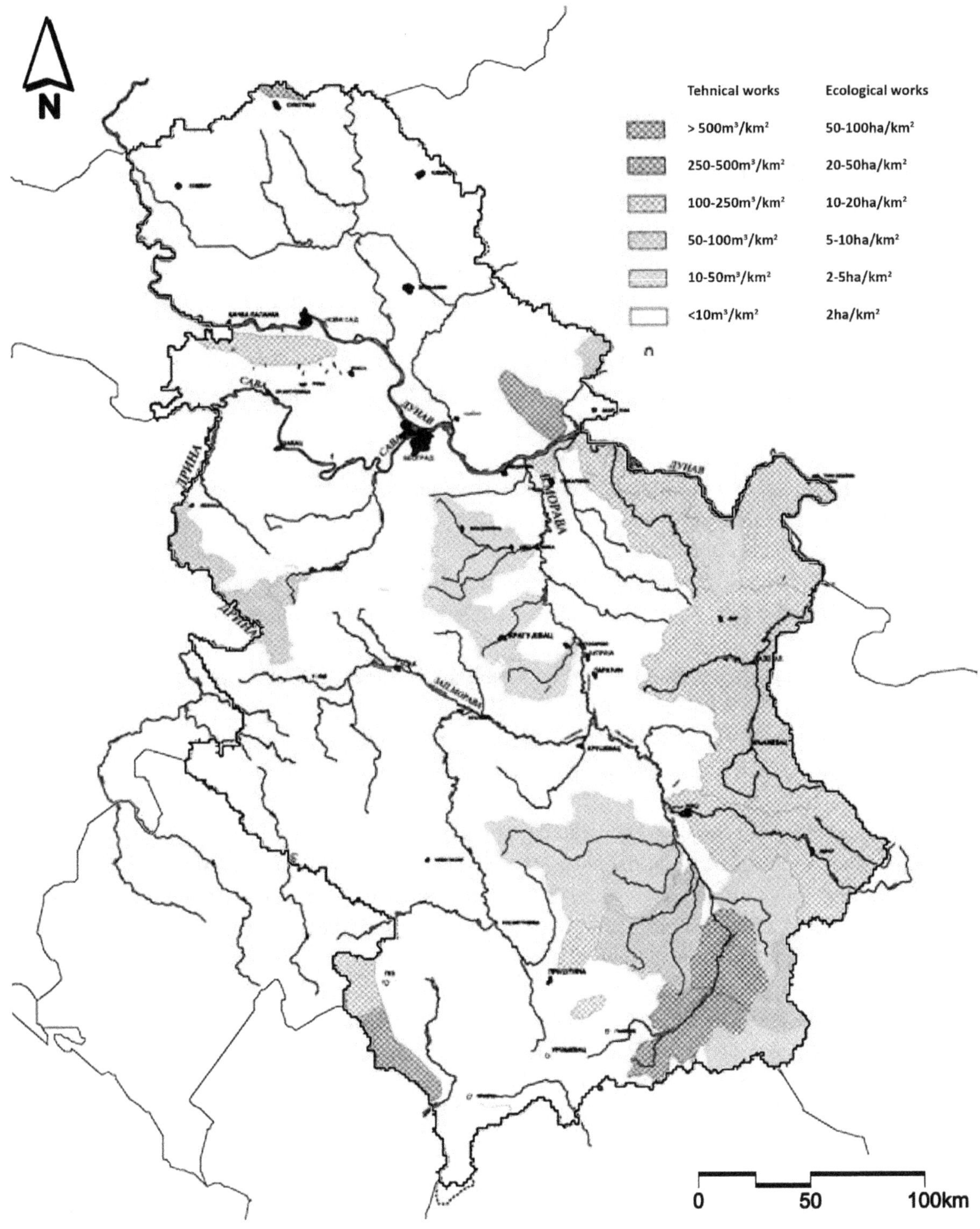

The Strategic and Legal Framework of Torrential Floods Control in North Macedonia

The provisions regulating torrent control are comprised of several sector laws focusing on various aspects related to torrent management, including erosion control. In practice, very often, torrents are not separated from river floods. There is no official delineation and cadaster of torrents in the country. Because of that, provisions for flood control are generally relevant for torrent floods.

The system for flood control encompasses elements of prevention of damage caused by floods, protection by taking measures to reduce the likelihood of floods, information system about flood risks and in the event of a flood, as well as emergency response and mitigation of the impacts on the affected population. The registered floods in the last 50 years in North Macedonia are shown in Figure 9. This text aims to give an overview and analyze the key national legal documents applicable to flood management.

Law on Waters -The core national legal instrument referring to the issue of flood management is the Law on Waters (hereinafter: LW) (Law on Waters, 2016). It incorporates the basic principles and procedures of water resources management. In general, the LW incorporates flood management in the overall river basin district management principles:

- the planning and management is based on the river basin district as a geographical unit for flood management,
- the river basin management plan encompasses the flood risk management and ensures efficiency of the implementation of measures, and the development of a program for protection against harmful effects of waters should be carried out in coordination with and is integrated into reviews of RBMPs,
- acknowledges the extreme floods as exceptional circumstances allowing deviation from the environmental objectives for a particular water body, and
- sets the competencies for planning and implementing the measures for protection against harmful effects of the waters within the same management body.

Within this law torrents, floods are separated then river and otter floods. The most relevant part of the law is the Chapter V: *Protection against harmful effects of waters* that contains provisions on activities and measures for protection and defense against floods, defense against erosion and torrents, defense against freezing of surface water bodies, as well as the elimination of the consequences from such harmful effects of waters, and competences thereof. The LW establishes references to other relevant *lex specialists* stipulating that relevant provisions of other laws determining the conditions, manner, and procedures for protection against the harmful effects of waters shall also be applicable. Within sub-chapter 4 – Protection from erosion and torrents, definitions of torrent and torrent bed are presented (article 139 and 140) as well as direction for torrent control and responsibilities (art.141).

The Law on Protection and Rescue - specifies the establishment and organization of a protection and rescue system, the construction of protection and rescue facilities, risk assessment for possible hazards, the creation of a protection and rescue plan, and spatial planning (Law on Protection and Rescue, 2019). The Law on Crisis Management governs the crisis management system in the Republic of North Macedonia such as the organization and functioning, decision-making and the use of the resources, communication, coordination and cooperation, assessment of the security jeopardy of the Republic of North Macedonia, planning and financing (Law on Crisis Management, 2015). The Law on Hydro-meteorological activities governs the functioning of the Hydro-meteorological activity and responsibilities of Administration for Hydro-meteorological activities and establishes a single meteorological and hydrological observation system, and also sets obligations for warning and notice of extreme weather conditions (Law on Hydro-meteorological activities, 2015). The Law on Local Self-government regulates inter alia the responsibility for the execution of preparations and undertaking of activities for the protection and rescuing of citizens and goods against war destructions, natural and other disasters, as well as against the consequences caused by them (Law on Local Self-government, 2002). They also have other competencies in flood management, which are delegated by other sector laws. Municipalities have inherent the Law on water economy - regulates the management, utilization, operation, and maintenance of hydro-systems and irrigation and drainage systems by the entities acting as water management activity providers. It provides information on some institutional and operational competencies related to the integrated flood management system (Law on water economy, 2015). The Law on forests regulates the management, utilization of silviculture, and protection of all forests in the country. Erosion and torrent control are not elaborated within the Law but some provisions for afforestation or those for protective forest and sustainable forest management significantly contribute to erosion and torrent control (Law on forests, 2014). Within the Rulebook for preparation forest management plans, there is a new article 2014 for protective forest (delineation of these forests) and it follows erosion and torrent control.

Erosion and torrent control are mentioned or touched (together with river floods or other hazards) in several national strategies or plans, as follows: Water strategy (mostly), National Action plan to combat Land degradation and desertification, National communication on climate changes, etc. Some consequences of torrent floods in North Macedonia are depicted in Figure 10.

Water Strategy (2012-2042). This document provides a direction for the country and sets the long-term vision of where the water sector should be by 2042. It is a document explaining the status of waters, inter alia the river training and protection against harmful effects of waters (river training, flood protection, erosion protection; water management objectives, inter alia, for protection against floods and other harmful effects of water (river training, flood protection, erosion protection, irrigation, surface water drainage); as well as a program of activities and measures, inter alia, for protection against floods and other harmful effects of water.

Within other national plans and strategies, torrent floods and other hazards are not mentioned separately but are recognizable as term natural hazards. All provisions are generally for natural hazards, not for any hazard separately. All provisions are generally for natural hazards, not for any hazard separately. An exception in the National Action Plan to Combat Desertification, where torrents (including erosion) are mentioned separately from river floods and consist of some measures against it.

The relevant national legislation assigns competent authorities which are responsible generally for flood management, including torrents. The main responsible body is the Ministry of Environment and Physical Planning, but also Local Self-governments and Water management enterprises. In a case of a crisis, the legislation includes additional competent public and private entities.

Legislation related to torrent control exists. There are strategies, national plans where torrents are separated from rivers, but provisions in some laws and documents are generally for floods even together for all-natural hazards.

Some provisions, especially within Law on Water as main legislation, are ambiguities and create difficulties in operative torrent control. According to the LW, municipalities are responsible for settled areas while water management enterprises are out of the urban area. For the biggest rivers, it is partially correct, while for torrents, it is incorrect, taking into consideration the responsibilities of various entities

Figure 9. Registered floods in last 50 years in North Macedonia
Source: http://globalfloodmap.org/ (Annotations for torrents by I. Blinkov, 2021)

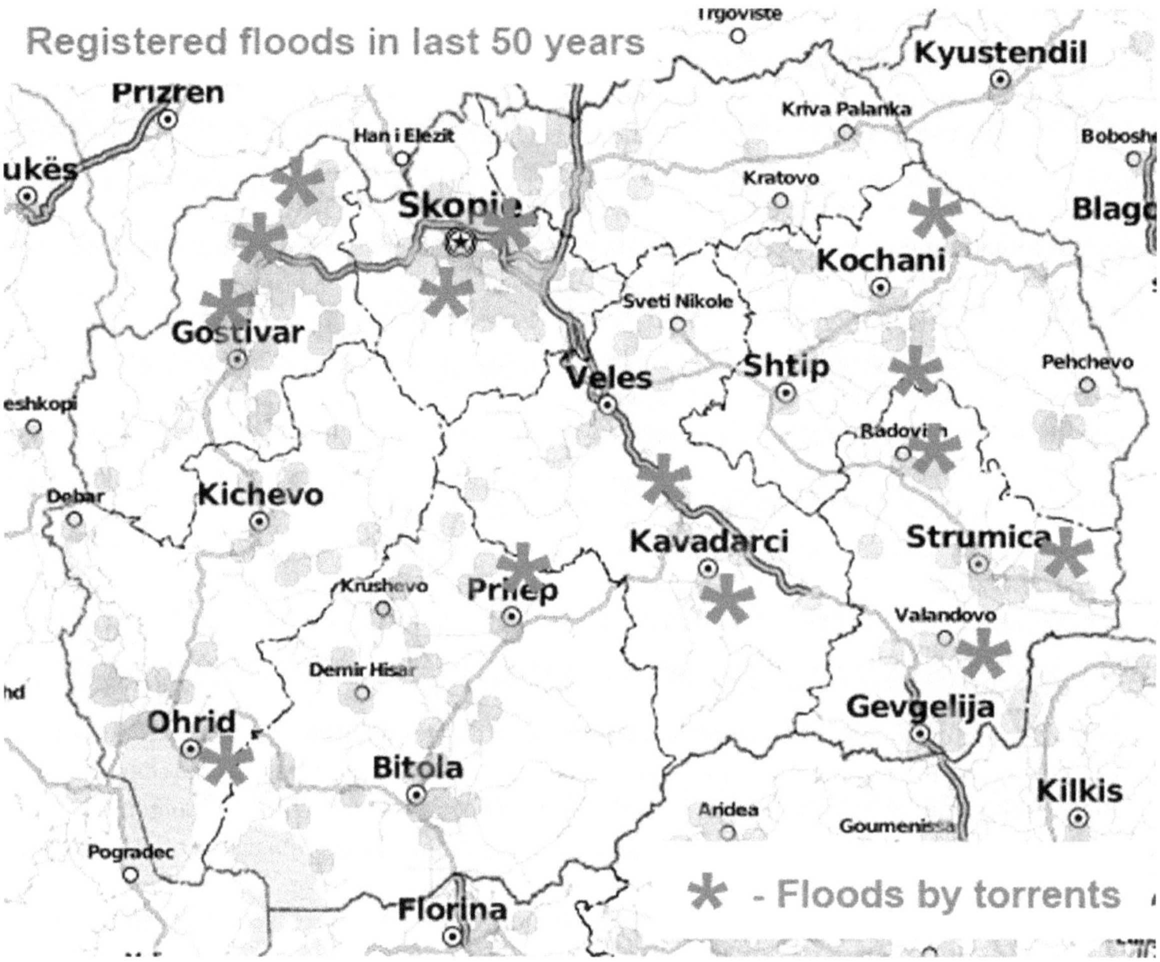

in any torrent catchment. The usually greater part of the torrent basins is in the mountain regions that are under the competencies of Public Enterprise "National Forests" and on the highest elevation PE for "Pastures management". Notorious fact but also according to the art., 141 of LW is that torrents should be integrally treated on the whole catchment with a focus on erosive areas. It means that mayors have an obligation for torrent control on the territory that belongs to their municipalities, but the basin could be mostly under competences of any otter subject (enterprise) on a national level (e.g. PE National Forest, PE Pastures Management) or part of the basin to be within other municipality. Besides, the misunderstanding of the problem by people from LMAs contributes to the preparation of designs for regulating the torrent within the settlement and without other measures - cross structures or measures on the catchment slopes. It resulted even in destroying the newly constructed channels.

Figure 10. Consequences of torrent floods 2015 and 2016 in North Macedonia

2015 - Tetovo region - 4 victims, 25 ME losses and damages

2016 - Skopje region - 23 victims , >50 ME losses and damages

COMPARATIVE ANALYSES OF STRATEGIC AND LEGAL FRAMEWORK OF TORRENTIAL FLOODS

In the EU countries considered in the study, Austria, and Italy, the area of flood protection is regulated by different laws. The basic law governing this area in Austria is the Austrian Forest Low. The Constitution of the Government of Austria determines the competence of the Federal Government, legislative and executive, for protection against torrents. This law regulates the development of hazard zone maps, planning, and execution of control measures, organization of torrential flood control services, etc. In addition to state-level laws, provincial laws have been enacted that regulate land use and buildings hazards. Management of larger river flows and performance of protective works on them is regulated by the Law on Water Management. Compared to Austria, in Italy, the legislative framework of torrential floods control is given through the Soil Protection Act. The aim of this law is the integrated planning and management of river basins in Italy. Based on the Soil Protection Act, the Department for Civil Protection was formed, and a network of advanced early warning and alerting centers was installed.

The Water Framework Directive and Floods Directive, which are binding on EU countries, were the umbrella laws in Italy for the adaptation of the existing Soil Protection Act and the enactment of the

Legislative Decree for the development of Flood hazard maps and Flood risk maps which are the basis for the Flood Risk Management Plans. By harmonizing the Soil Protection Act with the WFD and FD, the River Basin Authorities are adopted as the river basin management units responsible for drafting a Flood and landslide management plan.

The strategy in the field of flood control in Austria is "Protection against natural hazards".

In the WB countries, no special laws have been adopted that regulate the field of protection against torrents, but it is mainly regulated by the Law on Waters. The strategic and legislative framework of this area is specific in B&H due to its territorial division into three entities. However, in all three entities, the basis for legislation related to torrential floods control is the Law on Waters. In the Federation of B&H, in addition to the adopted Law on Waters at the federal level, these laws have been adopted at the cantonal level. Water management strategies have also been adopted in the Federation of B&H and the Republika Srpska, while Flood Defense Plans have been adopted in the Brcko District and the Republika Srpska.

The laws that regulate the area of torrential floods in Serbia are the Law on Waters, the Law on Disaster Risk Reduction and Emergency Management, and the Law on Meteorological and Hydrological Activities. The basic strategic document in this area in Serbia is the Water Management Strategy on the territory of the Republic of Serbia until 2034, adopted in 2017. Water Law prescribes that watercourses of the second-order which are all the torrential watercourses (more than 12,000) fall under the jurisdiction of the local self-government which cannot cope with floods neither in financial nor in personnel and technical sense. Thus, local communities have a "tie hands" in the sense of taking necessary steps in torrential flood risk reduction.

As in Serbia, the legal framework for flood protection in North Macedonia consists of the Law on Waters and the Law on Hydro-meteorological activities. In addition to these two laws, one-act regulates this area and Law on Protection and Rescue, Law on Crisis Management, and Law on water economy. The most important strategic document in the field of water in North Macedonia is the Water Strategy (2012-2042). An overview of the Legal and Strategic Framework of torrential floods control in the selected countries is presented in Table 1.

Following the floods that occurred in 2014, WB countries have intensified their work on the adoption of strategic, legal acts, and by-laws for flood protection, but there is still evident inconsistency between legal acts at the horizontal (other relevant state-level laws) and vertical levels (EU legislation). For effective defense against torrential floods, it is necessary to adopt all bylaws, and even more important is their implementation in practice.

Flood protection in WBC is organized on a different level, and different authorities and institutions are responsible for water management. All considered countries are aligning their water legislation with the requirements stipulated in the EU Water Framework Directive and Flood Directive.

Flood risk management still has to take into consideration some challenges, such as integrated water resource management and community participation in the decision-making process, including adaptation need regarding climate change.

Common challenges also include funding problems, limited capacity at the local level, and unclear allocation of roles and responsibilities in flood management across responsible authorities.

Table 1. Strategic and legal framework in some EU and WB countries

Country	Law	Strategy
Austria	• Austrian Forest Act • Water Management Act • Hydraulic Structures • Development Act	• Protection against natural hazards
Italy	• Soil Protection Act	
Bosna and Herzegovina	• Law on Waters	• Water Management Strategy of the Federation of Bosnia and Herzegovina 2010 – 2022 • Strategy of Integrated Water Management of the Republika Srpska up to 2024
Serbia	• Law on Waters • Law on Disaster Risk Reduction and Emergency Management • Law on Meteorological and Hydrological Activities	• Water management strategy on the territory of the Republic of Serbia until 2034
North Macedonia	• Law on Waters • Law on Protection and Rescue • Law on Hydro-meteorological activities • Law on Crisis Management • Law on water economy	• Water Strategy (2012-2042)

SOLUTIONS AND RECOMMENDATIONS

Torrential floods occur in a short period of several hours due to heavy rainfall. Whilst, the large river floods occur after one or more days. Torrential flood risk management is more complex due to the short arrival time, which makes the adoption of flood defense plans even more important. These plans are adopted at the local level and must be in accordance with the Flood risk management plan and the River basin management plan at the state level. The analysis of the legal framework for flood protection in the EU and WBC indicates its coverage by laws from various fields (water, forest, land, environment).

The multidisciplinarity of the factors influencing the occurrence of torrential floods is also reflected in the diversity of aspects in the legal and strategic framework. This leads to the phenomenon that this area is legally regulated by several laws instead of being defined organizationally, strategically, and legislatively in a single way. The recommendation is to create a strategy and legislation that will cover the whole problem of soil erosion and torrential floods.

FUTURE RESEARCH DIRECTIONS

Future research directions should focus on integrated water resource management, community participation in the decision-making process, as well as adaptation needs regarding climate change. Also, the funding, human and technical capacity at the local level and adequate allocation of roles and responsibilities across responsible authorities are the keys to the development and implementation of the torrential flood management plans.

During strategic planning, as well as during undertaking measures and works on torrential flood control, and following EU directives, it is necessary to consider environmental aspects (biological and hydro morphological quality of torrent flows). One of the important components of flood prevention is protection against soil erosion by undertaking biological works. For the success of these works, it is necessary to include in the future, forest management strategies measure to prevent the occurrence of climate change (invasive species, forest fires, etc.).

CONCLUSION

In recent decades, floods in Europe have had catastrophic consequences, and two EU countries (Austria and Italy) and three Western Balkan countries (Bosnia and Herzegovina, Serbia, and North Macedonia) have been considered for the analysis of the strategic and legislative framework for torrential floods. Binding laws in EU countries are the Water Framework Directive and Floods Directive, but in the observed EU countries, the area of control of torrential floods is regulated through Forest Law (Austria) and Soil Protection Law (Italy). In all analyzed countries of the Western Balkans, this area is legally regulated through the Law on Waters, which is relatively harmonized with EU directives. In addition to this law, torrential floods control in WB countries is also regulated by the Law on Meteorological and Hydrological Activities (Serbia and North Macedonia), the Law on Disaster Risk Reduction and Emergency Management (Serbia), the Law on Protection and Rescue (North Macedonia) and others. Flood management plans have been adopted in most countries, for the whole territory or individual catchment areas. Flood defense plans rely on these plans and local communities are responsible for their adoption in the WBC. A strategic flood defense framework has been established in the analyzed WB countries by developing Water Management Strategies for each of the countries.

ACKNOWLEDGMENT

This research was supported by the Erasmus+ Program of the European Union through the Capacity Building in Higher Education project entitled: Soil Erosion and Torrential Flood prevention: curriculum development at the universities of Western Balkan countries (SETOF), Project No. 598403-EPP-1-2018-1-RS-EPPKA2-CBHE-JP.

REFERENCES

Barbano, A., Braca, G., Bussettini, M., Dessì, B., Inghilesi, R., Lastoria, B., Monacelli, G., Morucci, S., Piva, F., Sinapi, L., & Spizzichino, D. (2012). Proposta metodologica per l'aggiornamento delle mappe di pericolosità e di rischio - Attuazione della Direttiva 2007/60/CE relativa alla valutazione e alla gestione dei rischi da alluvioni (Decreto Legislativo n.49/2010). Manuali e Linee Guida ISPRA, 82/2012 (in Italian).

Boardman, J., & Poesen, J. (2006). Soil Erosion in Europe. John Wiley & Sons Ltd. doi:10.1002/0470859202

Boeuf, B., & Fritsch, O. (2016). Studying the implementation of the Water Framework Directive in Europe: A meta-analysis of 89 journal articles. *Ecology and Society*, *21*(2), 19. doi:10.5751/ES-08411-210219

Dragović, N., & Ristić, R. (2013). Institutional Organisation and Financing of Works for the Protection of Soil from Erosion and Torrent Control in Serbia: Current Problems and Possible Solutions. *Wildbach- und Lawinenverbau, Journal of Torrent, Avalanche, Landslide and Rock Fall Engineering*, (171), 256–266.

Dragović, N., Vulević, T., Zlatić, M., Ristić, R., & Todosijević, M. (2018). Compliance of the Legal Framework for Soil and Water Resources Management in Serbia and Some Regional Countries with the EU Legislation. *Advances in Geoecology, 45*.

Dragović, N., Zlatić, M., & Petković, S. (2007). Comparative Analysis of Institutional Organization of Erosion and Torrent Control in Serbia and Same EU Countries. In *Conference Proceedings Erosion and Torrent Control as a Factor in Sustainable River Basin Management*. University of Belgrade.

Duile, J. (1826). *Ueber Verbauung der Wildbäche in Gebirgs-Ländern, vorzüglich in der Provinz Tirol, und Vorarlberg*. Rauchische Schriften.

Federal Ministry of Agriculture. Forestry, Environment, and Water Management. (2016). Austrian Service for Torrent and Avalanche Control. Author.

Federal Ministry of Agriculture. Water Management, and Forestry Water Management. (2012). Strategy of the Federation of Bosnia and Herzegovina 2010 – 2022. Author.

Flood Defense Plan, R. S. Official Gazette of RS, No. 9/19, (2019), https://ruczrs.org/wp-content/uploads/2019/09/Plan-odbrane-od-poplava-u-Republici-Srpskoj-u-2019.-godini.pdf

Fritsch, O., & Benson, D. (2020). Mutual Learning and Policy Transfer in Integrated Water Resources Management: A Research Agenda. *Water (Basel)*, *12*(1), 72. Advance online publication. doi:10.3390/w12010072

Galloway, G., Seminara, G., Blöschl, G., Garcia, M., Montanari, A., & Solari, L. (2017). Saving a World Treasure: Protecting Florence from Flooding. Firenze University Press.

Guzzetti, F., Cardinali, M., & Reichenbach, P. (1994). The AVI Project: A Bibliographical and Archive Inventory of Landslides and Floods in Italy. *Environmental Management*, *18*(4), 623–633. doi:10.1007/BF02400865

Guzzetti, F., Crosta, G., Marchetti, M., & Reichenbach, P. (1992). Debris flows triggered by the July, 17-19, 1987 storm in the Valtellina area Northern Italy. *Proceedings Interpraevent*, *1992*(2), 193–204.

Institute for Water Management. (2012). *The Strategy of integrated water management of the Republic of Srpska until 2024, Framework plan for the development of water management of the Republic of Srpska - Annex 3 Protection against erosion and torrents*. Author.

ISPRA. (2018). Landslides and floods in Italy: hazards and risk indicators. Summary Report 2018. ISPRA, Rapporti 287 bis/2018.

Kallis, G. (2005). Beyond limits and efficiency, what? Assessing developments in EU water policy. *International Journal of Water*, *3*(2), 121–145.

Lastoria, B., Simonetti, M. R., Casaioli, M., Mariani, S., & Monacelli, G. (2006). Socio-economic impacts of major floods in Italy from 1951 to 2003. *Advances in Geosciences*, *7*, 223–229.

Law of Water Protection, Official Gazette of the Brčko District of B&H, No. 1/05, 19/07 (2007), www. podaci.net/_gBiH/propis/Zakon_o_zastiti/Z-zvoda_05v0425-0719. html

Law on Crisis Management, Official Gazette of the Republic of Macedonia No. 29/05 and Amendments Official Gazette 36/11, 41/14, 104/15, (2015), https://www.refworld. org/cgi-in/texis/vtx/rwmain/opendocpdf.pdf?reldoc=y&docid=5d31a4344

Law on Disaster Risk Reduction and Emergency Management. Official Gazette of RS, No. 87/2018 (2018), https://www.paragraf.rs/propisi/zakon-o-smanjenju-rizika-od-katastrofa-i-upravljanju-vanrednim-situacijama.html

Law on Forests, Official Gazette of the Republic of Macedonia No. 64/09, and amendments 24/11, 53/11, 25/13, 79/13, 147/13 and 43/14, (2014), http://extwprlegs1.fao.org/docs/pdf/mac167487.pdf

Law on Hydro meteorological Activity, Official Gazette of the Republic of Macedonia No. 103/08; 53/11; 51/15, (2015) https://dejure.mk/zakon/zakon-za-hidrometeo roloshka-dejnost

Law on Local Self-Government, Official Gazette of the Republic of Macedonia No. 5/2002, (2002), https://aa.mk/WBStorage/Files/Zakon_lokalnata_samouprava.pdf

Law on Meteorological and Hydrological Activities. Official Gazette of RS" No. 88/2010, (20110), https://www.pravno-informacioni-sistem.rs/SlGlasnikPortal/eli/rep/sgrs/ skupstina/zakon/2010/88/9/reg

Law on Protection and Rescue, Official Gazette of the Republic of Macedonia No. 36/04) and amendments 49/04, 86/08, 124/1018/11, 41/14, editorially revised texts in 2019, https://www.ifrc.org/Global/Publications/IDRL/DM%20acts/Macedonia.pdf

Law on Public Administration of the BD of B&H. Official Gazette of the Brčko District of B&H", No. 36/09, 60/10, 37/14, 13/16 i 14/19 (2019), https://skupstinabd.ba/3-zakon/ba/Zakon%20o%20Javnoj%20upravi%20Brc-ko%20Distrikta%20BiH/07B 25-20%20Zakon%20o%20javnoj%20upravi%20-%20prec--is--c-eni%20tekst.pdf

Law on the Government of the BD of B&H, Official Gazette of the Brčko District of B&H, No. 49/18, 8/19, 10/19 i 32/19 (2019), https://www.paragraf.ba/propisi/ brcko/zakon-o-vladi-brcko-distrikta-bosne-i-hercegovine.html

Law on the organization and work of the bodies of the state administration, Official Gazette of the Republic of Macedonia No. 58/2000) and amendments 44/02, 82/08, 167/10, 51/11, 96/19, 110/19) (2019), https://mioa.gov.mk/sites/default /files/pbl_files/documents/legislation/Zakon_za_organizacija_i_rabota_na_organite_na_drzavnata_uprava_konsolidiran_15102015.pdf

Law on water economy - (Official Gazette of the Republic of Macedonia No. 51/2015) and 193/15 (2015) https://vodostopanstvo.mk/Zakoni/MK/.51-15.pdf

Law on Water. Official Gazette of RS No. 95/2018 (2018), https://www.paragraf.rs/ propisi/zakon_o_vodama.html

Law on Waters. Official Gazette of RS, number: 50/06, 92/09 and 121/12, (2012) http://www.voders. org/images/PDF/zakoni/Zakon121_12b.pdf

Law on Waters. Official Gazette of FB&H, No. 70/06, (2006) https://regeneracija.ba/wp-content/uploads/2016/12/Zakon-o-vodama-22Slu% C5%BEbene-novine-FBiH22-br-7006.pdf

Law on Waters, Official Gazette of the Republic of Macedonia No. 87/08; and amendments 6/2009; 161/2009; 83/2010; 51/2011 44/2012, 23/2013, 163/2013, 180/2014, 52/16, (2016), https://www.moepp.gov.mk/wp-content/uploads/2014/ 10/LAW-ON-WATERS.pdf

Ministry of Agriculture. (2006). *Framework Plan for the Development of Water Management of the Republic of Srpska*. Republic Water Directorate Bijeljina.

Mysiak, J., Testella, F., Bonaiuto, M., Carrus, G., De Dominicis, S., Ganucci Cancellieri, U., Firus, K., & Grifoni, P. (2013). Flood risk management in Italy: Challenges and opportunities for the implementation of the EU Floods Directive (2007/60/EC). *Natural Hazards and Earth System Sciences*, *13*, 2883–2890.

National Action Plan to Combat Desertification, version *prepared for adopting by the Government.*

National Water Strategy, 2012-2042, Official Gazette of the Republic of Macedonia no.122/2012, (2012)

Nikolić Popadić, S. (2020). Legal aspects of flood risk management. *Proceedings of the Faculty of Law in Nis, 59*(86), 201-221

Priest, S. J., Suykens, C., Van Rijswick, H. F. M. W., Schellenberger, T., Goytia, S. B., Kundzewicz, Z. W., Van Doorn-Hoekveld, W. J., Beyers, J.-C., & Homewood, S. (2016). The European Union approach to flood risk management and improving societal resilience: Lessons from the implementation of the Floods Directive in six European countries. *Ecology and Society*, *21*(4), 50. https://doi. org/10.5751/ES-08913-210450

Salvati, P., Bianchi, C., Rossi, M., & Guzzetti, F. (2010). Societal landslide and flood risk in Italy. *Natural Hazards and Earth System Sciences*, *10*, 465–483.

Statute of the BD of B&H, Official Gazette of the Brčko District of B&H", No. 39/09, (2009), https://skupstinabd.ba/images/dokumenti/ba/statut-brcko-distrikta.pdf

Wang, F. (1901). *Grundriss der Wildbachverbauung, S.* Hirzel.

Water management strategy on the territory of the Republic of Serbia until 2034, Official Gazette of RS No. 3/2017 (2017), https://www.paragraf.rs/propisi/strategija-upravljanja-vodama-u-srbiji-do-2034.html

ADDITIONAL READING

England, P. (2019, July). Trends in the Evolution of Floodplain Management in Australia: Risk Assessment, Precautionary and Robust Decision-Making. *Journal of Environmental Law*, *31*(2), 315–341. doi:10.1093/jel/eqz012

European Commission, European court of auditors (2018). Floods Directive: progress in assessing risks, while planning and implementation need to improve, Special Report, (pursuant to Article 287(4), second subparagraph, TFEU)

Mehryar, S., & Surminski, S. (2020). The role of national laws in managing flood risk and increasing future flood resilience. Centre for Climate Change Economics and Policy Working Paper 365/Grantham Research Institute on Climate Change and the Environment Working Paper 334. London: London School of Economics and Political Science

OECD. (2019). *Applying the OECD Principles on Water Governance to Floods: A Checklist for Action, OECD Studies on Water*. OECD Publishing., doi:10.1787/d5098392-

Rauter, M., Thaler, T., Attems, M. S., & Fuchs, S. (2019). Obligation or Innovation: Can the EU Floods Directive Be Seen as a Tipping Point Towards More Resilient Flood Risk Management? A Case Study from Vorarlberg, Austria. *Sustainability*, *11*(19), 5505. doi:10.3390u11195505

Todić, D. (2013). *Climate change in law and governance reform*. Institute for International Politics and Economics. (In Serbian)

Todić, D. (2015). Water Resources Management and European Integration of Serbia, Proceedings of the Faculty of Law in Nis, No. 71, Year LIV, 2015 (In Serbian) 10.5937/zrpfni1571149T

van Ruiten, J. L., & Hartmann, T. (2016). The spatial turn and the scenario approach in flood risk management—Implementing the European Floods Directive in the Netherlands. *AIMS Environmental Science*, *3*(4), 697–713. doi:10.3934/environsci.2016.4.697

KEY TERMS AND DEFINITIONS

Floods Directive: Directive 2007/60/EC is the EU flood directive on the assessment and management of flood risks. Protecting the community from the risk and impact of flooding is at the heart of the European Floods Directive (2007/60/EC). Introduced in 2009, the Floods Directive provides a new approach to managing flood risk on a catchment-wide scale.

Forest Law: This Law regulates the conservation, protection, planning, cultivation, use, and management of forests and forest land, including all necessary rules and requirements defining the control over the implementation of provided rules, monitoring, inspection, as well as other issues relevant to forests and forest land and areas.

Law: A normative act of the state which is passed by its legislative body according to the decision procedure. Laws regulate social relations in all areas of life. By law, social rules are transformed into norms, which are binding for all citizens.

Law on Water: The basic legal document in the field of water management, which regulates the legal status of water, integrated water management, management of water facilities and water land, sources, and methods of financing water activities as well as and other issues of importance to water management.

Soil Protection Act: This act defines important functions of soil and its degradation processes, such as soil erosion, reduction of organic matter content, loss of soil biodiversity, soil contamination, salinization, soil compaction, blocking of its functions, and high hydrogeological risks (floods and landslides). Also, it defines measures and responsibilities for reducing degradation processes.

Strategy: It is defined as a general plan of action with the purpose of achieving specific, clearly defined goals.

Torrential Flood: A flood caused by heavy or excessive rainfall in a short period of time, (generally less than 6 hours) that occurred in a small watershed characterized by a steep slope and intensive erosion processes.

Water Framework Directive: Directive 2000/60/EC of the European Parliament and Council is a legislative approach established for managing and protecting water, based not on national or political boundaries but on natural geographical and hydrological formations: river basins.

Watershed Management: The process of implementing land-use practices and water management practices to protect and improve the quality of the water and other natural resources within a watershed by managing the use of those resources in a comprehensive manner.

ENDNOTE

1 The Article is based on the flyer "Austrian Service for Torrent and Avalanche Control", published by the Federal Ministry of Agriculture, Forestry, environment and water management (2016), the proceeding "From practical experience to national guidelines for debris-flow mitigation measures in Austria", published at the 7th Debris flow hazard mitigation conference, Denver (2019) and on the homepage https://www.bmlrt.gv.at /english/forests/forest-and-natural-hazards/die-wildbach-in-austria.html

Chapter 4
Prioritization of Soil Erosion-Susceptible Sub-Watersheds Using Multi-Criteria Decision Method in the Lesser Himalayas

Shachi Pandey
Forest Research Institute, India

Parmanand Kumar
Forest Research Institute, India

Vijender Pal Panwar
Forest Research Institute, India

Raman Nautiyal
Indian Council of Forestry Research and Education, India

ABSTRACT

Watersheds in the Lesser Himalayan region are highly susceptible to natural hazards, particularly those instigated by action and movement of water, such as soil erosion, flood, and mass movements of lands. Hilly watersheds with diversified land use and fragile ecosystems are responsible for accelerating soil erosion. Soil erosion is one of the most implicit hazards as it degrades water and soil quality in a watershed. The study prioritizes the soil erosion-susceptible zones in the Tons river watershed (India) in the Lesser Himalayan region. The interrelationships and role of morphometry, soil quality, slope, and land use together as four components in soil erosion are studied. Remote sensing data and multi-criteria decision method (MCDM) framework has been used to estimate soil erosion susceptibility of sub-watersheds. Results showed that morphometric parameters like elongation ratio and slope of sub-watersheds play a major role in determining the state of erosion.

DOI: 10.4018/978-1-7998-8459-0.ch004

INTRODUCTION

Soil erosion is a threat to humankind as it degrades one of the essential natural resources i.e., soil. Soil erosion depends on combinations of factors such as the steepness of slope, land use & land cover, climate and ecological calamities like forest fires and landslides. On-site and off-site soil erosion have detrimental impacts on natural resources. Soil erodibility, erosivity and land use management practices play a key role in defining the status of the soil. Soil erosion leads to a reduction of protective cover, making the region more vulnerable. In a watershed, soil erosion leads to the degradation of the water quality of the streams as the eroded material is carried down to the lower reaches of the rivers, making it incompatible to carry an excess amount of water and sediment load. Vulnerability assessment towards soil erosion in a watershed plays a key role in identifying the extent of fragility in an area and supports in making appropriate plans for its conservation. Therefore, it is crucial for watershed management's point of view to identify the vulnerable zones towards erosion. Assessing soil erosion vulnerable zones based on various factors responsible for inducing erosion provides a basis for planning different strategies. There are several methods available to assess the region's susceptibility to soil loss and may be used depending upon the availability of data and other resources for assessment. Several qualitative and quantitative methods are being used worldwide to assess soil erosion and vulnerability zones towards erosion. Studies need to attempt methods based on the availability of accurate input for assessing soil erosion vulnerability. Among various qualitative methods, Multi-criteria decision methods (MCDM) based on different approaches like Više Kriterijumska Optimizacija i Kompromisno Rešenje (VIKOR), Technique for Order of Preference by similarity to Ideal Solution (TOPSIS) and Compound Factors (CF) have been used by researchers worldwide, however, CF approach is the most used method due to its flexibility about the inclusion of different components (Pandey et al., 2021).

The study identifies the most vulnerable sub-watersheds towards soil erosion using MCDM. It highlights the inclusion of various components which play a role in governing soil erosion. The study highlights the importance and application of the MCDM method based on CF value to prioritize zones susceptible to soil erosion. In addition, it suggests that methods like MCDM allow the inclusion of various components due their flexibility and that it may make the study more vigorous. The outcome of the study ranks and provides information on the sub-watersheds that need soil and water conservation. The related objective is to understand the usefulness of methods like MCDM to examine the vulnerability status of watersheds through ranking for strategizing protection and conservation of the watershed.

BACKGROUND

A watershed is the surface area drained by a part or the totality of one or several given water courses. Watersheds are naturally occurring hydrologic units characterized by a set of similar topographic, climatic and physical conditions (Javed et. al., 2009). Many watersheds in the fragile lesser Himalayan region are subject to water induced soil erosion of varying degrees leading to their degradation. Adoption of various land uses leads to an increase in soil vulnerability and adds to its proneness towards erosion. Therefore, there is a need to prioritize the zones prone to soil erosion in watersheds of the Lesser Himalayan region. To formulate the mitigation strategies effectively and implement conservation measures to deal with soil erosion, it is crucial to identify and quantify areas at risk (Blinkov, 2015). Many watersheds of the Tons river region are also facing problems of excessive runoff and pronounced soil erosion as it is located in the

Lesser Himalayan (Kumar et. al., 2014). Morphology and land use of the watershed plays a vital role in determining the degree of soil erosion in the watershed. Complex degradation processes like soil erosion require a multidisciplinary approach to the use of modern methods and techniques (Polovina et al., 2016). Among various MCDM based qualitative approaches, the VIKOR approach provides qualitative results and accuracy generally based on the number of parameters used (Pandey et al., 2021). In this approach, assigning weights are important and need extensive raw data, which is challenging in many cases. Another method, TOPSIS is a less popular method used to assess soil erosion vulnerability due to difficulty in selecting the most preferred alternative. In this approach less parameters lead to less accurate results (Pandey et. al., 2021). However, MCDM based on Compound Factors (CF) is the most used method due to its flexibility in inclusion of different components (Pandey et al, 2021). Multi-criteria decision analysis using the compound value method was performed for prioritizing the sub-watersheds for susceptibility to erosion. This approach is based on the principles of knowledge-driven modeling (Todorovski et al., 2006) and converts the qualitative understanding of a phenomenon based on scientific knowledge into a quantitative estimation. Many researchers have focused on multi-criteria decision-making (MCDM) techniques for solving complex decision-making (Ameri et al., 2018). Assigning a lumped value for a parameter of an entity as well as its use in only a comparative study are some drawbacks of this method. Further, it also imparts the same weightage to all the parameters involved, which in some cases can exaggerate the final output. However, this method relies on detailed estimation and parameterization of the processes involved and that is why it becomes one of the best approaches to compare processes in the watershed. The total number of ranks assigned is based on the number of watersheds. The average of the ranks of all the parameters for a particular watershed is designated as compound value (Cp) and represents the collective impact of all the parameters on the erosion susceptibility of a sub-watershed. It is determined from the following formula (Altaf et al., 2014):

$$C_p = \frac{1}{n}\sum_{i}^{n} R_i$$

where, R_i is the rank of a particular watershed for a parameter, and n is the number of parameters.

Multiple factors like geomorphology (Horton, 1945), soil erodibility parameters, land use and land cover (Ashmore, 1993; Ameri et al., 2018) play a crucial role in deciding the state of a watershed for its proneness towards erosion. Thus, morphometry, land use (Altaf et al, 2014), slope and soil quality with the help of various remote sensing and GIS tools and field study can provide a better solution to assess proneness of watersheds towards soil erosion. Morphometric parameters (Strahler, 1964) of a river basin provide a quantitative description of the drainage system, which is an important aspect of the characterization of basins. It involves evaluation of streams, analysis of various drainage parameters, namely ordering of the various streams and measurement of the area of the basin, perimeter of basin, length of drainage channels, drainage density (D_d), drainage frequency, bifurcation ratio (R_b), circulatory ratio (R_c), basin relief (B_h) and length of overland flow (L_g) to predict the approximate behaviour of the watersheds during periods of heavy rainfall (Parveen et al, 2012; Altaf et al., 2014). Land use and land cover and slope of a watershed help us to understand the status of types of land use in the watershed and provide us with the information about its contribution in erosion (Ameri et al, 2018). Soil erodibility depends primarily on parameters like organic matter, permeability and bulk density etc. and plays an important role in identifying erosion status (Ghosh et al., 2013). Moreover, soil properties have impact on sediment detachment and transport, infiltration (Ashmore, 1993) though use of soil quality parameters are restricted to few studies for qualitative assessment of vulnerability of watersheds to soil erosion (Pandey et al., 2021). In the present study, morphometric parameters have been grouped with land use and land cover, slope and soil quality parameters to assess soil erosion susceptibility of different sub-watersheds of the Tons river watershed.

MATERIAL AND METHODS

Study area

The watershed, lying in Sirmaur and Dehradun, lies in the Lesser Himalayan region within the Tons River catchment on the Himachal Pradesh and Uttarakhand boundary. It is situated between 30^0 40' 30" N to 77^0 43' 30'' E and 30^0 28' 30" N to 77^0 51' 0'' E (Figure 1) with elevation ranging from 431 to 1803 m above mean sea level (amsl) and covers an area of 91.91 km^2. The drainage point of the watershed is taken at its confluence in the mainstream Yamuna. The drainage pattern is mainly dendritic.

Data used

Drainage networks and other baseline data of the watershed were generated using the toposheets 53F/10 and 53F/14 from the Survey of India on a 1:50,000 scale. Watershed was delineated using ArcGIS tool with 20m contour lines provided on the toposheet. Advanced Spaceborne Thermal Emission and Reflection Radiometer (ASTER) 30 m digital elevation model (DEM) and LANDSAT 8 imagery were downloaded from United State Geological Survey (USGS) website was used for the analysis of streams of the watershed by using spatial analyst tool. With the help of ArcGIS, the toposheet was masked with DEM, and the whole catchment area was extracted from the DEM. After delineating the watershed, sub-watersheds were delineated. Twenty-one sub-watersheds (SWS) were identified (Figure 2) and taken up for further study (Figure 3). Area, elevation, and mean elevation of the whole watershed and 21SWS were calculated with the help of the ArcGIS tool. Land use classification was conducted using LANDSAT8 images using the tool of Arc GIS 10.3. Morphometric parameters were analysed for morphometric parameters using the ASTER DEM (30m).

Methodology

The assessment of the watershed is based on four components, namely, Morphometry, Slope, Land use and Soil quality. Total of 19 parameters were calculated using remote sensing and GIS and field study. The framework of the study is described below (Figure 3).

Morphometric parameters

The data were analysed for morphometry, and 10 variables were estimated using the data on stream and watershed boundaries obtained using Arc GIS 10.3. The selection of 10 important morphometric variables was based on Altaf et al. (2014). Further, each morphometric parameter was determined using the standard formula given in Table 1. Morphometric parameters having a direct relationship with erod-

Figure 1. Map showing the Tons river watershed
Source: Author

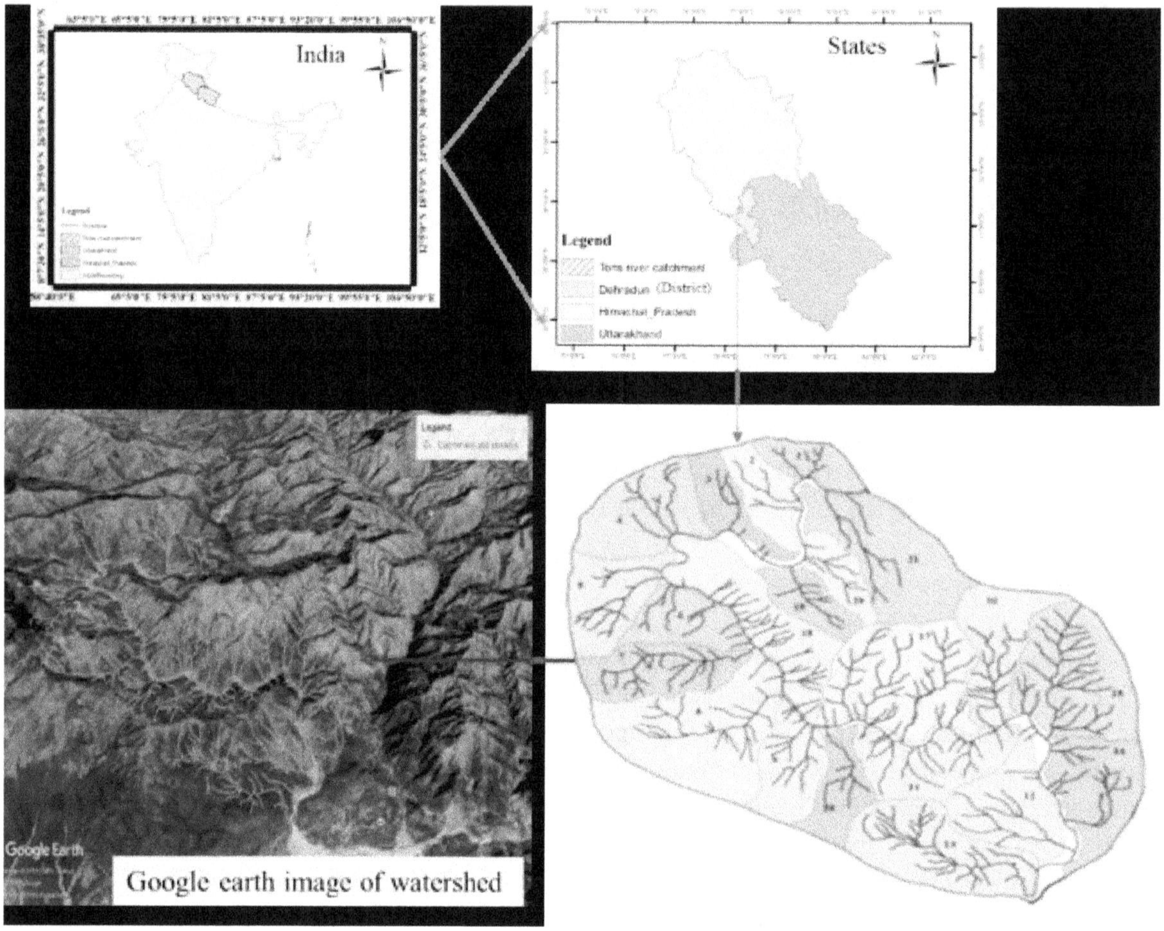

ibility, directly serve as indicators of soil erosion intensity and have been termed as 'erosion risk assessment parameters'. These include the linear morphometric parameters such as drainage density, stream frequency, mean bifurcation ratio, drainage texture, length of overland flow and areal morphometric parameters like basin relief. The shape morphometric parameters such as elongation ratio, circulatory ratio, form factor, basin shape and compactness coefficient have an inverse relation with erodibility (Ratnam et al., 2005). Based on the direct relationships for the linear and shape parameters, the highest value of a morphometric parameter and Hypsometric Integral was given rank 1; the immediate higher value was ranked 2, and so on. Whereas for the shape parameters, the lowest value of a morphometric parameter was given rank 1; the value lower than this was ranked 2, and so on.

Soil quality parameters

The soil quality parameters which affect erodibility of the soil were analyzed. Field study was conducted, and soil samples were collected from 10 watersheds selected based on land use of the sub-watersheds, Samples of soil were collected from each land use of all the 10 selected sub-watersheds. Standard proto-

Figure 2. Sub watersheds (SWS) of Tons watershed
Source: Author

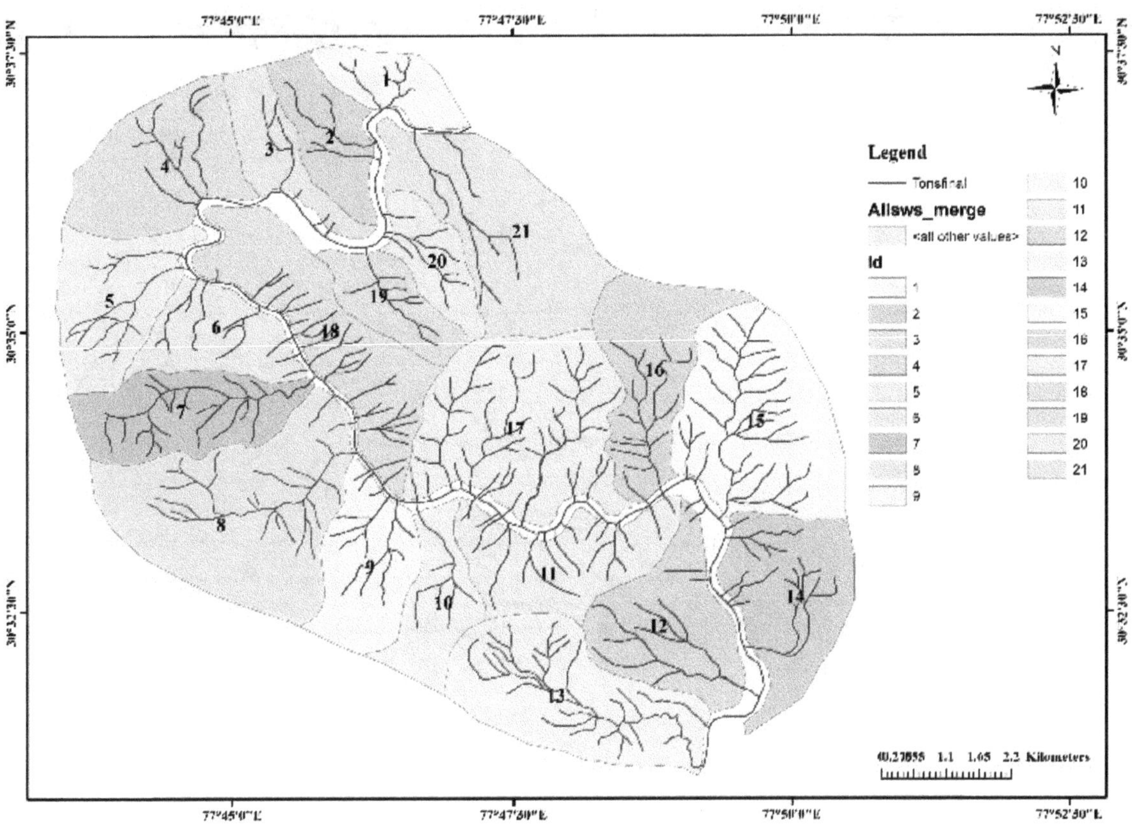

cols were used to test the samples for bulk density, organic carbon and water holding capacity. The soil quality parameters obtained for ten sub-watersheds were further used to predict and assign the value of soil properties for the remaining 11 sub-watersheds using Hierarchical Cluster Analysis (HCA) based on Euclidean distance (Biswas et al., 2015) under an assumption that sub-watersheds similar in morphological characteristics within the same watershed will have similar soil properties. Based on morphometric parameters, all 21 sub-watersheds were grouped into ten clusters using HCA, as given in Figure 6.

Slope

Weighted average slope for all the sub-watersheds was calculated using the Surface tool in ArcGIS 10.3.

Land use parameters

The sub-watersheds were classified into five classes i.e., forest, agriculture, settlement, barren and scrub. A combination of bands of LANDSAT8 images was used and supervised classification in ArcGIS 10.3 was performed for land use classification. The accuracy assessment was carried out using ground control points (GCP's) and was cross verified with the Google Earth tool. The kappa value for the accuracy was 82.3 percent.

Figure 3. Framework of the study
Source: Author

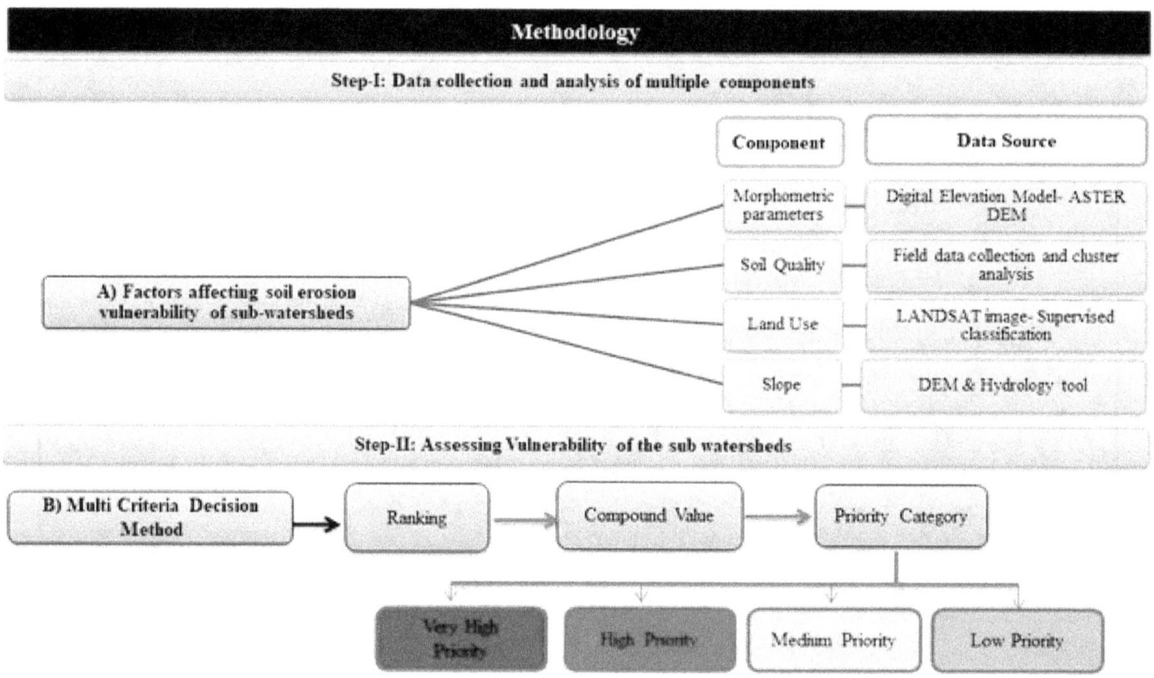

Multi-Criteria Decision Method

For this study, soil quality, slope, land use and morphometry criteria are included as they play a role in making a sub-watershed susceptible to soil erosion. Since 21 sub-watersheds are taken up for the

Table 1. Showing geomorphometric parameters of the watershed

S. No	Morphometric Parameters	Formula	Method
1	Mean Bifurcation Ratio		Horton (1932)
2	Form factor ratio (R_f)	$R_f = A / L_b^2$	Horton (1945)
3	Elongation ratio (R_e)	$R_e = 2/ L_b*(A/\pi)$	Schumm (1956)
4	Circularity ratio (R_c)	$R_c = 4* \pi (A/P^2)$	Strahler (1964)
5	Basin Relief (H)	$H = Z-z$	Schumm (1961)
6	Ruggedness Number (R_n)	$R_n = D_d*(H/1000)$	Strahler (1964)
7	Stream frequency (F_s)	$F_s = N_u / A$	Horton (1945)
8	Drainage density (D_d)	$D_d = L_u / A$	Horton (1945)
9	Drainage Intensity (D_i)	$D_i = F_s / D_d$	Horton (1945)
10	Length of overland flow (L_g)	$L_g = A/2* L_b$	Horton (1945)
Where, A= Basin Area, L_b= Basin Length, Z= Maximum Elevation, z=Minimum elevation, P= Perimeter			

Source: Author

study, ranks were assigned from 1 to 21. For morphometry, soil quality, slope and land use, rank 1 was assigned so that the value of the parameter represents a maximum contribution to the erosion and rank 21 represents minimum contribution. The average of the ranks of all the parameters for each watershed was designated as compound value (Cp) and represents the collective impact of all the parameters on erosion susceptibility of a sub-watershed using the formula by Altaf et al. (2014).

Based on Cp values, the sub-watersheds are categorized into four priority groups-very high priority (8.00-9.33), high priority (9.34 – 10.66), medium priority (10.70 – 11.99) and low priority (12.00-13.32).

RESULTS

The results obtained by multi-criteria decision method from the data generated of the watershed are discussed below.

Morphometric analysis

The measurement of various morphometric parameters (Table 2), namely bifurcation ratio (R_b), total relief (H), relief ratio (R_h), drainage density (D_d), stream frequency (F_s), elongation ratio (R_e), circularity ratio (R_c), form factor ratio (R_f), and length of overland flow (Lg) of the watershed has been carried out, and the result have been presented in Table 2. The results show that the river basin is not circular since the elongation ratio (R_e) varies from 0.64 (SWS7) to 1.91 (SWS15) (Table 2). The value of R_f lies towards the higher side as it varies from 0.32(SWS7) to 2.85 (SWS15), as given in Table 2. The circularity Ratio (R_c) of the watershed was observed from 0.36(SWS18) to 0.74(SWS15) (Table 2). Stream order (N_u) 4th is the highest order in this watershed. It is the trunk stream through which all discharge of water and sediment passes. The study area has a maximum number of 1st order streams. Stream number (N_u) provides the number of streams of different orders and the total number of streams in the basin. Total 481 streams segments were recognized in the watershed, out of which 298 are 1st order, 104 are 2nd order, 47 are 3rd order, and 32 are 4th order. The R_b values in the sub-watersheds of the area range from 1.75 (SWS10) to 6.73 (SWS18). Total Relief (H) aspects of the sub watershed play an important role in permeability, drainage development, surface and sub-surface water flow, landforms development and erosion properties of the terrain. The analysis reveals that the SWS19 has the maximum relief i.e., 1390m (Table 2). Ruggedness Number (R_n) is the product of basin relief and drainage density and defines the slope length and steepness. The value of ruggedness number for the watershed varied from 1.18 to 4.38 in SWS1 and SWS17, respectively (Table 2), which indicates the presence of steeper and longer slopes.

Drainage Density (D_d) indicates the closeness of spacing of channels (Table 2). It provides a numerical measurement of landscape dissection and runoff potential. Horton (1932) suggests that a high drainage density indicates the watershed is low permeable sub-soil and thin vegetation cover. The high value of D_d i.e. 3.58 (SWS 17) indicates high runoff potential of the watershed and impermeable subsoil. Drainage Intensity (D_i) indicates the period taken by surface runoff to get removed from the watershed, making it less or highly susceptible to flooding, gully erosion and landslides. The D_i in the area varies from 1.18 in SWS21 to 2.85 in SWS18. Stream Frequency (Fs) is indicative of low relief and high infiltration capacity of bedrock. In the present study, Fs exhibits a positive correlation with the drainage density values of the sub-watersheds, which indicates an increase in stream population with respect to an increase in

Table 2. Quantitative morphometry results of Tons sub watersheds

SWS	Rbm	Rf	Re	Rc	H	Rn	Fs	D_d	Di	Lg
SWS1	2.00	0.95	1.10	0.63	780.00	1.18	3.47	1.51	2.30	0.33
SWS2	2.00	0.50	0.80	0.52	971.00	1.66	2.61	1.71	1.53	0.29
SWS3	6.00	0.42	0.73	0.37	1067.00	1.74	3.42	1.63	2.10	0.31
SWS4	2.83	1.75	1.49	0.71	1079.00	1.64	2.61	1.52	1.71	0.33
SWS5	2.58	0.47	0.78	0.57	1151.00	2.47	4.30	2.15	2.00	0.23
SWS6	5.00	0.57	0.85	0.58	990.00	2.24	4.47	2.27	1.97	0.22
SWS7	3.30	0.32	0.64	0.61	1151.00	2.47	6.76	3.15	2.15	0.03
SWS8	1.76	0.67	0.93	0.63	1274.00	2.30	4.60	1.80	2.55	0.28
SWS9	2.00	0.51	0.81	0.62	724.00	1.66	5.51	2.30	2.40	0.22
SWS10	1.75	0.34	0.65	0.43	718.00	1.33	3.78	1.86	2.04	0.27
SWS11	4.75	2.02	1.60	0.46	781.00	2.10	5.82	2.69	2.16	0.19
SWS12	3.33	0.65	0.91	0.59	676.00	1.77	3.53	2.61	1.35	0.19
SWS13	2.44	0.37	0.69	0.68	857.00	2.70	6.43	3.15	2.04	0.16
SWS14	2.83	1.02	1.14	0.57	742.00	1.72	5.40	2.32	2.33	0.22
SWS15	2.40	2.85	1.91	0.74	1225.00	3.91	6.15	3.19	1.93	0.16
SWS16	2.55	1.61	1.43	0.43	1273.00	3.01	4.86	2.37	2.05	0.21
SWS17	2.40	2.44	1.76	0.62	1222.00	4.38	8.99	3.58	2.51	0.14
SWS18	6.73	2.61	1.82	0.36	867.00	2.22	7.30	2.57	2.85	0.19
SWS19	2.25	0.44	0.75	0.62	1390.00	3.08	5.54	2.22	2.50	0.23
SWS20	3.25	0.35	0.67	0.50	938.00	2.35	6.86	2.50	2.74	0.20
SWS21	2.50	0.42	0.73	0.54	1115.00	1.35	1.43	1.21	1.18	0.41
Min	1.75	0.32	0.64	0.36	676.00	1.18	1.43	1.21	1.18	0.03
Max	6.73	2.85	1.91	0.74	1390.00	4.38	8.99	3.58	2.85	0.41

*Where, Mean Bifurcation Ratio (Rbm), Form Factor (Rf), Elongation Ratio (Re) Circularity Ratio (Rc), Basin Relief (H), Ruggedness Number (Rn), Stream Frequency (Fs), Drainage Density (D_d), Drainage Intensity (Di), Length of overland flow (Lg)
Source: Author

drainage density. The highest stream frequency is for SWS17 (8.99) (Table 2). The value of Lg for all sub-watersheds in the watershed varies from 0.03 in SWS7 to 0.41 in SWS21.

Land use

The land use was identified and determined for each sub-watershed and it was observed that SWS 13 has a maximum forest area of 4.5 km^2 (Figure 4). At the same time, the least forest area is in SWS1 (0.16 km^2) (Figure 5). Similarly, SWS 8 has maximum agricultural land use (3.47 km^2) while the minimum agricultural land use is in SWS 19. The presence of both settlement and scrub is maximum in SWS12, while it was minimum in SWS6 and SWS 11, respectively. The barren area, which is an open rocky surface in the watershed, is maximum in SWS17 (4.17 km^2) and minimum in SWS 6. The scrub is found in a few sub-watersheds being maximum in SWS12 (0.96 km^2).

Figure 4. Land use (A), Slope (B) and Distribution of land use (C) in the watershed
Source: Author

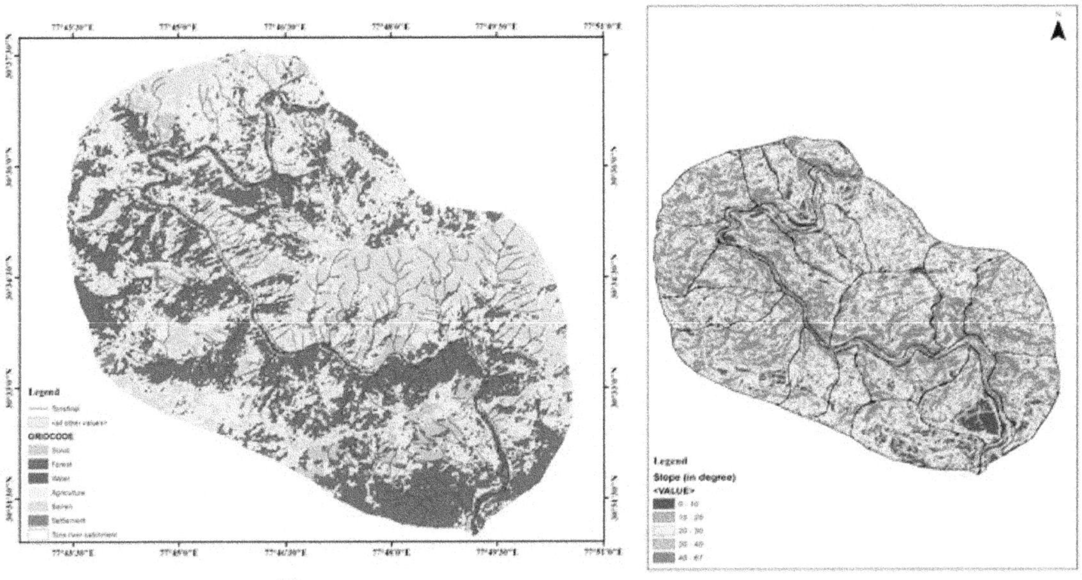

(A)

(B)

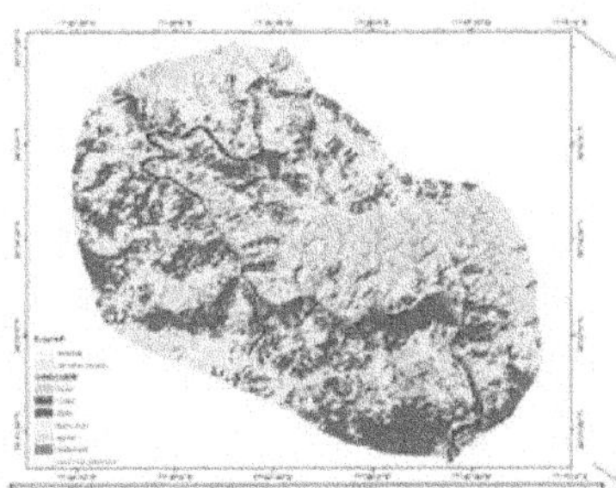

Land use and Land cover (km²) in percentage					
SWS	Forest	Agriculture	Barren	Settlement	Scrub
1	8.63	55.59	28.81	1.17	5.80
2	18.09	59.92	18.91	3.08	0.00
3	12.02	53.55	33.32	1.10	0.00
4	29.19	37.99	23.20	1.06	8.57
5	48.26	40.07	9.98	1.69	0.00
6	58.07	33.72	5.88	2.33	0.00
7	52.02	33.73	10.82	3.42	0.00
8	37.34	39.21	21.92	1.52	0.00
9	45.01	42.54	9.47	2.98	0.00
10	46.71	34.74	12.15	0.85	5.55
11	66.93	25.05	6.05	1.35	0.64
12	29.20	36.61	8.89	3.73	21.55
13	69.48	18.88	4.58	0.47	6.79
14	49.20	33.47	16.77	0.56	0.00
15	17.09	26.83	55.48	0.60	0.00
16	19.28	34.60	44.73	1.39	0.00
17	8.69	26.00	52.36	0.80	12.16
18	31.41	44.04	21.51	2.04	1.20
19	50.78	36.31	11.84	1.08	0.00
20	42.64	41.50	15.72	0.14	0.00
21	34.65	45.07	19.11	1.16	0.00

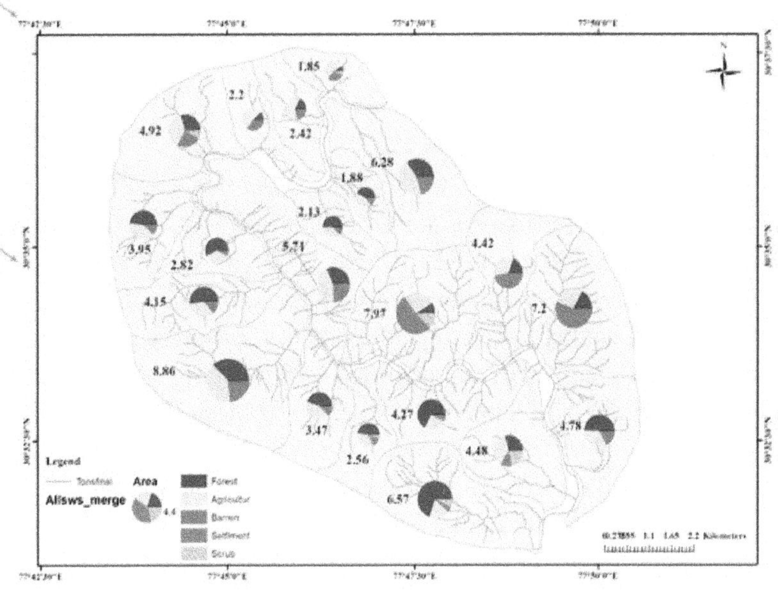

(C)

Slope

Most of the area has a steep slope going up to 74.76 percent (Figure 4). The weighted average slope of the sub-watersheds varied from 40.70 percent in SWS 13 to 74.76 percent in SWS 6. High slope was observed in SWS 19 (70.76%), SWS18 (73.08%), SWS 17(70.56%), and SWS 15 (68.61%) (Figure 5).

Figure 5. Distribution of various land use and slope across the different sub watersheds
Source: Author

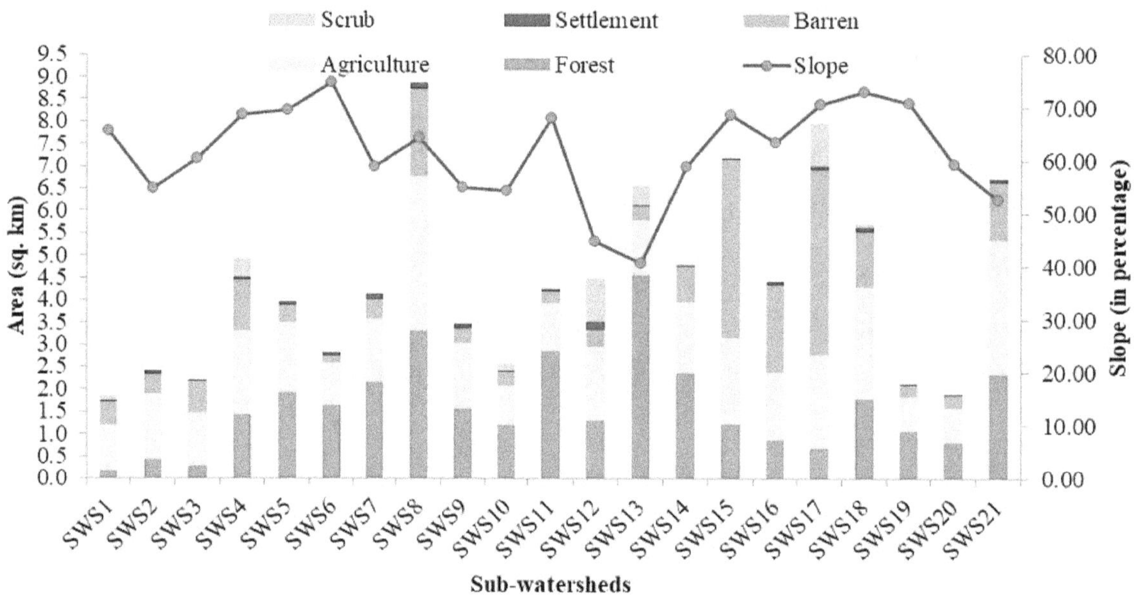

Soil parameters

The hierarchical cluster analysis identified homogenous clusters of sub-watersheds. The 10 sampled sub-watersheds were SWS14, SWS20, SWS4, SWS12, SWS13, SWS17, SWS8, SWS9, SWS15 and SWS 10. The sub-watersheds that clustered with the sampled sub-watersheds are given in Figure 6.

Out of 3 soil quality parameters, the organic carbon ranged from 0.66 percent to 1.17 percent, while the water holding capacity varied from 33.46 percent to 38.3 percent (Table 3). The bulk density was slightly high in SWS 17 and SWS 16 i.e. 1.45, and in the rest of the watershed, it varied from 1.38 to 1.45 g/ml. It was observed that, overall, the bulk density was high in all the sub-watersheds making them more prone to soil erosion. Low organic carbon content across the sub-watersheds indicated their less water infiltration capacity on the soil surface.

Multi-Criteria Decision Method Analysis

Multi-Criteria Analysis was performed by taking three different components (morphometry, soil quality, slope and land use) together to understand the susceptibility of sub-watersheds was analysed (Table 4). Results show that five sub-watersheds fall in the medium priority category and five in the low priority category, while four sub-watersheds are in the very high priority category, and seven are in the high priority category (Figure 7). The results showed that SWS13, SWS11, SWS15 and SWS1 are under very high vulnerable sub-watersheds.

Figure 6. Dendrogram for different sub-watersheds with explanations of their groups
Source: Author

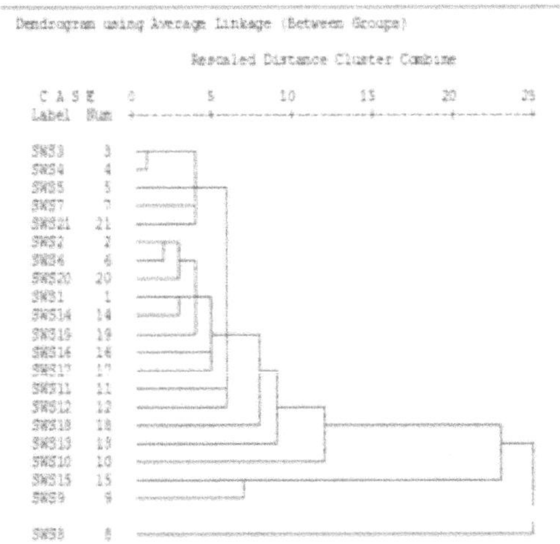

Groups	Sampled SWS	Clustered SWS	Groups	Sampled SWS	Clustered SWS
I	SWS 14	SWS 1	VI	SWS 17	SWS 16, SWS 19
II	SWS 20	SWS 2, SWS 6	VII	SWS 8	
III	SWS 4	SWS 3, SWS 5, SWS 7, SWS 21	VIII	SWS 9	
IV	SWS 12	SWS 11	IX	SWS15	
V	SWS 13	SWS 18	X	SWS 10	

DISCUSSION

The highest compound value and rank 1 in this group were determined to SWS13 (Figure 7). SWS13 has a low elongation ratio (0.7) and form factor (0.4) while a high circulatory ratio (0.68) and basin shape (2.7). Chow, 1964 had noted that strongly elongated basins have circularity ratios between 0.40 and 0.50. The form of the basin is wide in the heading part and narrow in the lower area of confluence,

Table 3. Soil Quality Parameters for sub-watersheds

	Soil Quality						
	OC (%)	WHC (%)	BD (g/ml)		OC (%)	WHC (%)	BD (g/ml)
Sub-watershed	Mean	Mean	Mean	Sub-watershed	Mean	Mean	Mean
SWS1	0.964	38.358	1.431	SWS12	0.776	36.783	1.384
SWS2	1.020	36.571	1.423	SWS13	0.903	38.279	1.383
SWS3	0.846	33.466	1.408	SWS14	0.964	38.358	1.431
SWS4	0.846	33.466	1.408	SWS15	1.008	36.299	1.415
SWS5	0.846	33.466	1.408	SWS16	0.666	33.986	1.450
SWS6	1.020	36.571	1.423	SWS17	0.666	33.986	1.450
SWS7	0.846	33.466	1.408	SWS18	0.903	38.279	1.383
SWS8	0.986	39.095	1.386	SWS19	0.666	33.986	1.450
SWS9	0.748	37.433	1.423	SWS20	1.020	36.571	1.423
SWS10	1.170	36.512	1.417	SWS21	0.846	33.466	1.408
SWS11	0.776	36.783	1.384				

*OC=Organic Carbon; WHC= Water Holding Capacity; BD=Bulk Density
(Source: Author)

Table 4. Ranking based on MCDM for watershed prioritization towards soil erosion susceptibility

SWS	Rbm	Rf	Re	Rc	H	Rn	Fs	D	Di	Lg	BD	WHC	OC	Sl	F	A	B	S	SB	Rank	Prt
SWS1	19	14	14	17	17	1	17	20	8	2	4	19	14	9	1	17	10	20	6	2	VH
SWS2	17	9	9	7	12	5	18	17	19	5	6	11	18	17	3	12	11	7	9	10	H
SWS3	2	5	6	2	10	8	18	18	11	4	12	1	7	12	2	15	9	17	9	20	L
SWS4	8	17	17	20	9	4	19	19	18	3	12	1	7	6	11	6	7	13	4	12	M
SWS5	10	8	8	9	6	15	14	14	15	8	12	1	7	5	15	9	14	8	9	14	M
SWS6	3	11	11	11	11	11	13	12	16	10	6	11	18	1	13	18	21	9	9	9	H
SWS7	6	1	1	13	7	16	4	3	10	21	12	1	7	15	16	13	12	2	9	19	L
SWS8	20	13	13	18	2	13	12	16	3	6	17	21	16	10	20	1	4	3	9	8	H
SWS9	18	10	10	16	19	6	9	11	6	11	6	16	4	16	12	11	15	5	9	11	H
SWS10	21	2	2	3	20	2	15	15	13	7	10	10	21	18	8	19	16	19	5	5	H
SWS11	4	18	18	5	16	10	7	5	9	17	18	14	5	8	19	16	19	12	8	3	VH
SWS12	5	12	12	12	21	9	16	6	20	16	18	14	5	20	10	7	13	1	2	7	H
SWS13	13	4	4	19	15	17	5	4	14	18	20	17	12	21	21	14	17	15	3	1	VH
SWS14	9	15	15	10	18	7	10	10	7	12	4	19	14	14	18	8	8	16	9	6	H
SWS15	14	21	21	21	4	20	6	2	17	19	11	9	17	7	9	5	2	14	9	3	VH
SWS16	11	16	16	4	3	18	11	9	12	13	1	6	1	11	6	10	3	11	9	18	L
SWS17	15	19	19	15	5	21	1	1	4	20	1	6	1	4	4	4	1	10	1	21	L
SWS18	1	20	20	1	14	11	2	7	1	15	20	17	12	2	14	3	6	4	7	17	L
SWS19	16	7	7	14	1	19	8	13	5	9	1	6	1	3	7	21	20	18	9	15	M
SWS20	7	3	3	6	13	14	3	8	2	14	6	11	18	13	5	20	18	21	9	13	M
SWS21	12	6	5	8	6	3	21	21	21	1	12	1	7	19	17	2	5	6	9	16	M

*Where, Mean Bifurcation Ratio (Rbm), Form Factor (Rf), Elongation Ratio (Re) Circularity Ratio (Rc), Basin Relief (H), Ruggedness Number (Rn), Stream Frequency (Fs), Drainage Density (D_d), Drainage Intensity (Di), Length of overland flow (Lg), Organic Carbon (OC); Water Holding Capacity (WHC); Bulk Density (BD), Slope (Sl), Forest (F), Agriculture (A), Barren (B), Settlement (S), Scrub (SB), Priority Status (Prt)

Source: Author

and the risk of flood is high during excessive rains based on stream order. The land use provides the sub-watershed an optimum condition i.e. highest forest cover (4.56 sq.km) but its very high vulnerability status indicates the influence of morphological behaviour of the sub-watershed playing an important role. Low elongation ratio leads to soil susceptibility towards erosion and are indicative of a wide variety of climate and geology of the region (Altaf et al., 2014). High drainage density in the SWS13 indicates the region to be affected by weak and impermeable subsurface material. Low R_b value (2.44) in SWS13 indicates the less structural disturbance and the drainage patterns have not been distorted. It indicates that the watershed is falling under not normal basin category (Strahler, 1957) and that there is high structural complexity and low permeability of the terrain. Sparse vegetation and mountainous relief, even in a small proportion of sub-watershed, can lead to high drainage density making it vulnerable to erosion (Horton 1945). The SWS under the high vulnerability category is SWS 2, SWS6, SWS8, SWS9, SWS10, SWS12 and SWS14. SWS6 has the highest rank among this category. It consists of a high forest area (1.63 km^2), but due to its high slope of 74.76 percent and low elongation ratio, it is highly susceptible to erosion. SWS12 has the highest area under settlement, contributing to its status of being highly prone to soil erosion. The relief of this sub-watershed is the lowest (676m), implying land use is not playing a role in deciding the vulnerability status of the region. Also, it has less ruggedness number (1.8), drainage intensity (1.4) and stream frequency (3.5), making it less susceptible to soil erosion. SWS2, SWS7, SWS16, SWS17 and SWS18 are least prone to erosion. These sub watersheds consist of different proportions of land use, among which barren is the major one. SWS17 is least vulnerable to soil erosion, although it has high barren (4.17 km^2) and scrub (0.97 km^2) areas. High length of overland flow (Lg), 0.41 in the sub-watershed indicates that the streamflow will take a longer time to travel through the sub-watershed resulting in increasing the susceptibility towards erosion, but high elongation ratio and circulatory ratio (Figure 7) makes the sub-watershed less susceptible towards erosion. Form Factor Ratio (R_f) lower values of form factor indicates a highly elongated shape. The elongated basin with a low form factor indicates that the watershed has a flatter flow peak for a longer duration than the circular basin. Circulatory ratio indicates the elongated and circular shape of few sub-watersheds. Since form factor ratio and elongation ratio indicate the presence of an elongated watershed, a circulatory ratio of 0.5 will be considered an elongated watershed. Thus, in case of SWS1,7 morphometric parameters are playing major role despite the presence of maximum barren land in the sub-watershed (Figure 5). In SWS 18, the low length of overland flow (0.14) along with low relief (867m) contributes to its less susceptibility. The study suggests that the presence of high value of favorable land use like forest and low value of factors like slope percentage is not alone sufficient to address vulnerability status of the sub-watersheds. A combined contribution of all different parameters provides a comprehensive way to understand the vulnerability of different sub-watersheds. The MCDM method is a reliable way to include all the parameters for vulnerability assessment (Pandey et al., 2021). This method can be applied to assess any watershed and rank the watersheds according to their vulnerability status towards erosion. The limitation of this method is as it may overestimate or underestimate the results depending upon the number of parameters and unavailability of weights to the parameter. The selection of the parameters for vulnerability assessment is crucial and should be reviewed thoroughly before selection.

FUTURE RESEARCH DIRECTIONS

Soil and water conservation is a matter of concern worldwide. Soil erosion degrades both soil and water quality of the region. In future, the detailed inventory on loss caused to natural resources due to soil erosion will help in forming strategies and policy for conservation. In addition, future research should focus on identifying the best methods to assess soil erosion in region with scare data. It should focus on identifying the methods, which provides best estimates towards the susceptibility of soil erosion. Research on soil conservation techniques needs to be conducted to promote and facilitate long-term conservation programs. In future research holistic management approach of soil and water resources should be considered. It should be linked with social and economic development with protection of natural ecosystem. There is need of conducting advance approach considering various parameters like water quality, which may be a good indicator in suggesting erosion status in the region. Last but not the least, frequency of studies should be increased in fragile regions like the Himalayan and coastal regions.

Figure 7. Grouped criteria watershed prioritization for erosion susceptibility
Source: Author

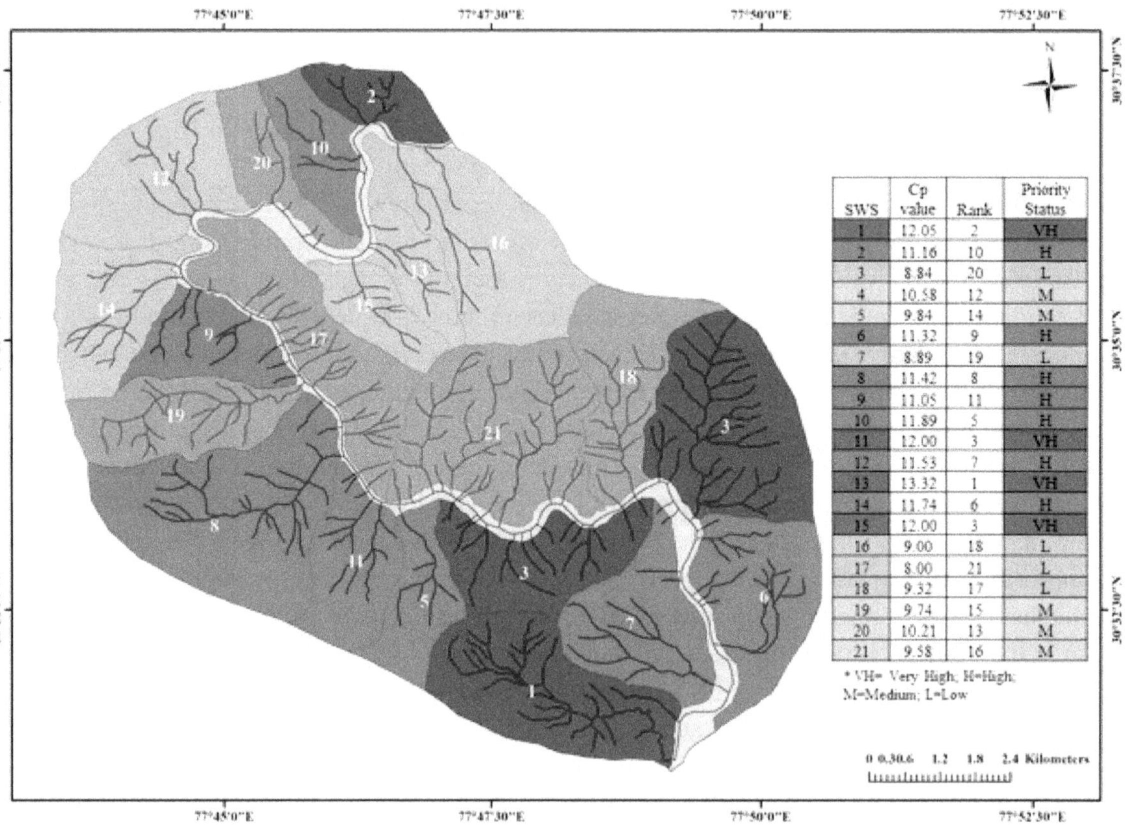

CONCLUSION

Watershed management is of utmost importance in prevention of degradation. Such management plans should include soil and water conservation measures and understanding of erosion-prone zones in a watershed. Assignment of relative status to sub-watersheds based on vulnerability to erosion becomes important when it comes to their management and conservation. For assessing the vulnerability of the Tons watershed towards soil erosion, MCDM qualitative approach is an efficient way, especially in regions lacking a substantial amount of data for quantitative approach. Morphometry alone is not sufficient to understand the priority status of any watershed. Quantitative morphometric characterization of a drainage basin, along with various other components like slope, soil quality, land use that affect soil erosion, should be considered for determining the vulnerability of a watershed to soil erosion. The study inferred that the vulnerability of a watershed could be understood in a better way with a combination of different components rather than considering a single component. The MCDM approach provided a better assessment of soil susceptibility as it is based on the combined role of morphometry, soil quality, slope and land use as grouped components categorizing them into low, medium-high and very high priority. With the help of satellite-based remote sensing, data to study watersheds can be accurately extracted. Their analysis for prioritizing the sub-watersheds may further help provide useful input in formulating conservation strategies of soil and water in watersheds.

RECOMMENDATION

It is suggested to combine satellite imagery, remote sensing technology and the field data to understand the hydrological processes. In addition, a combination of parameters may provide better scenarios to understand watersheds for their management and conservation.

REFERENCES

Altaf, S., Meraj, G., & Romshoo, S. A. (2014). Morphometry and land cover based multi-criteria analysis for assessing the soil erosion susceptibility of the western Himalayan watershed. *Environmental Monitoring and Assessment*, *186*(12), 8391–8412. doi:10.100710661-014-4012-2 PMID:25154685

Ameri, A. A., Pourghasemi, H. R., & Cerda, A. (2018). Erodibility prioritization of sub-watersheds using morphometric parameters analysis and its mapping: A comparison among TOPSIS, VIKOR, SAW, and CF multi-criteria decision making models. *The Science of the Total Environment*, *613–614*, 1385–1400. doi:10.1016/j.scitotenv.2017.09.210 PMID:29898506

Ashmore, P. (1993). Canadian Landscape Charter. *Progress in Physical Geography*, *17*(2), 190–204. doi:10.1177/030913339301700206

Biswas, H., Raizada, A., Mandal, D., Kumar, S., Srinivas, S., & Mishra, P. K. (2015). Identification of areas vulnerable to soil erosion risk in India using GIS methods. *Solid Earth*, *6*(4), 1247–1257. doi:10.5194e-6-1247-2015

Blinkov, I. (2015). The Balkans - The most erosive part of Europe? *Bulletin of the Faculty of Forestry*, *111*, 9-20.

Chow, V. T. (1964). *Handbook of applied hydrology: A compendium of water-resources technology*. McGraw-Hill.

Faniran, A. (1968). The Index of Drainage Intensity - A Provisional New Drainage Factor. *Australian Journal of Science*, *31*, 328–330.

Ghosh, K., Kumar De, S., Bandyopadhyay, S., & Saha, S. (2013). Assessment of soil loss of the Dhalai river basin, Tripura, India using USLE. *International Journal of Geosciences*, *04*(01), 11–23. doi:10.4236/ijg.2013.41002

Horton, R. E. (1932). Drainage Basin Characteristics. *Transactions - American Geophysical Union*, *13*(1), 350–361. doi:10.1029/TR013i001p00350

Horton, R. E. (1945). Erosional Development of Streams and Their Drainage Basins: Hydrophysical Approach to Quantitative Morphology. *Geological Society of America Bulletin*, *56*(3), 275–370. doi:10.1130/0016-7606(1945)56[275:EDOSAT]2.0.CO;2

Javed, A., Khanday, M. Y., & Ahmed, R. (2009). Prioritisation of sub watersheds based on Morphometric and Land use analysis using remote sensing and GIS techniques. *Photonirvachak (Dehra Dun)*, *37*(2), 261–274. doi:10.100712524-009-0016-8

Kumar, A., Devi, M., & Deshmukh, B. (2014). Integrated remote sensing and geographic information system based RUSLE Modelling for estimation of soil loss in Western Himalaya, India. *Water Resources Management, 28*(10), 3307–3317. doi:10.100711269-014-0680-5

Kumar, R., Kumar, S., Lohani, A. K., Nema, R. K., & Singh, R. D. (2000). Evaluation of Geomorphological Characteristics of a Catchment Using GIS. *GIS India, 9*(3), 13–17.

Pandey, S., Kumar, P., Zlatic, M., Nautiyal, R., & Panwar, V. P. (2021). Recent advances in assessment of soil erosion vulnerability in a watershed. *International Soil and Water Conservation Research*, 1–14.

Parveen, R., Kumar, U., & Singh, V. K. (2012). Geomorphometric Characterization of Upper South Koel Basin, Jharkhand: A Remote Sensing & GIS Approach. *Journal of Water Resource and Protection, 2012*(4), 1042–1050. doi:10.4236/jwarp.2012.412120

Polovina, S., Radic, B., Ristic, R., & Milcanovic, V. (2016). Spatial and temporal analysis of natural resources degradation in the Likodra river watershed. *Glasnik Sumarskog fakulteta, 144*, 169-188.

Ratnam, N. K., Srivastava, Y. K., Rao, V. V., Amminedu, E., & Murthy, K. S. R. (2005). Checkdam positioning by prioritization micro-watersheds using SYI model and morphometric analysis – remote sensing and GIS perspective. *Photonirvachak (Dehra Dun), 33*(1), 25–38. doi:10.1007/BF02989988

Schumm, S. A. (1956). Evolution of Drainage Systems and Slopes in Badlands at Perth Amboy, New Jersey. *Geological Society of America Bulletin, 6*(5), 597–646. doi:10.1130/0016-7606(1956)67[597:EODSAS]2.0.CO;2

Strahler, A. N. (1957). Quantitative analysis of watershed geomorphology. *American Geophysical Union Transactions, 38*(6), 913–920. doi:10.1029/TR038i006p00913

Strahler, A. N. (1964). Quantitative Geomorphology of Drainage Basin and Channel Networks. McGraw Hill.

Todorovski, L., & Džeroski, S. (2006). Integrating knowledge driven and data-driven approaches to modelling. *Ecological Modelling, 194*(1), 3–13. doi:10.1016/j.ecolmodel.2005.10.001

ADDITIONAL READING

Blanco-Canqui, H., & Lal, R. (2010). Soil and Water Conservation. In *Principles of Soil Conservation and Management*. Springer. doi:10.1007/978-1-4020-8709-7_1

Borrelli, P., Robinson, D. A., Panagos, P., Lugato, E., Yang, J. E., Alewell, C., Wuepper, D., Montanarella, L., & Ballabio, C. (2020, September). Land use and climate change impacts on global soil erosion by water (2015-2070). *Proceedings of the National Academy of Sciences of the United States of America, 117*(36), 21994–22001. doi:10.1073/pnas.2001403117 PMID:32839306

Issaka, S., & Ashraf, M. (2017). Impact of soil erosion and degradation on water quality: A review. *Geology Ecology Landscapes, 9508*(1), 1–11. Advance online publication. doi:10.1080/24749508.2017.1301053

Johnson, L. B., Richards, C., Host, G. E., & Arthur, J. W. (1997). Landscape influences on water chemistry in Midwestern stream ecosystems. *Freshwater Biology*, *37*(1), 193–208. doi:10.1046/j.1365-2427.1997.d01-539.x

Panagos, P., Ballabio, C., Poesen, J., Lugato, E., Scarpa, S., Montanarella, L., & Borrelli, P. (2020). A Soil Erosion Indicator for Supporting Agricultural, Environmental and Climate Policies in the European Union. *Remote Sensing*, *12*(9), 1365. doi:10.3390/rs12091365

Panagos, P., Borrelli, P., Poesen, J., Ballabio, C., Lugato, E., Meusburger, K., Montanarella, L., & Alewell, C. (2015). The new assessment of soil loss by water erosion in Europe. *Environmental Science & Policy*, *54*, 438–447. doi:10.1016/j.envsci.2015.08.012

Stathopoulos, N., Lykoudi, E., Vasileiou, E., Rozos, D., & Dimitrakopoulos, D. (2017). Erosion Vulnerability Assessment of Sperchios River Basin, in East Central Greece—A GIS Based Analysis. *Open Journal of Geology*, *7*(5), 621–646. doi:10.4236/ojg.2017.75043

Todorovski, L., & Džeroski, S. (2006). Integrating knowledge driven and data-driven approaches to-modeling. *Ecological Modelling*, *194*(1), 3–13. doi:10.1016/j.ecolmodel.2005.10.001

KEY TERMS AND DEFINITIONS

Hierarchical Cluster Analysis (HCA): It is a method to build a tree diagram where the groups that are most similar based on characteristics in the study are placed on branches that are close together.

Infiltration Capacity: Infiltration capacity is the maximum rate at which a soil is capable of absorbing water in a given condition. Several factors control infiltration capacity. Antecedent rainfall and soil-moisture conditions.

Morphometry: In geomorphology, morphometry is a quantification of morphology. Indices of watershed morphometry can interpret the shape and hydrological characteristics of a river basin. Morphometric analysis of watershed is the best method to identify the relationship of various aspects in the area. It is a comparative evaluation of different watersheds in various geomorphological and topographical conditions.

Soil Erodibility: It is defined as a resistance to two energy sources defines the erodibility of a soil as a material with a greater or lesser degree of coherence: the impact of raindrops on the soil surface, and the shearing action of runoff between clods in grooves or rills.

Soil Erosion: Soil erosion is one of the ten major soil threats, identified in the Status of the World's Soil Resources Report. It is defined as the accelerated removal of topsoil from the land surface through water, wind and tillage.

Soil Erosivity: Erosivity is the term used to describe the potential of raindrop impact, runoff from snowmelt, or water applied with an irrigation system rainstorm to detach and erode soil.

Watershed: A watershed is the geographical area drained by a watercourse. The concept applies at various scales – from, for example, a farm drained by a creek (a "micro-watershed") to a large river basin (or a lake basin).

Chapter 5
Prevention of Soil Erosion and Torrential Floods

Bhavya Kavitha Dwarapureddi
GITAM University (Deemed), India

Nihanth Soury Garika
GITAM University (Deemed), India

Swathi Dash
GITAM University (Deemed), India

Ankit Kumar
GITAM University (Deemed), India

Aman Raj
GITAM University (Deemed), India

Saritha Vara
https://orcid.org/0000-0003-0473-993X
GITAM University (Deemed), India

ABSTRACT

Climatic conditions, precise relief features, variations of soil, flora cover, socio-economic conditions together lead to torrential flood waves as a result of current soil erosion processes. Erosion and torrential floods are aggravated due to over exploitation of agricultural and forest land along with urbanization. Effects of soil erosion include nutrient loss, land use changes, reduced productivity, siltation of water bodies, among other effects like affecting livelihood of marginal communities dependent on agriculture globally and public health. Nearly 11 million km2 of soil is impacted by erosion precisely by water. Other factors like intensified agriculture and climate change contribute to and aggravate the erosion rate. Contemporary torrential floods are characterized by their increased destruction and frequency unlike the pre-development periods when their occurrence was rare. The focus of this review is to compile and aid as a data base for understanding methods of preventing erosion of soil and torrential floods as put forth by various researchers.

INTRODUCTION

Lithosphere is one of the most vital yet fragile resources on earth, playing essential roles in all living things. In the process of achieving sustainable development goals (SDGs), an enhanced pressure is being exerted on SDGs related to water, food, climate and health (Keesstra et al., 2018). About sixty percent

DOI: 10.4018/978-1-7998-8459-0.ch005

increase in the demand of fiber, feed and food is projected during the coming years due to the population growth (FAO, 2015). In order to fulfil certain aspects of sustainable generation, including zero hunger, no poverty, well-being and good health, it is required that land be searched for satisfying these needs via development and agricultural activities, which result in land degradation problems. Hence, land and soil conservation need immediate attention.

Soil science, among other land-associated disciplines, is significantly related to many of SDGs. Moreover, land degradation directly affects lithosphere and hydrosphere. Among several concerns related to land degradation, one of the important that attracts scientists and stakeholders is soil erosion (Baveye et al. 2016; Keesstra et al., 2016). Soil erosion has become a prevalent issue in several countries apart from being a global problem. Soil erosion can be understood as destruction of soil due to natural phenomena involving wind, snow and water and anthropogenic factors like extensive and intensive agriculture, which are happening simultaneously. Depending on the intensity, soil erosion can be a natural process taking place for millions of years, resulting in the formation of new soils, or it can be an accelerated process due to anthropogenic activities comprising of overgrazing, deforestation and farming practices that are not sustainable leading to loss of soil greater than its formation (Holy, 1980).

Several studies have presented adverse effects of soil erosion on water quality, agricultural production and the health of the ecosystem (Fayas et al., 2019). Some of the aspects leading to soil erosion include terrain, land cover, climate erosivity and soil erodibility. Agricultural activities are impacted due to soil erosion which results in reduced productivity. The impact of erosion on soil productivity can be seen both on-site and as off-site. This is a consequence of 3 interaction effects - long-term, small productivity and reduced soil quality (Lal, 2001).

Soil particles in agricultural lands detach and segregate from soil mass when rain water droplets hit the soil surface. Such unceasing exposure to heavy rainfall significantly deteriorates the soil. Several researchers have presented the effects of cropping on erosion rates of soil, resulting in denudation of topsoil reducing soil fertility (Nearing et al., 2017; Ahmad et al., 2020).

Soil erosion not only affects productivity and results in loss of nutrients, siltation of water bodies, but it also affects the public health of marginal communities across the globe that are contingent on agriculture (Meena et al., 2017). Human activities like the expansion of agricultural lands, conversion of forest lands, deforestation, shifting cultivation on steep slopes, rapid urbanization, road construction and other progressive activities followed by poor soil preservation, high rainfall and enhanced soil erosivity are the reasons for higher rates of soil loss and sedimentation of water bodies like rivers and lakes which result in reduced water quality and recreation (Markose et al., 2016; Al-Abadi et al., 2016; Dutta, 2016; Barman et al., 2020).

It is known that precise characteristics of relief, climate, variations in vegetation and soil cover, along with social conditions, usually result in the torrential floods, which is one of the prime reasons for soil erosion. As they appear in numerous forms, floods are the utmost recurrent natural catastrophic events occurring across the world (Barredo, 2007). Among types of floods, torrential floods are categorized under hydrological hazards that manifest as a sudden occurrence with maximum water discharge resulting in a greater rate of sediments in torrential flood trend, usually concurring with movements of landslides and clay flows (Foulds et al., 2014, Petrovic, 2015; Kostadinov et al., 2014; Garambois et al., 2014; Tiranti et a., 2014; Liste et al., 2014). The phenomenon of torrential floods is due to dangerous rainfall episodes having a rapid reaction from areas of small watersheds, intensive soil erosion and steep slopes. Changes due to urbanization and activities like deforestation and agriculture bring about changes in the higher and lower parts of the watershed along with natural processes and conditions. Further, these activities tend to disturb the natural structure of the soil, resulting in conditions for intensified soil erosion, sedimentation in the river bed and sudden runoff on slopes.

In contrast to great rivers floods, torrential floods occurrence is sudden, leaving no reaction time. Significantly this phenomenon is characterized by short duration, localized havoc aftermath and severe causalities. The frequency of waterborne diseases is higher in emergency management conditions after the occurrence a torrential flood.

Apart from physical consequences, psychological consequences like trauma and related diseases also occur (Petrovic, 2015). Erosivity due to rainfall is the chief factor for the loss of nutrients and sediments across the world, leaving farmers susceptible to failure of crops leading to unstable equilibrium in various landscapes (Wuepper et al., 2020). Earth's surface exposure to destructive precipitation is a crucial parameter governing erosion due to water on terrestrial ecosystems along with other destructive hydrological events like flash floods and floods (li et al., 2016; Diodato et al., 2017). Incidences of hydrological extremes leading to accompanying sediment loss during precipitation are perceived to be dominant features in the global climate system since, across the world, variation in precipitation and temperature arrays produce consistent variations in the development of natural hazards (Brönnimann et al., 2018). It has been presumed that rainfall-runoff erosivity and extreme storms are becoming more recurrent owing to climate change (Yin et al., 2018). Highly susceptible areas might result in disastrous regime changes related to the occurrence of such destructive hydrological events (Diodato et al., 2019). This explains the constant interest of engineers and scientists towards the hydrological response of landscapes, ranging from an elementary understanding of the phenomenon to predictions of altering situations, that are driven by enhanced recognition of both environmental and financial cost, neglecting hydroclimatic forcing parameters in association to the preservation of soil systems as well as land-use planning.

REASONS FOR ENHANCED SOIL EROSION DUE TO TORRENTIAL FLOODS

Mismanagement and over-exploitation of agricultural, forest lands and urbanization aggravated the phenomenon of torrential floods and soil erosion. Erosion of soil leads to several changes in land use mainly focusing on abandoning arable land due to reduced productivity (Bakker et al., 2005). Soil erosion gets more complicated with the local economic development. Also, with watershed development, the hydrological regime changes resulting in the enhanced volume of the torrential flood (Ristić et al., 2010). The occurrence of torrential floods was very rare during the pre-development era, but these have now become frequent, more disparaging owing to the alteration of the watershed from land uses of rural to urban. Clearing forests and their vegetation, inadequate agricultural measures and unrestrained urbanization are among few negative human activities, enhancing the occurrence of torrential floods with a frequency or recurrence interval of 100 to reduce to 20 years (Ristić et al., 2012).

EXTENT OF SOIL EROSION

On an average, the soil erosion rates across the globe are approximately 12 – 15 tonnes/hectare/year, implying that nearly 0.90 – 0.95mm of topsoil is lost every year (FAO, 2015; Ashiagbor et al., 2013). It is estimated that about 11 million kilometers of the area is impacted by water-related erosion. Erosion of soil is also aggravated due to climate change and intensive agricultural practices leading to a loss in soil fertility, soil quality degradation, siltation of water bodies and dams resulting in endangered food production, degrading water quality, eutrophication, etc. (Bou-Imajjane et al., 2020; Belasri et al., 2016; Aiello et al., 2015). Influences of erosion of soil can be understood as on-site and off-site erosion. On-site soil erosion comprises of decreased nutrient and water holding capacity of the soil, decreased organic matter, together with soil depth that can support organisms and roots. While off-site impacts reflect the fate of soil particles after dislodging from the crop land, adverse effects include economic consequences owing to deposition of large quantities of sediments into infrastructure, settlements and residences in terms of muddy floods and associated environmental damage. Further, the soil particles tend to adsorb chemicals and cause eutrophication of water bodies (Lal et al., 1990; Langdale et al., 1992; Pimentel et al., 1995 and 2006; Wardle et al., 2004; Boardman, 2010; Mullan, 2013). Further, nearly 84% of soil erosion is estimated to be caused by wind and water (Sayl et al., 2016). Moreover, it is estimated that the average soil loss due to water is greater than 2000 t-km-2y-1 (Vaezi et al., 2017; Shit et al., 2020).

SOIL EROSION AND RESERVOIRS

The estimates say that nearly 0.5 to 1% of deposit every year affects reservoir volume limit across the world. It is projected that by 2050 several dams will lose half of their capacity owing to human activities like deforestation (Chuenchum et al., 2020). In Asia alone, the deposit has affected nearly 40% of the total storage of reservoirs (Walling., 2011). Emerging countries are at high risk owing to deposit as it affects the long-term sustainability of storage structures. Nearly 30 - 32.8 Mha of land is affected by soil erosion due to water in India. However, in Pakistan, about 16 Mha of land is affected by the loss of soil via various processes, and nearly 70% of soil is lost i.e., 11.2 Mha is lost through water erosion. Trends across the globe show that nearly 20 billion tons of deposit are collected in the ocean discharged from rivers (Gelagay et al., 2016; Jozaghi et al., 2018; Atoma et al., 2020; Munir et al., 2021).

SOIL EROSION AND SOIL BIODIVERSITY

Lack of detailed spatial assessments for depicting various processes contributing to soil-driven ecosystem services globally enhances the challenges for addressing interactions among soil biodiversity (Costanza et al., 2017). Erosion of soil, among other effects, degrade conditions for soil biodiversity resulting in a negative influence on biogeochemical cycles (Wall et al. 2015; Quinton et al. 2010). Drivers of soil erosion combine to bring changes in the ecosystem and contribute to soil condition degradation for several human livelihoods as well as soil biodiversity (Jónsson and Davíðsdóttir, 2016; Bardgett and van der Putten 2014). According to the recent meta-analyses and assessments, IPBES, 2018, has recognized a related optimistic link between soil biodiversity and decline in soil biodiversity. However, global susceptibility of soil biodiversity to processes of soil degradation is still understudied owing to the contemporary sub-surface conservation strategies stressing primarily on ecosystem processes without representation on how sub-surface diversity is related to them (Nielsen et al., 2015). Nonetheless, several earlier regional and global risk assessments related to soil erosion have not included quantification of the straight influence of natural systems on deterrence of erosion of soil, nor have they considered this process to oversee long-term, static variations. Along with the non-inclusion of several temporal and spatial dimensions of risk from soil erosion, these assessments also ignore potential spatial matches among soil biodiversity and risk of erosion, specifically at a global scale.

In similar climatic and environmental conditions, enhanced vegetation cover led to a significant risk reduction in water-driven erosion of soil, resulting in increased supply of ecosystem service (Guerra et al., 2016). In the framework of contemporary scenario where data availability for input and process-based physical models are not sufficiently adequate for applications at a global scale, using physical empirical methods towards prediction of risk from soil erosion would provide rationally precise estimates (García-Ruiz et al. 2015; Borrelli et al. 2017). Such experimental models enable users to reason vigorously for impacts of land cover, and climate change through modeling unceasing changes in vegetation cover and erosivity from rainfall (Guerra et al., 2020).

MANAGEMENT AND ASSESSMENT OF TORRENTIAL FLOODS

Ephemeral mountainous streams originating from mountains and several other zones of the world carrying exhaustive bedload that cause flash floods are Torrents. Recurrent changes in the course of river migration and meandering cause widespread impairment to the environment in alluvial and valley lands, where consequences are have washing their beds, eroding their banks and destroying valuable arable lands. Torrential watersheds vary in dimensions experiencing unanticipated utmost discharge with high sediment volume. Thus, it is vital to understand torrential streams and their hydraulic and hydrological characteristics in order to take measures for conservation and minimizing negative impacts on the surrounding environment (Ristić and Malošević, 2011).

Enrichment of soil with minerals and transport of sediment by water are the two significant benefits of flooding. The sediments that are transported enhance soil fertility replacing long-lasting with novel soil; also, floods are a great normal way for recharging groundwater (Gioti et al., 2013). Nevertheless, floods also cause economic and human loss and are viewed as catastrophes. Researchers across the globe are constantly trying to find ways and means to manage floods, utilize flood water and control assets loss.

Previous studies presented non-structural as well as structural management of floods. Structural management takes into account physical measures comprising of construction of floodwalls, dams, raised buildings, cleansing of aquatic environments and flood proving characteristics (Munir et al., 2020). While non-structural methods comprise of planning for flood plain partitioning, disaster and early warning schemes.

During the current eras, hydrological models have been demonstrated as efficient procedures for predicting and monitoring flash floods. Hydrometeorological methods globally have been categorized as the finest management customs available for flash floods. These are classified into distributed, semi-distributed and lumped schemas. Distributed models incorporate inferential and distributed data like temperature, solar radiation, soil moisture, precipitation etc., while performing calculations on both chronological and chorological scales. On the other hand, semi-distributed models' sub-catchment is considered the smallest unit of a watershed, while lumped modeling encompasses assumptive conditions that are uniform. It has been reported that semi-distributed and distributed models have been equally adopted and implemented for monitoring flash floods as they have greater performance over lumped models (Jia et al., 2019). In general, various uncertainties existing in cause input data tend to disturb hydraulic models, specifically available secondary data that regulates the output of the model (Walega et al., 2017).

Few attempts for estimation of torrent areas and their affected zones have been taken up across the world. Land degradation induced by upstream torrential activities and their impacts on human settlements and natural resources in the downstream call for periodic assessment and monitoring of extent and nature of hazards through the systematic acquisition of data using satellite remote sensing. This provides a great input for hazard assessment of torrential river systems. Several torrents seam together, forming larger rivers with a moderate slope, braiding patterns creating an overflow of river water causing floods that lead to loss of forest and agricultural production and destruction of human settlements. Risk assessment and hazard mapping as a consequence of torrents in Aswan Governorate, Egypt, was carried out by Omran, 2020, who reported a greater risk of lower elevation over higher. HEC-RAS (Hydrologic Engineering Center River Analysis System) and Personal Computer SWMM models were used to carry out flash flood analysis of Wadore hill torrent by Munir et al., 2020 in Pakistan (Kumar et al., 2021). Adopting various models for assessment enhances user-defined scenarios portraying the reality of systems having sufficient accuracy (Xie et al., 2017; Cheng et a., 2017; Wu et al., 2017; Iosub et al., 2020).

CONTROL OF SOIL EROSION

It is well recognized that vegetation reduces erosion of soil. Vegetation intercepts rainfall, while some water drops are stored in the canopy remaining evaporates or reach the soil surface directly as raindrops or indirectly via leaf drainage or stem flow (Puigdefábregas, 2005). Water that reaches the surface of soil gets stored in small depressions or infiltrates into the soil, while the remaining soil moves downhill as surface runoff gets entrain in soil particles and transport them, they may also get concentrated, forming gullies and rills. Soil erosion is governed by several factors like slope angle, erosivity of eroding agent, erodibility of soil and nature of plant cover (Figure 1). Significance of vegetation cover towards erosion control has been established by several studies from bench-scale/plot to basin-scale (Zhao et al., 2013; Quinton et al., 1997) and have presented that vegetation cover plays an imperative role in the recovery of degraded land (Zhao et al., 2013). Linear to Exponential reduction in the rates of erosion has been observed among vegetation-associated lands, as there is a variation among the kind of relationship for canopy and ground covers (Boer and Puidgefábregas, 2005). Optimistic effects of vegetation towards erosion control are attributed to the reduced kinetic energy of raindrops and the reduction in velocities of surface runoff (Puigdefábregas, 2005).

Moreover, vegetation changes the inherent properties of soil that determine soil erodibility (Gyssels et al., 2005); it makes conditions for the supply of organic matter and microclimate, provide the nutrients, and ultimately enhances the activity of microorganisms leading to better plant productivity.

Organic matter, along with secretions from bacteria and fungi, enhances the development of stable accumulates (Vásquez-Méndez et al., 2010) that affect the hydraulic conductivity and thus water storage, increasing the shear strength of soil (De Baets et al., 2008). It has been clearly demonstrated that soil loss is reduced due to the combined effect of below and above-ground biomass. Reduction to nearly zero within 0 – 10% of cross-sectional grass-root occupation was found whole, and only 25 – 50% reduction with aerial cover for the same area (De Baets et al., 2006). Zhou and Shangguan, 2007 conducted a rainfall simulation experiment after 27 weeks of sowing ryegrass demonstrated in erosion pans and presented that this species account 90% reduction in soil detachment below ground (Ola et al., 2015).

Figure 1. Controlling Methods of Soil Erosion

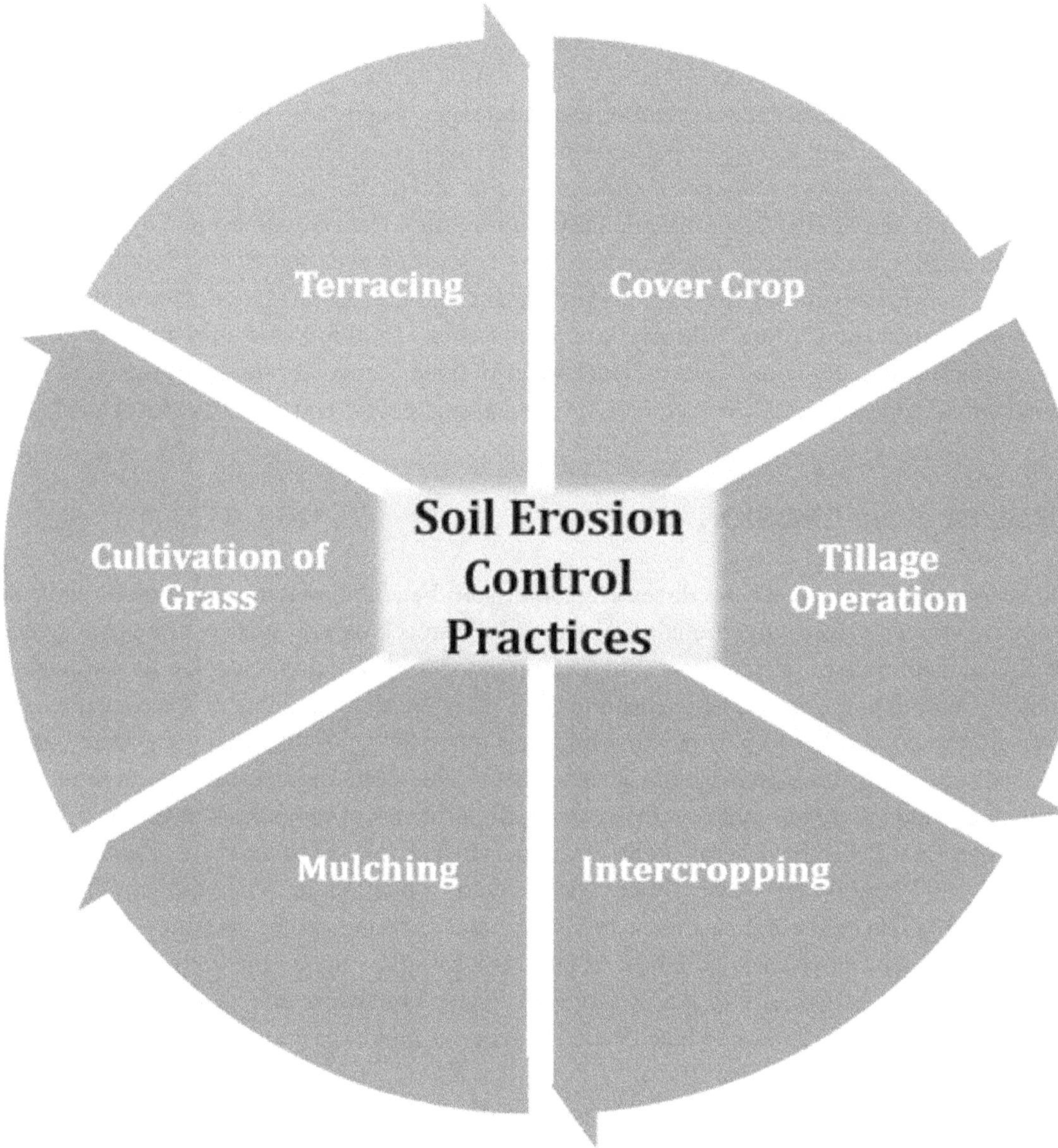

ASSESSMENT OF SOIL EROSION

Efforts towards quantifying water ruled soil erosion through measuring gully growth rates can be drawn back to nearly 4000 years, but it was only lately that concern regarding the issue had received attention resulting in significant attempts being made for obtaining widespread data related to water ruled soil erosion as baseline data for the conservation of soil (Sang et al., 2002). During 1939, Jacks and Whyte had highlighted the problem of human-induced soil erosion at a global scale. As per estimates, nearly 10 million ha of arable land is lost every year as a consequence of soil erosion (Shikangalah et al., 2016). Swift expansion of urban functions and population explosion led to variations in land use and land cover, resulting in the transformation of croplands, semi-natural and natural areas into non-porous lands (Radic et al., 2020). These changes resulted in the development of various kinds of erosion processes which, along with the current conditions of climate change, have become the key catalysts for the occurrence of torrential floods (Kusky, 2020).

Probably the most significant environmental problems associated with the combined effects of urbanization, soil erosion, and torrential floods are the negative effect on quantity, quality and spatial distribution of water as well as soil resources (Pribadi et al., 2018). Certain changes in the soil, such as retention capacity and infiltration, directly affect the magnitude and extent of erosion leading to surface runoff (Ristic et al., 2012). Besides being affected by the intensity and frequency of natural disasters, environmental risks in urban areas are increasing as a consequence of population explosion and the effects of socio-economic activities on land use and land cover (Herslund et al., 2016).). Hydrological events triggered by intense and short storms or lengthy duration rainfall result in sustained erosive force and lead to significant loss of soil in several zones of the globe (Duulatov et al., 2021; Wuepper et al., 2020; Diodato et al., 2020). These events might increase in severity and frequency along with climate system variations in relation to global warming (Wei et al., 2020; Rineau et al., 2019; Harris et al., 2018). Significant challenges are faced in quantifying undercurrents and present expansions of climate extremes with rainfall erosivity and their impact on landscapes. Soil erosivity due to rainfall is significant towards understanding dynamics of surface processes such as degradation of soil as well as other stressors of landscape like landslides and flooding (Reimann et al., 2018; Schmidt et al., 2016).

Understanding and interpreting these changes have become an essential part of research as it helps propose and implement management measures towards the control of soil erosion. Numerous methods have been proposed and used to estimate soil erosion, including monitoring and measurement, plot studies, modeling and use of tracers (Parsons, 2019). Understanding these dynamics also presents an opportunity to capture the fingerprint of climate changes happening recently (Zittis et al., 2019; Diodato et al., 2021). Among these approaches, modelling methods are significant towards gaining insights into these processes and the impacts of different factors on the genesis, occurrence and erosion intensity and predicting production of erosion material (De Vente et al., 2005; Boardman, 2006). While modeling is the most appropriate solution, this also comes with the advantages and disadvantages of its application. Modeling comes with a certain degree of uncertainty attributed to the quality of input data used necessary for applying a specific model. The advantages of modeling lie in their wide applications towards calculating erosion production due to the impact of changes in land use, climate change leading to changes in land cover (Renschler, 2003). The spatial and temporal dynamic modeling methods application enabled deeper speedy and dependable conclusion regarding causes of erosion processes (Mitasova et al., 2013). Modern cities have the most noticeable dynamics of change among urban zones and their surrounding semi-natural and natural elements and agricultural areas (Pribadi et al., 2018; Polovina et al., 2021).

Powerful rainfall, specifically on sheer slopes of hill torrents, stimulates flash flooding having a brief lag time causing intense economic loss (Petrovic, 2015). The runoff behavior in such sheer hilly catchment is understood as a composite process that is exclusively contingent on physical parameters like size, shape, type of stream of the catchment (Fan et al., 2013). Accurate measurement of discharge in these catchments remains a challenge owing to precise calculations required for physical parameters. Recently, advancements and improvements in studies related to changes in the natural environment and water balance have boosted hydraulic as well as hydrological studies (Moynihan et al., 2014). Researchers have developed various strategies for integrating physical and man-induced factors on runoff from storm events (Gutiérrez et al., 2014; Jiang et al., 2015). Ahmad et al., 2015 presented that among several available methods SCS- CN (Soil Conservation Service Curve Number) has proven to be sustaining method towards quantifying voluminous discharges from flash floods in catchments (Fan et al., 2013).

FUTURE RESEARCH DIRECTION

- Long-term and large-scale monitoring is required, which allows the researchers to analyze the impacts of land management policies and practices on erosion.
- Interdisciplinary research for developing erosion predictions and controlling technologies to ensure the new technologies' adoption at regional/ local levels. Land managers and end-users must be involved in the entire process to enhance the rates of the adaptation.
- More practical, well organized, and essential methods for collecting data related to soil erosion and torrential flood are required to develop the tools for enabling more productive data sharing.
- Efforts to significantly increase our understanding of sediment transport by wind or water and the off-site impacts on air and water quality are required, and research is necessary to conduct these lines.

CONCLUSION

Being an indispensable resource for life, water might as well act as a land-disturbing aspect due to its erosive force of precipitation, expressed as storm erosivity. Hydrological extremes tend to change the structure of soil which triggers the erosion. Ecological consequences like modifications in characteristics as a consequence of climate change and precipitation extremes are usually understood poorly. Population explosion coupled with a rise in income resulted in enhanced challenges of meeting the world's demand for products from agriculture and ecosystem services. There has been a need for the production of more energy and food; therefore, soil protection worldwide and prevention of soil erosion have become mandatory with high priority. The above discussion presents direct relation between floods and soil erosion which enhances with the duration of the flood. Also, a relation between land-use changes and spatial and temporal patterns of accelerated erosion has also been presented. Since anthropogenic activities altered the source-to-sink system, there is a need for a holistic approach towards understanding novel pathways and stores that are created along the landscapes, towards ensuring the transport of natural amount of particulates. Hence, novel methods and models of assessment and measurements are required for enabling appropriate management strategies. Along these lines, accessibility of widespread high-temporal resolution archives of rainfall for huge areas with the progress of climate models has unlocked novel prospects towards using these approaches for hazard prevention and large-scale planning (Table 1).

Table 1. Recent studies carried out on Torrential Floods and Soil Erosion

S.No	Author	Country	Objective	Results
1.	Křeček et al., 2021	Czechia	To study the defensive actions of the forest of the Jizera Mts.	The projected ecological changes cannot significantly deny the great possibility of water recharge or inverse actions in new rescue after acidification and fundamentally disturb the forest cover.
2.	Zhou et al., 2021	China	To recognize the spatial form and foundation of gullies.	The spatial form of gullies in north-eastern China is a mixed type, and varied ecological factors were observed.
3.	Bombino et al., 2021	Italy	To evaluate various parameters in 4 variations of management practices i.e., Mechanical tillage (MT), total soil cover with a net (SP), and mulching in the grove of Southern Italy.	The various influence of soil mulching at dissimilar doses in the residue in Mediterranean olive groves.
4.	Gwapedza et al., 2021	Africa	To examine Modified Universal Soil Loss Equation (MUSLE).	Modified Universal Soil Loss Equation (MUSLE) could estimate the sediment yield, which can be observed by calibrating it.
5.	de Almeida et al., 2021	Brazil	To identify various rainfall patterns and their impacts on the sediments.	The study showed various technologies for designing the water erosion experiments by considering the duration, intensity.
6.	Ivica et al., 2018	Croatia	Soil losses due to water were studied in Stagnosols in central lowland Croatia, under different tillage treatments that vary in depth and tillage direction.	Tillage treatments like tillage across the slope (PAS, VDPAS, SSPAS) efficiently store the soil nutrients.
7.	Asfaw et al., 2019	Amhara National Regional State	The study mainly focused on implementing water and soil conservation practices by quantifying the morphometric characteristics.	The author mainly focused on enhancing/strengthening the water storage capacity and preventing sediment loss from the watershed by practicing various conservation projects related to soil and water.
8.	Gholami et al., 2021	Iran	This study aims to evaluate the pattern of the storm and its effect on surface runoff, erosion of soil and sediment intensity on rangeland soil slope.	GLM test was conducted to evaluate the pattern of storm influence on surface runoff, erosion of soil and sediment concentration on rangeland soil slope significantly at a level of 99%.
9.	Boardman et al., 2020	Belgium	The study focused on the Muddy runoff from agricultural fields and their effects on properties, freshwater systems in Europe.	The study focused on the protection of measures for muddy water from agriculture fields.
10.	Kaboli et al., 2021	Iran	Temporal and spatial disparities in rainfall were ascertained using DRCI as well MRPI in Iran by collecting 30 years of data from eighty meteorological stations.	The increasing trend was observed in DRCI values at eighty percent of meteorological stations and was important at thirty seven percent of tested stations.
11.	Pant et al., 2020	India	Morphometric, morphotectonic, statistical, and hazard analyses were studied.	Future considerations for the Alaknanda river basin were analyzed by soil erosion, and flash flood possibilities by relating watershed management to develop the area.
12.	Du et al., 2021	China	Analyzing spatial and temporal disparities of sediment load and runoff in Malian River basin.	Anthropogenic activity effects on runoff along with sediment load decrease using the double-mass curve method resulting in ninety and seventy-eight percent, respectively, and rainfall changes were also observed, i.e., 9.3% and 21.3%, respectively.
13.	Yavari et al., 2021	Australia	To know the erosion process by relating runoff, rainfall, and suspended and sediment bedload export by assessing yearly, inter-event, seasonal, and intra-event periods in the trial plot.	An experimental study was conducted in a trial plot to understand the sediment dynamics and develop predictive models by improving landform models.
14.	Munir et al., 2021	Pakistan	The author mainly focused on emerging a method to recognize the appropriate sites for dams using RDSA. The author focused primarily on the soil erosion estimation using the RUSLE model.	The study showed that the yearly average sediment yield (SY) would feed the dam A and B were approximately 298,073 and 318,000 tons, respectively, and expected that there would be no harm for 87 and 90 years for both the dams.
15.	Mohammadi et al., 2021	Iran	The study evaluated the land-use changes between 1991-2014 in the Talar watershed by using GIS.	The study prioritizes the improper management and planning of land and how it leads to the flow discharge of watershed.
16.	Kumar et al., 2021	India	The author used API to study the temporal Landsat data in GEE to know the land use and land cover pattern in Tangri River, India.	Models TOPSIS and ELECTRE were utilized to evaluate trends of vulnerable sections in torrential regimes and mainly focused on the planning, monitoring, and management.
17.	Polovina et al., 2021	Serbia	The author focused on the study is to evaluate the reasons for soil erosion by analyzing with the G2 erosion model.	The study depicted an increase in soil erosion in semi-natural and natural areas, i.e., forty-three and sixteen percent, respectively, in the land of Master Plan of Belgrade.
18.	Petrović et al., 2021	Serbia	To study the severe torrential flood that occurred in Serbia on 15th September 2014, with the magnitude, compensations and losses caused.	Ancient events of torrential floods should be considered to plan risk mitigation measures, as of now present incident was used to form mitigation and prevention measures.
19.	Stefanidis et al., 2018	Greece	To measure the changes by soil erosion in catchment areas beneath climate change.	The decline in rainfall and temperature will be seen till the 21st century yearly, possibly these variations could lower the soil erosion probability according to the results.
20.	Luino et al., 2018	Italy	To study both critical floods that happened in September 1993 and October 2000 in the Soana Valley.	The results point out the land use planning and management that happened on both critical flood days, with the errors in protection organization.

REFERENCES

Ahmad, I., Verma, V., & Verma, M. K. (2015). Application of curve number method for estimation of runoff potential in GIS environment. *2nd International Conference on Geological and Civil Engineering*, *80*(4), 16-20.

Ahmad, N. S. B. N., Mustafa, F. B., & Didams, G. (2020). A systematic review of soil erosion control practices on the agricultural land in Asia. *International Soil and Water Conservation Research*, *8*(2), 103–115. doi:10.1016/j.iswcr.2020.04.001

Aiello, A., Adamo, M., & Canora, F. (2015). Remote sensing and GIS to assess soil erosion with RUSLE3D and USPED at river basin scale in southern Italy. *Catena*, *131*, 174–185. doi:10.1016/j.catena.2015.04.003

Al-Abadi, A. M. A., Ghalib, H. B., & Al-Qurnawi, W. S. (2016). Estimation of soil erosion in northern Kirkuk governorate, Iraq using rusle, remote sensing and gis. *Carpathian Journal of Earth and Environmental Sciences*, *11*(1), 153–166.

Asfaw, D., & Workineh, G. (2019). Quantitative analysis of morphometry on Ribb and Gumara watersheds: Implications for soil and water conservation. *International Soil and Water Conservation Research*, *7*(2), 150–157. doi:10.1016/j.iswcr.2019.02.003

Ashiagbor, G., Forkuo, E. K., Laari, P., & Aabeyir, R. (2013). Modeling soil erosion using RUSLE and GIS tools. *Int J Remote Sens Geosci*, *2*(4), 1–17.

Atoma, H., Suryabhagavan, K. V., & Balakrishnan, M. (2020). Soil erosion assessment using RUSLE model and GIS in Huluka watershed, Central Ethiopia. *Sustainable Water Resources Management*, *6*(1), 1–17. doi:10.100740899-020-00365-z

Bakker, M. M., Govers, G., Kosmas, C., Vanacker, V., Van Oost, K., & Rounsevell, M. (2005). Soil erosion as a driver of land-use change. *Agriculture, Ecosystems & Environment*, *105*(3), 467–481. doi:10.1016/j.agee.2004.07.009

Bardgett, R. D., & Van Der Putten, W. H. (2014). Belowground biodiversity and ecosystem functioning. *Nature*, *515*(7528), 505–511. doi:10.1038/nature13855 PMID:25428498

Barman, B. K., Rao, K. S., Sonowal, K., Prasad, N. S. R., & Sahoo, U. K. (2020). Soil erosion assessment using revised universal soil loss equation model and geo-spatial technology: A case study of upper Tuirial river basin, Mizoram, India. *AIMS Geosciences*, *6*(4), 525–544. doi:10.3934/geosci.2020030

Barredo, J. I. (2007). Major flood disasters in Europe: 1950–2005. *Natural Hazards*, *42*(1), 125–148. doi:10.100711069-006-9065-2

Baveye, P. C., Baveye, J., & Gowdy, J. (2016). Soil "ecosystem" services and natural capital: Critical appraisal of research on uncertain ground. *Frontiers in Environmental Science*, *4*, 41. doi:10.3389/fenvs.2016.00041

Belasri, A., & Lakhouili, A. (2016). Estimation of soil erosion risk using the universal soil loss equation (USLE) and geo-information technology in Oued El Makhazine Watershed, Morocco. *Journal of Geographic Information System*, *8*(01), 98–107. doi:10.4236/jgis.2016.81010

Boardman, J. (2006). Soil erosion science: Reflections on the limitations of current approaches. *Catena, 68*(2-3), 73–86. doi:10.1016/j.catena.2006.03.007

Boardman, J. (2010). A short history of muddy floods. *Land Degradation & Development, 21*(4), 303–309. doi:10.1002/ldr.1007

Boardman, J., & Vandaele, K. (2020). Managing muddy floods: Balancing engineered and alternative approaches. *Journal of Flood Risk Management, 13*(1), e12578. doi:10.1111/jfr3.12578

Boer, M., & Puigdefábregas, J. (2005). Effects of spatially structured vegetation patterns on hillslope erosion in a semiarid Mediterranean environment: A simulation study. *Earth Surface Processes and Landforms: The Journal of the British Geomorphological Research Group, 30*(2), 149–167. doi:10.1002/esp.1180

Bombino, G., Denisi, P., Gómez, J. A., & Zema, D. A. (2021). Mulching as best management practice to reduce surface runoff and erosion in steep clayey olive groves. *International Soil and Water Conservation Research, 9*(1), 26–36. doi:10.1016/j.iswcr.2020.10.002

Borrelli, P., Robinson, D. A., Fleischer, L. R., Lugato, E., Ballabio, C., Alewell, C., & Panagos, P. (2017). An assessment of the global impact of 21st century land use change on soil erosion. *Nature Communications, 8*(1), 1–13. doi:10.103841467-017-02142-7 PMID:29222506

Bou-Imajjane, L., & Belfoul, M. A. (2020). Soil loss assessment in Western High Atlas of Morocco: Beni Mohand watershed study case. *Applied and Environmental Soil Science*, 2020.

Brönnimann, S., Pfister, C., & White, S. (2018). In S. White, C. Pfister, & F. Mauelshagen (Eds.), *The Palgrave Handbook of Climate History* (pp. 27–36). Palgrave Macmillan. doi:10.1057/978-1-137-43020-5_3

Cheng, T., Xu, Z., Hong, S., & Song, S. (2017). Flood risk zoning by using 2D hydrodynamic modeling: A case study in Jinan City. *Mathematical Problems in Engineering, 2017*, 2017. doi:10.1155/2017/5659197

Chuenchum, P., Xu, M., & Tang, W. (2020). Estimation of soil erosion and sediment yield in the Lancang–Mekong river using the modified revised universal soil loss equation and GIS techniques. *Water (Basel), 12*(1), 135. doi:10.3390/w12010135

Costanza, R., De Groot, R., Braat, L., Kubiszewski, I., Fioramonti, L., Sutton, P., Farber, S., & Grasso, M. (2017). Twenty years of ecosystem services: How far have we come and how far do we still need to go? *Ecosystem Services, 28*, 1–16. doi:10.1016/j.ecoser.2017.09.008

de Almeida, W. S., Seitz, S., de Oliveira, L. F. C., & de Carvalho, D. F. (2021). Duration and intensity of rainfall events with the same erosivity change sediment yield and runoff rates. *International Soil and Water Conservation Research, 9*(1), 69–75. doi:10.1016/j.iswcr.2020.10.004

De Baets, S., Poesen, J., Gyssels, G., & Knapen, A. (2006). Effects of grass roots on the erodibility of topsoils during concentrated flow. *Geomorphology, 76*(1-2), 54–67. doi:10.1016/j.geomorph.2005.10.002

De Baets, S., Poesen, J., Reubens, B., Wemans, K., De Baerdemaeker, J., & Muys, B. (2008). Root tensile strength and root distribution of typical Mediterranean plant species and their contribution to soil shear strength. *Plant and Soil, 305*(1), 207–226. doi:10.100711104-008-9553-0

De Vente, J., & Poesen, J. (2005). Predicting soil erosion and sediment yield at the basin scale: Scale issues and semi-quantitative models. *Earth-Science Reviews*, *71*(1-2), 95–125. doi:10.1016/j.earscirev.2005.02.002

Diodato, N., Borrelli, P., Fiener, P., Bellocchi, G., & Romano, N. (2017). Discovering historical rainfall erosivity with a parsimonious approach: A case study in Western Germany. *Journal of Hydrology (Amsterdam)*, *544*, 1–9. doi:10.1016/j.jhydrol.2016.11.023

Diodato, N., Borrelli, P., Panagos, P., Bellocchi, G., & Bertolin, C. (2019). Communicating hydrological hazard-prone areas in Italy with geospatial probability maps. *Frontiers in Environmental Science*, *7*, 193. doi:10.3389/fenvs.2019.00193

Diodato, N., Ljungqvist, F. C., & Bellocchi, G. (2020). Historical predictability of rainfall erosivity: a reconstruction for monitoring extremes over Northern Italy (1500–2019). *NPJ Climate and Atmospheric Science, 3*(1), 1-10.

Diodato, N., Ljungqvist, F. C., & Bellocchi, G. (2021). Climate Patterns in the World's Longest History of Storm-Erosivity: The Arno River Basin, Italy, 1000–2019 CE. *Frontiers of Earth Science*, *9*, 108.

Du, M., Mu, X., Zhao, G., Gao, P., & Sun, W. (2021). Changes in Runoff and Sediment Load and Potential Causes in the Malian River Basin on the Loess Plateau. *Sustainability*, *13*(2), 443. doi:10.3390u13020443

Duulatov, E., Chen, X., Issanova, G., Orozbaev, R., Mukanov, Y., & Amanambu, A. C. (2021). Introduction and background of rainfall erosivity processes and soil erosion. In *Current and future trends of rainfall erosivity and soil erosion in Central Asia* (pp. 1–7). Springer. doi:10.1007/978-3-030-63509-1_1

Fan, F., Deng, Y., Hu, X., & Weng, Q. (2013). Estimating composite curve number using an improved SCS-CN method with remotely sensed variables in Guangzhou, China. *Remote Sensing*, *5*(3), 1425–1438. doi:10.3390/rs5031425

FAO. (2015). Status of the world's soil resources (SWSR)–main report. Food and agriculture organization of the United Nations and intergovernmental technical panel on soils, Rome, Italy.

Fayas, C. M., Abeysingha, N. S., Nirmanee, K. G. S., Samaratunga, D., & Mallawatantri, A. (2019). Soil loss estimation using rusle model to prioritize erosion control in KELANI river basin in Sri Lanka. *International Soil and Water Conservation Research*, *7*(2), 130–137. doi:10.1016/j.iswcr.2019.01.003

Foulds, S. A., Griffiths, H. M., Macklin, M. G., & Brewer, P. A. (2014). Geomorphological records of extreme floods and their relationship to decadal-scale climate change. *Geomorphology*, *216*, 193–207. doi:10.1016/j.geomorph.2014.04.003

Garambois, P. A., Larnier, K., Roux, H., Labat, D., & Dartus, D. (2014). Analysis of flash flood-triggering rainfall for a process-oriented hydrological model. *Atmospheric Research*, *137*, 14–24. doi:10.1016/j.atmosres.2013.09.016

García-Ruiz, J. M., Beguería, S., Nadal-Romero, E., González-Hidalgo, J. C., Lana-Renault, N., & Sanjuán, Y. (2015). A meta-analysis of soil erosion rates across the world. *Geomorphology*, *239*, 160–173. doi:10.1016/j.geomorph.2015.03.008

Gelagay, H. S., & Minale, A. S. (2016). Soil loss estimation using GIS and Remote sensing techniques: A case of Koga watershed, Northwestern Ethiopia. *International Soil and Water Conservation Research*, *4*(2), 126–136. doi:10.1016/j.iswcr.2016.01.002

Gholami, L., Darvishan, A. K., Spalevic, V., Cerdà, A., & Kavian, A. (2021). Effect of storm pattern on soil erosion in damaged rangeland; field rainfall simulation approach. *Journal of Mountain Science*, *18*(3), 706–715. doi:10.100711629-019-5633-2

Gioti, E., Riga, C., Kalogeropoulos, K., & Chalkias, C. (2013). A GIS-based flash flood runoff model using high resolution DEM and meteorological data. *EARSeL eProceedings*, *12*(1), 33–43.

Guerra, C. A., Maes, J., Geijzendorffer, I., & Metzger, M. J. (2016). An assessment of soil erosion prevention by vegetation in Mediterranean Europe: Current trends of ecosystem service provision. *Ecological Indicators*, *60*, 213–222. doi:10.1016/j.ecolind.2015.06.043

Guerra, C. A., Rosa, I. M., Valentini, E., Wolf, F., Filipponi, F., Karger, D. N., & Eisenhauer, N. (2020). Global vulnerability of soil ecosystems to erosion. *Landscape Ecology*, *35*(4), 823–842. doi:10.100710980-020-00984-z PMID:32587435

Gutiérrez, F., Parise, M., De Waele, J., & Jourde, H. (2014). A review on natural and human-induced geohazards and impacts in karst. *Earth-Science Reviews*, *138*, 61–88. doi:10.1016/j.earscirev.2014.08.002

Gwapedza, D., Nyamela, N., Hughes, D. A., Slaughter, A. R., Mantel, S. K., & van der Waal, B. (2021). Prediction of sediment yield of the Inxu River catchment (South Africa) using the MUSLE. *International Soil and Water Conservation Research*, *9*(1), 37–48. doi:10.1016/j.iswcr.2020.10.003

Gyssels, G., Poesen, J., Bochet, E., & Li, Y. (2005). Impact of plant roots on the resistance of soils to erosion by water: A review. *Progress in Physical Geography*, *29*(2), 189–217. doi:10.1191/0309133305pp443ra

Harris, R. M., Beaumont, L. J., Vance, T. R., Tozer, C. R., Remenyi, T. A., Perkins-Kirkpatrick, S. E., & Bowman, D. M. J. S. (2018). Biological responses to the press and pulse of climate trends and extreme events. *Nature Climate Change*, *8*(7), 579–587. doi:10.103841558-018-0187-9

Herslund, L. B., Jalayer, F., Jean-Baptiste, N., Jørgensen, G., Kabisch, S., Kombe, W., & Vedeld, T. (2016). A multi-dimensional assessment of urban vulnerability to climate change in Sub-Saharan Africa. *Natural Hazards*, *82*(2), 149–172. doi:10.100711069-015-1856-x

Holy, M. (1980). *Erosion and environment- Environmental sciences and application* (Vol. 9). Pergamon Press Ltd.

Iosub, M., Minea, I., Chelariu, O. E., & Ursu, A. (2020). Assessment of flash flood susceptibility potential in Moldavian Plain (Romania). *Journal of Flood Risk Management*, *13*(4), e12588. doi:10.1111/jfr3.12588

Ivica, K., Igor, B., Zeljka, Z., & Darija, B. (2018). Effects of soil erosion by water under different tillage treatments on distribution of soil chemical parameters. *Soil and Water Research*, *13*(1), 36–43. doi:10.17221/25/2017-SWR

Jia, P., Liu, R., Ma, M., Liu, Q., Wang, Y., Zhai, X., & Wang, D. (2019). Flash flood simulation for ungauged catchments based on the distributed hydrological model. *Water (Basel)*, *11*(1), 76. doi:10.3390/w11010076

Jiang, C., Xiong, L., Wang, D., Liu, P., Guo, S., & Xu, C. Y. (2015). Separating the impacts of climate change and human activities on runoff using the Budyko-type equations with time-varying parameters. *Journal of Hydrology (Amsterdam)*, *522*, 326–338. doi:10.1016/j.jhydrol.2014.12.060

Jónsson, J. Ö. G., & Davíðsdóttir, B. (2016). Classification and valuation of soil ecosystem services. *Agricultural Systems*, *145*, 24–38. doi:10.1016/j.agsy.2016.02.010

Jozaghi, A., Alizadeh, B., Hatami, M., Flood, I., Khorrami, M., Khodaei, N., & Ghasemi Tousi, E. (2018). A comparative study of the AHP and TOPSIS techniques for dam site selection using GIS: A case study of Sistan and Baluchestan Province, Iran. *Geosciences*, *8*(12), 494. doi:10.3390/geosciences8120494

Kaboli, S., Hekmatzadeh, A. A., Darabi, H., & Haghighi, A. T. (2021). Variation in physical characteristics of rainfall in Iran, determined using daily rainfall concentration index and monthly rainfall percentage index. *Theoretical and Applied Climatology*, *144*(1), 507–520. doi:10.100700704-021-03553-9

Keesstra, S., Mol, G., De Leeuw, J., Okx, J., De Cleen, M., & Visser, S. (2018). Soil-related sustainable development goals: Four concepts to make land degradation neutrality and restoration work. *Land (Basel)*, *7*(4), 133. doi:10.3390/land7040133

Keesstra, S. D., Bouma, J., Wallinga, J., Tittonell, P., Smith, P., Cerdà, A., Montanarella, L., Quinton, J. N., Pachepsky, Y., van der Putten, W. H., Bardgett, R. D., Moolenaar, S., Mol, G., Jansen, B., & Fresco, L. O. (2016). The significance of soils and soil science towards realization of the United Nations Sustainable Development Goals. *Soil (Göttingen)*, *2*(2), 111–128. doi:10.5194oil-2-111-2016

Kostadinov, S., Borisavljević, A., & Mlađan, D. (2014). Torrents and torrential floods in Serbia: Characteristics and possibilities of its control. *Advances in GeoEcology*, *41*, 278–285.

Křeček, J., Nováková, J., Palán, L., Pažourková, E., & Stuchlík, E. (2021). Role of forests in headwater control with changing environment and society. *International Soil and Water Conservation Research*, *9*(1), 143–157. doi:10.1016/j.iswcr.2020.11.002

Kumar, P., Garg, V., Mittal, S., & Murthy, Y. K. (2021). GIS-based hazard and vulnerability assessment of a torrential watershed. *Environment, Development and Sustainability*, 1–31. doi:10.100710668-021-01476-z

Kusky, M. T. (2020). Encyclopedia of Earth and Space Science. Academic Press.

Lal, R., & Stewart, B. A. (1990). *Soil Degradation*. Springer-Verlag.

Lal, R. A. T. T. A. N. (2001). Soil degradation by erosion. *Land Degradation & Development*, *12*(6), 519–539.

Langdale, G. W., West, L. T., Bruce, R. R., Miller, W. P., & Thomas, A. W. (1992). Restoration of eroded soil with conservation tillage. *Soil Technology*, *5*(1), 81–90.

Liste, M., Grifoll, M., & Monbaliu, J. (2014). River plume dispersion in response to flash flood events. Application to the Catalan shelf. *Continental Shelf Research*, *87*, 96–108.

Luino, F., Turconi, L., Paliaga, G., Faccini, F., & Marincioni, F. (2018). Torrential floods in the upper Soana Valley (NW Italian Alps): Geomorphological processes and risk-reduction strategies. *International Journal of Disaster Risk Reduction*, *27*, 343–354.

Markose, V. J., & Jayappa, K. S. (2016). Soil loss estimation and prioritization of sub-watersheds of Kali River basin, Karnataka, India, using RUSLE and GIS. *Environmental Monitoring and Assessment*, *188*(4), 225.

Meena, N. K., Gautam, R., Tiwari, P., & Sharma, P. (2017). Nutrient losses in soil due to erosion. *Journal of Pharmacognosy and Phytochemistry*, *SP1*, 1009–1011.

Mitasova, H., Barton, C. M., Ullah, I., Hofierka, J., & Harmon, R. S. (2013). GIS-based soil erosion modeling. In *Remote Sensing and GIScience in Geomorphology* (pp. 228–258). Elsevier Inc.

Mohammadi, M., Khaledi Darvishan, A. K., Spalevic, V., Dudic, B., & Billi, P. (2021). Analysis of the Impact of Land Use Changes on Soil Erosion Intensity and Sediment Yield Using the IntErO Model in the Talar Watershed of Iran. *Water (Basel)*, *13*(6), 881.

Moynihan, K. P., & Vasconcelos, J. G. (2014). SWMM modeling of a rural watershed in the lower coastal plains of the United States. *J. Water Manag. Model*, 1-12.

Mullan, D. (2013). Managing soil erosion in Northern Ireland: A review of past and present approaches. *Agriculture*, *3*(4), 684–699.

Munir, B. A., Ahmad, S. R., & Hafeez, S. (2020). Integrated hazard modeling for simulating torrential stream response to flash flood events. *ISPRS International Journal of Geo-Information*, *9*(1), 1.

Munir, B. A., Ahmad, S. R., & Rehan, R. (2021). Torrential Flood Water Management: Rainwater Harvesting through Relation Based Dam Suitability Analysis and Quantification of Erosion Potential. *ISPRS International Journal of Geo-Information*, *10*(1), 27.

Nearing, M. A., Xie, Y., Liu, B., & Ye, Y. (2017). Natural and anthropogenic rates of soil erosion. *International Soil and Water Conservation Research*, *5*(2), 77–84.

Nielsen, U. N., Wall, D. H., & Six, J. (2015). Soil biodiversity and the environment. *Annual Review of Environment and Resources*, *40*, 63–90.

Ola, A., Dodd, I. C., & Quinton, J. N. (2015). Can we manipulate root system architecture to control soil erosion? *Soil (Göttingen)*, *1*(2), 603–612.

Pant, N., Dubey, R. K., Bhatt, A., Rai, S. P., Semwal, P., & Mishra, S. (2020). Soil erosion and flood hazard zonation using morphometric and morphotectonic parameters in Upper Alaknanda river basin. *Natural Hazards*, *103*(3), 3263–3301.

Parsons, A. J. (2019). How reliable are our methods for estimating soil erosion by water? *The Science of the Total Environment*, *676*, 215–221.

Petrović, A. M., Novković, I., & Kostadinov, S. (2021). Hydrological analysis of the September 2014 torrential floods of the Danube tributaries in the Eastern Serbia. *Natural Hazards*, 1–15.

Pimentel, D., Harvey, C., Resosudarmo, P., Sinclair, K., Kurz, D., McNair, M., ... Blair, R. (1995). Environmental and economic costs of soil erosion and conservation benefits. *Science*, *267*(5201), 1117–1123.

Polovina, S., Radić, B., Ristić, R., Kovačević, J., Milčanović, V., & Živanović, N. (2021). Soil Erosion Assessment and Prediction in Urban Landscapes: A New G2 Model Approach. *Applied Sciences (Basel, Switzerland)*, *11*(9), 4154.

Pribadi, D. O., Vollmer, D., & Pauleit, S. (2018). Impact of peri-urban agriculture on runoff and soil erosion in the rapidly developing metropolitan area of Jakarta, Indonesia. *Regional Environmental Change*, *18*(7), 2129–2143.

Puigdefábregas, J. (2005). The role of vegetation patterns in structuring runoff and sediment fluxes in drylands. *Earth Surface Processes and Landforms: The Journal of the British Geomorphological Research Group*, *30*(2), 133–147.

Quinton, J. N., Edwards, G. M., & Morgan, R. P. C. (1997). The influence of vegetation species and plant properties on runoff and soil erosion: Results from a rainfall simulation study in south east Spain. *Soil Use and Management*, *13*(3), 143–148.

Quinton, J. N., Govers, G., Van Oost, K., & Bardgett, R. D. (2010). The impact of agricultural soil erosion on biogeochemical cycling. *Nature Geoscience*, *3*(5), 311–314.

Reimann, L., Vafeidis, A. T., Brown, S., Hinkel, J., & Tol, R. S. (2018). Mediterranean UNESCO World Heritage at risk from coastal flooding and erosion due to sea-level rise. *Nature Communications*, *9*(1), 1–11.

Renschler, C. S. (2003). Designing geo-spatial interfaces to scale process models: The GeoWEPP approach. *Hydrological Processes*, *17*(5), 1005–1017.

Rineau, F., Malina, R., Beenaerts, N., Arnauts, N., Bardgett, R. D., Berg, M. P., & Vangronsveld, J. (2019). Towards more predictive and interdisciplinary climate change ecosystem experiments. *Nature Climate Change*, *9*(11), 809–816.

Ristić, R., Kostadinov, S., Abolmasov, B., Dragićević, S., Trivan, G., Radić, B., ... Radosavljević, Z. (2012). Torrential floods and town and country planning in Serbia. *Natural Hazards and Earth System Sciences*, *12*(1), 23–35.

Ristić, R., Kostadinov, S., Radić, B., Trivan, G., & Nikić, Z. (2012). *Torrential floods in Serbia–man made and natural hazards. 12th congress interpraevent.*

Ristić, R., Radivojević, S., Radić, B., Vasiljević, N., & Bjedov, I. (2010). Restoration of eroded surfaces in ski resorts of Serbia. *Global change: Challenges for soil management*, 165-174.

Sang, G. S., Gan, Z. M., & Yue, D. P. (2002). Landform Evolution and Soil Erosion on Zhouyuan During Human History Period. *Journal of Mountain Research*, 6.

Sayl, K. N., Muhammad, N. S., Yaseen, Z. M., & El-shafie, A. (2016). Estimation the physical variables of rainwater harvesting system using integrated GIS-based remote sensing approach. *Water Resources Management*, *30*(9), 3299–3313.

Schmidt, S., Alewell, C., Panagos, P., & Meusburger, K. (2016). Regionalization of monthly rainfall erosivity patterns in Switzerland. *Hydrology and Earth System Sciences*, *20*(10), 4359–4373.

Shikangalah, R. N., Jeltsch, F., Blaum, N., & Mueller, E. N. (2016). A review on urban soil water erosion. *Journal for Studies in Humanities and Social Sciences*, 163-178.

Shit, P. K., Pourghasemi, H. R., & Bhunia, G. S. (Eds.). (2020). *Gully erosion studies from India and surrounding regions*. Springer International Publishing.

Stefanidis, S., & Stathis, D. (2018). Effect of climate change on soil erosion in a mountainous mediterranean catchment (Central Pindus, Greece). *Water (Basel)*, *10*(10), 1469.

Tiranti, D., Cremonini, R., Marco, F., Gaeta, A. R., & Barbero, S. (2014). The DEFENSE (debris Flows triggEred by storms–nowcasting system): An early warning system for torrential processes by radar storm tracking using a Geographic Information System (GIS). *Computers & Geosciences*, *70*, 96–109.

Vaezi, A. R., Abbasi, M., Keesstra, S., & Cerdà, A. (2017). Assessment of soil particle erodibility and sediment trapping using check dams in small semi-arid catchments. *Catena*, *157*, 227–240.

Vásquez-Méndez, R., Ventura-Ramos, E., Oleschko, K., Hernández-Sandoval, L., Parrot, J. F., & Nearing, M. A. (2010). Soil erosion and runoff in different vegetation patches from semiarid Central Mexico. *Catena*, *80*(3), 162–169.

Walega, A., Cupak, A., Amatya, D. M., & Drozdzal, E. (2017). Comparison of direct outflow calculated by modified SCS-CN methods for mountainous and highland catchments in upper Vistula Basin, Poland and lowland catchment in South Carolina, USA. *Acta Scientiarum Polonorum. Formatio Circumiectus*, *16*(1), 187–207.

Wall, D. H., & Six, J. (2015). *Give soils their due*. Academic Press.

Walling, D. E. (2011). Human impact on the sediment loads of Asian rivers. *Sediment Problems and Sediment Management in Asian River Basins. IAHS Publ*, *349*, 37–51.

Wardle, D. A., Bardgett, R. D., Klironomos, J. N., Setälä, H., Van Der Putten, W. H., & Wall, D. H. (2004). Ecological linkages between aboveground and belowground biota. *Science*, *304*(5677), 1629–1633.

Wei, X., Cai, S., Ni, P., & Zhan, W. (2020). Impacts of climate change and human activities on the water discharge and sediment load of the Pearl River, southern China. *Scientific Reports*, *10*(1), 1–11.

Wu, X., Wang, Z., Guo, S., Liao, W., Zeng, Z., & Chen, X. (2017). Scenario-based projections of future urban inundation within a coupled hydrodynamic model framework: A case study in Dongguan City, China. *Journal of Hydrology (Amsterdam)*, *547*, 428–442.

Wuepper, D., Borrelli, P., & Finger, R. (2020). Countries and the global rate of soil erosion. *Nature Sustainability*, *3*(1), 51–55.

Xie, J., Chen, H., Liao, Z., Gu, X., Zhu, D., & Zhang, J. (2017). An integrated assessment of urban flooding mitigation strategies for robust decision making. *Environmental Modelling & Software*, *95*, 143–155.

Yavari, S., McIntyre, N., & Baumgartl, T. (2021). An Empirical Analysis of Sediment Export Dynamics from a Constructed Landform in the Wet Tropics. *Water (Basel)*, *13*(8), 1087.

Zhao, G., Mu, X., Wen, Z., Wang, F., & Gao, P. (2013). Soil erosion, conservation, and eco-environment changes in the Loess Plateau of China. *Land Degradation & Development*, *24*(5), 499–510.

Zhou, Y., Zhang, B., Qin, W., Deng, Q., Luo, J., Liu, H., & Zhao, Y. (2021). Primary environmental factors controlling gully distribution at the local and regional scale: An example from Northeastern China. *International Soil and Water Conservation Research*, *9*(1), 58–68.

Zittis, G., Hadjinicolaou, P., Klangidou, M., Proestos, Y., & Lelieveld, J. (2019). A multi-model, multi-scenario, and multi-domain analysis of regional climate projections for the Mediterranean. *Regional Environmental Change*, *19*(8), 2621–2635.

ADDITIONAL READING

Carrasco, R. M., Pedraza, J., Martin-Duque, J. F., Mattera, M., Sanz, M. A., & Bodoque, J. M. (2003). Hazard zoning for landslides connected to torrential floods in the Jerte Valley (Spain) by using GIS techniques. *Natural Hazards*, *30*(3), 361–381. doi:10.1023/B:NHAZ.0000007171.13298.01

Durlević, U. (2021). Assessment of torrential flood and landslide susceptibility of terrain: Case study-Mlava River Basin (Serbia). *Glasnik Srpskog Geografskog Drustva*, *101*(1), 49–75. doi:10.2298/GS-GD2101049D

Gurung, N., Fort, M., Bell, R., Arnaud-Fassetta, G., & Maharjan, N. R. (2021). Hydro-torrential hazard vs. anthropogenic activities along the Seti valley, Kaski, Nepal: Assessment and recommendations from a risk perspective. *Journal of Nepal Geological Society*, *62*, 58–87. doi:10.3126/jngs.v62i0.38695

Marković, M., Lukić, S., & Baumgerte, A. (2021). Use of Flash Flood Potential Index (FFPI) Method for Assessing the Risk of Roads to the Occurrence of Torrential Floods-Part of the Danube Basin and PEK River Basin. *IJTTE. International Journal for Traffic and Transport Engineering*, *11*(4).

Petrović, A. M., Novković, I., & Kostadinov, S. (2021). Hydrological analysis of the September 2014 torrential floods of the Danube tributaries in the Eastern Serbia. *Natural Hazards*, *108*(1), 1–15. doi:10.100711069-021-04737-2

Shaaban, F., Othman, A., Habeebullah, T. M., & El-Saoud, W. A. (2021). An integrated GPR and geoinformatics approach for assessing potential risks of flash floods on high-voltage towers, Makkah, Saudi Arabia. *Environmental Earth Sciences*, *80*(5), 1–15. doi:10.100712665-021-09454-4

Stolf, P., Berthet, J., Bosc, C., Bremond, P., Ceyte, A., Champonnois, V., . . . Sayah, A. (2021). Towards Flash Floods Management Using Temporal Loops (Doctoral dissertation, IRIT-Institut de Recherche en Informatique de Toulouse).

KEY TERMS AND DEFINITIONS

Biodiversity: Variety of flora and fauna precisely in a habitat.

Climate Change: Change in global and regional climatic patterns specifically apparent from middle to late of the 20th century.

Lithosphere: Rigid outer part of the earth.

Management: Process of dealing with issues.

Over-Exploitation: Excessive use of a resources.

Soil Erosion: The process of gradual movement and transport of upper layer of soil through agents like water, wind etc. is called soil erosion.

Torrential Floods: Extremely heavy rainfall or downpour.

Chapter 6
Environmental Consequences of Soil Erosion

Tatjana D. Golubović
University of Niš, Serbia

ABSTRACT

Soil is one of the most valuable natural resources. Despite soil importance, the pressures on soil have increased in recent decades. Soil degradation is a critical and growing problem, whereby soil erosion presents a prevailing process compared to other degradative processes. The intensity of erosion depends on the topography, climate conditions, soil characteristics, human activities, and the presence of vegetation. In this chapter, the diverse factors that cause soil erosion have been evaluated. The level of damage associated with soil erosion has been analyzed, with emphasis on the impacts they may have on the global carbon cycle, phosphorus loss, dust emissions, eutrophication, and soil biodiversity.

INTRODUCTION

Soil, the biologically active medium that has been formed in the uppermost layer of Earth's crust, plays a vital role in the functioning of terrestrial ecosystems. Soil is essential for humans because it presents the basis for food production. Besides, soil acts as a filter for contaminants, a reservoir of water and nutrients, participates in the cycling of carbon and other elements through the global ecosystem, serves as a habitat for soil organisms, as well as a landscaping and engineering medium.

Despite its importance, the pressure on soil is increasing. As an element of the land degradation process, soil degradation refers to a decline in soil productivity and quality. Several physical, chemical, and biological processes are responsible for the degradation of soil. The physical methods include deterioration of soil structure, compaction, crust formation, erosion, and desertification. Among all chemical processes, leaching, soil fertility decline, acidification, salinization and pollution are significant. The biological processes of soil degradation include carbon reduction and a decrease in soil biodiversity (Osman, 2014).

Several natural and anthropogenic causes can lead to soil degradation. Among the natural ones, the most important are frequent floods, high-velocity wind, high-intensity rains, leaching in humid regions,

DOI: 10.4018/978-1-7998-8459-0.ch006

drought in arid areas, etc. As regards anthropogenic causes, the most significant are deforestation, overgrazing, inadequate use of agrochemicals, monocropping, shifting cultivation (Dragović & Vulević, 2020; Osman, 2014).

Many studies confirm the connection between deforestation and soil degradation. Based on research, conversion of the forest area into the pasture and the subsequent unregulated grazing have altered the erosion rate, nutrient amount and soil characteristics (Rhoades, 2011). Hajabbasi et al. (1997) were investigated deforestation effects on soil physical and chemical properties in the Lordegan region of central Zagros Mountain, Iran. Deforestation and clear-cutting of the forests resulted in lower soil quality. Namely, deforestation and subsequently tillage practices resulted in a 50% decrease in organic matter and total nitrogen, a 10 to 15% decrease in soluble ions comparing to the undisturbed forest soil. Also, the tilth index coefficient of the forest site was significantly higher (0.717) than the cultivated forest and the deforested areas (0.633 and 0.573, respectively). The consequences of deforestation vary from soil erosion and sedimentation to loss of biodiversity, climate change, flood and drought, the decline in water quality, etc.

Overgrazing is one of the most severe problems on a global scale in terms of soil degradation. Depending on grazing intensity, soil organic matter, nitrogen content, microbial activity and soil moisture vary. All the above-mentioned factors and loss of vegetation cover lead to soil erosion and desertification. The results obtained by studying a dry Mediterranean agro-forest landscape (Crete, Greece) indicate that soil erosion increases and soil moisture decrease on overgrazed land and highlight the crucial role of sustainable grazing in land degradation mitigation (Kairis et al., 2015). According to research conducted in eastern Hovsgol, Mongolia, soil fertility decreases and soil chemical properties change due to overgrazing. In overgrazed areas, soil organic matter content was 30-50% lower and exchangeable calcium 40-60% lower than non-grazing areas. Nitrogen content was also lower in overgrazed valleys. Overgrazing was also found to affect soil physical characteristics in terms of increasing topsoil temperature, decreasing moisture, and increasing bulk and particle density (Wang & Batkhishig, 2014).

Modern agriculture relies on the extensive use of agrochemicals, pesticides, fertilizers and soil conditioners, to increase crop productivity by controlling harmful pests, pathogens, and weeds (Mandal et al., 2020). Excessive applications of nitrogen and phosphorus fertilizers lead to the disruption of the natural nitrogen and phosphorus cycles. Also, the use of fertilizers introduces heavy metals and radionuclides from mineral fertilizers and pathogens, veterinary pharmaceuticals, and endocrine disrupters from organic fertilizers into the soil. Their inappropriate use hurts the soil quality and represents one of the basic forms of degradation.

Monocropping is the agricultural practice of growing a single crop (corn, soybeans, wheat) year after year on the same soil, in the absence of rotation through other crops or growing multiple crops (polyculture) on the same soil. The monocropping can lead to many adverse effects, including soil degradation. Long-term utilization of the same culture leads to loss of soil nutrients, some types of insects and pests proliferate, so farmers become dependent on fertilizers, pesticides and other agrochemicals. Henry (1995) found that a deterioration in soils' physical and chemical properties causes cane yield decline (during monocropping).

Shifting cultivation is an old farming practice, where the "slash-and-burn" technique is applied to clear land, followed by a long fallow period important for the restoration of soil fertility. Many studies indicate that crop burning has harmful effects on soil, such as increased susceptibility to soil erosion and a reduction in nutrients (Dragović & Vulević, 2020).

According to the Global Assessment of Soil Deterioration (GLASOD), there are five principal types of soil degradation: water erosion, wind erosion, chemical deterioration, physical deterioration, and degradation of biological activity (Oldeman, 1991).

Soil erosion presents an overall process compared to other degradative processes. The slow geologic erosion is a constructive process, which has created fertile soils of alluvial flood plains and loess plateaus around the world (Lal, 2003). In contrast, intensive soil erosion is a global problem with several environmental impacts. Erosion is generally characterized by three phases, including separation of individual particles from the soil mass, transport of eroded material, and deposition. These processes usually result in the relocation of the top layer of soil rich in nutrients and organic matter. Organic matter affects the improvement of soil structure, soil water retention, aeration, soil thermal regime, as well as adsorption properties. Also, organic matter has the role of a buffer, participates in the global carbon cycle, serves as an energy source for soil microorganisms, etc. Soil erosion hinders the growth of plants, agricultural yields, and quality of water. Erosion may lead to reduced crop production by reducing nitrogen, phosphorus, and soil organic matter reserves, varying soil clay content, depleting available water capacity, and reducing soil aggregation. Much of the phosphorus from erosive soils are transported to water bodies together with eroded soil causing eutrophication, or significant growth of algae and other aquatic plants in nutrient-enriched waters that lowers dissolved oxygen levels. As these algae and other plants decompose, the result is increased fish kills, increased turbidity, and shifts in aquatic flora and fauna populations.

Without appropriate erosion prevention and control, the soil may lose its ability to hold nutrients, regulate water flow, filtrate pollutants, and other significant properties and functions.

This chapter aims to point out causes and types of soil degradation, emphasizing soil erosion and providing future guidelines for soil degradation prevention.

BACKGROUND

Soil and Sustainable Development

The United Nations' adoption of the 17 sustainable development goals (SDGs), under the 2030 Agenda for Sustainable Development, supported the scientific and academic community to generate accurate information to support planning and monitor socio-economic development by interlinking them with environmental sustainability dimensions (UN 2015; Tóth et al., 2018). Some SDGs, directly or indirectly, consider soil resources through the issues of food security, food safety, urban development and sustainability of terrestrial ecosystem services. SDG target 15.3 on land degradation neutrality point out "by 2030 to combat desertification, restore degraded land and soil, including land affected by desertification, drought and floods, and strive to achieve a land degradation-neutral world".

As mentioned, sustainable development goal 15 considers sustainable use of terrestrial ecosystems, combat desertification, and halt and reverse land degradation. Many ecosystem services and soil functions are related to this goal. Appropriate land management is crucial for sustainability. The importance of the Agenda for Sustainable Development results from the fact that 75% of the land is degraded (IPBES, 2018).

Figure 1. a) Aeolian erosion in the desert area Ischigualasto, San Juan, Argentina b) Water erosion, Đavolja varoš, Kuršumlija, Serbia c) Barren landscape (granite), the monastery of St. Archangel Michael, Prilep, Northern Macedonia d) Fluvial erosion of the karst area in the valley of the river Cijevna, Podgorica, Montenegro e) Fluvial erosion in the valley of the river Pčinja, Trgovište, Srbija f) Gully erosion in the valley of the river Struma, Sandanski, Bulgaria
(Source: Author)

The Causes and Effects of Soil Erosion

Soil erosion is considered the most widespread form of land degradation that may cause environmental and property damages and social and economic disruption. Erosion is a process of detachment of soil particles by different agents (e.g., runoff, raindrop, wind, gravity, etc.) and their displacement and deposition (Figure 1).

The erosion process lowers the quantity and quality of soils on-site; however, significant sediment-related problems off-site can notice. The content of soil loss by natural erosion is easily recoverable by the biological processes of soil formation. Erosion caused by anthropogenic influences, known as accelerated soil erosion, is a much bigger problem. Solving the problem of soil erosion requires an understanding of the complexity of erosion processes, their interactions, and their spatial extent.

Several erosion processes occur simultaneously in many environments and may interact, resulting in amplification or compensation of soil loss rates (Poesen, 2018).

The intensity of soil erosion depends on many factors, such as the configuration of the terrain (slopes), climate conditions, soil characteristics, human activities, and the presence of vegetation. Depending on the dominance of the above factors, soil erosion may be more or less harmful. Soil erosion is a pronounced process on steep land, where soil material is easily transported. Land with a high slope will relieve the rainwater flow rate or runoff saturation in the area, primarily due to the faster movement of the water downhill (Chen et al., 2011; Nenadović et al., 2013). The wind is one of the most important factors for decreasing soil quality and quantity through erosion, especially for sandy or lighter soil. It causes the loss of the most fertile part of the soil and significantly reduces soil productivity (Osman, 2014). Since wind erosion is affected by climatic factors, wind velocity, precipitation, temperature, as well as evaporation and soil moisture content, wind erosion poses a severe threat in the arid and semi-arid regions with rare vegetation, low rainfall and high temperature. The evaporation is more increased than precipitation over a long time, which causes depletion of soil humidity and organic matter and deteriorates the soil structure. Wind erosion not only affects the land but also harms crops and infrastructure. The most important soil characteristics that affect soil erodibility by wind are texture, structure, and water content. When it comes to vegetative cover, it has been proven that it slows the wind velocity near the ground and decreases the eroding capacity.

Rainfall and rainwater runoff cause the dispersion of the soil organic matter and sand particles, but in heavy rainfalls, more significant soil components also get affected, causing the erosion processes. In areas with lush vegetation, natural erosion is moderate, gradual, and harmless. In these conditions, the rate of soil loss is less than the rate of soil formation (Osman, 2014). Crops and grasses improve the structure of soils; thereby, a decrease in soil erosion is significant. Areas with less naturally occurring vegetation are more susceptible to soil erosion. Moreover, anthropogenic activities that are responsible for soil erosion are mining, agriculture, deforestation, and overgrazing. There are numerous water erosion's on-site effects upon agricultural soils, which can be reflected in soil loss, loss of nutrients and organic matter, soil compaction, the decline in soil quality, reduction in growth and yield of crops, the reduced capability of ecosystem functions, etc. (Hicks & Anthony, 2001; Bogunovic et al., 2018; Alletto & Coquet, 2009; Bongiorno et al., 2019). Soil erosion causes land degradation and leads to many off-site environmental problems such as flooding, water siltation, and pollution (Al-Wadaey & Ziadat, 2014; Issaka & Ashrafa, 2017; Poleto & Beier, 2012). Off-site effects of soil erosion are not always easily noticed because eroded material is transported to distant places and deposited in streams, rivers, lakes, agricultural land, roads, etc.

In addition to the above effects caused by soil erosion, it is essential to mention the economic consequences.

Cost estimates for soil erosion have been made worldwide. Telles et al. (2011) estimated the costs of soil erosion, as an issue of fundamental importance in view of the current discussions on sustainability. The on-site costs were estimated based on soil loss, nutrients, pesticides, organic matter, productivity and yield. On-site effects directly influence productive land, primarily through decline, degradation, or reduction of nutrients and organic matter. Also, they lead to a drop in fertile soil depth and moisture available for plants. As a result, the use of plant crops is limited. The side effect of these difficulties is instability in food production, land devaluation, additional costs for irrigation, loss of investments in improved production systems that become non-effective in soils with accelerated erosion, and labor costs to repair the damage caused by soil erosion (Telles et al., 2011; Telles et al., 2013).

Off-site costs were estimated in various ways. The significant off-site consequence is sedimentation (Telles et al., 2013; Montanarella, 2007). The content of sediment in water bodies causes the functionality of hydroelectric power plants and water treatment stations. The costs incurred by this process were estimated in terms of water treatment (Foster et al., 1987) and increased expenditure on electrical energy generation (Marques, 1998). It is known that sedimentation can adversely affect irrigation, navigation, recreation and water storage and distribution, which may cause an increase in costs concerning reparation

(Hansen et al., 2002). Off-site costs can also be related to the increased price of agricultural products, resulting in macroeconomic instabilities (Cohen et al., 2006).

SOIL EROSION: ENVIRONMENTAL EFFECTS

Soil is a complex of solid components (about 45% mineral and 5% organic matter), gases comprise about 20–30%, and water (20–30%). The amount of gases and water in the pore space can change quickly depending on weather conditions and other factors. The inorganic components of soils represent more than 90% of the solid components. The major elements C, N, P, and S (macronutrients) are essential to the life cycles of organisms and are absorbed by them in significant amounts. The global biogeochemical cycles of these elements are therefore of major interest, especially because of the large anthropogenic influence they experience. Soil organic matter (SOM) includes the entire organic compounds in soils, excluding undecayed plant and animal tissues, their "partial decomposition" products, and the soil biomass (FAO, 2005).

Various environmental problems are directly or indirectly related to soil composition and soil-related processes. Significant losses of carbon (C) from soils occur due to erosion, leaching and runoff. Also, part of the carbon is emitted into the atmosphere in the form of CO_2 or CH_4.

Emissions to the atmosphere are an important pathway of nitrogen (N) loss (N_2, N_2O and NH_3). Cultivated soils, unlike natural, often show exceeding leaching and runoff losses of nitrogen due to the addition of high contents of nitrate or ammonium fertilizers that rise inorganic nitrogen amounts.

Leaching losses are minimal for soil phosphorus (P), while gaseous phosphorus emissions to the atmosphere do not occur from natural soils. Phosphorus can be absorbed by living organisms or lost with soil particles on which it is adsorbed. Fertilizer and organic waste applications to soils can lead to phosphorus losses, mostly by erosion, posing an environmental threat (Sposito, 2008).

As mentioned above, soil erosion is one of the most serious environmental problems facing modern society. Namely, the soil is being lost from land areas 10 to 40 times faster than the rate of soil renewal, which significantly endangers environmental quality.

As the major terrestrial carbon pool, the soil carbon pool is approximately 3.3 times the size compared to atmospheric and 4.5 times the size of the biotic pool (Lal, 2004a).

Soil erosion and transport of sediments by rivers are basic processes for soil carbon lateral transfer at the land surface. As a consequence of these processes, a significant carbon budget of terrestrial ecosystems occurs (Li et al., 2019; Wang et al., 2019). Nevertheless, this view was not in line with the opinion of another group of researchers. Jacinthe, Lal and co-workers have concluded that soil erosion leads to a terrestrial carbon loss due to soil aggregates' decomposition, causing a decrease in soil nutrients (Jacinthe et al., 2002; Lal, 2004b).

One of the significant impacts of soil erosion on the environment relates to carbon dynamics and possible emissions of greenhouse gases (GHGs). Namely, the rapid increase in the atmospheric concentration of CO_2 and other greenhouse gases into the atmosphere has elevated to a higher level concerns regarding the identification of sources and sinks of these gases. It isn't easy to precisely estimate local, regional and global emissions of CO_2 and other GHGs originating from soil cultivation, fertilizers, deforestation, biomass burning, etc. It is crucial to identify the "unknown" sinks and sources (e.g., terrestrial including soils and vegetation, industrial, aquatic, biotic) of CO_2 and other GHGs in order to reduce GHGs concentration in the atmosphere. The decline in the SOC pool happens due to transport by soil erosion processes, and leaching into groundwater and mineralization. Limited soil cultivation can significantly reduce C losses from soils owing to reducing mineralization and erosion. As mentioned, the impact of soil erosion on carbon dynamics needs to be accurately evaluated.

Soil C is affected by erosion processes (water and wind erosion). The level of interactions depends on different factors such as soil characteristics (the content of mineral and organic matter, soil texture, amount of water and soil gasses, soil moisture and temperature regimes, porosity, etc.) and climatic conditions vegetative cover. Soil erosion affects SOC dynamics through its impact on disruption of aggregates formations, removal of C in runoff water or dust storms, mineralization of soil organic matter on-site, mineralization of SOC displaced and redistribution over the landscape and transported in rivers and dust storms, reaggregation of soil through the formation of organo-mineral complexes at the depositional/protected sites and deep burial of C-enriched sediments in depositional sites, flood plains and reservoirs and ocean floor (Lal, 2003). Conversion of natural to agricultural ecosystems depletes SOC pool because C inputs in managed ecosystems are usually lower than in natural ecosystems due to differences in soil moisture and temperature regimes caused by ploughing, drainage, biomass burning, or residue removal (Lal, 2001). The movement of soil particles and redistribution of SOC by erosion is a pedogenic process because eroded and depositional soils differ from the original soils. The SOC is easily transported by runoff water or wind because it is relatively low density and is concentrated in the topsoil. The SOC pool in eroded soils is much lower than the uneroded ones, especially on sloping terrain. The erosion causes slaking, disruption and decomposition of soil aggregates. The soil organic carbon, especially particulate organic carbon, is susceptible to mineralization. The degree of mineralization depends on temperature (an increase in high temperature) and the presence of microorganisms. Some research shows that a significant amount of carbon dioxide is sequestered in aquatic ecosystems (there is no mineralization of C on the route from eroded sites to aquatic ecosystems, and all terrestrial C is transported into aquatic ecosystems is covered with sediments). The river water contains particulate organic carbon (POC), dissolved organic carbon (DOC) and dissolved inorganic carbon (DIC). The total organic carbon (TOC) is a sum of particulate organic carbon and dissolved organic carbon (DOC). It is presumed that rivers could be a sink of atmospheric CO_2 depending on the $P(CO_2)$. Rivers receiving large volumes of agricultural and urban effluent containing a high concentration of DOC. It can be concluded that the fate of C translocated by erosional processes is determined by numerous complex and interacting processes.

One of the significant effects of soil erosion on the environment is the loss of phosphorus and, consequently, the process of eutrophication. Phosphorus (P) is an essential element for plant and animal growth, and its presence is necessary to crops and animal production. Soil P exists in organic (undecomposed residues, microbes, and organic matter) and inorganic forms. Phosphorus needs to be added to most soils to provide the required level of this element for optimum crop yield. Due to its pronounced adsorption capacity, P can be rapidly fixed in relatively insoluble forms and unavailable to plants. Availability depends on soil pH and soil type (presence of Al, Fe, and Ca). Agriculture is a significant source of phosphorus in the environment. The necessary amount of phosphorus is achieved by the addition of fertilizer or manure. In most soils, the P content of topsoil is greater than that of the subsoil due to sorption of added P, pronounced biological activity, cycling of P from roots to aboveground plant biomass, and more organic material in surface layers. Permanent long-term application of fertilizer or manure at levels exceeding crop needs increases soil P content. If a significant amount of P is added, it results in the accumulation of P in soil. Phosphates are not water-soluble (unlike nitrates), so they move with soil movement, adsorbed on soil particles. A major problem related to increasing P content in soil is that any factors that increase soil erosion (water or wind) will also increase the runoff of P with soil to streams, rivers, lakes, and coastal regions. The loss of P in agricultural runoff occurs in sediment-bound and dissolved forms (Sharpley et al., 2003). Phosphorus, which binds to the soil, is usually lost through sheet and rill erosion from agricultural lands. Sharpley et al. (1992) have suggested that sediment P includes P associated with soil particles and organic material eroded during flow events and constitutes about 80 percent of P transported in surface runoff from most cultivated land. The author's point is that surface runoff from grass, forest, or noncultivated soils carries a small amount of sediment. A significant fraction is dissolved P origins from the soil and plant material. The greatest amount of dissolved P is available for biological uptake, unlike sediment P, which is considered a long-term P source for aquatic biota. Usually, surface runoff is dominant compared to subsurface flow in terms of phosphorus loss from the soil. Many studies report that the loss of P in surface runoff as well as in subsurface flow depends on soil P concentration.

Phosphorus loss depends on soil characteristics, the form of phosphorus applied, rainfall, uptake by plants and water movement. Phosphorus Retention Index expresses the capacity of soils to sorb P

(PRI). For example, soils with high P retention decrease P loss by leaching and intensify P loss by erosion. Soils with high P retention are prone to storing considerable amounts of P in the soil, which can then be lost through erosion when surface runoff removes soil particles that have been enriched with nutrients (Weaver, 2001).

The over-enrichment of aquatic ecosystems with nutrients (phosphorus, nitrogen, etc.) leads to algal blooms and anoxic events. This phenomenon is called eutrophication and presents a serious environmental problem. Eutrophication is essentially a natural process. In that case, water quality is generally acceptable, and there is a diverse biological community and little algae. However, human activities (extensive agriculture, deforestation, development of industry) accelerate these natural processes. As mentioned, agriculture is the major source of nonpoint water pollution. Extensive use of fertilizers results in an essential amount of nutrients in agricultural runoff. If eroded soil reaches the water bodies, both phosphorus and nitrogen in the soil contribute to eutrophication. Possible effects of eutrophication, caused by excessive inputs of phosphorus and nitrogen to lakes, rivers and coastal oceans include:

- Increased biomass of phytoplankton resulting in algal blooms
- Toxic or inedible phytoplankton species (harmful algal blooms)
- Increased blooms of gelatinous zooplankton
- Decreases in water transparency (increased turbidity)
- Dissolved oxygen depletion or hypoxia
- Eutrophication-driven acidification
- Dead zones
- Species biodiversity decreases and the dominant biota changes
- Green tides (http://www.coastalwiki.org).

Eutrophication and related algal blooms (a short-lasting substantial increase of an algal population) seriously impact ecological functions and water quality. Harmful algal blooms (HABs) are recognized as one of the most damaging aspects of cultural eutrophication. Many cyanobacterial genera have optimal growth rates and bloom potentials at relatively high water temperatures; therefore, global warming plays a crucial role in their boost and persistence. Bloom-forming cyanobacterial taxa can be harmful to the environment, humans, and other organisms by outcompeting useful phytoplankton, depleting oxygen, and producing a variety of toxic secondary metabolites (e.g., cyanotoxins). (Paerl & Otten, 2013). Algal blooms operate by photosynthesis and therefore produce a lot of oxygen. The reverse happens when the bloom decays; oxygen is extracted from the water when the organic material is mineralized. Namely, bacteria decompose algal remains, a process that consumes dissolved oxygen contained in the water. When the oxygen consumption rate in the aquatic environment exceeds resupply, oxygen content can rapidly collapse to levels inadequate to sustain most animal life, producing hypoxic conditions. Algal blooms also reduce water clarity and inhibit sunlight from penetrating the water and reach submerged aquatic vegetation and benthic (seafloor) microalgae, causing them to release less oxygen that would generally help replenish the water's oxygen supply.

Except for hypoxia, acidification of coastal waters is a significant problem related to microbial degradation of organic matter. CO_2 released in microbial degradation of organic matter in the water forms carbonic acid (H_2CO_3), which causes acidification of coastal water. Coastal ecosystems can also be acidified through:

- atmospheric carbon dioxide fluxes
- the introduction of acidic river water
- upwelling of CO2 enriched deep water (Wallace et al., 2014).

It may be noted that due to a combination of these processes in some coastal zones, seasonal decrease of pH and/or high $P(CO_2)$ was detected, while in some coastal regions, the continuous decline of pH value in recent decades was noticed (Feely et al., 2010; Cai et al., 2011; Waldbusser et al., 2011; Duarte et al., 2013; Melzner et al., 2013).

Eutrophication increases phytoplankton biomass and reduces water transparency. Water transparency is an indicator of water quality. Actually, water transparency is a function of the concentrations of chlorophyll-a, total suspended matter, and colored dissolved organic matter. Knowledge of water transparency could ensure useful information concerning aquatic environmental changes and the marine ecosystem (Bai et al., 2020). Qiang and co-workers gave an overview of research related to lake eutrophication and its ecosystem response. One aspect concerned the relationship between total phosphorus concentration and transparency. It was found the correlation between total phosphorus concentration and transparency in 13 eutrophic lakes in Denmark indicates that water transparency decreases as the total phosphorus concentration increases, and the sensory water quality decreases with transparency (BoQiang et al., 2013).

Biotic and abiotic changes associated with eutrophication influence the direction and strength of natural selection and, thereby, the shape of the ecological fitness landscape experienced by component populations (Alexander al., 2017). Rapidly increasing of nutrients in a short time resulted in anthropogenic eutrophication in the majority of aquatic ecosystems. To avoid risk of hyper-eutrophication, some species have developed mechanisms of adaptation to the mentioned environmental conditions (behavioural adaptations). Frisch et al. (2014), using the resurrection ecology approach, environmental, genetic and phenotypic data spanning pre- and post-industrialized agricultural areas reveal the evolutionary consequences of anthropogenic environmental change. They concluded that younger genotypes of *Daphnia pulicaria* of South Center Lake, Minnesota, grew less efficiently under low-phosphorus conditions as they had adapted to take advantage of the high-phosphorus diet while the lake was eutrophic. Some zooplankton species (*Acartia tonsa*) also show behavioural adaptations to the enhanced eutrophication. Modifications in the environment often indicate that the well-adapted native species are not competitive to non-native species. Changes in the environment very often display the reverse outcomes of competitive interactions among species. Anthropogenic influences may substantially impact ecosystem changes during eutrophication, which may result in the fact that invasive species (usually arriving from more phosphorus-rich habitats) have an ecological advantage over native species (Byers, 2002). The modified environmental conditions related to eutrophication can influence visual signalling between mates and between competitors. If the fact that sexual selection is mediated by vision-based mate selection takes into account among many species, it is evident that eutrophication-induced turbidity weakens sexual selection. Modifying natural and sexual selection and reproductive isolation and gene flow, eutrophication affects basic processes that cause and sustain biodiversity.

As mentioned above, erosion is one of the most serious threats to soil, primarily from the aspect associated with land loss, which is reflected in the quantity and diversity of soil organisms. Soil biodiversity implies a variety of organisms, including micro-organisms (e.g., bacteria, fungi, protozoa and nematodes), mesofauna (e.g., acari and springtails), as well as macro-fauna (e.g., earthworms and termites). Causal relationships between soil biodiversity and erosion can be considered through different aspects. Microorganisms are widespread in nature because of their small size, adaptation to different habitats, capability to utilize a wide range of nutrients etc. The bacterial population is prevalent and participates in soil structure dynamics, cycling processes, regulation of different plant communities, soil aeration, and soil fertility. The loss of a considerable content of soil cause a decrease, disperse and mix of a tremendous resource of soil organisms. As a result, erosion has a direct impact on soil population quantity and dynamics.

The influence of water and wind erosion on organisms replacement is previously described. It is known that rivers are important for the movement, migration, and dispersal of different aquatic and riparian organisms. Namely, soil runoff can move organisms' long distances and lead to invasion effect. Many studies have shown that invasive species can change biodiversity and ecosystem functioning. Merritt & Wohl (2002) have investigated the correlation between flow regime, channel morphology, dispersal phenology, and seed deposition. Systematic delivery of seeds under different environmental conditions (mentioned above) in repeated processes suggested that hydrochory may function as a form of dispersal for some coastal plant species. The results of the investigation have shown that knowledge of hydrograph characteristics may be used to accommodate or inhibit species using hydrochory. Some soil animals are subject to similar processes (e.g., nematodes). Based on the above, soil erosion can be considered as an invasion facilitator.

The intensity of erosive processes as well as agriculture and desertification may change and destroy, partially or entirely, communities of soil organisms. This problem is even more important when it comes to rare, vulnerable or endangered species. It is assumed that many soil organisms have developed protection mechanisms to reduce the effects of soil erosion as a type of stressor.

Microbial biomass carbon and microbial abundance immediately after, and two years after soil erosion were investigated by Mabuhay et al. (2004). This study aimed to estimate the degree of damage, the rate of recovery of microorganisms, the determination of the community diversity, and establish relationships between microbial biomass, microbial abundance, and the physicochemical properties of the soil. The research area was Hiroshima Prefecture (three different sites). This study pointed out that soil erosion has a damaging effect on microbial biomass, abundance and composition, by changing natural soil characteristics and removing vegetative protection. Immediately after the occurrence of erosion, a decline in microbial amount was noted, with the exception of a short period in spring and autumn. The difference in microbial biomass carbon between the study areas was noticed. The area in Kagamiyama (undisturbed area; pine trees) showed the highest biomass carbon, followed by the area in Yasumiyama (naturally eroded; mixed, primarily needle-leaved, vegetation) and Itsukaichi (naturally eroded; secondary mixed-pine forest). The microbial abundance showed a significant difference between the Kagamiyama site and the eroded sites. The three groups of microorganisms (Gram-positive bacteria, Gram-negative bacteria and fungi) were significantly affected by erosion. As can be seen, the eroded areas showed poor microbial diversity compared to the undisturbed area. This is in line with the assumption that microbial diversity declines with soil depth.

Regarding microorganism diversity, the analysis showed that the eroded sites (Yasumiyama and Itsukaichi) were most similar in composition, whereas the Kagamiyama site was the most different (Mabuhay et al., 2004). Based on the presented results, it can be seen that almost half of the microorganisms present in the undisturbed habitat are not present in the eroded sites, which indicates that many groups of microorganisms were removed during erosion processes. Also, some groups of microorganisms were present in the area two years after erosion (Yasumiyama) and in the undisturbed area (Kagamiyama), but not in the area immediately after erosion (Itsukaichi). These groups of microorganisms were probably removed with the eroded soils and had not been re-established. Finally, microorganisms in the soil (immediately after erosion) are not present in the undisturbed soil and soil two years after erosion. This is probably a new colonizing species in the area.

The conclusions of Fierer et al. (2003) support the above results. Namely, they examined variations in microbial community composition through two soil depth profiles. The results showed that the composition of the soil microbial communities changes significantly with soil depth. The differentiation of microbial communities within the two profiles coincides with an overall decline in microbial diversity. Gram-negative bacteria, fungi, and protozoa were highest in the topsoil, unlike Gram-positive bacteria and actinomycetes, which are dominant in the subsurface. The vertical distribution of these microorganisms may be conditioned by carbon availability (lower at greater depth).

For the carbon analysis (Mabuhay et al., 2004) the values greatly vary from one area to another. The Kagamiyama area showed the highest percentage of carbon and nitrogen content. This was followed by the area at Yasumiyama, while the lowest value was in Itsukaichi.

As stated earlier, soil biodiversity and soil erosion are closely interlinked. Namely, not only does erosion intensely transform soil organisms, but it also affects soil characteristics and loss. For example, mycorrhizal fungi that inhabit either a forest or grassland will probably reduce the soil eroded by rain or wind, keeping soil aggregates compacted. On the contrary, the extensive excavation activity of some soil-living mammals (e.g., moles) may distract soil structure and, thus, accelerate the erosion rate (Orgiazzi, 2018). A group of soil organisms, earthworms, has been confirmed to affect soil structure and hydrology significantly. In most cases, this is a positive impact reflected in better soil porosity, aeration, water infiltration etc. The burrowing activity of earthworms ensures a better run of water, favoring water infiltration (Edwards et al., 1990). However, according to the literature data, water infiltration and erosion processes depend on the dominant species (e.g., anecic *L. terrestris*, endogeic *A. tuberculata*) as well as on the abundance of earthworms (Bohlen et al., 1997; Edwards et al., 1990; Shuster, 2001).

Another very significant impact of erosion on the environment can be seen from the aspect of air pollution. Soil erosion by wind causes dust emission to the atmosphere. The major compounds of global aerosols are dust particles origins in different soil sources (sediment-deposition mostly in arid regions, playa-dry lake etc.). Global dust emissions from soils are estimated to be about 3000 million tons per year. A significant fraction includes particles less than 10 micrometers in diameter (PM10). The significance of PM10 is reflected in fact that they originate from soil organic carbon and clay material responsible for the adsorption capacity of the soil. Also, they are crucial air pollutants associated with health risks. As mentioned, wind erosion is an important problem in the arid and semiarid regions with poor vegetation, low rainfall and high temperature. The most critical factors affecting wind erosion are soil structure, climate, field width and slop, and vegetation. Soil erodibility refers to the ease of detachment and transport by wind. Climate influences refer to wind velocity, precipitation, and temperature that result in evaporation and soil moisture content. Strong winds and soil dryness increase the susceptibility of the topsoil to erosion. Vegetative covers reduce the wind velocity at the soil surface and also generally decrease the soil erodibility.

Anthropogenic activities change the soil's properties and its capacity to resist soil erosion. The soil erosion by wind can cause considerable nutrient loss (particularly phosphorus loss) and may lead to soil degradation and a decrease in fertility. Li et al. (2015) suggested that soil tillage and wind erosion represent two major emission sources of particulate matter (PM2.5 and PM10) from cultivated soils. The results of experiments with cultivated and uncultivated land showed that no-till soils had consistently lower PM2.5 and PM10 emissions than paired conventional tilled soils for uncrushed samples. One of the most important soils in overall dust emission is loess (cover about 10% of the Earth's land surface) as primary or secondary deposits. Dust emission of loess is due to the high availability of suspended sediments in silt and clay fractions (Katra, 2020). The destruction of the top layer of soil by erosion reduces soil nutrients, decreasing soil productivity and stability. Dust emissions, caused by erosion processes, can have both positive and negative effects. Namely, increased dust emissions from agricultural soils may intensify the phosphorus inputs to aquatic systems and accelerate eutrophication. On the other hand, Saharan dust that is transported over the Atlantic Ocean provides a vital input of phosphorus to the oligotrophic waters of the ocean and the P-depleted rain forests of America (Swap et al., 1992). The study conducted by Katra et al. (2016) aimed to focus on phosphorus cycling via dust emissions under standard land-use practices in an arid environment. The experiments indicate significant phosphorus fluxes by PM10 dust due to agricultural land use. The emission of phosphorus from soil dust sources has significant implications for soil nutrient resources and management strategies in agricultural regions as well as for loading to the atmosphere and global biogeochemical cycles.

Some soil from damaged land enters suspension and becomes part of the atmospheric dust load. Dust obscures visibility and pollutes the air, impacts water quality, and imperils animal and human health.

EROSION PREVENTION AND CONTROL: SOLUTIONS AND RECOMMENDATIONS

Soil erosion control measures have proven to reduce erosion potential and therefore preserve environmental quality, primarily soil quality. The most significant principles which should be applied in order to reduce soil erosion (by water) imply:

- Reducing detachment of soil particles from the surface of the soil. In this way, the possibility of splash erosion is reduced. This can be realized by providing a cover on the soil during the pluvial season. Tillage and cropping management systems are significant elements for reducing raindrop impact on soil particles due to crop residue availability to protect the soil surface.
- Improving the structure and texture of the soil, especially the surface layer (topsoil). A well-structured soil improves water drainage after irrigation or heavy rain and minimizes runoff. Soil

organic carbon is important as a binding medium in the formation of soil aggregates and soil aggregate stability (Bronick & Lal, 2005).

- Increasing the movement of water into the soil due to gravity and capillary forces (infiltration) and reducing surface runoff. The slopes, orientation of the slopes, topsoil properties, and the microtopography of the soil surface affect soil infiltration and erosion processes. Namely, the average erosion value increases at a higher slope. The erosion was observed to be high at higher slope steepness due to the increase in the surface flow velocity which is greater than the infiltration capacity (Duhita et al., 2021).
- Reducing runoff velocity leads to an increased infiltration rate. Since the degree and length of the slope affect the runoff velocity, it is necessary to correct them with appropriate techniques. Commonly used engineering approaches to reduce runoff velocity include the formation of terraces and waterways.
- Minimal soil destruction ensured by adequate land resources management (no-tillage, minimum tillage, and subsoil tillage). In contrast, intensive tillage increases the possibility of erosion.
- Wind erosion is a dominant type of erosion in the arid and semiarid regions characterized by poor vegetation, small precipitation, and high temperature. As a result of the notable differences in soil types, climate conditions, and used crops, there is no uniform erosion control method. However, by combining several different methods, wind erosion can be reduced to a tolerable level. Some of the most important measures in controlling wind erosion can be seen in the following:
- The primary method of reducing consequences caused by wind erosion is land coverage with surface residues. The residue, in general, originates from previous crops and can affect the maintenance of soil moisture by reducing evaporation. Apart from the benefits that are achieved by using this type of soil protection, there are also negative effects. Namely, after residues decomposition, significantly fewer residues remain in the soil.
- In areas with inadequate residue amounts to protect the soil, proper cultivation can highly reduce the erodibility of the soils. Tillage disturbs the natural soil structure, leading to the soil aggregates' decomposition into smaller particles. The small soil particles become susceptible to aeolian erosion, why tillage must be avoided or minimized.
- Timing of the tillage practice is most important to produce the combination of surface clods and roughness needed to control wind erosion. Rough cloddy surfaces increased infiltration and decreased soil erosion. Overtilling reduced surface roughness and cloddiness.
- Crop strips, crop barriers, or shelterbelts reduce field width and erosion from perpendicular winds. Barriers placed in the path of prevailing winds decrease wind velocity and reduce the intensity of aeolian erosion. Shelterbelts permit some flow of the air which removes the unwanted turbulence that occurs behind solid barriers. It has been shown that the yield was higher in the presence of shelterbelts.
- Provide a cover of growing vegetation. Perennial grasses and legumes cover the soil densely and ensure the best long-term protection against wind erosion. Grasses and legumes can also be used in combination with other crops such as different types of trees.

Soil erosion can be easily controlled on-site if the appropriate methods are used.

FUTURE RESEARCH DIRECTIONS

Soil resource is crucial to the survival of the human population, but its degradation is permanent. Preservation of this resource is important to maintaining agricultural productivity and environmental quality. Improving knowledge of the possible future rates of soil erosion is important for policymakers engaged in land-use decision-making and for earth-system modelers seeking to reduce uncertainty on global predictions. Also, increased endeavor, both from the research and financial aspect, is needed to develop new technologies to reduce soil erosion and sediment transport.

CONCLUSION

The pressure on soils will continue in the future, and a clearly defined regulation is needed. The United Nations' adoption of the 17 sustainable development goals (SDGs), under the 2030 Agenda for Sustainable Development, supported the scientific and academic community to generate accurate information to support planning and monitor socio-economic development by interlinking them with environmental sustainability dimensions. Sustainable development goal 15 considers sustainable use of terrestrial ecosystems, combat desertification, and halt and reverse land degradation. Many ecosystem services and soil functions are related to this goal. To achieve a sustainable situation, appropriate land management is crucial. The importance of the Agenda for Sustainable Development results from the fact that 75% of the land is degraded. Soil erosion is considered the major and the most widespread form of land degradation that may cause environmental and property damages and social and economic disruption. One of the significant impacts of soil erosion on the environment relates to carbon dynamics and possible emissions of greenhouse gases (GHGs). Also, important effects of soil erosion on the environment are the loss of phosphorus and, consequently, the process of eutrophication. Eutrophication is one of the serious threats to the environment, primarily for freshwater systems, in terms of reducing water quality and altering ecosystem structure and function. A solution to eutrophication is acute. Governments should perform more beneficial policies to regulate the industrial and agricultural sectors to reduce activities that contribute to eutrophication. Erosion is one of the most serious threats to the soil from the aspect associated with land loss, which is reflected in the quantity and diversity of soil organisms (micro-organisms, mesofauna, and macro-fauna). Causal relationships between soil biodiversity and erosion can be considered through different aspects. Another, very significant, impact of erosion on the environment can be seen from the aspect of air pollution. Soil erosion by wind causes dust emission to the atmosphere. The major compounds of global aerosols are dust particles origins in different soil sources (sediment-deposition mostly in arid regions, playa - dry lake, etc.). Global dust emissions from soils are estimated to be about 3000 million tons per year. A significant fraction includes particles less than 10 micrometers in diameter (PM10).

Solving the problem of soil erosion requires an understanding of the complexity of erosion processes, their interactions, and their spatial extent.

ACKNOWLEDGMENT

This research was supported by the European Commission under the Project "Soil Erosion and Torrential Flood Prevention: Curriculum Development at the Universities of Western Balkan Countries" (SETOF) [grant number 598403-EPP-1-2018-1-RS-EPPKA2-CBHE-JP], and the Ministry of Education, Science and Technological Development of the Republic of Serbia - Agreements on the implementation and financing of scientific research in 2021 [number 451-03-9/2021-14/200148].

REFERENCES

Al-Wadaey, A., & Ziadat, F. (2014). A participatory GIS approach to identify critical land degradation areas and prioritize soil conservation for mountainous olive groves (case study). *Journal of Mountain Science, 11*(3), 782–791. doi:10.100711629-013-2827-x

Alexander, T. J., Vonlanthen, P., & Seehausen, O. (2017). Does eutrophication-driven evolution change aquatic ecosystems? *Philosophical Transactions ofThe Royal Society B: Biological Sciences, 372*(1712), 20160041. doi:10.1098/rstb.2016.0041

Alletto, L., & Coquet, Y. (2009). Temporal and spatial variability of soil bulk density and near-saturated hydraulic conductivity under two contrasted tillage management systems. *Geoderma, 152*(1-2), 85–94. doi:10.1016/j.geoderma.2009.05.023

Bai, S., Gao, J., Sun, D., & Tian, M. (2020). Monitoring Water Transparency in Shallow and Eutrophic Lake Waters Based on GOCI Observations. *Remote Sensing, 12*(1), 163. doi:10.3390/rs12010163

Bogunovic, I., Pereira, P., Kisic, I., Sajko, K., & Sraka, M. (2018). Tillage management impacts on soil compaction, erosion and crop yield in Stagnosols (Croatia). *Catena, 160*, 376–384. doi:10.1016/j.catena.2017.10.009

Bohlen, P. J., Parmelee, R. W., McCartney, D. A., & Edwards, C. A. (1997). Earthworm effects on carbon and nitrogen dynamics of surface litter in corn agroecosystems. *Ecological Applications, 7*(4), 1341–1349. doi:10.1890/1051-0761(1997)007[1341:EEOCAN]2.0.CO;2

Bongiorno, G., Bünemann, E. K., Oguejiofor, C. U., Meier, J., Gort, G., Comans, R., Mader, P., Brussaard, L., & de Goede, R. (2019). Sensitivity of labile carbon fractions to tillage and organic matter management and their potential as comprehensive soil quality indicators across pedoclimatic conditions in Europe. *Ecological Indicators, 99*, 38–50. doi:10.1016/j.ecolind.2018.12.008

Bronick, C. J., & Lal, R. (2005). Soil structure and management: A review. *Geoderma, 124*(1–2), 3–22. doi:10.1016/j.geoderma.2004.03.005

Byers, J. E. (2002). Impact of non-indigenous species on natives enhanced by anthropogenic alteration of selection regimes. *Oikos, 97*(3), 449–458. doi:10.1034/j.1600-0706.2002.970316.x

Cai, W. J., Hu, X., Huang, W.-J., Murrell, M. C., Lehrter, J. C., Lohrenz, S. E., Chou, W.-C., Zhai, W., Hollibaugh, J. T., Wang, Y., Zhao, P., Guo, X., Gundersen, K., Dai, M., & Gong, G.-C. (2011). Acidification of subsurface coastal waters enhanced by eutrophication. *Nature Geoscience*, *4*(11), 766–770. doi:10.1038/ngeo1297

Chen, X. A., Cai, Q., Zheng, M. G., & Li, J. (2011). Empirical soil erosion model for single rainstorm in Chabagou drainage basin. *Progress in Geography*, *30*, 325–329.

Cohen, M. J., Brown, M. T., & Shepherd, K. D. (2006). Estimating the environmental costs of soil erosion at multiple scales in Kenya using energy synthesis. *Agriculture, Ecosystems & Environment*, *114*(2-4), 249–269. doi:10.1016/j.agee.2005.10.021

Dragović, N., & Vulević, T. (2020). Soil Degradation Processes, Causes, and Assessment Approaches. In Life on Land, Encyclopedia of the UN Sustainable Development Goals. Springer Nature Switzerland AG 2020 W. doi:10.1007/978-3-319-71065-5_86-1

Duarte, C. M., Losada, I. J., Hendriks, I. E., Mazarrasa, I., & Marbà, N. (2013). The role of coastal plant communities for climate change mitigation and adaptation. *Nature Climate Change*, *3*(11), 961–968. doi:10.1038/nclimate1970

Duhita, A. D. P., Rahardjo, A. P., & Hairani, A. (2021). The Effect of Slope on the Infiltration Capacity and Erosion of Mount Merapi Slope Materials. *Journal of the Civil Engineering*, *7*(1), 71–84. doi:10.22146/jcef.58350

Edwards, W., Shipitalo, M., Owens, L., & Norton, L. (1990). Effect of *Lumbricus terrestris* L. burrows on hydrology of continuous no-till corn fields. *Geoderma*, *46*(1-3), 73–84. doi:10.1016/0016-7061(90)90008-W

FAO. (2005). Organic matter decomposition and the soil food web. In *The importance of soil organic matter: Key to drought-resistant soil and sustained food and production*. Food and Agriculture Organization of the United Nation.

Feely, R., Alin, S., Newton, J., Sabine, C., Warner, M., Devol, A., Krembs, C., & Maloy, C. (2010). The combined effects of ocean acidification, mixing, and respiration on pH and carbonate saturation in an urbanized estuary. *Estuarine, Coastal and Shelf Science*, *88*(4), 442–449. doi:10.1016/j.ecss.2010.05.004

Fierera, N., Schimela, J. P., & Holden, P. A. (2003). Variations in microbial community composition through two soil depth profiles. *Soil Biology & Biochemistry*, *35*(1), 167–176. doi:10.1016/S0038-0717(02)00251-1

Foster, D. L., Bardos, C. P., & Southgate, D. D. (1987). Soil erosion and water treatment costs. *Journal of Soil and Water Conservation*, *42*, 349–352.

Frisch, D., Morton, P. K., Chowdhury, P. R., Culver, B. W., Colbourne, J. K., Weider, L. J., & Jeyasingh, P. D. (2014). A millennial-scale chronicle of evolutionary responses to cultural eutrophication in Daphnia. *Ecology Letters*, *17*(3), 360–368. doi:10.1111/ele.12237 PMID:24400978

Hajabbasi, M., Jalalian, A. A., & Karimzadeh, H. R. (1997). Deforestation effects on soil physical and chemical properties, Lordegan, Iran. *Plant and Soil*, *190*(2), 301–308. doi:10.1023/A:1004243702208

Hansen, L. T., Breneman, V. E., Davison, C. W., & Dicken, C. W. (2002). The Cost of Soil Erosion to Downstream Navigation. *Journal of Soil and Water Conservation*, *57*(4), 205–212.

Henry, P. C. (1995). Soil as a factor in sugarcane ratoon yield decline on an irrigated estate in Swaziland. *Proc S Afr Sug Tech Ass*, *69*, 122-126.

Hicks, D. H., & Anthony, T. (2001). *Soil conservation technical handbook*. The Ministry for the Environment. http://www.coastalwiki.org

IPBES. (2018). The IPBES assessment report on land degradation and restoration. In L. Montanarella, R. Scholes, & A. Brainich (Eds.), *Secretariat of the Intergovernmental Science-Policy Platform on Biodiversity and Ecosystem Services*. doi:10.5281/zenodo.3237392

Issaka, S., & Ashrafa, M. A. (2017). Impact of soil erosion and degradation on water quality: a review. *Geology, Ecology, and Landscapes, 1*(1), 1–11. doi:10.1080/24749508.2017.1301053

Jacinthea, P. A., Lal, R., & Kimble, J. M. (2002). Carbon dioxide evolution in runoff from simulated rainfall on long-term no-till and plowed soils in southwestern Ohio. *Soil & Tillage Research*, *66*(1), 23–33. doi:10.1016/S0167-1987(02)00010-7

Kairis, O., Karavitis, C., Salvati, L., Kounalaki, A., & Kosmas, K. (2015). Exploring the Impact of Overgrazing on Soil Erosion and Land Degradation in a Dry Mediterranean Agro-Forest Landscape (Crete, Greece). *Arid Land Research and Management*, *29*(3), 360–374. doi:10.1080/15324982.2014.968691

Katra, I. (2020). Soil Erosion by Wind and Dust Emission in Semi-Arid Soils Due to Agricultural Activities. *Agronomy (Basel)*, *10*(1), 89. doi:10.3390/agronomy10010089

Katra, I., Gross, A., Swet, N., Tanner, S., Krasnov, H., & Angert, A. (2016). Substantial dust loss of bioavailable phosphorus from agricultural soils. *Scientific Reports*, *6*(1), 24736. doi:10.1038rep24736 PMID:27095629

Lal, R. (2001). World cropland soils as a source or sink for atmospheric carbon. *Advances in Agronomy*, *71*, 145–191. doi:10.1016/S0065-2113(01)71014-0

Lal, R. (2003). Soil erosion and the global carbon budget. *Environment International*, *29*(4), 437–450. doi:10.1016/S0160-4120(02)00192-7 PMID:12705941

Lal, R. (2004a). Soil carbon sequestration impacts on global climate change and food security. *Science*, *304*(5677), 1623–1627. doi:10.1126cience.1097396 PMID:15192216

Lal, R. (2004b). Soil carbon sequestration to mitigate climate change. *Geoderma*, *123*(1-2), 1–22. doi:10.1016/j.geoderma.2004.01.032

Li, H., Tatarko, J., Kucharski, M., & Dong, Z. (2015). PM2.5 and PM10 emissions from agricultural soils by wind erosion. *Aeolian Research*, *19*, 171–182. doi:10.1016/j.aeolia.2015.02.003

Li, T., Zhang, H., Wang, X., Cheng, S., Fang, H., Liu, G., & Yuan, W. (2019). Soil erosion affects variations of soil organic carbon and soil respiration along a slope in Northeast China. *Ecological Processes*, *8*(28), 1–10. doi:10.118613717-019-0184-6

Mabuhay, J. A., Nakagoshi, N., & Isagi, Y. (2004). Influence of erosion on soil microbial biomass, abundance and community diversity. *Land Degradation & Development*, *15*(2), 183–195. doi:10.1002/ldr.607

Mandal, A., Sarkar, B., Mandal, S., Vithanage, M., Patra, A. K., & Manna, M. C. (2020). Impact of agrochemicals on soil health. In Agrochemicals Detection, Treatment and Remediation: Pesticides and Chemical Fertilizers (pp. 161-187). Elsevier.

Marques, J. F. (1998). Costs of soil erosion due to their on-site and off-site effects in the area of agricultural production. *Revista Brasileira de Economia e Sociologia Rural*, *36*, 61–79.

Melzner, F., Thomsen, J., Koeve, W., Oschlies, A., Gutowska, M. A., Bange, H. W., Hansen, H. P., & Kortzinger, A. (2013). Future ocean acidification will be amplified by hypoxia in coastal habitats. *Marine Biology*, *160*(8), 1875–1888. doi:10.100700227-012-1954-1

Merritt, D., & Wohl, E. E. (2002). Processes governing hydrochory along rivers: Hydraulics, hydrology, and dispersal phenology. *Ecological Applications*, *12*(4), 1071–1087. doi:10.1890/1051-0761(2002)012[1071:PGHARH]2.0.CO;2

Montanarella, L. (2007). Climate and land degradation. In M. V. K. Sivakumar & N. Ndiang'ui (Eds.), *Trends in land degradation in Europe* (pp. 83–104). Springer.

Nenadović, S., Kljajević, L., Nenadović, M., Milanović, M., & Markovi, S. (2013). Physico-chemical soil analysis of Rudovci region. *Geonauka*, *1*(2), 1–8. doi:10.14438/gn.2013.07

Oldeman, L. R., Hakkeling, R. A., & Sombroek, W. G. (1991). *Global Assessment of Soil Degradation (GLASOD): World Map of the Status of Human-Induced Soil Degradation*. United Nations Environment Programme, International Soil Reference and Information Centre.

Orgiazzi, A., & Panagos, P. (2018). Soil biodiversity and soil erosion: It is time to get married. *Global Ecology and Biogeography*, *27*(10), 1155–1167. doi:10.1111/geb.12782

Osman, K. T. (2014). *Soil Degradation, Conversation and Remediation*. Springer Science+Business Media.

Pearl, H., & Otten, T. (2013). Harmful Cyanobacterial Blooms: Causes, Consequences, and Controls. *Microbial Ecology*, *65*(4), 995–1010. Advance online publication. doi:10.100700248-012-0159-y PMID:23314096

Poesen, J. (2018). Soil erosion in the Anthropocene: Research needs. *Earth Surface Processes and Landforms*, *43*(1), 64–84. doi:10.1002/esp.4250

Poleto, C., & Beier, E. V. (2012). Siltation and erosion processes on a tributary of Lake Itaipu due a dam reservoir. *Lakes, Reservoirs and Ponds*, *6*(2), 108–119.

Qin, B. Q., Gao, G., Zhu, G. W., Zhang, Y. L., Song, Y. Z., Tang, X. M., Xu, H., & Deng, J. M. (2013). Lake eutrophication and its ecosystem response. *Chinese Science Bulletin*, *58*(9), 961–970. doi:10.100711434-012-5560-x

Rhoades, J. L., Demyan, M. S., & Orr, B. (2011). Impacts of Deforestation and Land Cover Change on Mountain Soils in Hrazdan, Armenia. *Journal of Sustainable Forestry*, *30*(7), 677–696. doi:10.1080/10549811.2011.567965

Sharpley, A. N., Smith, S. I., Iones, O. R., Berg, W. A., & Coleman, G. A. (1992). The transport of bioavailable phosphorus in agricultural runoff. *Journal of Environmental Quality*, *21*(1), 30–35. doi:10.2134/jeq1992.00472425002100010003x

Sharpley, A. N., Weld, J. L., Beegle, D. B., Kleinman, P. J. A., Gburek, W. J., Moore, P. A., & Mullins, G. (2003). Development of phosphorus indices for nutrient management planning strategies in the United States. *Journal of Soil and Water Conservation*, *58*(3), 137–152.

Shuster, W., Subler, S., & McCoy, E. (2001). Deep-burrowing earthworm additions changed the distribution of soil organic carbon in a chisel-tilled soil. *Soil Biology & Biochemistry*, *33*(7-8), 983–996. doi:10.1016/S0038-0717(01)00002-5

Sposito, G. (2008). *The Chemistry of Soils* (2nd ed.). Oxford University Press Inc.

Swap, R., Garstang, M., Greco, S., Talbot, R., & Kållberg, P. (1992). Saharan dust in the Amazon Basin. *Tellus. Series B, Chemical and Physical Meteorology*, *44*(2), 133–149. doi:10.3402/tellusb.v44i2.15434

Telles, T. S., de Fátima Guimarães, M., & Dechen, S. C. F. (2011). The costs of soil erosion. *Revista Brasileira de Ciência do Solo*, *35*(2), 287–298. Advance online publication. doi:10.1590/S0100-06832011000200001

Telles, T. S., Dechen, S. C. F., de Souza, L. G. A., & de Fátima Guimarães, M. (2013). Valuation and assessment of soil erosion costs. *Scientia Agrícola*, *70*(3), 209–216. doi:10.1590/S0103-90162013000300010

Tóth, G., Hermann, T., Ravina da Silva, M., & Montanarella, L. (2018). Monitoring soil for sustainable development and land degradation neutrality. *Environmental Monitoring and Assessment*, *190*(57), 57. Advance online publication. doi:10.100710661-017-6415-3 PMID:29302746

UN. (2015). Transforming our world: the 2030 Agenda for Sustainable Development. Resolution adopted by the General Assembly on 25 September 2015. United Nations.

Waldbusser, G. G., Voigt, E. P., Bergschneider, H., Green, M. A., & Newel, R. I. E. (2013). Biocalcification in the Eastern Oyster (*Crassostrea virginica*) in Relation to Long-term Trends in Chesapeake Bay pH. *Estuaries and Coasts*, *34*(2), 221–231. doi:10.100712237-010-9307-0

Wallace, R. B., Baumann, H., Grear, J. S., Aller, R. C., & Gobler, C. J. (2014). Coastal ocean acidification: The other eutrophication problem. *Estuarine, Coastal and Shelf Science*, *148*, 1–13. doi:10.1016/j.ecss.2014.05.027

Wang, Q., & Batkhishig, O. (2014). Impact of Overgrazing on Semiarid Ecosystem Soil Properties: A Case Study of the Eastern Hovsgol Lake Area, Mongolia. *Journal of Ecosystem & Ecography*, *04*(01). Advance online publication. doi:10.4172/2157-7625.1000140

Wang, X., Quine, T. A., Zhang, H., Tian, G., & Yuan, W. (2019). Redistribution of soil organic carbon induced by soil erosion in the nine river basins of China. *Journal of Geophysical Research. Biogeosciences*, *124*(4), 1018–1031. doi:10.1029/2018JG004781

Weaver, D., & Summers, R. (2001). Soil factors influencing eutrophication. In A handbook for understanding and managing agricultural soils. Department of Agriculture and Food.

ADDITIONAL READING

Cowie, A. L., Waters, C. M., Garland, F., Orgill, S. E., Baumber, A., Cross, R., O'Connell, D., & Metternicht, G. (2019). Assessing resilience to underpin implementation of land degradation neutrality: A case study in the rangelands of western new south wales, Australia. *Environmental Science & Policy, 100*, 37–46. doi:10.1016/j.envsci.2019.06.002

Liniger, H., Harari, N., Lynden, G., Fleiner, R., Leeuw, J., Bai, Z., & Critchley, W. (2019). Achieving land degradation neutrality: The role of SLM knowledge in evidence-based decision-making. *Environmental Science & Policy, 94*, 123–134. doi:10.1016/j.envsci.2019.01.001

Sun, C., Hou, H., & Chen, W. (2021). Effects of vegetation cover and slope on soil erosion in the Eastern Chinese Loess Plateau under different rainfall regimes. *PeerJ, 9*(11226), e11226. Advance online publication. doi:10.7717/peerj.11226 PMID:33954047

Tsymbarovich, P., Kust, G., Kumani, M., Golosov, V., & Andreeva, O. (2020). Soil erosion: An important indicator for the assessment of land degradation neutrality in Russia. *International Soil and Water Conservation Research, 8*(4), 418–429. doi:10.1016/j.iswcr.2020.06.002

Wunder, S., & Bodle, R. (2019). Achieving land degradation neutrality in Germany: Implementation process and design of a land use change based indicator. *Environmental Science & Policy, 92*, 46–55. doi:10.1016/j.envsci.2018.09.022

KEY TERMS AND DEFINITIONS

Air Pollution: The presence in or introduction into the air of a substance with harmful or poisonous effects.

Biodiversity: The variability among living organisms from all sources, including terrestrial, marine, and other aquatic ecosystems and the ecological complexes of which they are part; this includes diversity within species, between species, and ecosystems.

Carbon Cycle: The continuous movement of carbon between different living organisms on earth and between living organisms and the environment, through natural processes like photosynthesis, respiration, and decomposition in the soil, and also the burning of fossil fuels.

Environmental Quality: A state of ecological conditions in environmental media, expressed in terms of indicators or indices related to environmental quality standards.

Eutrophication: The process by which a body of water becomes enriched in dissolved nutrients that stimulate the growth of aquatic plant life, usually resulting in the depletion of dissolved oxygen.

Soil Degradation: Element of the land degradation process which refers to a decrease in soil's productivity and quality.

Soil Erosion: Detachment and movement of topsoil or soil material from the upper part of the profile by the action of wind or running water, especially due to changes brought about by human activity (such as unsuitable or mismanaged agricultural methods).

Sustainable Development: The practice of maintaining productivity by replacing resources used with resources of equal or greater value without degrading or endangering natural biotic systems.

Chapter 7
The Role of Polyculture in Sustainable Agricultural Development and Prevention of Land Degradation

Milan Marković
https://orcid.org/0000-0002-9617-6697
Innovation Center, University of Niš, Serbia

Ivana Marjanović
Faculty of Economics, University of Niš, Serbia

ABSTRACT

The aim of the chapter is to show the possible impact of policulture farming on some determinants of sustainable agricultural development, especially from the point of view of economic viability, biodiversity, and land degradation. Increasing the area under polyculture is one of the main solutions to the present environmental problems. The key constraints are economic pressures due to the question of the cost-effectiveness of such a mode of production and the need to provide sufficient food for a growing population, especially in developing countries. The results of the research show that policulture (organic agriculture) should be favored, while monoculture farming must be adequately directed and put in the function of achieving ecological goals of sustainable development as much as possible. In addition, on the example of European countries, it was assessed that there are good conditions for further "greening" of agriculture, bearing in mind the movement of the analyzed indicators.

INTRODUCTION

The agricultural sector faces many challenges. The main one is the need to specialize production without disturbing the ecological balance. Soil pollution, biodiversity loss and land erosion are leading problems,

DOI: 10.4018/978-1-7998-8459-0.ch007

especially in monoculture production (Alcon, 2020). Therefore, economic and environmental goals should be harmonized through the principles of sustainable development.

At the global level, the risk of biodiversity loss, abandonment of agricultural land and its degradation has become one of the important issues of sustainable agricultural and rural development. Also, a responsible attitude towards soil fertility is important for overall sustainable economic development. Without the major resource on which agriculture is based, its long-term sustainable development will not be possible. Soil erosion and excessive use of chemicals are particularly dangerous for the preservation of agricultural land. Land is a renewable natural resource. However, restoration is much slower than land degradation and fertility loss (Kelam, 2016).

Monoculture is a form of production that is often seen in collision with the principles of sustainability. This is the practice of growing the same culture in the same field, from year to year. The risks of monoculture farming arise because the obtained products are used mainly in industrial production. Intensive agricultural practices such as monoculture and simplified two-crop rotation systems threaten ecosystem services essential for crop production, reduce agricultural productivity and cause adverse effects on the environment (Wang et al., 2021). Furthermore, monoculture represents one of the practices of intensive agricultural production, which endangers land quality and leads to land degradation. Due to the industrialization of agricultural production, which includes mechanization of farms, intensive use of pesticides and fertilizers, specialization of farms and improved transport network, there has been a spatial concentration of certain types of crops and loss of crop diversity (Crossley et al., 2020). This production depletes nutrients from the soil making it less productive, reduces organic matter in the soil and can cause significant erosion. In addition to soil degradation, monocultures have a detrimental effect on surface and groundwater, as they can cause contamination due to runoff of nutrients and agricultural chemicals (Wang et al., 2021). Monoculture production is in conflict with certain economic principles too. It is clear that above-average profitability can be achieved in this way, but it is not in line with the principles of risk diversification.

The aim of the chapter is to provide insight into the possibilities of polyculture production in achieving the goals of sustainable development and preventing the loss of land quality. The secondary goal of this paper is to point out the potential shortcomings of monoculture, and to provide potential solutions so that it would not be considered an alternative to polyculture farming. The chapter provides answers to some open questions and discussions in the scientific literature related to this topic. At the same time, data at the level of the European Union are used to assess sustainable agricultural development.

BACKGROUND: AGRICULTURAL SUSTAINABILITY AS AN INTEGRAL PART OF SUSTAINABLE SOCIO-ECONOMIC DEVELOPMENT

The considerable importance of agriculture in the socio-economic development of European countries can be viewed from several aspects. From the economic aspect, the benefit of agriculture is reflected primarily in its role in food production but also in ensuring food security, production of raw materials for other industries, and foreign trade. From the social aspect, the agricultural sector affects the demographic and social development of a country. The development of the agricultural sector enables the creation of conditions reduce poverty in the country. Christiaensen et al. (2011) emphasise key factors that determine the ability of a particular sector to contribute to poverty reduction: (i) the extent to which poor people participate in the sector; (ii) the relative importance of that sector in the economy; (iii) growth and growth potential of the sector; and (iv) indirect impact of the sector on growth in other sectors. Numerous studies have proven the great power of the agricultural sector in reducing poverty (Liu et al., 2020; Warr & Suphannachart, 2021), which in many cases surpasses other economic sectors (Christiaensen et al., 2011). Having in mind the importance of agriculture, it is recommended to include and analyse the effect of the agricultural sector in growth strategies, where the emphasis is placed on the analysis of the role of increasing agricultural productivity. The mechanism behind the impact of agricultural productivity on poverty reduction is as follows: as labour productivity increases, the amount of food produced per worker increases, which causes falling prices, withdraws real wages, reduces absolute poverty and allows farmers to shift to more productive sectors, thus increasing overall productivity (Dorinet et al., 2021).

Given the availability of significant natural and human resources, agricultural production is one of the most important economic activities of many European countries. Agriculture is a sector that is almost completely supported at the European level, unlike most other economic sectors for which national governments are responsible. The primary reason for forming a unified policy at the European Union level is the fact that it is necessary to ensure the existence of a uniform public policy whose basic responsibility is to enable food security, responsible and rational use of natural resources and encourage the economic development of rural areas.

The Common Agricultural Policy of the European Union (CAP) was launched in 1962 and represents a partnership between agriculture and society, and between Europe and its farmers. Its main objectives are (European Commission [EC], 2021a):

- to support farmers and improve agricultural productivity, while ensuring a stable supply of affordable food;
- to protect the farmers of the European Union and to provide them with opportunities to earn an adequate living;
- help tackle climate change and the sustainable management of natural resources;
- maintain rural areas and landscapes across the European Union;
- keep the rural economy alive by promoting jobs in farming, agri-foods industries and associated sectors.

The CAP has undergone numerous reforms over the years. However, today, after almost 60 years of implementation, it faces perhaps the most significant challenges for enabling sustainable food production that would help feed the world's population, whose significant growth is expected in the future. Another challenges pertain to responding to the climate change and providing sustainable management of natural resources, as well as protecting the countryside throughout the European Union and maintaintaining the rural economy (Pe'er et al., 2020). In 2018 the European Commission has proposed the new, reformed CAP aimed at providing the conditions to achieve the objectives of the European Green Deal. The reformed CAP should incentivise, empower and reinforce European farmers, helping them contribute more resolutely to tackling climate change, protecting the environment and moving to more sustainable and resilient food systems (EC, 2021b). With planned implementation in 2023, the new CAP reform aims to create and support the conditions for the transition to more sustainable agriculture with a special focus on climate and the environment.

Given the growing problems facing the agricultural sector in recent years, it is not surprising that one of the strategic directions of the reformed CAP is related to achieving the goals of the European Green Deal. Contemporary tendencies, primarily related to demographic trends globally, have led to increased pressures on the agricultural sector. Specifically, meeting the growing needs of the world's population has led to the expansion and intensification of agriculture. Yet, these processes are key determinants of biodiversity loss as well as climate change. Generally, certain agricultural practices had favoured biodiversity. However, those practices have gradually been halted or substituted by agricultural systems that capitalize on yields through unsustainable use of natural resources and to the detriment of biodiversity (Pe'er et al., 2020). The consequence of the dynamic growth of the world's population in the last hundred years is an increased demand for food, which leads to increased pressures on the agricultural sector and more intensive agricultural production, resulting in the conversion of new land into agricultural, which as a consequence has the degradation of the environment (Kolasa-Więcek, 2018).

In recent years, a number of challenges have been imposed on the agricultural sector. Not only must the agricultural sector produce more food for humans, animals and other raw materials to meet the needs of a growing human population, but it must also contribute to the socio-economic prosperity, while ensuring the protection of natural resources (Trivino-Tarradas et al., 2019).

Due to the need to transform the agricultural sector in accordance with the principles of sustainable development, more and more developed and developing countries are facing various environmental and socio-economic problems. Implementation of the concept of sustainable development in the context of agriculture inevitably implies consideration and creation of appropriate strategies for balancing crop and livestock production with protection of the quality of national soil and water resources, as well as consideration of socio-economic consequences of such activities (Wilson & Maliszewska-Kordybach, 2000). The implementation of sustainability in agriculture has led to the emergence of the concept of agricultural sustainability.

Sustainable development in the agricultural sector requires the establishment of food and fiber production and processing practices that protect or enhance natural resources, that are compatible with the environment, contribute to the socio-economic well-being of all people, ensure a secure and high-quality agricultural supply, and provide well-being of agricultural and agri-food enterprises, workers and their families (McRae et al., 2000). Agricultural sustainability is a comprehensive priority of the European Union, which is reflected in the preservation of ecosystems, improvement of biodiversity, preservation of water and soil quality in ecosystems related to agriculture (Trivino-Tarradas et al., 2019). Hansen (1996) states that agricultural sustainability denotes meeting the needs of the present without compromising on the needs of the future generations with regard to food, feed and fibre production. Agricultural sustainability is a narrower concept than environmental sustainability; however, it is an important element of environmental sustainability. The main contributions of sustainable agriculture practice to the preservation of the environment are reflected in maintaining soil quality, reducing erosion, and preserving water.

Agricultural sustainability is a multidimensional concept that requires consideration of environmental, economic and social factors. Since the achievement of agricultural sustainability is an integral part of the strategies of most countries in the world, it is desirable to monitor appropriate indicators to determine the level and trends of agricultural sustainability of an individual country. Numerous indicators for monitoring agricultural sustainability have been developed. Some of them are based on analysing spatial dimension (Sabiha et al., 2016; Trivino-Tarradas et al., 2019), while others evaluate agricultural sustainability at the local level (Wezel et al., 2014; Rao et al., 2019). Three groups of agri-environmental indicators used for measuring agricultural sustainability can be identified (Bockstaller et al., 2008): (i) indicators based on the use of one type of variable obtained by survey, databases and not directly measured; (ii) indicators based on calculation; and (iii) indicators based on one or several measurements. In recent years, the application of multicriteria analysis methods when creating composite sustainability indices has also become important (Diaz-Balteiro et al., 2017; Stanković et al., 2021a; Stanković et al., 2021b; Stanković et al., 2021c) and represents a methodology that can significantly contribute to measuring and monitoring agricultural sustainability (Talukder et al., 2018; Tzouramani et al., 2020). The purpose of agri-environmental indicators is to (McRae et al., 2000):

- inform agricultural and other decision-makers about environmental performance in agriculture;
- determine how environmental conditions within agriculture are changing over time;
- provide information on the impact of the adoption of stewardship principles and the use of environmentally sound practices;
- support the development of strategies and actions targeted at areas and resources that remain at environmental risk; and
- facilitate the environmental analysis of agricultural policies and programs and the monitoring of their performance.

The European Commission has taken a number of actions to align with the UN Sustainable Development Goals in order to enhance climate change mitigation and foster sustainable agricultural practices (Schiavon et al., 2021). Reducing and mitigating land degradation is an integral part of the United Nations Sustainable Development Goals for 2030 (SDG), primarily SDG15 Protect, restore and promote sustainable use of terrestrial ecosystems, sustainably manage forests, combat desertification, and halt and reverse land degradation and halt biodiversity loss. However, land use care is also part of several other goals, e.g., SDG02 End hunger, achieve food security and improved nutrition, and promote sustainable agriculture, as well as SDG 11 Make cities and human settlements inclusive, safe, resilient and sustainable.

Consequences of Inadequate Agricultural Practices

Agriculture is often identified as one of the greatest threats to biodiversity, as cultivated land requires more water, increases soil erosion and water pollution, preserves less carbon, emits more greenhouse gases and maintenances considerably less habitat and biodiversity than the ecosystems it replaces (Zhang et al., 2011). Although agriculture is an economic sector that provides multiple benefits, such as food security and the provision of resources for other sectors, it also generates negative externalities, such as air, water and soil degradation (Adegbeye et al., 2020). It is evident that there is an interrelationship between agriculture and the natural environment; however, the nature of that connection is complex. Despite being one of the most sensitive sectors when it comes to the negative effects of climate change, the agricultural sector is simultaneously one of the main causes of climate change (Agovino et al., 2019). Although agriculture contributes to the maintenance of valuable habitats in Europe, intensive agricultural activity and inappropriate agricultural practices can lead to depletion of natural resources, create negative impacts on the ecosystem and worsen the impact of climate change (Bosco & Simeoni, 2018).

Almost half of local non-CO2 greenhouse gas emissions in the European Union are generated by agriculture and almost half of land for economic use is agricultural land (Renner et al., 2020). In addition to greenhouse gas emissions, one of the problems facing agricultural production to ensure sustainability is land degradation. The use of agrochemical inputs, such as fertilizers and pesticides, has intensified in recent decades to increase production in response to a number of market and economic factors, but has consequently led to an increase in soil and water pollution (Expósito & Velasco, 2020).

Land worldwide faces worldwide growing threats due to population growth, economic development and climate change (Montanarella et al., 2016). Bearing in mind that a significant part of food is produced on land, while slightly more than 1% of food is obtained from water systems, it can be concluded that land degradation also poses a serious threat to food safety (Pimentel, 2006). As some projections predict that the world's population will reach over 9 billion in 2050, food security for that population requires a significant increase in world crop production in the coming decades (Foley et al., 2011). In response to the growing needs of the population, it is assumed that there will be an expansion of agricultural land to the detriment of natural ecosystems. Another unfavorable trend is unsustainable agricultural intensification that can accelerate land degradation, which currently affects many productive lands around the world (Prăvălie, 2021).

Land degradation can be defined as a loss of the land system's biological and/or economic resilience and adaptive capacity (Holling, 2001). Land degradation and environmental sustainability are interrelated processes. While environmental sustainability is determined by the resilience and solidity of the exploited resources, their sensitivity to modification, and the ability of a system to adjust to change, land degradation arises when the resilience and adaptive capacity of the land are compromised (Baartman et al., 2007). Over 75% of the world's soil is facing moderate to high degradation, and this could surpass 90% by 2050, which threatens global food supplies, increases carbon emissions and conditions potential migratory movements due to the abandonment of infertile soil (Perović et al., 2021). Soil degradation is the process of reducing the quality of the soil and its ability to support animals and plants. An integral part of soil degradation is soil erosion that occurs when the topsoil and nutrients are lost either naturally (for example, due to wind erosion), or due to human actions (such as poor land management). Although soil degradation can occur as a natural process, it has been considerably accelerated in the last few decades by intensive agricultural practices such as deforestation, intensive tillage, overgrazing, and forest fires.

Baartman et al. (2007) have listed several land degradation characteristics: (i) it is a human-induced phenomenon that cannot be caused by natural processes alone; (ii) land degradation decreases the capacity of the land system to meet its user demands; and (iii) land degradation threatens the long-term biological and/or economic resilience and adaptive capacity of the ecosystem.

The United Nations (UN, 2013) have issued a document regarding the importance of soil sustainability and responsible land management according to which (Gomiero, 2016):

- soil sustainability is a prerequisite for meeting the needs of a growing population,
- sustainable land management can contribute to the maintenance of healthy land, which in turn contributes to achieving food security for the world's population, as well as ensuring environmental sustainability,
- good land management generates a number of socio-economic benefits, which are reflected in the contribution to economic growth, biodiversity conservation, sustainable agriculture and food security and poverty reduction,
- issues such as climate change, water availability, land degradation and desertification, and droughts are the most important challenges facing the world and to be addressed as soon as possible,
- raising awareness about the sustainable use of limited resources is an important task in order to achieve sustainability.

European Commission identifies several causes of land and soil degradation (EC, 2021c): poor land management, such as unsustainable farming and forestry practices, land use change, construction activities and soil sealing, as well as pollution from industrial emissions to the atmosphere, poor waste management or contaminants present in fertilisers or sewage sludge that are applied to the soil. Land degradation negatively affects 3.2 billion people and causes an economic loss of about 10% of the annual global gross product (Scholes et al., 2018). As a phenomenon closely related to soil degradation, soil erosion generates both direct and indirect costs (Panagos et al., 2018). Direct costs are reflected in the loss of fertile land and are borne by farmers, since due to soil erosion, they have losses in production, reduced yields, as well as a reduction in arable land. Indirect costs are borne by society and are reflected in floods, landslides, water pollution, habitat loss and damage to biodiversity.

Bearing in mind the consequences of poor land management, attention is given to the consequences of intensive agricultural production practices on the long-term sustainability of crops and to impending environmental costs reflected in a shortfall of wildlife habitat, diminishing water quality, degraded biodiversity and greenhouse gas emissions (McRae et al., 2000).

Since monoculture is an agricultural practice that can jeopardize agricultural sustainability, and thus the environmental sustainability, more details about the practice of monoculture will be provided in the next section. One of the basic risks of monoculture is the reduction of the quality and degradation of the soil, which undoubtedly leads to its erosion in the future.

MONOCULTURE VS. POLYCULTURE

Monoculture is the production of only a certain type of product (usually from plant production) on a certain soil over a long period of time. Monoculture farming mainly means plant production, although it can also refer to livestock production. In the second case, only one breed of cattle is raised on certain land and in a certain period of time.

Products made from monoculture can meet the growing demand for food. Higher demand exists due to the rapid increase in the world population. Monoculture is just trying to solve this problem. At the same time, rice and wheat, as the most important food sources, are the most common crops covered by monoculture farming. As the majority of the world's population lives in developing and underdeveloped countries, it is imperative to provide affordable food prices on a global level, especially given the global trend of increasing prices of agricultural and food products due to market monopolization (Radukić & Marković, 2015).

The economic benefits of monoculture, which make the products obtained from this production cheaper, should not be overlooked. The best reason for growing one type of plant on larger arable land is to reduce costs. The savings can be more efficient: tillage with agricultural machinery, the use of pesticides, herbicides, as well as the final harvest. There is a reduced need for labor, which affects the reduction of fixed costs per product unit.

In monoculture, it is possible to apply some solutions of modern agriculture without higher costs. The concept of smart agriculture mainly finds its application in this form of production. The use of drones, meteorological stations, as well as sensors that are placed in the ground is significantly facilitated in that case. Monitoring plant development via drones and some forms of satellite monitoring can reduce the use of pesticides in monoculture-based agricultural production. Also, as the application of modern information and communication technology carries with it high fixed costs (Rađenović et al., 2020), they can be significantly reduced per unit of product. This facilitates the control of plant diseases; there is no delay in applying certain pesticides, while the constant monitoring of crops is at an exceptional level because there is an up-to-date early warning system for infections. These systems use numerous parameters that make it easier for managers to act with the help of modern software solutions. Air temperature, humidity, moisture on the leaves, soil temperature determine the probability of crop infection and optimize the use of fertilizers. This has a twofold effect: reducing production costs and protecting the environment, and greening agricultural production by reducing pesticide consumption. Lower production costs are directly correlated with lower market prices that can be expected in this case.

The essence of the increasingly frequent land consolidation processes is in favor of the aspiration for the existence of monocultures. This provides higher productivity, economic efficiency, yield, as well as the easier process of managing the entire production (the same method of pruning, the same planting period, the same harvest period, uniform application of the same or similar chemicals, the same climate, rainfall and number of hours when the sun is active). The specialization of production leads to a radical reduction of costs so that the increased free funds can be used for new investments. Monitoring crop development is much simpler and requires less time. However, the crop that will be most profitable given the existing climatic and other natural conditions and resources must be selected. On the other hand, without crop insurance, the risks of production on large plantations increase significantly.

The economic effects of adequate management of monoculture farming will be drastically higher than polyculture if the appropriate crop, suitable land, and correct application of mechanical and chemical means are chosen. Also, a lot depends on the climatic conditions in that area (average amount of precipitation, average daily exposure to the sun or light, number of days with frost, terrain disposition, etc.). Therefore, before planting, the soil that is most suitable for a specific plant culture must be carefully selected. Since in that case, there is no possibility of risk diversification, it is necessary to manage the production process well. Producers who do not diversify their production will face many challenges if there is a price disturbance in a particular cultivated crop market.

All of this would be ideal if there were no other side to the coin. These are potential disadvantages of this form of agriculture. In the first place, it is about the quality of the land that remains after the end of the production cycle. The composition of the soil is becoming evidently poorer. As the reduction of necessary minerals and other elements in the soil occurs, the question arises of whether the initial richness of the soil will be maintained. Namely, pests and parasites will multiply more easily in the soil where the same type of agricultural crop has been planted for many years. On the other hand, biological pest control, as well as the regulation of the entire ecosystem, are significantly more present in non-specialized production (Klasen et al., 2016). In polyculture, there are often certain wild plants that can regulate the appearance of parasites. In addition, farmers can sow certain plants themselves to protect the main crops. Sustainable agriculture often means replacing a good portion of conventional chemicals with plants. For example, parsley is suitable in vineyards because it can protect the vine from various diseases, garlic if planted between rows of strawberries has a positive effect on the firmness and size of strawberries, chamomile found among cereals also favors the development and resistance of some cereals, etc. (Đekić, 2010).

On the opposite side of monoculture production is polyculture. Such production is environmentally sustainable, but it can bring an economic loss to agricultural producers. At the level of the national economy, it can cause inefficient management of farms, as well as an increase in the price of basic foodstuffs. It dominates on smaller parcels that most often appeared due to decades of inheritance by several owners and division of land. This is inevitably associated with higher labor costs. In polyculture, lands of the same owners are often distant from each other, which leads to unprofitable farming.

Table 1. Distinguishing between monoculture and polyculture farming

Characteristics	Monoculture	Polyculture
Production risk	Increased, but also the possibility of making significant profits is higher	Decreased
Production costs	Significantly lower	Significantly higher
Risk diversification	Impossible due to production specialization, crop insurance is necessary, which increases costs	Possible
Profitability of investments	Significant savings are made, so large investments are possible	Investments are often unprofitable, which makes it impossible to improve the production process
Sustainability of production	Only possible with adequate management of production processes	Possible
Danger of soil erosion	There is a danger	There is reduced risk, except in areas where there is excessive deforestation and conversion of forest to agricultural land
Use of phytosanitary preparations	Usually increased	Usually reduced
Biodiversity conservation	There is a risk of disturbing natural biodiversity	There are good conditions for preserving agrobiodiversity, especially in organic farming systems
Preservation of the landscape	Not possible	Possible

Source: (Authors' contribution)

Table 1 shows the distinction between monoculture and polyculture based on some basic parameters, such as: production risk, production costs, risk diversification, the profitability of investments, production sustainability, risk of soil erosion, use of pesticides and other phytosanitary preparations, biodiversity conservation and preservation of the agricultural landscape.

The problem of increased risk in monoculture farming has already been highlighted. In contrast, risk diversification stands out as an advantage in polyculture, because the yield and market price of certain crops can compensate for the negative circumstances of yield decline due to natural or some other factors, as well as market price disturbances or changes in demand. However, traditional agriculture increases the costs of organizing production since the landholdings are reduced. Each crop has its own specific requirements, which brings with it increased certain costs. These costs can be drastically higher than in monoculture due to:

- higher labor costs,
- use of different types of pesticides,
- use of a different method of land cultivation,
- higher costs of agricultural machinery per unit of product.

Diversification of production, in addition to the impact on reducing risk, can also have effects on increasing yields. For example, the diversity of plant species on grasslands affects the increase in milk production and, consequently, increases farmers' incomes (Schaub et al., 2020).

Soil erosion stands out as one of the problems with monoculture. In fact, monoculture land can quickly become susceptible to erosion. This will reduce the fertility of the land, but also prevent crop production in the future. Soil erosion and pollution are especially present if there is intensive and irrational use of land, as the main natural resource needed for agricultural activity. In that case, arable agricultural land may lose its purpose. It is known that soils with the same crops and the same types of the root system are primarily exposed to soil erosion. This prevents the natural balance of the land, especially if there is deforestation of certain forests and trees in larger areas. Soil erosion often occurs due to disturbances of moisture and the amount of water in the soil due to the inadequate irrigation process on larger parcels.

An additional problem for land degradation is created by excessive deforestation. Deforestation is directly related to soil erosion processes. It is especially dangerous when it is done on a large area of land growing certain cereals with the application of pesticides or for the purpose of converting agricultural land into construction land for housing (in suburban areas or weekend zones) or industrial activity. On the example of European countries, there are positive trends, bearing in mind that there has been a constant increase in the share of forest areas in total land. Since 2009, from which data are available, there has been a permanent increase in forest areas, which is fully in line with the proclaimed goals of sustainable development. In the countries of the European Union, the share of forests in the total area is about 43%, according to the latest available data (Eurostat, 2021). There is evidence in the literature that agroforestry increases biodiversity, soil quality and enables landscape heterogeneity (Beillouin et al., 2019), so that afforestation, in the context of soil conservation and sustainable development, achieves a number of benefits for socio-economic development.

A special concern in monoculture is the ecological dimension of sustainable development. Loss of soil fertility exists primarily because the same plants on a larger agricultural area prevent beneficial microorganisms from multiplying and surviving in such a disturbed environment. The reduction of land fertility affects the accelerated degradation of land, which raises the question of the possibility of making a profit in the future.

As monoculture farming grows the same crops in a certain area, the quality of these products may decrease over time. In addition, the inability to adequately control pests creates additional pressure for farmers to use pesticides. Such intensive agriculture production associated with increased pesticide use can often mean a reduction in of products' quality and nutritional properties..

Table 2 shows the change in sales of pesticides (in kilograms) in the countries of the European Union in percentages compared to 2011. The data indicate a declining trend in the use of almost all forms of pesticides except Insecticides and acaricides. Sales of Molluscicides were particularly reduced. The slowest decline in sales of Plant growth regulators was recorded. It is a change that is in line with the proclaimed goals written in sustainable development strategies.

In addition, monoculture farming can impair the quality of the environment as well as natural biodiversity. A richer ecosystem usually means the diversity of the living world, which includes plants, animals and microorganisms. For example, as the most important pollinators, bees look for different plants in order to improve their immunity and ultimately survive. The potential danger arises from the fact that such production calls into question the achievement of the environmental goals of agriculture. Therefore, space is opened for violating the goal of biodiversity conservation, which is highly ranked in the sustainable development strategies of many countries.

All these problems associated with the loss of biodiversity cause the loss of landscape and land degradation. Among them are: intensive agriculture, land abandonment, fires due to human negligence, expansion of urban areas, as well as the accelerated emergence of commercial tourist destinations (Bajocco et al., 2012). The loss of the natural landscape is often considered inevitable since conventional industrial agriculture is often necessary for the nutrition of the population and the economic development

Table 2. Sales of pesticides in the countries of the European Union (as a percentage compared to 2011 as a reference year)

PESTICIDE	2015.	2019.
Fungicides and bactericides	+2,95%	-11,78%
Herbicides, haulm destructors and moss killers	-2,66%	-13,53%
Insecticides and acaricides	+1,93%	+26,11%
Molluscicides	+63,14%	-43,68%
Plant growth regulators	+15,60%	-10,79%
Other plant protection products	-13,47%	-31,29%

Source: (Eurostat, 2021)
Note: Countries for which data were not available for all years were not taken into analysis

of poor countries (Russo Lopes et al, 2021). However, on the other hand, there are huge risks to the environment, land and other natural resources. Given that agricultural landscapes have become increasingly monotonous, farmers who will adequately manage the diversification of production should play a key role in order to prevent the loss of the landscape (Peltonen-Sainio et al., 2020).

Organic Agriculture and Sustainable Development

Profitability of production is the goal of every economic entity in a market economy. However, in the context of achieving the goals of sustainable development, economic goals are a necessary condition but not a sufficient condition. In fact, everything that is economically justified does not have to be desirable for sustainable socio-economic development. That is why environmental goals are playing an increasingly important role. Market-oriented production can often lead to environmental degradation and, in the long run, prevent further development due to overexploitation of natural resources. The degradation of natural resources on which agriculture rests will affect the impossibility of dynamic growth of agricultural production. As the main part of agricultural production is carried out in the open, it is necessary to adequately manage the environment in order to protect it.

Organic agriculture is a form of sustainable agriculture (Marković, 2018b). This method of farming implies the principles of good production practice, and the preservation of biodiversity and agricultural resources. The ultimate goal is to get healthy, quality and environmentally friendly products. Given the growing demand for such products from consumers, it can provide producers with significant economic benefits. The prices of organic products can be several times higher than conventional agricultural products. In addition, the price of organic food is far more stable, with a tendency to constantly increase, especially processed products from organic farming. Food products of organic origin and with little added value can bring significantly higher incomes for farmers. Organic farming also has its social significance because it affects the development of entire rural areas, both from an economic point of view and from the point of view of increasing arable land and preventing land degradation. Sustainable, organic agriculture means a process in which quality products will come out as an output. Such production strives to preserve the environment and natural resources through their rational use, while respecting certain ecological principles.

The concept of sustainable agriculture enables adequate land management, tillage, crop rotation, as well as the use of chemical preparations in order to achieve stable agricultural production, but with the preservation of natural resources (Turek, 2013). The goals of sustainable agriculture are (Marković, 2018a, p. 413):

- ensuring food security of the population,
- maintaining the economic value of production, and
- preserving the existing and improving the future quality of the environment and natural resources.

Globally, more and more large farmers are shifting to organic farming, both because of significant environmental incentives and because of the market price of these products. Thus, the areas under organic production in the European Union are increasing. Compared to 2013, in 2019, there was an increase of 2 percentage points (Eurostat, 2021). According to the latest available data, the share of organic agriculture in the total utilized agricultural area in the European Union is about 8%. As the use of chemical preparations in monoculture can significantly disrupt the existing biodiversity, so agrobiodiversity is one of the most prominent goals of organic agriculture. Such diversity of flora and fauna encourages diversity and is important for agricultural landscapes.

Avoiding the negative effects of unsustainable agricultural practices can be achieved by using regenerative agricultural techniques focused on building soil health through ecosystem-focused techniques. Achieving sustainable agricultural production involves the use of various techniques to create and maintain a healthy soil structure, rich in nutrients, without the use of synthetic fertilizers or pesticides. These techniques include composting, crop rotation, no-till or low-till techniques, zero pesticide use and sustainable pest management techniques and similar (Sherwood & Uphoff, 2000).

MONOCULTURE FARMING MANAGEMENT FOR SUSTAINABLE DEVELOPMENT: SOLUTIONS AND RECOMMENDATIONS

The above clearly indicates that the practice of monoculture is one of the leading causes of land degradation. Land degradation is a reversible process but can become an irreversible phenomenon of desertification when it is accompanied by deteriorating climate conditions, and also can lead to the degradation of ecosystems that communicate with the soil layer, such as groundwater and atmosphere (Salvati & Zitti, 2009). Potential solutions for solving the problem of land degradation and desertification are different and diverse and depend on the specific situation. The most general classification of measures includes (Bowyer et al., 2009): (i) Measures to avoid and minimize land degradation (prevention measures and those aimed at increasing soil resilience); (ii) Measures to reduce the impact of ongoing land degradation processes; (iii) Measures for recovery, remediation and management of degraded land.

However, there is simply no one-size-fits-all solution to the problem of land degradation. The problem is complex, caused by numerous factors and processes, and it is necessary to adequately analyze the reasons for the occurrence of land degradation, and create an appropriate response. Potential solutions must be flexible and adapted to the given circumstances. Certain practices can be applied in order to prevent, mitigate or even avoid soil degradation. The simplest way is to leave the vegetation on the ground so that the nutrients can return to the soil. In addition, it is possible to educate stakeholders on sustainable practices in order to preserve biodiversity. Education about the harmful effects of monoculture practice is extremely important. Possible recommendations for tackling the issue of land degradation can be systematized as follows (Baartman et al., 2007):

- Using incentives, financial (subsidy) or material (say in the form of seeds or other material), to encourage local actors to apply practices that help preserve the soil;
- Raising awareness of the importance of land conservation, reducing land degradation and implementing sustainable agricultural practices;
- Education and training of farmers and other stakeholders on the benefits of implementing sustainable agricultural practices;
- Giving greater security and responsibility to land tenants, in terms of providing the possibility of buying land after a certain number of years of use;
- Encouraging research related to new sustainable agricultural practices.

Addressing land degradation involves systemic changes at the macroeconomic level, including joint efforts to improve the sustainability of both production systems and consumer lifestyles, while fostering a socio-economic environment conducive to low population growth rates and per capita consumption (Scholes et al., 2018). Often, policies related to the problem of land degradation are adopted after the problem occurs, in terms of land mitigation measures, which do not bring adequate results. The focus of policymakers should be on ex-ante policies that will not only mitigate but also contribute to avoiding land degradation. Evading, reducing and reversing land degradation is key to human well-being, as short-term gains from unsustainable land management often translate into long-term losses, and on average, the benefits of land restoration are ten times greater than the costs of land degradation (Scholes et al., 2018).

Disruption of natural biodiversity and excessive exploitation of natural resources are the main risks faced by the concept of monoculture. Therefore, in order to protect ecosystems in agriculture, sustainable agriculture systems are being applied in practice. One of the main forms of sustainable agriculture is organic agriculture, which aims to increase agrobiodiversity. In addition, the goal is to obtain quality and health-safe products whose production is strictly controlled. Having in mind the respect of ecological principles, this production strives to protect the quality of the environment. Increasing the area under polyculture is one of the main solutions to the present environmental problems and improving the function of the ecosystem. The key constraint is economic pressures due to the cost-effectiveness of such a mode of production and the provision of sufficient food for the market.

In parallel with favoring organic production, it is necessary to adequately manage monoculture farming, bearing in mind its potential economic benefits. It is primarily about increased economic and production efficiency, bearing in mind the effects of economies of scale. Industrial agriculture should not completely replace organic food production, which is present mainly in small, local areas. Thanks to this production, a large number of the population in rural and suburban areas is fed. Although it carries with it certain risks, monoculture should not be suppressed, but adequately directed. Industry-oriented monoculture production must not conflict with the environmental and social goals of sustainable economic development. Otherwise, after a certain period of time, the process of soil erosion will begin, which will endanger the basic agricultural resource. A certain compromise must be made between the economic benefits of production specialization and the environmental costs of monoculture farming (Klasen, 2016).

However, it should be noted that adequate management of monoculture practices can lead to a reduction and gradually to a cessation of land degradation. For example, there are certain crops that are suitable for planting in monoculture, such as corn or wheat. However, it is necessary to rotate such crops annually on specific land, so managers must keep this in mind in order to maximize yields. In this way, through crop rotation, the negative effects of monoculture on land and biodiversity can be mitigated. Some plant species (e.g., parsley, onion) often fail if they are planted again in the same field a year after harvest. In some cases, such a practice is needed that in a period of several years, the same culture must not be planted on the same land, and even sometimes such planting must be several hundred meters away from the previous one. In practice, the case of rotation of corn and wheat planting is common.

In polyculture, it is necessary for the farmer to organize production, planting schedule, maintenance time and harvest in a good way. As unit costs are often increased, it is necessary to unite farmers, not only horizontally, but also vertically. The ultimate goal is to create strong rural family farms. This is because small farmers often have limited access to agroeconomic and market information, as well as a low ability to take significant risks (Van Zonneveld et al., 2020).

Greening of agriculture, encouraging biodiversity, as well as reduced use of pesticides are solutions to the growing problems of preventing land degradation and sustainable agricultural development. In the European Union, through various reforms of the common agricultural policy since 1992, the aim was to ensure self-sufficiency in food, but also to reduce soil pollution, preserve animal welfare standards, as well as food safety (Marković, 2015). Various programs for greening agricultural production have been initiated, in which farmers have often been conditioned to receive certain payments to apply good agricultural practice from an environmental point of view. The key goal was to prevent soil erosion, land degradation and respect principles of sustainable development. The data showed that the European Union is going in the right direction, having in mind the increase of areas under organic production, reduction of the use of almost all types of pesticides, as well as the increase of areas under forest land.

FUTURE RESEARCH DIRECTIONS

Further research in this area can be directed towards creating a composite index that would initially enable monitoring of the level of land degradation. In contrast, further improvement and inclusion of other indicators would make it possible to create a composite index of agricultural sustainability.

The development of composite indices assumes the application of methodology suitable for quantifying regional differences in terms of land quality. The stated methodology should be able to evaluate the complex phenomena of land degradation from the point of view of several relevant criteria, by summarizing individual indicators into a composite index. Composite indices are widely used in research of various social, environmental and economic phenomena, whenever it is necessary to observe more variables to obtain a complex assessment. Such indices are used as a means for benchmarking countries, regions or areas in many domains. The growing number of constructed composite indices in the world is an obvious sign of their importance and usefulness in policy-making. These indices are increasingly recognized as a useful tool in the analysis of economic policy and public communication. In order to avoid large amounts of data and indicators when creating development strategies or policies, policymakers will certainly benefit if this data is replaced by a composite index. It is often easier to interpret the composite index to the general public than to identify common trends in a large number of specific indicators.

The need to create composite indices arises in situations when individual indicators cannot adequately represent a complex or multidimensional concept, such as in this case land degradation or agricultural sustainability. With the help of such an index, it is possible to compare areas, regions and countries by taking into account several different dimensions or performances (for example, environmental, economic and social dimensions of agricultural sustainability) at the same time. However, composite indices can give a deceptive assessment of a phenomenon if they are not adequately constructed. The quality of a composite index depends not only on the quality of the data but also on the methodology used to construct it. One of the scientifically accepted methodologies, which leads to the creation of non-compensatory or semi-compensatory composite indices, is multicriteria analysis. The application of multicriteria analysis methods to create an index for monitoring land degradation and agricultural sustainability can be extremely useful since multicriteria analysis is used to evaluate complex, multidimensional phenomena, such as the concept of land degradation, but also agricultural sustainability. Therefore, the focus of future research in this area will be primarily on developing a model for assessing land degradation, with the aim of further improving the model for assessing agricultural sustainability.

CONCLUSION

Easier process management, as well as lower resource use costs, are cited as the main benefits of monoculture. The costs of using the machines are lower because of the same irrigation conditions, the same pesticides are needed, as well as a smaller number of agricultural machinery. However, at the national level, monoculture farming can often require increased costs due to environmental pollution. In addition, it poses a greater risk to small farmers. Therefore, the specialization of production is possible only for large companies in the field of the agricultural sector with higher capital.

In order to achieve the goals of sustainable development from the aspect of environmental protection, the prevention of soil erosion stands out as a priority. Soil erosion is one of the leading determinants of land degradation. It is especially dangerous in agricultural systems that use frequent tillage. The chapter has shown certain advantages of the polycultural way of agricultural production in preserving the quality of the soil. Summarizing the most important topics in the paper, the recommendation is to increase biodiversity, afforestation, as well as the controlled use of pesticides and agro-technical measures. It is necessary to raise farmers' awareness of the benefits of manure use in sustainable agricultural systems. The favorable circumstance is that the use of pesticides is decreasing (in the analyzed countries), while the areas under organic production, as a specific form of agriculture, are growing. Future research should also examine contemporary sustainable agricultural land management practices in soil erosion prevention. It is about the integration of crop and livestock production, application of agroforestry and the concept of a circular economy.

Monoculture production should be directed in the right way in order to extract the maximum economic effects, at the same time without depleting fertility and nutrients in the soil. In practice, this is usually achieved by crop rotation. The major goal is to find a balance between the economic and environmental goals of sustainable agricultural development. The success of the creators of agrarian, economic and environmental policy will depend on this.

The ultimate goal of sustainable development in the modern economy should be the creation of economically viable production, which will not disturb the environment and which will ensure overall socio-economic development. The remaining population in rural areas is a much-needed condition for sustainable agricultural and rural development. There is also an increasing number of abandoned fertile lands due to unresolved problems regarding inheritance or low interest of young people in this type of activity.

In the following period, food prices on the world market are expected to rise due to the existing health and economic crisis, threatening to persist for years. Therefore, the return of people to the countryside and their own production can ensure food self-sufficiency, nutrition with quality products, as well as the preservation of arable agricultural land. Otherwise, there may be an abandonment of agricultural land and additional socio-economic problems. Also, care must be taken to ensure that natural production is rational, which is often not difficult given the price and profit those organic products can make on the market.

ACKNOWLEDGMENT

This research was supported by the Ministry of Education, Science and Technological Development of the Republic of Serbia - Agreements on the implementation and financing of scientific research in 2021 [number 451-03-9/2021-14/200371 and number 451-03-9/2021-14/200100].

REFERENCES

Adegbeye, M. J., Reddy, P. R. K., Obaisi, A. I., Elghandour, M. M. M. Y., Oyebamiji, K. J., & Salem, A. Z. M., Morakinyo-Fasipe, O. T., Cipriano-Salazar, MCamacho-Díaz, L. M. (2020). Sustainable agriculture options for production, greenhouse gasses and pollution alleviation, and nutrient recycling in emerging and transitional nations - An overview. *Journal of Cleaner Production*, *242*, 118319. doi:10.1016/j.jclepro.2019.118319

Agovino, M., Casaccia, M., Ciommi, M., Ferrara, M., & Marchesano, K. (2019). Agriculture, climate change and sustainability: The case of EU-28. *Ecological Indicators*, *105*, 525–543. doi:10.1016/j.ecolind.2018.04.064

Alcon, F., Marín-Miñano, C., Zabala, J. A., de-Miguel, M. D., & Martínez-Paz, J. M. (2020). Valuing diversification benefits through intercropping in Mediterranean agroecosystems: A choice experiment approach. *Ecological Economics*, *171*, 106593. doi:10.1016/j.ecolecon.2020.106593

Baartman, J. E., van Lynden, G. W., Reed, M. S., Ritsema, C. J., & Hessel, R. (2007). *Desertification and land degradation: origins, processes and solutions. DESIRE Report series: Scientific Report*. DESIRE.

Bajocco, S., De Angelis, A., Perini, L., Ferrara, A., & Salvati, L. (2012). The impact of land use/land cover changes on land degradation dynamics: A Mediterranean case study. *Environmental Management*, *49*(5), 980–989. doi:10.100700267-012-9831-8 PMID:22419398

Beillouin, D., Ben-Ari, T., & Makowski, D. (2019). Evidence map of crop diversification strategies at the global scale. *Environmental Research Letters, 14*(12).

Bockstaller, C., Guichard, L., Makowski, D., Aveline, A., Girardin, P., & Plantureux, S. (2008). Agri-environmental indicators to assess cropping and farming systems. A review. *Agronomy for Sustainable Development*, *28*(1), 139–149. doi:10.1051/agro:2007052

Bosco, G., & Simeoni, L. (2018). Environmental impact of intensive farming in sloping areas. In *MATEC Web of Conferences* (Vol. 174, p. 01001). EDP Sciences. 10.1051/matecconf/201817401001

Bowyer, C., Withana, S., Fenn, I., Bassi, S., Lewis, M., Cooper, T., Benito, P., & Mudgal, S. (2009). *Land Degradation and Desertification*. European Parliament. https://www.europarl.europa.eu/RegData/etudes/etudes/join/2009/416203/IPOL-ENVI_ET(2009)416203_EN.pdf

Christiaensen, L., Demery, L., & Kuhl, J. (2011). The (evolving) role of agriculture in poverty reduction-An empirical perspective. *Journal of Development Economics*, *96*(2), 239–254. doi:10.1016/j.jdeveco.2010.10.006

Crossley, M. S., Meier, A. R., Baldwin, E. M., Berry, L. L., Crenshaw, L. C., Hartman, G. L., Lagos-Kutz, D., Nichols, D. H., Patel, K., Varriano, S., Snyder, W. E., & Moran, M. D. (2020). No net insect abundance and diversity declines across US Long Term Ecological Research sites. *Nature Ecology & Evolution*, *4*(10), 1368–1376. doi:10.103841559-020-1269-4 PMID:32778751

Đekić, S. (2010). *Agrarni menadžment (Agrarian management)*. University of Niš, Faculty of Economics.

Diaz-Balteiro, L., González-Pachón, J., & Romero, C. (2017). Measuring systems sustainability with multi-criteria methods: A critical review. *European Journal of Operational Research*, *258*(2), 607–616. doi:10.1016/j.ejor.2016.08.075

Dorinet, E., Jouvet, P. A., & Wolfersberger, J. (2021). Is the agricultural sector cursed too? Evidence from Sub-Saharan Africa. *World Development*, *140*, 105250. doi:10.1016/j.worlddev.2020.105250

EC. (2021a). *Political agreement on new Common Agricultural Policy: fairer, greener, more flexible*. https://ec.europa.eu/commission/presscorner/detail/en/IP_21_2711

EC. (2021b). *Working with Parliament and Council to make the CAP reform fit for the European Green Deal*. https://ec.europa.eu/info/sites/default/files/food-farming-fisheries/key_policies/documents/factsheet-cap-reform-to-fit-european-green-deal_en.pdf

EC. (2021c). *Soil and Land*. https://ec.europa.eu/environment/soil/index_en.htm

Eurostat. (2021). [Data set]. https://ec.europa.eu/eurostat/data/database

Expósito, A., & Velasco, F. (2020). Exploring environmental efficiency of the European agricultural sector in the use of mineral fertilizers. *Journal of Cleaner Production*, *253*, 119971. doi:10.1016/j.jclepro.2020.119971

Foley, J. A., Ramankutty, N., Brauman, K. A., Cassidy, E. S., Gerber, J. S., Johnston, M., Mueller, N. D., O'Connell, C., Ray, D. K., West, P. C., Balzer, C., Bennett, E. M., Carpenter, S. R., Hill, J., Monfreda, C., Polasky, S., Rockström, J., Sheehan, J., Siebert, S., ... Zaks, D. P. (2011). Solutions for a cultivated planet. *Nature*, *478*(7369), 337–342. doi:10.1038/nature10452 PMID:21993620

Gomiero, T. (2016). Soil degradation, land scarcity and food security: Reviewing a complex challenge. *Sustainability*, *8*(3), 281. doi:10.3390u8030281

Hansen, J. W. (1996). Is agricultural sustainability a useful concept? *Agricultural Systems*, *50*(2), 117–143. doi:10.1016/0308-521X(95)00011-S

Holling, C. S. (2001). Understanding the complexity of economic, ecological, and social systems. *Ecosystems (New York, N.Y.)*, *4*(5), 390–405. doi:10.100710021-001-0101-5

Kelam, I. (2016). Odgovorno upravljanje poljoprivrednim zemljištem (Responsible management of agricultural land). *JAHR*: *Europski časopis za bioetiku, 7*(2), 203-215.

Klasen, S., Meyer, K. M., Dislich, C., Euler, M., Faust, H., Gatto, M., Hettig, E., Melati, D. N., Jaya, I. N. S., Otten, F., Pérez-Cruzado, C., Steinebach, S., Tarigan, S., & Wiegand, K. (2016). Economic and ecological trade-offs of agricultural specialization at different spatial scales. *Ecological Economics*, *122*, 111–120. doi:10.1016/j.ecolecon.2016.01.001

Kolasa-Więcek, A. (2018). Neural modeling of greenhouse gas emission from agricultural sector in European Union member countries. *Water, Air, and Soil Pollution*, *229*(6), 1–12. doi:10.100711270-018-3861-7

Liu, Y., Amin, A., Rasool, S. F., & Zaman, Q. U. (2020). The role of agriculture and foreign remittances in mitigating rural poverty: Empirical evidence from Pakistan. *Risk Management and Healthcare Policy*, *13*, 13–26. doi:10.2147/RMHP.S235580 PMID:32021516

Marković, M. (2015). Adjustment of EU agricultural policy to the new environment. [Economics of Agriculture]. *Ekonomika Poljoprivrede*, *62*(4), 1031–1044. doi:10.5937/ekoPolj1504031M

Marković, M. (2018a). The economic dimension of sustainable agricultural development of Serbia. In D. Karabašević, S. Vukotić, & M. Maksimović (Eds.), *International Thematic Monograph - Thematic proceedings: Innovation as an initiator of the development "Innovations – Basis for development"* (pp. 410–425). Faculty of Applied Management, Economy and Finance.

Marković, M. (2018b). Organic agriculture as a form of sustainable agricultural development in Serbia. *Economics of Sustainable Development*, 2(2), 29–36.

McRae, T., Smith, C. A. S., & Gregorich, L. J. (2000). *Environmental sustainability of Canadian agriculture. Report of the Agri-Environmental Indicators Project.* Agriculture and Agri-Food Canada.

Montanarella, L., Pennock, D. J., McKenzie, N., Badraoui, M., Chude, V., Baptista, I., Mamo, T., Yemefack, M., Singh Aulakh, M., Yagi, K., Young Hong, S., Vijarnsorn, P., Zhang, G.-L., Arrouays, D., Black, H., Krasilnikov, P., Sobocká, J., Alegre, J., Henriquez, C. R., ... Vargas, R. (2016). World's soils are under threat. *Soil (Göttingen)*, *2*(1), 79–82. doi:10.5194oil-2-79-2016

Panagos, P., Standardi, G., Borrelli, P., Lugato, E., Montanarella, L., & Bosello, F. (2018). Cost of agricultural productivity loss due to soil erosion in the European Union: From direct cost evaluation approaches to the use of macroeconomic models. *Land Degradation & Development*, *29*(3), 471–484. doi:10.1002/ldr.2879

Pe'er, G., Bonn, A., Bruelheide, H., Dieker, P., Eisenhauer, N., Feindt, P. H., Hagedorn, G., Hansjürgens, B., Herzon, I., Lomba, Â., Marquard, E., Moreira, F., Nitsch, H., Oppermann, R., Perino, A., Röder, N., Schleyer, C., Schindler, S., Wolf, C., ... Lakner, S. (2020). Action needed for the EU Common Agricultural Policy to address sustainability challenges. *People and Nature*, *2*(2), 305–316. doi:10.1002/pan3.10080 PMID:32626843

Peltonen-Sainio, P., Jauhiainen, L., & Latukka, A. (2020). Interactive tool for farmers to diversify high-latitude cereal-dominated crop rotations. *International Journal of Agricultural Sustainability, 18*(4), 319–333. doi:10.1080/14735903.2020.1775931

Perović, V., Kadović, R., Đurđević, V., Pavlović, D., Pavlović, M., Čakmak, D., Mitrović, M., & Pavlović, P. (2021). Major drivers of land degradation risk in Western Serbia: Current trends and future scenarios. *Ecological Indicators, 123*, 107377. doi:10.1016/j.ecolind.2021.107377

Pimentel, D. (2006). Soil erosion: A food and environmental threat. *Environment, Development and Sustainability, 8*(1), 119–137. doi:10.100710668-005-1262-8

Prăvălie, R., Patriche, C., Borrelli, P., Panagos, P., Roşca, B., Dumitraşcu, M., Nita, I.-A., Săvulescu, I., Birsan, M.-V., & Bandoc, G. (2021). Arable lands under the pressure of multiple land degradation processes. A global perspective. *Environmental Research, 194*, 110697. doi:10.1016/j.envres.2020.110697 PMID:33428912

Rađenović, Ž., Krstić, B., & Marković, M. (2020). Smart farming in agricultural industry: Mobile technology perspective. *Ekonomika Poljoprivrede, 67*(3), 925–938. doi:10.5937/ekoPolj2003925R

Radukić, S., & Marković, M. (2015). Limitation of trade margins as a measure of food price controls: Experience of Serbia. *Ekonomika Poljoprivrede, 62*(1), 193–205. doi:10.5937/ekoPolj1501193R

Rao, C. S., Kareemulla, K., Krishnan, P., Murthy, G. R. K., Ramesh, P., Ananthan, P. S., & Joshi, P. K. (2019). Agro-ecosystem based sustainability indicators for climate resilient agriculture in India: A conceptual framework. *Ecological Indicators, 105*, 621–633. doi:10.1016/j.ecolind.2018.06.038

Renner, A., Cadillo-Benalcazar, J. J., Benini, L., & Giampietro, M. (2020). Environmental pressure of the European agricultural system: Anticipating the biophysical consequences of internalization. *Ecosystem Services, 46*, 101195. doi:10.1016/j.ecoser.2020.101195

Russo Lopes, G., Bastos Lima, M. G., & dos Reis, T. N. (2021). Maldevelopment revisited: Inclusiveness and social impacts of soy expansion over Brazil's Cerrado in Matopiba. *World Development, 139*(C), 105316. doi:10.1016/j.worlddev.2020.105316

Sabiha, N. E., Salim, R., Rahman, S., & Rola-Rubzen, M. F. (2016). Measuring environmental sustainability in agriculture: A composite environmental impact index approach. *Journal of Environmental Management, 166*, 84–93. doi:10.1016/j.jenvman.2015.10.003 PMID:26492465

Salvati, L., & Zitti, M. (2009). The environmental "risky" region: Identifying land degradation processes through integration of socio-economic and ecological indicators in a multivariate regionalization model. *Environmental Management, 44*(5), 888–898. doi:10.100700267-009-9378-5 PMID:19787396

Schaub, S., Buchmann, N., Lüscher, A., & Finger, R. (2020). Economic benefits from plant species diversity in intensively managed grasslands. *Ecological Economics, 168*, 106488. doi:10.1016/j.ecolecon.2019.106488

Schiavon, E., Taramelli, A., Tornato, A., & Pierangeli, F. (2021). Monitoring environmental and climate goals for European agriculture: User perspectives on the optimization of the Copernicus evolution offer. *Journal of Environmental Management, 296*, 113121. doi:10.1016/j.jenvman.2021.113121 PMID:34217939

Scholes, R. J., Montanarella, L., Brainich, E., Barger, N., Ten Brink, B., Cantele, M., Erasmus, B., Fisher, J., Gardner, T., Holland, T. G., Kohler, F., Kotiaho, J. S., Von Maltitz, G., Nangendo, G., Pandit, R., Parrotta, J., Potts, M. D., Prince, S., Sankaran, M., & Willemen, L. (2018). *IPBES (2018): Summary for policymakers of the assessment report on land degradation and restoration of the Intergovernmental Science-Policy Platform on Biodiversity and Ecosystem Services*. Academic Press.

Sherwood, S., & Uphoff, N. (2000). Soil health: Research, practice and policy for a more regenerative agriculture. *Applied Soil Ecology, 15*(1), 85–97. doi:10.1016/S0929-1393(00)00074-3

Stanković, J. J., Janković-Milić, V., Marjanović, I., & Janjić, J. (2021a). An integrated approach of PCA and PROMETHEE in spatial assessment of circular economy indicators. *Waste Management, 128*, 154–166. doi:10.1016/j.wasman.2021.04.057 PMID:33991754

Stanković, J. J., Marjanović, I., Papathanasiou, J., & Drezgić, S. (2021b). Social, Economic and Environmental Sustainability of Port Regions: MCDM Approach in Composite Index Creation. *Journal of Marine Science and Engineering, 9*(1), 74. doi:10.3390/jmse9010074

Stanković, J. J., Marjanović, I., & Stojković, N. (2021c). DEA Assessment of Socio-economic Development of European Countries. *Management Journal of Sustainable Business & Management Solutions in Emerging Economies, 26*(1).

Talukder, B., Hipel, K. W., & vanLoon, G. W. (2018). Using multi-criteria decision analysis for assessing sustainability of agricultural systems. *Sustainable Development, 26*(6), 781–799. doi:10.1002d.1848

Trivino-Tarradas, P., Gomez-Ariza, M. R., Basch, G., & Gonzalez-Sanchez, E. J. (2019). Sustainability assessment of annual and permanent crops: The Inspia model. *Sustainability, 11*(3), 738. doi:10.3390u11030738

Turek, A. (2013). Sustainable Agriculture: Between Sustainable Development and Economic Competitiveness. In A. Jean-Vasile, T. Adrian, J. Subic, & D. Dusmanescu (Eds.), *Sustainable Technologies, Policies, and Constraints in the Green Economy* (pp. 219–235). IGI Global. doi:10.4018/978-1-4666-4098-6.ch012

Tzouramani, I., Mantziaris, S., & Karanikolas, P. (2020). Assessing sustainability performance at the farm level: Examples from greek agricultural systems. *Sustainability, 12*(7), 2929. doi:10.3390u12072929

UN. (2013). *Draft Resolution Submitted by the Vice-Chair of the Committee, Ms. Farrah Brown (Jamaica), on the Basis of Informal Consultations on Draft Resolution A/C.2/68/L.21*. United Nations General Assembly. https://www.fao.org/fileadmin/user_upload/GSP/docs/iys/World_Soil_Day_and_International_Year_of_Soils__UNGA_Resolution_Dec._2013.pdf

Van Zonneveld, M., Turmel, M. S., & Hellin, J. (2020). Decision-making to diversify farm systems for climate change adaptation. *Frontiers in Sustainable Food Systems, 4*, 32. doi:10.3389/fsufs.2020.00032

Wang, T., Jin, H., Fan, Y., Obembe, O., & Li, D. (2021). Farmers' adoption and perceived benefits of diversified crop rotations in the margins of US Corn Belt. *Journal of Environmental Management, 293*, 112903. doi:10.1016/j.jenvman.2021.112903 PMID:34102504

Warr, P., & Suphannachart, W. (2021). Agricultural Productivity Growth and Poverty Reduction: Evidence from Thailand. *Journal of Agricultural Economics, 72*(2), 525–546. doi:10.1111/1477-9552.12412

Wezel, A., Casagrande, M., Celette, F., Vian, J. F., Ferrer, A., & Peigné, J. (2014). Agroecological practices for sustainable agriculture. A review. *Agronomy for Sustainable Development, 34*(1), 1–20. doi:10.100713593-013-0180-7

Wilson, M., & Maliszewska-Kordybach, B. (Eds.). (2000). *Soil quality, sustainable agriculture and environmental security in Central and Eastern Europe* (Vol. 69). Springer Science & Business Media. doi:10.1007/978-94-011-4181-9

Zhang, Y., Li, Y., Jiang, L., Tian, C., Li, J., & Xiao, Z. (2011). Potential of perennial crop on environmental sustainability of agriculture. *Procedia Environmental Sciences, 10*, 1141–1147. doi:10.1016/j.proenv.2011.09.182

ADDITIONAL READING

Baldwin, R. F., Trombulak, S. C., Leonard, P. B., Noss, R. F., Hilty, J. A., Possingham, H. P., Scarlett, L., & Anderson, M. G. (2018). The future of landscape conservation. *Bioscience, 68*(2), 60–63. doi:10.1093/biosci/bix142 PMID:29599547

Kumar, A., & Singh, K. (2019). An analysis of crop diversification in Rup Nagar district of Punjab. *International Journal of Recent Technology and Engineering, 7*, 269–287.

Montgomery, D. R. (2007). Soil erosion and agricultural sustainability. *Proceedings of the National Academy of Sciences of the United States of America, 104*(33), 13268–13272. doi:10.1073/pnas.0611508104 PMID:17686990

Palma, I. P., Toral, J. N., Vázquez, M. R. P., Fuentes, N. F., & Hernández, F. G. (2015). Historical changes in the process of agricultural development in Cuba. *Journal of Cleaner Production, 96*, 77–84. doi:10.1016/j.jclepro.2013.11.078

Peltonen-Sainio, P., Jauhiainen, L., & Latukka, A. (2020). Interactive tool for farmers to diversify high-latitude cereal-dominated crop rotations. *International Journal of Agricultural Sustainability, 18*(4), 319–333. doi:10.1080/14735903.2020.1775931

Pogutz, S., & Winn, M. I. (2016). Cultivating ecological knowledge for corporate sustainability: Barilla's innovative approach to sustainable farming. *Business Strategy and the Environment, 25*(6), 435–448. doi:10.1002/bse.1916

Rausser, G., Sexton, S., & Zilberman, D. (2019). The economics of the naturalist food paradigm. *Annual Review of Resource Economics, 11*(1), 217–236. doi:10.1146/annurev-resource-100516-053623

Uddin, K., Abdul Matin, M., & Maharjan, S. (2018). Assessment of land cover change and its impact on changes in soil erosion risk in Nepal. *Sustainability*, *10*(12), 4715. doi:10.3390u10124715

KEY TERMS AND DEFINITIONS

Agrobiodiversity: Diversity of living world (microorganisms, bacteria, parasites) necessary for sustainable agricultural production and conservation of soil quality.

Economies of Scale: A term used to denote the cost-effectiveness of increased production volumes, where as production volumes increase, unit costs decrease.

Land Degradation: The comprehensiveness of processes related to the reduction of soil fertility, land erosion or the reduction of useful living world in arable land.

Monoculture Farming: A form of production in which only one crop is selected to be planted, or only one type of livestock is raised, usually on a larger agricultural area. The choice of plant production is made in accordance with the basic climatic predispositions of the soil, the type of soil and its position.

Organic Agriculture: A traditional agricultural practice that involves the use of as many natural preparations as possible in the production process in order to preserve the quality of food, food nutrients, as well as human health in the final instance.

Polyculture: A method of production that includes the diversification of cultivated crops, harmonized with the principles of biodiversity, organic or ecological agriculture and preservation of soil fertility.

Risk Diversification: One of the key economic principles based on risk reduction through the production and sale of various outputs.

Sustainable Development: Economic development that will not have negative consequences for the environment and people's quality of life in the future.

Chapter 8
Wind Erosion, Climate Change, and Shelterbelts

Aleksandar Baumgertel
Faculty of Forestry, University of Belgrade, Serbia

Sara Lukić
Faculty of Forestry, University of Belgrade, Serbia

Milica Caković
Faculty of Forestry, University of Belgrade, Serbia

Radovan Savić
Faculty of Agriculture, University of Novi Sad, Serbia

Atila Bezdan
Faculty of Agriculture, University of Novi Sad, Serbia

Boško Blagojević
Faculty of Agriculture, University of Novi Sad, Serbia

ABSTRACT

Wind erosion is a widespread phenomenon causing serious soil degradation. It is estimated that about 28% of the global land area suffers from this process. Global climate changes are expected to accelerate land degradation and significantly affect the intensity of wind erosion. Shelterbelts are linear multi-row planting strips of vegetation (trees or shrubs) established for numerous environmental purposes. Shelterbelts are a specific type of agroforestry system which could reduce soil degradation (soil erosion). Shelterbelts mitigate greenhouse gas through trees storing carbon (C) in their above- and below-ground biomass, wherefore they are highlighted as one of the potential ways to mitigate climate change. The purpose of this chapter is to present wind erosion as a land degradation problem, especially in line with climate changes and the present concept of vegetation establishment in the form of shelterbelts for long-term multi-functional provision of ecosystem services, in particular carbon sequestration.

DOI: 10.4018/978-1-7998-8459-0.ch008

INTRODUCTION

Land degradation in arid, semi-arid, and dry sub-humid areas results from climatic variations and human activities and commonly leads to desertification (The United Nations Convention to Combat Desertification). Climate change has a significant influence on land degradation through an increase in air temperature, decrease in precipitation which results in frequent and long-term droughts, and frequent extreme events, such as devastating heavy rain and stormy winds.

During the last several decades, the range and intensity of desertification have risen in dryland areas. Inappropriate land management, particularly when coupled with droughts and reduced and forest vegetation cover, has contributed to high wind erosion.

Vegetation could protect soil from wind erosion processes. In that sense, shelterbelts as vegetation objects can supply a long-term and very efficient option. Likewise, a well-designed shelterbelt can protect crops, reduce nutrient loss, change a microclimate, and provide other ecosystem services.

Although their primary role is directly or indirectly related to reducing wind speed, they have recently been recognized as carbon reservoirs that could contribute to climate change mitigation.

This chapter discusses the problem of wind erosion in line with climate change and concepts of vegetation establishment in shelterbelts for long-term multifunctional provision of ecosystem services, in particular carbon sequestration.

BACKGROUND

Wind erosion leads to serious soil degradation (Lal, 1994; Sterk, 2002). This process has a negative impact on soil quality (Wang et al., 2020), agricultural production (Santra et al., 2017), human health (Brevik & Burgess, 2014), etc. Wind erosion has mainly become evident in arid and semi-arid areas (Zhibao et al., 2000). A large concentration of GHG in the atmosphere leads to global warming and climate changes, which could accelerate wind erosion processes in the future all over the world (Sharratt et al. 2015; Li et al., 2019). As an appropriate tool for carbon sequestration, agroforestry systems are recognized worldwide (Montagnini and Nair 2004; Ramachandran Nair, 2009). Shelterbelts represent one of the agroforestry systems and need special attention due to their environmental services, including the potential to mitigate the greenhouse effect (GHG) (Mayrinck et al., 2019). In this regard, numerous studies on shelterbelts' carbon sequestration potential have been performed worldwide (Kort and Turnock, 1998; Czerepowicz et at., 2012; Possu et al., 2017). In line with the modern concept of ecosystem services promoted by Millennium Ecosystem Assessment (MEA) (Reid, 2005), the potential of shelterbelts to sequester carbon could be considered an ecosystem service (Xie et al., 2018).

SOIL EROSION BY WIND, CLIMATE CHANGES INFLUENCE AND EFFICIENCY OF SHELTERBELTS

Wind Erosion Problem and Mechanism of the Process

Wind erosion is a natural process of transportation and deposition of soil in conditions of high-speed winds, reduced humidity, and the absence of vegetation cover (Weinan & Fryrear, 1996; Laity, 2016).

Figure 1. Wind erosion in Serbia
Source: Letić et al. (2009)

Wind erosion is a widespread phenomenon causing serious soil degradation. It is estimated that about 28% of the global land area and about 42 million ha of European agricultural land experiencing land degradation suffer from this process (JRC, 2010) (Figure 1 and Figure 2). On the European continent, some research reported high intensity of wind erosion processes in parts of Germany (Schmidt et al., 2017), Hungary (Mezösi et al., 2015), Spain (Gomes et al., 2003), Serbia (Baumgertel et al., 2019), UK (Riksen & De Graaff, 2001), etc. Wind erosion is also often associated with desertification, a degradation process that most commonly occurs in arid, semi-arid, and dry sub-humid areas where soil productivity depends on water availability (UNCCD, 1994).

Wind erosion is a dynamic process in which soil particles are detached and moved under wind erosive force. Wind erosion is caused by two basic processes, deflation and abrasion (Laity, 2016). Deflation is the dominant process that causes wind erosion. During deflation, incoherent, small particles are blown off the surface. As a result of deflation, smaller or larger depressions in the relief may occur. The intensity of deflation depends on wind speed and particle size. The wind speed is directly proportional to the deflation, while the particle size is inversely proportional (Gonzales et al., 2017). Abrasion is caused by the mechanical action of particles that move by salting on a coherent material. Abrasion leads to damage to both rock material and plants in the transport of erosion material (Gardiner et al., 2016). The degree of impact of abrasion depends on the composition (density and strength), shape, and particle size of the erosion sediment (Laity, 2016). Particle density directly affects the total kinetic energy, while strength determines whether the kinetic energy will damage the object/plant or deform the particle itself.

Figure 2. Wind erosion in Netherland
Source: Borrelli et al. (2015)

Soil Erodibility and Influence of Vegetation and Land-use Change on Wind Erosion

Soil erodibility is the ability of soil to resist the erosive force of the wind. The inherent physical and chemical properties of the soil affect its ability to resist wind erosion. These are texture, organic matter content (Colazo & Buschiazzo, 2010; Stredova et al., 2015), CaCO3 content, hydraulic properties, density, structure, as well as the current soil condition in terms of moisture content, compaction, and presence of crust (Fan et al., 2008). In terms of texture, the most erodible are soils with a high content of sand fraction. Sandy soils dry out quickly and have stable structural aggregates (Kisić, 2017). Soils with a higher content of clay-loamy soils are more resistant than sandy soils, but they create a higher dust emission in the erosion process, while clay soils are the most resistant to wind erosion. According to Woodruff & Siddowey (1965), all particles and aggregates larger than 0.84 mm are considered non-erodible particles, and the erodibility of soil depends on their percentage.

The content of organic matter in the soil directly influences soil structure stability, which determines soil resistance to erosion and water availability to vegetation. According to Zobeck et al., (2013), the density of particles of the erodible fraction (> 0.106 mm) is inversely proportional to the content of organic matter and with increasing content of organic matter, the density of initiated particles of the erodible fraction decreases. According to recent research, the biotechnological process of introducing carbonates with microbial activity (microbial-induced carbonate precipitation - MICP) stimulates the formation of carbonate deposits and thus increases the resistance of naturally erodible soils to wind erosion (Tang et al., 2020). This method of reducing erodibility has been well demonstrated in sandy soils (Maleki et al., 2016).

The key role of vegetation in reducing erosion is to absorb the wind energy before it can detach soil particles. The vegetation cover is under the constant influence of natural factors, mostly climate and human activities. The degree of protection that the vegetation cover provides to the soil depends on the vegetation characteristics, condition, and height. From the aspect of wind erosion control, the most important is that the soil surface is not exposed to the direct action of the erosive wind force. The critical speed of movement also increases with the height of the vegetation cover so that forest ecosystems provide the best protection against wind erosion and agroecosystems and grass ecosystems (Zhang et al., 2018).

A crop does not provide equal protection against erosion in all crop phenophases, which means that its efficiency depends on the number of days from sowing/planting. This implied the development of equations for determining crop sets for summer, winter, crops with narrower and wider leaves (Fryrear et al., 2001). Standing crop residues function similarly to growing crops. Flattened crop residues are also effective in reducing soil erosion. The degree of protection of the land by lying harvest residues is expressed through the content of non-erodible lying remains on the surface. In addition to protecting the soil surface from the direct action of the wind, harvest residues also contribute to the return of organic matter to the soil. In the process of decomposition, the lying remains are incorporated into the soil and cease to perform a protective function as a cover on the soil surface, while the standing remains to pass into the lying ones and take over their protective function as a surface covering of the soil (Blanco & Lal, 2008). In addition to the number of crop residues, their orientation has a significant impact on wind erosion. On the other side, forest vegetation in the form of shelterbelts on the agricultural area is very effective in controlling wind erosion.

The impact of land use/cover change, soil overexploitation, deforestation, and intensive tillage is just some of the human activities that can lead to increased susceptibility to wind erosion. However, at the same time, appropriate management methods can reduce sensitivity and control degradation due to wind erosion. In Inner Mongolia, the impact of human activity on the dynamics of wind erosion predominates at 11.33% of the total area (Zhang et al., 2018). The same authors considered that human activity consists mainly of grazing, human-caused land use/cover change, and ecological restoration programs.

Agriculture as a way of using land requires applying certain tillage operations, which often leave the area without any protection for a certain time. In this way, the surface becomes susceptible to wind erosion due to the collapse of structural aggregates of the soil, destruction of crop residues, and drying of the soil. In the absence of protection, the crushed particles of drained soil are easily torn off and transported along the field. In areas where agriculture is the dominant land use, degradation caused by wind erosion can be successfully controlled by good land management strategies and planning and management, applying best practices and implementing ecological restoration programs. Some of the strategies applied to control wind erosion in areas where agriculture is the dominant use are: reduction of cultivation during fallow, the establishment of field shelterbelts, reduction of intensive grazing, minimal tillage or its elimination if possible, reduction of tillage speed, and leaving crop residues on the surface, the introduction of belt cultivation of crops and application of mulch, application of stabilizers and polymers to the soil, increasing the roughness of the surface layer of the soil and reducing the length of the field in the wind direction, etc. For successful erosion assessment and planning of wind erosion control measures, it is necessary to know exactly the condition of the soil surface that causes a certain way of tillage. Tillage operations modify the crust on the soil surface, change the arrangement of lumps of soil on the surface, orient the ridge furrow roughness, and spread and burying plant remains. In this way, some tillage operations can influence the reduction of wind erosion, indicating the importance of tillage inputs in assessing wind erosion.

Negative Impact of Wind Erosion and Damages

Wind erosion has been neglected in the past as a process of degradation. However, when it was recognized as a cause of a sharp decrease in soil fertility, and as a source of atmospheric pollution, more attention was paid to it (Oldemann et al., 1990; Gobin et al., 2003; Funk & Reuter, 2006). The effects of wind erosion are present at the location where the deflation process takes place (on-site effects) and outside these locations on the path of erosion sediment transport and at the places of particle deposition (off-site effects). The main on-site effect of wind erosion is the decline in soil quality caused by removing the surface layer of soil with the highest organic matter content and nutrients (Buschiazzo & Funk 2015; Négyesi et al., 2019), as well as the reduction of retention capacity. Due to the drying effect of wind, the stability of structural aggregates decreases in organic soils, and the content of dry erodible material increases, while organic matter content decreases (Zobeck et al., 2013; Kučera & Podhrazska, 2016). Agricultural lands can, to some extent, mitigate the effect of fertility loss, but this can cause an off-site effect associated with groundwater and surface water pollution. The off-site effects of wind erosion are reflected in adverse health effects caused by dust storms and damage to traffic, communications, and irrigation and drainage infrastructure (Shi et al., 2004). On the way of transport of erosion sediment particles, plants are damaged due to abrasion. Moreover, erosion particles damage crops and natural vegetation, especially young plants (Gonzales et al., 2017). In addition, turbulent airflow increases evapotranspiration and further affects the reduction of soil moisture. In locations where the wind's kinetic energy is not sufficient to transport the erosion sediment, off-site effects are also observed in the deposition process, which is manifested by the backfilling of arable land and young crops on them. The pace of land degradation has shifted the focus from on-site effects, which are mainly related to economic effects, to off-site effects related to environmental impact (Funk & Reuter, 2006).

Climate Change and Its Influence on Wind Erosion

According to the UN Framework Convention on Climate Change (UNFCCC, 1994), climate change is defined as a change of climate that is directly or indirectly attributed to human activities that alter the composition of the global atmosphere. Climate change is characterized by an increase in temperature, precipitation regime changes, annual distribution, and an increased frequency of extreme weather events and periods with extreme climatic conditions (Weart, 1997).

Climate models are dynamic models that represent a simplified representation of reality and operate on a global scale, even in a limited region of the world (von Storch, 2005). Climate models are the primary tools for researching the climate system's response, enabling climate predictions on seasonal to decadal time scales and predicting future climate during the coming century and beyond (Kattsov et al., 2013).

Based on a reliable climate model, we can examine the impacts of different scenarios. The scenarios are described through possible events, i.e., events that will significantly affect climate development in the future. The climate scenario should be realistic, consistent, and convincing. For climate research, the Intergovernmental Panel on Climate Change (IPCC) (https://www.ipcc.ch/) has established a Representative Concentration Pathways scenario (RCPs) (Kwak et al., 2017). There are four such scenarios (Table 1.) that predict temperature changes based on the total influence of external factors (greenhouse gases and aerosols that affect the radiation balance) (Huang et al., 2013).

Figure 3 presents the temperature change according to the optimistic scenario (RCPs 2.6) and the pessimistic scenario (RCPs 8.5). It is noticed that the temperature does not increase evenly and that the

Table 1. Description of Representative Concentration Pathways scenarios (RCPs)

Scenarios	CO_2 Concentration (ppm)	Global Warming until 2100 (Mean and Likely Range)
RCP 2.6	490	1.0 (0.3 – 1.7) °C
RCP 4.5	650	1.8 (1.1 – 2.6) °C
RCP 6.0	850	2.2 (1.4 –3.1) °C
RCP 8.5	1370	3.7 (2.6 – 4.8) °C

Source: (Kwak et al., 2017)

warming is higher above the continents and the Arctic in relation to the warming above the ocean. The reason is that the oceans can absorb more heat.

IPCC (2013) estimates average global warming of 1.4 to 5.8°C for 1990 to 2100. IPCC (2013) states that global mean temperatures will increase in most land areas, which will affect the occurrence of more frequent hot and less cold temperature extremes within daily and seasonal time frames.

Figure 3. Change in average surface temperature
Source: (IPCC, 2013)

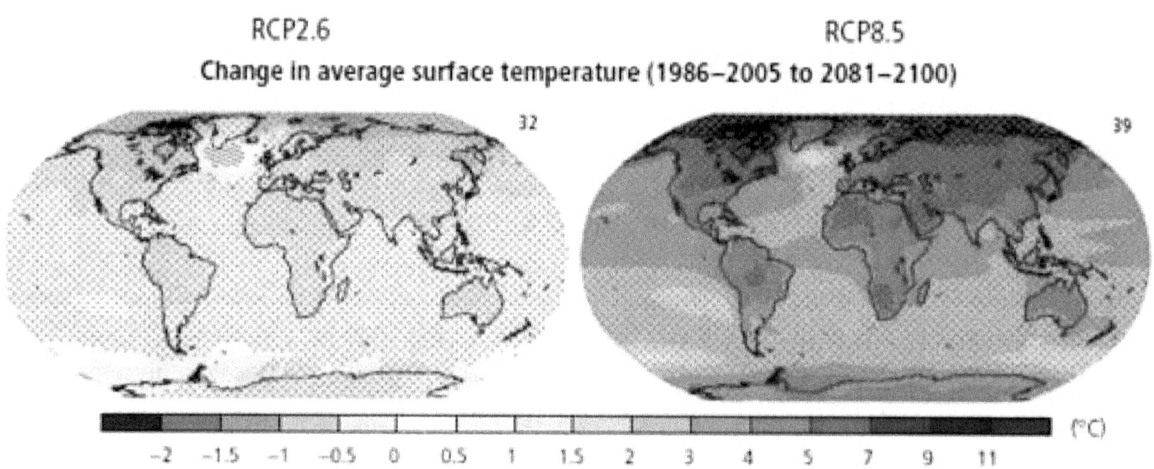

In addition to changing temperatures, climate change will affect the change in precipitation dynamics. Based on IPCC projections, we can expect more precipitation in some regions and less in others. Figure 4 shows the change in mean annual precipitation. A land area in southern Europe and northern Africa will have less rain, while much more precipitation will be in northern Eurasia and central Africa. An increase in the frequency and intensity of extreme precipitation is expected, and the reason for that is that warmer air can carry more moisture. Changes in mean precipitation (such as annual precipitation) reflect only one side of climate change, while the other side represents the more significant variability of individual rainfall episodes. It is assumed that there will be more extreme events in the future, such as heavy rains over a short period, which will increase the risk of floods. Also, long periods with very little rain can lead to more dry days. However, such changes can already be noticed.

Figure 4. Change in average precipitation
Source: (IPCC, 2013)

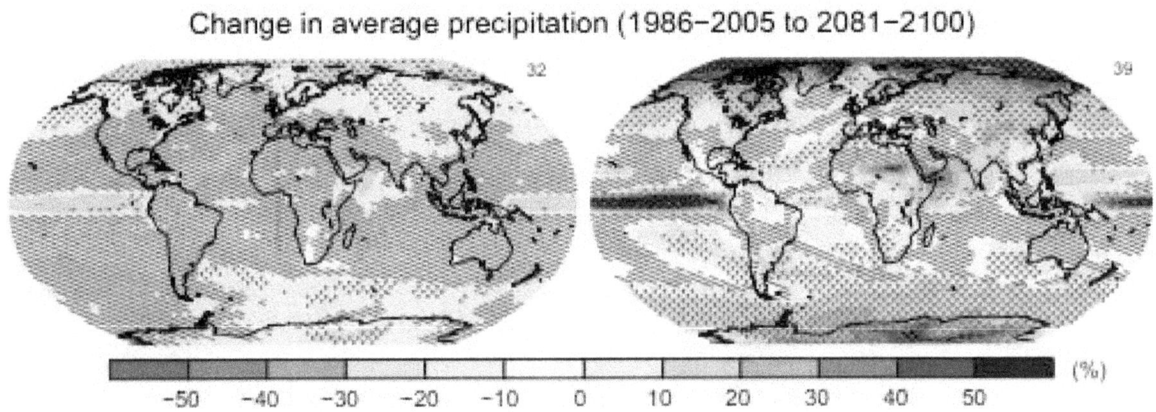

Global climate changes are expected to accelerate land degradation (Huang et al., 2015) and significantly affect the intensity of wind erosion (Sharratt et al., 2015; Li et al., 2020; Liddicoat et al., 2012; Mezosi et al., 2016). In conditions of reduced humidity, especially if longer periods of time have moisture deficit, wind erosion develops rapidly and can leave permanent consequences.

Climatic indicators key to the wind erosion process are wind speed, precipitation, and temperature (Yang & Lu, 2016). The erosive strength of the wind is determined by the direction and speed of the wind, while changes in precipitation and temperature directly affect the soil water content and soil retention function of an ecosystem (Sharratt et al., 2013). Since the beginning of the 21st century, there has been growing concern about the effects of climate change and its negative impact on many processes and components of the global ecosystem. Li et al. (2020) predicted that in Central Asia, spring soil wind erosion would increase by 10.34% (RCP4.5) to 10.71% (RCP8.5) and that winter soil wind erosion will increase by 23.32% (RCP4.5) to 33.74% (RCP8.5) in the late 21st century. Liddicoat et al. (2012) predicted that the temperature increases by 0.4 °C to 1.8 °C, average annual precipitation might decrease by 15%, farmlands in South Australia would suffer from increased soil wind erosion by 2030. Similarly, with the increase in temperature by 2 °C in grasslands in Asia, wind erosion will increase by about 25% (Gao et al., 2002). Also, the same authors further predict that the increase in annual wind erosion will likely be more pronounced in the areas with sandier soils.

The influence of wind speed as a driver of wind erosion is equally significant. According to research by Lee et al. (1999), aeolian erosion on agricultural areas under maize will increase by 20% under conditions of unchanged temperature and precipitation.

SHELTERBELTS AS A SOLUTION AND RECOMMENDATION FOR WIND EROSION CONTROL AND ECOSYSTEM SERVICES

Shelterbelts are linear multi-row planting strips of vegetation (of trees or shrubs) established for numerous environmental purposes. In different parts of the world, they are also known as Protective Forest Belts (PFB), windbreaks, or shelterbelts. They affect the reduction of wind speed on the leeward and

Figure 5. Ecosystem services provided by shelterbelts
Source: (Xie et al., 2018)

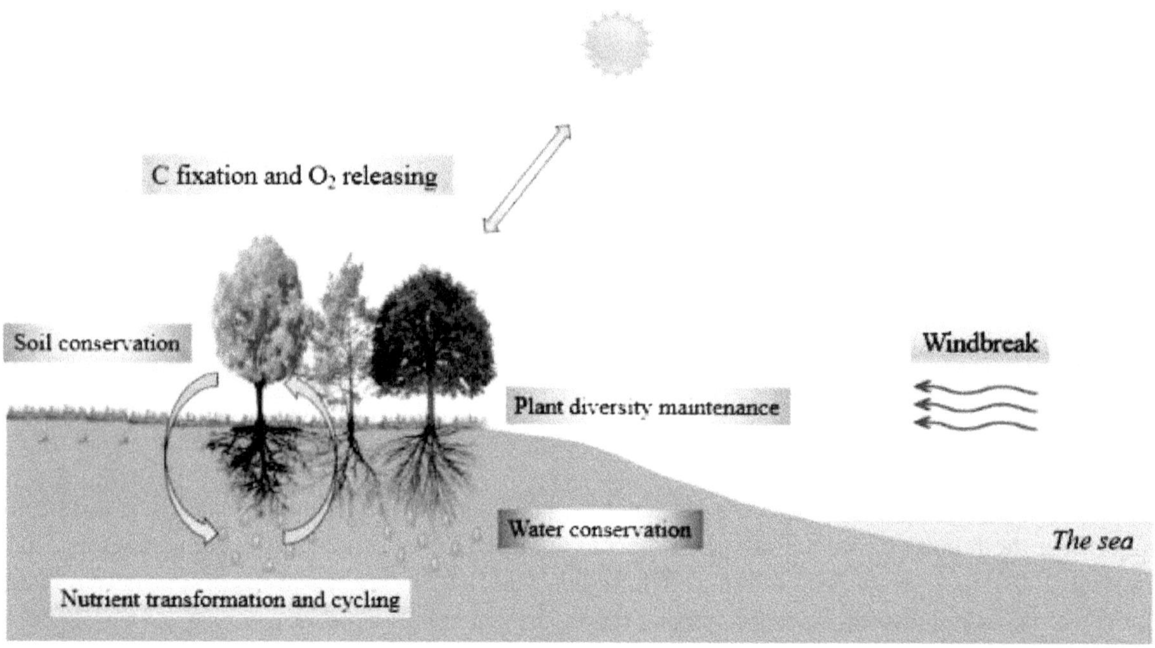

windward, and hereupon they represent one of the most used practices for soil protection from wind erosion (Baumhardt & Blanco-Canqui, 2014). In addition to their primary function (erosion protection), shelterbelts ensure numerous benefits for humans and the environment. Shelterbelts have a positive effect on increasing plant and animal biodiversity (Alemu, 2016), energy saving, increasing yields (Zheng et al., 2016), air quality (Chen et al., 2015), water conservation (Thevs et al., 2017), road (Chu et al., 2013) and noise (Stojanović et al., 2016) protection, improving landscape ecological network (Podhraska et al., 2021), etc. Lately, shelterbelts have been increasingly recognized as an appropriate tool for mitigating climate change through CO2 sequestration from the atmosphere.

Because of the numerous functions that shelterbelts provide, they are considered multifunctional objects (Mize, 2008). In recent times, by introducing a relatively new concept of ecosystem services (MEA, CICES), defined as a direct benefit for humans, some of the beforementioned functions could be considered as ecosystem services provided by shelterbelt created from Miscanthus giganteus. In this regard, Littlejohn et al. (2019) identified and quantified key ecosystem services groups (provisioning, regulatory and cultural). On the other hand, Xie et al., (2018) measured the forest structure, species diversity and major individual ecosystem services (Figure 5) of four single species dominated shelterbelts in Eastern China. Weninger et al. (2021) systematically analyzed published studies regarding ecosystem services (according to CICES) provided by shelterbelts. They pointed out most often mentioned positive functions, such as soil protection, biodiversity, and pest control, whereas for biomass production, nutrient, and water balance. Considering the wide spectrum of functions that shelterbelts provide, they could perform most ecosystem functions (according to MEA) (Table 2). Due to their multifunctionality and capability to improve environmental conditions, shelterbelts represent one of the most used agroforestry practices (Brendle et al., 2009). Agriculture production produces almost one-third of total emitted CO2 in the atmosphere (Gilbert et al., 2012); consequently, special attention should be on evaluating the carbon storage potential for agroforestry systems (Possu, 2015).

Table 2. Ecosystem services provided by shelterbelts

	Provisioning Services	Regulating Services	Cultural Services	Supporting Services
	Products obtained from ecosystems	*Benefits obtained from the regulation of ecosystem processes*	*Nonmaterial benefits obtained from ecosystem*	*Necessary services for the needs of other ecosystem services*
Service	**Food**	**Climate regulation**	Spiritual and religious	**Soil formation**
	Water	Disease regulation	**Recreation and Ecotourism**	**Nutrient cycling**
	Fuelwood	**Water regulation**	**Aesthetic**	Primary production
	Fiber	**Water purification**	Inspirational and Educational	
	Biochemicals	**Pollination**	Sense of place	
	Genetic resources		**Cultural heritage**	

Source: (Authors)

Erosion Protection by Shelterbelts

Shelterbelts reduce wind speed and decrease erosion intensity within the leeward zone (from 5 up to 30 shelterbelts heights). Wind speed reduction is mostly affected by internal and external shelterbelt structures (Heisler and De Walle, 1988). According to Yang et al. (2016) and Wu et al. (2018), two main parameters influencing wind speed reduction are shelterbelts height and porosity. Porosity can be used as a descriptor of internal structure, and it is defined as the ratio of perforated area to total area (Heisler and De Walle, 1988). Based on the results of research performed by Rahaček et al. (2017), there is a strong correlation ($r = 0.842$ to 0.936, $p < 0.05$) between optical porosity and band efficiency in terms of incoming wind speed reduction. Similar results are obtained by Stredova (2012) and Thuyet et al. (2013).

There are some general claims about how shelterbelts should be designed to perform different functions. In this regard, according to the document Shelterbelts – Design Guidelines (Agriculure and Agri-Food Canada, 2010) suggest denser shelterbelts, a mixture of tree and shrub species, native and herbicide-tolerant species and trees with an upright and narrow growth habit. Other similar guidelines published within different regions worldwide give similar instructions and emphasize the choice of mainly domestic woody species.

In addition to the direct impact on wind speed reduction, shelterbelts change other microclimatic indicators in the protection zone (Figure 6). This primarily refers to the change in solar radiation, air temperature, and relative humidity, which, together with the reduced wind speed, affect the change of evapotranspiration. Reduced evapotranspiration can lead to soil moisture conservation which further positively affects the soil's ability to resist erosion processes. Also, through the change of evapotranspiration, shelterbelts could affect increase of yield (Osorio et al., 2018).

In China and the United States, large-scale afforestation projects have been implemented on agricultural land using shelterbelts with the aim of reducing physical damage to plants and reducing the intensity of

Figure 6. Shelterbelts effect on microclimate conditions
Source: (Marshall, 1967)

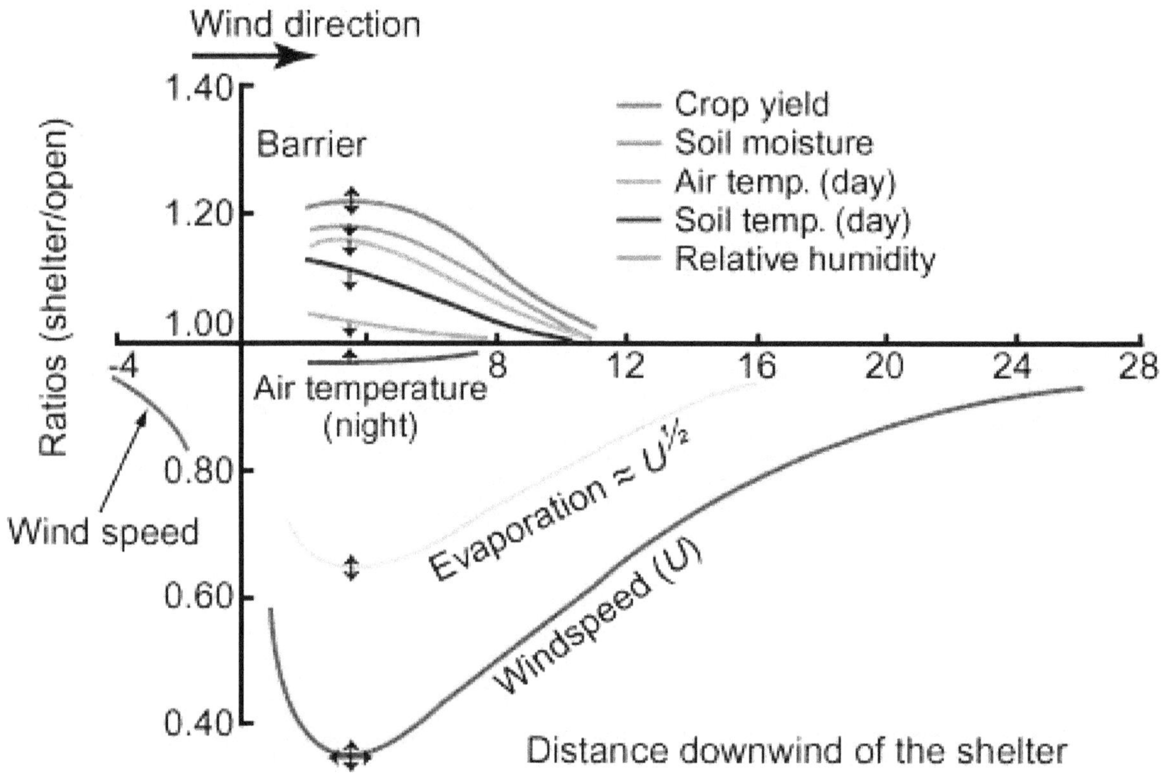

wind erosion. One of the largest projects undertaken to increase afforestation and desertification is the "Three-North" shelterbelts project in China. Since its inception 30 years ago, important achievements have been made concerning that forest cover increased from 5.05% in 1978 to 10.51% in 2008 (Li et al., 2012). In the United States (U.S.), a shelterbelt program was performed by the Prairie States Forestry Project (PSFP), which resulted in 30,000 km of shelterbelts planted from 1935 to 1942 across six the area of Great Plains states (Dunlop, 2000). One example from Eastern Europe is the establishment of shelterbelts in former Yugoslavia which started before the Second World War. In the period after had got wider extent including shelterbelts establishment in development strategies such as so-called "Five-year plan", as one of the activities of high priority for economic growth and development. The research of the positive effects of shelterbelts' standard types and the usual crops in a sheltered zone on wind speed reduction was conducted on shelterbelts in the area of Pančevo and Bačka Palanka (Lukić 2006; Lukić and Dožić 2006). Similar projects were carried out in some other parts of Eastern European countries, such as Hungary (Szigeti et al., 2020), Bulgaria (Shahanov & Cirella, 2021) and Russia (Chendev et al., 2015).

Nowadays, new strategies and projects on establishing new shelterbelts rely more on stimulation and subventions for local farmers in order to plant new shelterbelts on their land. For example, Prairie Farm Rehabilitation Administration (PFRA) includes a free tree distribution program for local farmers to purchase trees for shelterbelts (Kulshreshtha & Rempel 2014). Also, one of the mechanisms to introduce shelterbelts is through land commission where shelterbelts could be considered as "common facilities", which means that their purpose will provide benefits for the whole community.

Carbon Sequestration by Shelterbelts

One-third of the total carbon dioxide emissions into the atmosphere, according to Gilbert (2012), come from intensive agriculture, primarily due to the use of pesticides. In order to increase forest cover in agricultural systems, shelterbelts represent a highly acceptable measure of the introduction of woody components. Through a direct result of the growth of trees, shelterbelts storing carbon (C) in their above- and belowground biomass, wherefore they are highlighted as one of the potential ways to help mitigate climate change (Possu et al., 2017). In addition to using on agricultural land in the urban landscape, shelterbelts provide numerous environmental services (Li et al., 2007), which include avoiding carbon emissions indirectly (Nowak et al. 2017) and bringing to an energy reduction of 20% for heating (Ko, 2018).

In recent times, studies about the carbon sequestration potential of shelterbelts have become more represented worldwide. These studies are mostly performed in North America, mainly in Canada and (Kort & Turnock, 1998; Possu, 2015; Dhillon & Van Rees, 2017; Krobel et al., 2020; Mayrinck et al., 2019; Rudd et al., 2020; Amichev et al., 2016b). Moreover, the above studies assess carbon sequestration in shelterbelts. As a result, Amichev et al. (2016a) found that the carbon potential of the six shelterbelts species in Canada ranges from 1,78 to 6,54 Mg C km-1 yr-1. On the other hand, Possu (2015) analyzed carbon storage potential for different field windbreaks across the USA and reported 5.8 Mg C km-1 yr-1 for a three-row tall-deciduous windbreak. Also, similar research has been performed in different parts of China (Chu et al., 2019; Wang et al., 2013, and others). According to (Mayrinck et al., 2019) research in which shelterbelts are considered as agroforestry practices have also been done in Poland, New Zeeland, Russia, etc. Additionally, studies on carbon sequestration potential are performed in the Republic of Serbia, in the predominantly agricultural area of AP Vojvodina, where Lukić et al. (2018) determine the potential of some specific shelterbelts. According to the results, in Bačka Palanka (Serbia), 20-year-old shelterbelts consisted of the most commonly used species (Populus x euramericana, Ulmis pumila, Robinia pseudoacacia), estimated carbon stock was 0.300; 0.333 and 0.111 t per tree, respectively. Research about shelterbelts' carbon sequestration potential could serve as a basis for succeeding studies where their impact on climate change mitigation could be analyzed in more detail.

In the studies mentioned above, standard allometric equations are mostly used to assess carbon sequestration. In some cases, specific allometric equations do not exist, whereas available equations for another tree species were used (Czerepowicz et al., 2012). In that sense, there is a space for future research which could be oriented to establish allometric equations for given specific tree species. Also, in some studies, remote sensing techniques have been used. That includes using optical satellite images, point-cloud network, or/and vegetation indices obtained by UAV.

Although their carbon sequestration function is recognized as an essential element of smart agriculture and sustainable development, there was a lack of shelterbelt spatial distribution data and tree growth models that are needed for carbon inventory analyses (Amichev et al., 2016b). Likewise, according to Possu (2015), accessible and valid data for estimating windbreak contributions from local to regional C assessments in most areas do not exist.

Local farmers represent a substantial factor regarding accepting and implementing shelterbelts into their agricultural production systems. To provide practical knowledge for local farmers on how to design shelterbelts, numerous guidelines with that purpose are provided within different countries and regions worldwide. In the context of adaptation to climate change, the Net zero-emission concept, which promotes that all man-made greenhouse gases should be eliminated from the atmosphere, is a top priority. Considering the potential of shelterbelts for carbon sequestration, they represent an appropriate tool to achieve set goals.

Agroforestry systems are promoted through different strategies and programs which promote sustainable agriculture production. Also, there are numerous guidelines published in different countries worldwide with practical information about how to design and establish shelterbelt in an appropriate way. However, a broad interdisciplinary view on these shelterbelt functions lacks scientific literature and common knowledge (Weninger et al., 2021).

FUTURE RESEARCH DIRECTIONS

The assessment of carbon reserves in shelterbelts is usually performed using by IPCC methodology developed for typical forests. However, the conditions in the shelterbelts are somewhat different because the undergrowth is not present to the extent that it can be present in the forest. Consequently, further research could be addressed to revise the factor for the conversion of wood volume into biomass – BEF factor. Also, remote sensing techniques confirm the potential for studying different vegetation, from grassland to forest ecosystems, but also in agriculture. For further research, the development of specific allometric equations in areas where they do not exist is necessary. Also, the application of remote sensing techniques for assessing carbon storage in shelterbelts in combination with data collected from field should be put forward in further research.

Further research could be oriented to creating a national or regional database on shelterbelt carbon sequestration using the same methodology. That could cover the areas which are not limited just on country borders but covers up larger areas.

CONCLUSION

Due to climate change, there is a risk of wind erosion process acceleration in the future. Shelterbelts have the potential to help fight climate change because of their function in carbon sequestration. Shelterbelts have been established mainly for the function of wind speed reduction, but these days there is a need for their establishment in the context of smart agriculture within agroforestry systems. Within that system, shelterbelts could provide numerous ecosystem services.

ACKNOWLEDGMENT

This research was supported by the Erasmus+ Program of the European Union through the Capacity Building in Higher Education project entitled: Soil Erosion and Torrential Flood prevention: curriculum development at the universities of Western Balkan countries (SETOF), Project No. 598403-EPP-1-2018-1-RS-EPPKA2-CBHE-JP and carried out as part of the Agreement on financing the scientific research work of Scientific Research Organizations in 2021, No. 451-03-9/2021-14/200169 05.02.2021.

REFERENCES

Agriculture and Agri-Food Canada. (2010). *Design Guidelines for Farmyard, Field, Roadside, Livestock, Wildlife, and Riparian Buffer Plantings on the Prairies*. Author.

Alemu, M. M. (2016). Ecological Benefits of Trees as Windbreaks and Shelterbelts. *International Journal of Ecosystem*, *6*(1), 10–13. doi:10.5923/j.ije.20160601.02

Amichev, B., Murray, B., Kulshreshtha, S., Laroque, C. P., Piwowar, J. M., & Van Rees, K. (2016a). Carbon sequestration and growth of six common tree and shrub shelterbelts in Saskatchewan, Canada. *Canadian Journal of Soil Science*, *97*(3), 368–381. doi:10.1139/CJSS-2016-0107

Amichev, B., Murray, B., Kurz, W., Laroque, C. P., Kulshreshtha, S., Piwowar, J. M., & Van Rees, K. (2016b). Carbon sequestration by white spruce shelterbelts in Saskatchewan, Canada: 3PG and CBM-CFS3 model simulations. *Ecological Modelling*, *325*, 35–46. doi:10.1016/j.ecolmodel.2016.01.003

Baumgertel, A., Lukić, S., Belanović Simić, S., & Kadović, R. (2019). Identifying Areas Sensitive to Wind Erosion—A Case Study of the AP Vojvodina (Serbia). *Applied Sciences (Basel, Switzerland)*, *9*(23), 5106. doi:10.3390/app9235106

Baumhardt, R. L., & Blanco-Canqui, H. (2014). Soil: Conservation Practices. In Encyclopedia of Agriculture and Food Systems. Academic Press. doi:10.1016/B978-0-444-52512-3.00091-7

Blanco-Canqui, H., & Lal, R. (2008). Principles of Soil Conservation and Management. Springer Science+Business Media B.V.

Borrelli, P., Panagos, P., Hessel, R., Riksen, M., & Stolte, J. (2015). *Soil erosion by wind in Europe*. EU RECARE Project. doi:10.2788/488050

Brandle, J., Hodges, L., Tyndall, J., & Sudmeyer, R. (2009). Windbreak Practice. In North American Agroforestry: An Integrated Science and Practice (2nd ed.). American Society of Agronomy. doi:10.2134/2009.northamericanagroforestry.2ed.c5

Brevik, E. C., & Burgess, L. C. (2014). The Influence of Soils on Human Health. *Nature Education Knowledge*, *5*(12), 1.

Buschiazzo, D., & Funk, R. (2015). Wind Erosion of Agriculutral Soils and the Carbon Cycle 2015. In S. A. Banwart, E. Noellemeyer, & E. Milne (Eds.), *Soil Carbon: Science, Management and Policy for Multiple Benefits*. CAB International. doi:10.1079/9781780645322.0161

Chen, J., Yu, X., Sun, F., Lun, X., Fu, Y., Jia, G., Zhang, Z., Liu, X., Mo, L., & Bi, H. (2015). The Concentrations and Reduction of Airborne Particulate Matter (PM_{10}, $PM_{2.5}$, PM_1) at Shelterbelt Site in Beijing. *Atmosphere*, *6*(5), 650–676. doi:10.3390/atmos6050650

Chendev, Y., Sauer, T., Ramirez, G., & Burras, C. (2015). History of East European Chernozem Soil Degradation; Protection and Restoration by Tree Windbreaks in the Russian Steppe. *Sustainability*, *7*(1), 705–724. doi:10.3390u7010705

Chu, C.-R., Chang, C.-Y., Huang, C.-J., Wu, T.-R., Wang, C.-Y., & Liu, M.-Y. (2013). Windbreak protection for road vehicles against crosswind. *Journal of Wind Engineering and Industrial Aerodynamics*, *116*, 61–69. doi:10.1016/j.jweia.2013.02.001

Chu, X., Zhan, J., Li, Z., Zhang, F., & Qi, W. (2019). Assessment on Forest Carbon Sequestration in the Three-North Shelterbelt Program Region, China. *Journal of Cleaner Production*, *215*, 382–389. Advance online publication. doi:10.1016/j.jclepro.2018.12.296

Colazo, J. C., & Buschuazzo, D. E. (2010). Soil dry aggregate stability and wind erodible fraction in a semiarid environment of Argentina. *Geoderma, Elsevier*, *159*(1–2), 228–236. doi:10.1016/j.geoderma.2010.07.016

Czerepowicz, L., Case, B. S., & Doscher, C. (2012). Using satellite image data to estimate aboveground shelterbelt carbon stocks across an agricultural landscape. *Agriculture, Ecosystems & Environment*, *156*, 142–150. doi:10.1016/j.agee.2012.05.014

Dhillon, G. S., & Van Rees, K. C. J. (2017). Soil organic carbon sequestration by shelterbelt agroforestry systems in Saskatchewan. *Canadian Journal of Soil Science*, 1–16. doi:10.1139/cjss-2016-0094

Dunlop, A. (2000). *Spatial and Temporal Aspects of Saskatchewan Field Shelterbelts* (Ph.D. Thesis). University of Saskatchewan.

Canada European Environmental Agency (JRC). (2010). *The State of Soil in Europe*. European Commission.

Fan, Y., Liu, J., & Cai, Q. (2008). The effects of wetting rate on aggregate stability in three soils. In *15th ISCO Congress "Soil and Water Conservation, Climate Change and Environmental Sensitivity", 18– 23 May 2008 Budapest* (p. 4). Geographical Research Institute, Hungarian Academy of Science.

Fryrear, D. W., Sutherland, P. L., Davis, G., Hardee, G., & Dollar, M. (2001). Wind erosion estimates with RWEQ and WEQ. In *Sustaining the Global Farm*. Pardue University and USDA-ARS National Soil Erosion Research Laboratory.

Funk, R., & Reuter, H. I. (2006). Wind Erosion. In J. Boardman & J. Poesen (Eds.), *Soil Erosion in Europe*. John Wiley & Sons. doi:10.1002/0470859202.ch41

Gao, Q., Ci, L., & Yu, M. (2002). Modeling wind and water erosion in northern China under climate and land use changes. *Journal of Soil and Water Conservation*, *57*, 46–55.

Gardiner, B., Berry, P., & Moulia, B. (2016). Review: Wind impacts on plant growth, mechanics and damage. *Plant Science*, *245*, 94–118. doi:10.1016/j.plantsci.2016.01.006 PMID:26940495

Gilbert, N. (2012). (2012). One-third of our greenhouse gas emissions come from agriculture. *Nature*. Advance online publication. doi:10.1038/nature.2012.11708

Gobin, A., Govers, G., Jones, R., Kirkby, M., & Kosmas, C. (2003). *Assessment and Reporting on Soil Erosion*. Technical Report 94. European Environment Agency.

Gomes, L., Arrúe, J. L., López, M. V., Sterk, G., Richard, D., Gracia, R., Sabre, M., Gaudichet, A., & Frangi, J. P. (2003). Wind erosion in a semiarid agricultural area of Spain: The WELSONS project. *Catena*, *52*(3-4), 235–256. doi:10.1016/S0341-8162(03)00016-X

Gonzales, H.B., Casada, M.E., Hagen, L.J., Tatarko, J., & Maghirang, R.G. (2017). Sand Transprt and Abrasion within Simulated Standing Vegetation. *The American Society of Agricultural and Biological Engineers*, 791-802.

Heisler, G., & Dewalle, D. (1988). Effects of Windbreaks Structure on Wind Flow. *Agriculture, Ecosystems & Environment*, *22/23*, 41–69. doi:10.1016/0167-8809(88)90007-2

Huang, J., Mendoza, B., Daniel, J. S., Nielsen, C. J., Rotstayn, L., & Wild, O. (2013). Anthropogenic and natural radiative forcing. *Climate Change 2013 the Physical Science Basis: Working Group I Contribution to the Fifth Assessment Report of the Intergovernmental Panel on Climate Change*, 659–740.

Huang, J., Yu, H., Guan, X., Wang, G., & Guo, R. (2015). Accelerated dryland expansion under climate change. *Nature Climate Change*, *6*(2), 166–171. doi:10.1038/nclimate2837

IPCC. (2013). Summary for Policymakers. In Climate Change 2013: The Physical Science Basis. Cambridge University Press.

Kattsov, V., Federation, R., Reason, C., Africa, S., Uk, A. A., Uk, T. A., Baehr, J., Uk, A. B., Catto, J., Canada, J. S., & Uk, A. S. (2013). Evaluation of climate models. *Climate Change 2013 the Physical Science Basis: Working Group I Contribution to the Fifth Assessment Report of the Intergovernmental Panel on Climate Change*, 741–866.

Kisić, I. (2017). Erozija vjetrom. *Hrvatske Vode*, *25*(99), 1–12.

Ko, Y. (2018). Trees and vegetation for residential energy conservation: A critical review for evidence-based urban greening in North America. *Urban Forestry & Urban Greening*, *34*, 318–335. doi:10.1016/j.ufug.2018.07.021

Kort, J., & Turnock, R. (1998). Carbon reservoir and biomass in Canadian prairie shelterbelts. *Agroforestry Systems*, *44*(2/3), 175–186. doi:10.1023/A:1006226006785

Kröbel, R., Moore, J., Ni, Y. Z., McPherson, A., Poppy, L., Soolanayakanahally, R. Y., Amichev, B. Y., Ward, T., Laroque, C., Van Rees, K. C. J., & Akhter, F. (2020). Demonstration and Testing of the Improved Shelterbelt Component in the Holos Model. *Frontiers in Environmental Science*, *8*, 149. Advance online publication. doi:10.3389/fenvs.2020.00149

Kučera, J., & Podhrázská, J. (2016). *Analysis of Soil Aggregate Degradation in Heavy Soils Situated in Localities at Risk of Wind Erosion*. Conference MendelNet 2016, Brno, Chez Republic.

Kulshreshtha, S., & Rempel, J. (2014). Shelterbelts on Saskatchewan farms: an asset or nuisance. In S. Lac & M. McHenry (Ed.) Climate change and forest ecosystems (pp. 37–54). Nova Publishers.

Kwak, J., St-Hilaire, A., Chebana, F., & Kim, G. (2017). Summer season water temperature modeling under the climate change: Case study for fourchue river, Quebec, Canada. *Water (Basel)*, *9*(5), 346. doi:10.3390/w9050346

Laity, J. E. (2016). Eolian Erosional Processes and Landforms. In The International Encyclopedia of Geography. John Wiley & Sons, Ltd.

Lal, R. (1994). *Soil Erosion Research Methods* (2nd ed.). Soil and Water Conservation Society. doi:10.1201/9780203739358

Lee, J. J., Phillips, D. L., & Benson, V. W. (1999). Soil erosion and climate change: Assessing potential impacts and adaptation practices. *Journal of Soil and Water Conservation*, *54*, 529–536.

Letić, Lj., Savić, R., & Nikolić, V. (2009). *Wind Erosion Hazard in Western Bačka, Šumarstvo*. Academic Press.

Li, J., Ma, X., & Zhang, C. (2019). Predicting the spatiotemporal variation in soil wind erosion across Central Asia in response to climate change in the 21st century. *The Science of the Total Environment*, *709*, 136060. doi:10.1016/j.scitotenv.2019.136060 PMID:31905572

Li, M. M., Liu, A. T., Zou, C. J., Xu, W. D., Shimizu, H., & Wang, K. Y. (2012). An overview of the "Three-North" Shelterbelt project in China. *Forestry Studies in China*, *14*(1), 70–79. doi:10.100711632-012-0108-3

Li, W., Wang, F., & Bell, S. (2007). Simulating the sheltering effects of windbreaks in urban outdoor open space. *Journal of Wind Engineering and Industrial Aerodynamics*, *95*(7), 533–549. doi:10.1016/j.jweia.2006.11.001

Liddicoat, C., Hayman, P., Alexander, B., Rowland, J., Maschmedt, D., Young, M.A., Hall, J., Herrmann., T, & Sweeney, S. (2012). *Climate change, wheat production and erosion risk in South Australia's cropping zone: Linking crop simulation modelling to soil landscape mapping*. Government of South Australia, through Department of Environment, Water and Natural Resources.

Littlejohn, C. P., Hofmann, R. W., & Wratten, S. D. (2019). Delivery of multiple ecosystem services in pasture by shelter created from the hybrid sterile bioenergy grass Miscanthus x giganteus. *Scientific Reports*, *9*(1), 5575. doi:10.103841598-019-40696-2 PMID:30944349

Lukić, S. (2006). *The effect of standard windbreaks and usual agricultural crops on wind velocity* (Master's thesis). University of Belgrade, Faculty of Forestry.

Lukic, S., Belanovic Simic, S., Pantić, D., Beloica, J., Baumgertel, A., Miljkovic, P., Borota, D., & Kadović, R. (2018). Carbon storage in shelterbelts in the agroforestry systems of the Bačka Palanka area (Serbia). *AGROFOR*, *3*(2). Advance online publication. doi:10.7251/AGRENG1802080L

Lukić, S., & Dožić, S. (2006). Influence of black walnut belt along the road Pančevo - Kovin on the change of wind speed. Thematic collection of works. *Melioration - Land and Water - Novi Sad.*

Maleki, M., Ebrahimi, S., Asadzadeh, F., & Tabrizi, M. E. (2016). Performance of microbial-induced carbonate precipitation on wind erosion control of sandy soil. *International Journal of Environmental Science and Technology*, *13*(3), 937–944. doi:10.100713762-015-0921-z

Marshall, J. K. (1967). The effect of shelter on the productivity of grasslands and field crops. *Field Crop Abstracts*, *20*, 1–14.

Mayrinck, C. M., Laroque, C. P., Amichev, B., & Van Rees, K. (2019). Above- and Below- Ground Carbon Sequestation in Shelterbelts Trees in Canada: A Review. *Forests*, *10*(10), 922. Advance online publication. doi:10.3390/f10100922

Mezösi, G., Blanka, V., Bata, T., Kovacs, F., & Meyer, B. (2015). Estimation of regional differences in wind erosionsensitivity in Hungary. *Natural Hazards and Earth System Sciences*, *15*(1), 97–107. doi:10.5194/nhess-15-97-2015

Mezősi, G., Blanka, V., Bata, T., Ladányi, Z., Kemény, G., & Meyer, B. C. (2016). Assessment of future scenarios for wind erosion sensitivity changes based on ALADIN and REMO regional climate model simulation data. *Open Geosciences, 8*(1), 465-477.

Mize, C. W., Brandle, J. R., Schoeneberger, M. M., & Bentrup, G. (2008). Ecological Development and function of Shelterbelts in Temperate North America. *Toward Agroforestry Design*, 27–54. doi:10.1007/978-1-4020-6572-9_3

Montagnini, F., & Nair, P. K. R. (2004). Carbon sequestration: An underexploited environmental benefit of agroforestry systems. In New Vistas in Agroforestry. Springer. doi:10.1007/978-94-017-2424-1_20

Négyesi, G., Lóki, J., Buró, B., Bertalan-Balázs, B., & Pásztor, L. (2019). Wind erosion research in Hungary – past, present and future possibilities. Hungarian Geographical Bulletin, 68(3), 223-240.

Nowak, D. J., Appleton, N., Ellis, A., & Greenfield, R. (2017). Residential building energy conservation and avoided power plant emissions by urban and community trees in the United States. *Urban Forestry & Urban Greening*, *21*, 158–165. doi:10.1016/j.ufug.2016.12.004

Oldemann, L. R., Hakkeling, R. T. A., & Sombroek, W. G. (1990). World Map of Human-induced Soil Degradation: an Explanatory Note. Global Assessment of Soil Degradation (GLASOD), ISRIC and United Nations Environment Program (UNEP), FAO-ITC. International Soil Reference and Information Centre (ISRIC), Wageningen. United Nations Environment Programme (UNEP), in cooperation with Winand Staring Centre, Food and Agriculture Organization of the United Nations (FAO). International Institute for Aerospace Survey and Earth Science (ITC).

Osorio, R., Barden, C., & Ciampitti, A. (2018). GIS approach to estimate windbreak crop effect in Kansas-Nebraska. *Agroforestry Systems*, *93*(4), 1567–1576. doi:10.100710457-018-0270-2

Podhrázská, J., Kučera, J., Doubrava, D., & Doležal, P. (2021). Functions of Windbreaks in the Landscape Ecological Network and Methods of Their Evaluation. *Forests*, *12*(1), 67. doi:10.3390/f12010067

Possu, B. (2015). *Carbon storage potential of windbreaks on agricultural lands of the continental United States* (Doctoral Dissertation). University of Nebraska-Lincoln.

Possu, W. B., Brandle, J. R., Domke, G. M., Schoenberger, M., & Blankenship, E. (2017). Estimating carbon storage in windbreak trees on U.S. agricultural lands. *Agroforestry Systems*, *90*(5), 889–904. doi:10.100710457-016-9896-0

Ramachandran Nair, P. K., Mohan Kumar, B., & Nair, V. D. (2009). Agroforestry as a strategy for carbon sequestration. *Journal of Plant Nutrition and Soil Science*, *172*(1), 10–23. doi:10.1002/jpln.200800030

Rehacek, D., Khel, T., Kucera, J., Vopravil, J., & Petera, M. (2017). Effect of Windbreaks on Wind Speed Reduction and Soil Protection against Wind Erosion. *Soil and Water Research*, *12*(2), 128–135. doi:10.17221/45/2016-SWR

Reid, W. V. (2005). Ecosystems and human well-being: a report on the conceptual framework working group of the Millenium Ecosystem Assessment. Academic Press.

Riksen, M. J. P. M., & De Graaff, J. (2001). On-site and off-site effects of wind erosion on European light soils. *Land Degradation & Development*, *12*(1), 1–11. doi:10.1002/ldr.423

Rudd, L. (2020). *Carbon cycle assessment of shelterbelts in Saskatchewan* (Master's dissertation). University of Saskatchewan.

Santra, P., Moharana, P. C., Kumar, M., Soni, M. L., Pandey, C. B., Chaudhari, S. K., & Sikka, A. K. (2017). Crop production and economic loss due to wind erosion in hot arid ecosystem of India. *Aeolian Research*, *28*, 71–82. doi:10.1016/j.aeolia.2017.07.009

Schmidt, S., Meusburger, K., de Figueiredo, T., & Alewell, C. (2017). Modelling Hot Spots of Soil Loss by Wind Erosion (SoLoWind) in Western Saxony, Germany. *Land Degradation & Development*, *28*(3), 1100–1112. doi:10.1002/ldr.2652

Shahanov, V. M., & Cirella, G. T. (2021). Shelterbelt Planning in Agriculture: Application from Bulgaria. In G. T. Cirella (Ed.), *Human Settlements. Advances in 21st Century Human Settlements*. Springer. doi:10.1007/978-981-16-4031-5_8

Sharratt, B., Vaddella, V. K., & Feng, G. (2013). Threshold friction velocity influenced by wetness of soils within the Columbia Plateau. *Aeolian Research*, *9*, 175–182. doi:10.1016/j.aeolia.2013.01.002

Sharratt, B. S., Tatarko, J., Abatzoglou, J. T., Fox, F. A., & Huggins, D. (2015). Implications of climate change on wind erosion of agricultural lands in the Columbia plateau. *Weather and Climate Extremes, 10*(A), 20-31. doi:10.1016/j.wace.2015.06.001

Shi, P., Yan, P., Yuan, Y., & Nearing, M. A. (2004). Wind erosion research in China: Past, present and future. *Progress in Physical Geography*, *28*(3), 366–386. doi:10.1191/0309133304pp416ra

Sterk, G. (2002). Causes, consequences and control of wind erosion in Sahelian Africa: A review. *Land Degradation & Development*, *14*(1), 95–108. doi:10.1002/ldr.526

Stojanović, N., Ocokoljić, M., Anastasijević, V., Ralević, N., Veselinović, M., & Vasiljević, N. (2016). The Impact of Roadside Green Spaces on Ecological Conditions in the Urban Environment. *Poljoprivreda i Sumarstvo*, *62*(3), 37. doi:10.17707/AgricultForest.62.3.02

Stredova, H., Podhrazska, J., Litschmann, T., Streda, T., & Roznovsky, J. (2012). Aerodynamic parameters of windbreak based on its optical porosity. *Contributions to Geophysics and Geodesy*, *42*(3), 213–226. doi:10.2478/v10126-012-0008-5

Stredova, H., Spacilova, B., Podhrazska, J., & Chuchma, F. (2015). A universal meterorologicak method to identify potential risk of wind erosion on heavy textured soils. Moravian Geographical Report 23(2), Czech Academy of Sciences, Institute of Geonics.

Szigeti, N., Frank, N., & Vityi, A. (2020). The Multifunctional Role of Shelterbelts in Intensively Managed Agricultural Land – Silvoarable Agroforestry in Hungary. *Acta Silvatica et Lignaria Hungarica*, *16*(1), 19-38.

Tang, C. S., Yin, L., Jiang, N., Zhu, C., Zeng, H., Li, H., & Shi, B. (2020). Factors affecting the performance of microbial-induced carbonate precipitation (MICP) treated soil: A review. *Environmental Earth Sciences*, *79*(5), 94. doi:10.100712665-020-8840-9

The Intergovernmental Panel on Climate Change (IPCC). (n.d.). https://www.ipcc.ch/

Thevs, N., Strenge, E., Aliev, K., Eraaliev, M., Lang, P., Baibagysov, A., & Xu, J. (2017). Tree Shelterbelts as an Element to Improve Water Resource Management in Central Asia. *Water (Basel)*, *9*(11), 842. doi:10.3390/w9110842

Thuyet, D. V., Van Do, T., Sato, T., & Hung, T. T. (2014). Effects of species and shelterbelt structure on wind speed reduction in shelter. *Agroforestry Systems*, *88*(2), 237–244. doi:10.100710457-013-9671-4

UNCCD. (1994). United Nations Convention to Combat Desertification. A/AC.214/27. Paris, October 1994.

United Nations Treaty Collection. United Nations Framework Convention on Climate Change (1992). 1771, p. 107, https://treaties.un.org/pages/ViewDetailsIII.aspx?src=TREATY&mtdsg_no=XXVII-7&chapter=27&Temp=mtdsg3&clang=_en

von Storch, H. (2005). Models of Global and Regional Climate. Encyclopedia of Hydrological Sciences. doi:10.1002/0470848944.hsa035

Wang, F., Xu, X., Zou, B., Guo, Z., Li, Z., & Zhu, W. (2013). Biomass Accumulation and Carbon Sequestration in Four Different Aged Casuarina equisetifolia Coastal Shelterbelt Plantations in South China. *PLoS One*, *8*(10), e77449. doi:10.1371/journal.pone.0077449 PMID:24143236

Wang, Y., Dong, Y., Su, Z., Mudd, S., Zheng, Q., Hu, G., & Yan, D. (2020). Spatial distribution of water and wind erosion and their influence on the soil quality at the agropastoral ecotone of North China. *International Soil and Water Conservation Research*, *8*(3), 253–265. Advance online publication. doi:10.1016/j.iswcr.2020.05.001

Weart, S. R. (1997). The discovery of the risk of global warming. *Physics Today*, *50*(1), 34–40. doi:10.1063/1.881664

Weinan, C., & Fryrear, D. W. (1996). Grain-size distributions of wind-eroded material above a flat bare soil. *Phys. Geography, 17*, 554–58.

Weninger, T., Scheper, S., Lackóová, L., Kitzler, B., Gartner, K., King, N. W., Cornelis, W., & Michel, K. (2021). Ecosystem services of tree windbreaks in rural landscapes: A systematic review. *Environmental Research Letters*, *16*(10), 103002. Advance online publication. doi:10.1088/1748-9326/ac1d0d

Woodruff, N. P., & Siddoway, F. H. (1965). A wind erosion equation. *Soil Science Society of America Proceedings, 29*, 602–608.

Wu, T., Zhang, P., Zhang, L., Wang, J., Yu, M., Zhou, X., & Wang, G. (2018). Relationships between shelter effects and optical porosity: A meta-analysis for tree windbreaks. *Agricultural and Forest Meteorology*, *259*, 75–81. doi:10.1016/j.agrformet.2018.04.013

Xie, H., Wang, G. G., & Yu, M. (2018). Ecosystem multifunctionality is highly related to the shelterbelt structure and plant species diversity in mixed shelterbelts of eastern China. *Global Ecology and Conservation*, *470*. Advance online publication. doi:10.1016/j.gecco.2018.e00470

Yang, F., & Lu, C. (2016). Assessing changes in wind erosion climatic erosivity in China's dryland region during 1961–2012. *Journal of Geographical Sciences, 26*, 1263-1276.

Zhang, H., Fan, J., Cao, W., Harris, W., Li, Y., Chi., W., & Wang, S. (2018). Response of wind erosion dynamics to climate change and human activity in Inner Mongolia, China during 1990 to 2015. *Sci Total Environ, 639*, 1038-1050.

Zheng, X., Zhu, J., & Xing, Z. (2016). Assessment of the effects of shelterbelts on crop yields at the regional scale in Northeast China. *Agricultural Systems*, *143*, 49–60. Advance online publication. doi:10.1016/j.agsy.2015.12.008

Zhibao, D., Xunming, W., & Lianyou, L. (2000). Wind Erosion in Arind and Semiarid Chin: An Overview. *Journal of Soil and Water Conservation*, *55*(4), 439–444.

Zobeck, T.M., Baddock, M., Van Pelt, R.S., Tatarko, J., & Acosta-Martinez, V. (2013). Soil property effects on wind erosion of organic soils. *Aeolian Research, 10*, 43–51.

ADDITIONAL READING

Bao, Y., Li, H., & Yang, J. (2011, May). Notice of Retraction: Effects of Shelterbelts on Wind Erosion Control in the Desertified Cropland of North-Western Shandong Province, China. In 2011 5th International Conference on Bioinformatics and Biomedical Engineering (pp. 1-5). IEEE.

Bärring, L., Jönsson, P., Mattsson, J. O., & Åhman, R. (2003). Wind erosion on arable land in Scania, Sweden and the relation to the wind climate—A review. *Catena*, *52*(3-4), 173–190. doi:10.1016/S0341-8162(03)00013-4

Bartus, M., Barta, K., Szatmári, J., & Farsang, A. (2017). Modeling wind erosion hazard control efficiency with an emphasis on shelterbelt system and plot size planning. *Zeitschrift für Geomorphologie*, *61*(2), 123–133. doi:10.1127/zfg/2017/0406

Begimova, M. (2021). Climate indicators for forest landing and evaluation of forest shelterbelts. In *E3S Web of Conferences* (Vol. 227, p. 02004). EDP Sciences. doi:10.1051/e3sconf/202122702004

Rempel, J. C., Kulshreshtha, S. N., Amichev, B. Y., & Van Rees, K. C. (2017). Costs and benefits of shelterbelts: A review of producers' perceptions and mind map analyses for Saskatchewan, Canada. *Canadian Journal of Soil Science*, *97*(3), 341–352. doi:10.1139/CJSS-2016-0100

Wang, C. L., Liu, J., Sun, X., Zhang, X., Zhang, Y. L., & Yao, X. J. (2011). Simulating efficiency of resistance to wind erosion in area of complex erosion by wind and water. *Advanced Materials Research*, *183*, 1807–1811. doi:10.4028/www.scientific.net/AMR.183-185.1807

Zhang, F., Zou, X., Mao, R., Gong, D., & Feng, X. (2020). Wind Erosion Climate Change in Northern China During 1981–2016. *International Journal of Disaster Risk Science*, *11*(4), 484–496. doi:10.100713753-020-00291-w

KEY TERMS AND DEFINITIONS

Agroforestry: Is a mutual name for land-use systems and practices in which woody perennials are meaningfully integrated with crops and/or animals on the same land-management unit. The agroforestry system is the interaction of agriculture and trees with productivity, sustainability, and adaptability.

Carbon Sequestration: Is when atmospheric carbon dioxide is taken up by trees, grasses, and other plants through photosynthesis and stored as carbon in biomass and soils.

Climate Change: Includes global warming driven by human-induced greenhouse gas emissions and the resulting large-scale shifts in weather patterns.

Ecosystem Services: Are the several different benefits to humans provided by the natural environment and healthy ecosystems.

Shelterbelts: Are linear plantings of multiple rows of trees or shrubs established for environmental purposes such as protecting farmsteads and livestock areas, reduce wind speed and alter, wind field, saving energy, and to enhance wildlife habitat.

Soil Degradation: Is the loss of the intrinsic physical, chemical, and biological qualities of soil either by natural or anthropic processes.

Wind Erosion: Is a naturally occurring process that causes the transfer of soil particles by wind energy.

Chapter 9
The Importance of Proper Dam Maintenance Due to the Increase of Torrential Floods in the Face of Climate Change

Vesela Tanaskovic Gassner
Afforest for Future, Austria

ABSTRACT

In this chapter, the author discusses the importance of mitigation and adaptation actions needed to be taken from an environmental and engineering standpoint in regards to dams, reservoirs they form, the river basins they serve, and how this can benefit these systems in the future. One of the main problems identified for the mid-21st century will be the availability of fresh water. Currently, appx. 20% of the world's freshwater is stored in manmade reservoirs. However, these reservoirs sediment over time. This "sediment phenomena" adversely affects the water volume in reservoirs and their sustainable maintenance, potentially jeopardizing water supply and lives. To answer the "sediment phenomena," this chapter will explore a new approach to a no less devastating problem of land degradation, developed at the Technical University of Vienna. In the Balkan region, sediments are mostly composed of alluvial soil, decomposing organic matter, and sands, making them indeed a perfect soil amendment for degraded lands and barren topsoil terrains destroyed during torrential floods and landslides.

INTRODUCTION

Climate Change is a very complex issue of our time, and as such many international, intergovernmental and academic institutions worldwide have been trying to propose the best actionable advice to be considered by authorities when dealing with disaster aftermath, preparing for them and or risk assessment for such events.

Climate Change will have various consequences on Europe and the Balkans as its integral part, such as droughts, forest fires, river floods, which will all be felt throughout our economies.

DOI: 10.4018/978-1-7998-8459-0.ch009

Droughts are expected to be more severe and persistent in most of Europe, except for northern parts of Europe, including Baltic states and most northeast. South of Europe, including the Balkans will face strong reductions in low flows and experience agricultural droughts that can be expected to increase up to 7-fold in the various simulations, reaching 700,000 km2/year, almost twice the area of Germany.

The largest increase in the exposed area would be in Southern Europe (reaching nearly 60% of the total EU-affected area, compared to 30% today). People affected by droughts would also largely increase from today's levels by a factor of seven, reaching 153 million/year in the Reference simulation. Again, half of the overall population affected would be in the Southern Europe region. The multimodel ensemble projections of more cropland and people affected by drought in the south and the opposite signal in the north are statistically highly significant and robust amongst the Reference simulation members, while the projected changes are more dissonant in a transition zone in between. (Ciscar JC, 2014)

Looking into the same simulation references, forest fires are expected to more than double in Southern Europe, reaching more than 3 times the size of ex-Yugoslavia, almost 800,000 ha. However, if we were to achieve only a 2°C temperature increase, this would be around 400,000 ha, 50% less.

Torrential floods and soil erosion are among the top concerns when looking into the future of Climate Change, which increased or evoked disasters across the south and central Europe. The Balkans, as an integral part of Europe, has already been experiencing this increase in devastating effects, intensity and casualties.

Climate change is projected to largely change the frequency and magnitude of river floods. Flood damages could more than double with the 2080s climate under the Reference simulation, reaching around €11 billion/year. The largest increase would occur in the UK & Ireland and Central Europe South regions. The number of people affected by floods per year could almost double to 290,000. Under the 2°C simulation, the effects would be smaller, with annual economic damage of around €10 billion and 240,000 people/year affected by floods. If the 2080s economy is simulated (i.e. allowing for economic and population growth, instead of shocking the current economy as of today), then the damages would be much greater, reaching €98 billion/year and €68 billion/year under the A1B and E1 scenario, respectively. This difference is due to the much higher value of assets at risk because of economic and demographic developments. (Ciscar JC, 2014)

The adaptation measures to river flooding have been a focal study point in many simulation studies, with the most common objections to maintain a 1 in the 100-year level of flood protection across Europe manageable. However, it is important to note that flood damage simulations are usually subject to a high degree of uncertainty, partly because of the uncertainty in the extreme precipitation projections. The uncertainty of costs can be translated into all economic aspects, however, there is great recurrence that the overall economic costs will be around 2% of the total EU GDP today.

Most of the previous biophysical impacts (barring droughts and habitat suitability) have been integrated into an economic model to assess the effects in terms of household welfare losses. The economic effects consider the direct climate effects (as measured by the biophysical models) and the indirect effects in the economy (as calculated with the general equilibrium economic model). Under the Reference simula-

tion, the annual total damages would be around €190 billion, almost 2% the of EU GDP today. (Ciscar JC, 2014)

Unfortunately, when we look at the costs of climate change, it is clear that the global south will have to pay the price that was set by "industrial countries or so-called global west". The same will be with Europe and "poor" southern Europe, including the Balkans, will have to pay a major price in the very near future.

The geographical distribution of the climate damages is very asymmetric, with a clear bias towards the southern European regions. Relative to GDP (see next Figure), the welfare losses range from 0.2% in Northern Europe to 3% in both the Central Europe South and Southern Europe regions, i.e. fifteen times higher than the damage in Northern Europe. The highest welfare losses would occur in Southern Europe (€74 billion), Central Europe South (€58 billion) and Central Europe North (€45 billion). The damage in the two regions in the south accounts for 70% of the overall EU damages in monetary terms. (Ciscar JC, 2014)

In this chapter, we will look at trends that are expected to hit the region mid towards the end century and the effects they may have specifically on dams, reservoirs they make, downstream and upstream floodplains, river channels, "tamed rivers" and natural waterways specifically across Balkan region. We'll be taking into account the importance of hydropower in the region, but also the more pronounced land-use change, intensified urbanization, unplanned suburban sprawls, the legacy of the Yugoslavian era and the future projections of current states of action or inaction in the face of Climate Change.

BACKGROUND

For the past two decades of the 21st century, the Balkan region has been experiencing difficult torrential flood events for numerous consecutive years. In the face of Climate Change, these events are only expected to exacerbate, bringing more soil erosion, increased topsoil and vegetation losses together with environmental degradation, different levels of property damage, socio-economic losses and ultimately the irreversible loss of human lives.

A very short recollection of flood events from the region brings us to the May of 2014 and the flooding that happened across the Balkan peninsula, starting from Croatia, Bosnia and Herzegovina to Serbia and beyond rising the Danube levels along its course, causing human casualties on its way. Thankfully, this event led to greater empowerment of already existing cross border structures and cooperations, and to the formation of new relevant organizational cross border structures, which have till today produced a number of guidelines and recommendations, combining the data and values from all the countries of the basin, and producing the recommendations and actions looking at the Sava River Basin as a whole and addressing it as such, not cut up by borders.

In the report from the International Sava River Basin Commission from 2014 titled "Floods in May 2014 in the Sava River Basin - Brief overview of key events and lessons learned," we can find detailed reports on the weather and meteorological events that preceded the catastrophic torrential floods in May of 2014, and the aftermath, all neatly put together by each countries representatives.

The whole of April 2014 can be characterized as a period of unstable weather, with intense cyclonic activity. Several cyclonic systems passed over the region during April, bringing significant temperature drops, frequent rainfalls and even snow in some mountainous areas. Continuous precipitation was recorded at some meteorological stations (i.e. Banja Luka, Doboj) from 14th April till 4th May. Recorded rainfalls were significantly above the long-term average on almost all measuring stations in the middle and lower part of the basin. Unusually rainy weather caused the saturation of soils on large areas in the Sava River Basin. The soil moisture content in some parts of the basin was between 60 and 100% in mid-May. (International Commission for the Protection of the Danube River [ICPDR] & International Sava River Basin Commission [ISRBC], 2015)

Aside from extreme rainfall and snow that year, we can notice that the soil moisture was very high but not unorthodox for that type of rainfall. This came coupled with

Days of extensive rainfall on pre-saturated soil caused devastating floods in Balkans in May 2014. Due to a specific cyclone in mid-May 2014, a large portion of the Sava River Basin within Croatia, Bosnia and Herzegovina (B&H) and Serbia was hit by continuous, heavy rainfall. Intensive precipitation in the second half of April and the beginning of May preceded this event and caused a high soil saturation. This combination caused flash floods, erosion and landslides along small watercourses, but also disastrous flooding along the Sava River main course and its right tributaries. (ICPDR & ISRBC, 2015)

There were several factors at play during this particular torrential flood events as always; however, ones that are of relevance to this chapter and future forecasting are the land use, river basin maintenance - under which fall: dams, dikes, barges, spillways, channels, floodplains, regulated river flows and meandering planes, further, soil erosion, soil cover, urbanization and sewage systems (as closing up of soils in urban areas becomes more and more predominant trend). The ISRBC addresses most of these factors in its report, however, the urgency and the importance of addressing the outdated dam monitoring and controlling systems and also outdated dam structures themselves, of which across the Balkans most fall under the ex-Yugoslavian legacy, have not been mentioned. Perhaps the reason behind not taking into account the dams and their effect on the whole basin during this catastrophic event was the fact that most of the dams on river Sava are located in Slovenia, which did not suffer tremendously during this particular torrential flooding event.

Nevertheless, the dams as such present an obstacle and an opportunity for sustainable river basin development, especially in terms of climate change adaptation and mitigation measures in the Balkan region.

Why? Most of the dam structures in this region do not fall under major dams. Still, they are considered small to medium-sized dams, and as such, do not fall under the regulatory body of the International Commission on Large Dams, or ICOLD, meaning that their capacities, their maintenance and their productivity are left to be accounted for by local authorities and at best the local governments, most of the dams in the Balkans are still public property, as well as the reservoirs and water bodies they form. Calculating, forecasting and keeping track of the dead zone in the reservoir or the amount of sedimentation accumulated are key factors in assessing the risks in times of climate change.

When dam structures, most importantly, the systems that are in place to keep them safe and operable in the future are jeopardized, so are dams. Dams in the Balkans are at risk as well.

In the UK, for example, earthfill embankments are most likely to be vulnerable due to climate change:

increased erosion, more extreme fluctuations in water levels, changes in vegetation and prolonged drying during hot weather could combine to exploit existing weaknesses that may exist in the dam design or construction. The form of non-erodible structures (concrete, masonry, etc.) is unlikely to be particularly vulnerable to climate change, but there are exceptions, particularly at dams where existing climatic variation is known to cause problems associated with cracking or joint movement. Overflow structures and spillways may also be vulnerable due to the increasing frequency and size of flows and catchment impacts that might increase debris and vegetation.

Auxiliary structures such as valves or draw-off towers may be vulnerable to similar effects and can be prone to other factors such as siltation or heat-induced expansion. (Atkins, 2013)

Changes in hydrology or water quality of a reservoir are easy to notice and usually first to be noticed as such. However, various other less noticeable reservoir and dam functions, like increasing water level oscillations that can lead to a deterioration of inshore vegetation and therefore bankside fisheries, are far less likely to be observed without having a proper observation or monitoring system in place, and these are in direct correlation with climate change. These should be assessed using a different blend of approaches, such as: predictive modelling, trend analysis, or simple analysis of change factors by operators that are familiar enough with the catchment area, the reservoir and the dam.

Because of the variety of effects and timescales involved with climate change impacts at dams and reservoirs, the adaptation measures that have been recommended generally follow a 'plan' format, which involves an escalation of measures over time. Typically, this would range from monitoring to preventative or reactive maintenance, through to capital works and finally possible decommissioning or change of use. (Atkins, 2013)

Another problem that is at large present across the globe and in the Balkans is the aging of dam structures as the dam boom happened in the second half of the last century. For a bigger chunk of the world's dam structures, climate change was never considered in original plans. Nor do we have many dam plans that have set climate change adaptation measures in place.

The aging dam landscape faces new temperatures, snow, discharge, and flood patterns that increase the risk of hydrological failure. New dam release operations will be required to maintain historical levels of flood protection in the face of climate change. In addition, precise and reliable hydro-meteorological forecasts will be invaluable for maximizing flood protection and avoiding untimely and excessive outflows. (Boulange et al., 2021)

DAMS AND THEIR LEGACY IN TIMES OF CLIMATE CHANGE

There are numerous different classification styles and variables when talking about dam structures, this is perhaps the best described as follows:

Civil Engineering and its sub-branches like Hydraulics and Water Resource Engineering and/or Management, Structural Engineering, Coastal Engineering are usually the ones to deal with dam engineering.

This branching of civil engineering science is widely diverse across the globe and education systems. This is why today we have so many various dam classifications as we have different engineering groups in charge of their design. The most widely accepted classification system is provided by the International Coalition on Large Dams (ICOLD). As mentioned in the literature review, this system recognizes two basic types of dams:

- large dams - which they track and
- small dams - which they do not monitor.

In this classification, there are a few nuances, and so the large dams are seen by ICOLD as Large Dam - Grand barrage A dam with a height of 15 meters or greater from lowest foundation to crest or a dam between 5 meters and 15 meters impounding more than 3 million cubic meters, and defined in greater detail in the World Register of Dams.

While there are over 100 000 smaller dams with storage over 100 000 m3 and millions under 100 000 m3, many are unaccounted for by international organizations.

The large dams are further classified by ICOLD as single or multipurpose dams. In both cases, irrigation is the most common with 50% and 24%, respectively, reaffirming the analogy with agriculture. (Tanaskovic, 2018)

It is important to note that across the Balkans, there are a handful of large dams monitored by other institutions outside of the country or countries of placement, like Djerdap I. However, even these dams are extremely poorly maintained and often suffer turbine malfunctions due to inappropriate maintenance or in some cases, no maintenance at all. Debris is an overarching problem within dams and their reservoir maintenance, which cascades from environmental to engineering creating problems in each sector, and at some points, debris can cause a red alert in terms of jeopardizing the whole dam structure. This is not solely the case in the Balkans, but perhaps it is very easily visible here as large dams are not taken care of. Figure 1 presents the aforementioned problem.

Looking at medium or small dams, the overall importance of dam maintenance just becomes more predominant. Namely, most of the large dams have a so-called dead zone planned in the initial design of the reservoir and the dam itself. This zone is used to store the incoming silt or sediments, depending on the region, and to be safety storage to prevent the sediments from obstructing turbines during the first decades of the dam construction. This mainly depends on the geomorphology of the region and the location of the dam, climate conditions, incoming rainfall, frequency of flesh floods, vegetation cover, soil composition and similar activities that have an impact on turbidity and caring power of the river, which creates the reservoir, as well as the actual distance from the river entering the reservoir and the new dam. Nevertheless, dead zones are not bulletproof security. More often than not, dead zones get sedimented much faster than the initial projections calculated.

Siltation reduces the storage capacity of reservoirs, first reducing the dam's ability to produce hydropower, as with each cleaning of the turbines, production needs to stop and their efficiency before cleaning would have been reduced. Not only this, but siltation directly influences the water availability for the local communities. There is a less available volume for water and hence there is less available water for consumption. The same goes for irrigation. This has led many communities to abandon their villages once the dam has been heavily silted. With no access to potable and irrigation water, they were forced

Figure 1. Potpecko lake, Serbia (Reuters) – "Almost as far as the eye can see, trash spreads out over Serbia's Potpecko lake, lapping against the dam that crosses it"
Source: Reuters,2021

to relocate in large numbers, becoming the "first climate refugees". Such has often been the case in the developing world. After the siltation has taken over most of the dead zone, it can become a structural threat, sometimes even before that, causing the most dreaded of all the consequences of dam building, the dam failure. China has, on average, 68 dam failures each year on account of many factors, out of which siltation is always one of them. (Tanaskovic, 2018)

The problem with visible - debris and invisible - sedimentation dam and reservoir maintenance is that there is little to no literature for small and medium dams. Even the literature on sediment maintenance of large dams is obscured, as it has often been sponsored or written by dam engineers and dredging companies who have obviously vested interest in maintaining the status quo and repetitive dredging cycles. This is coupled with the fact that most of the dams are a legacy from the 20th century and the time when climate change has not been yet introduced nor taking into account when future forecasting the events an average 150 yr old dam should withstand, it is easy to see the eminent problem.

When looking into the Balkans in particular, the fact that the region is still recovering from the wars in the 90s and economic stagnation, it is clear that the dam infrastructures, reservoirs and the whole maintenance of the river basins don't come as the top priority. A similar situation is across the globe within all developing countries. Dams were built by one of the UN bodies financed by some extension or program of the World Bank, in a somewhat binding agreement with an unstable local government. Maintenance was rarely, if ever, put in place. This is why we have an increasing number of reservoir and dam sedimentation and abandonment issues. The effects of dam abandonment, reservoir sedimentation, and migration of peoples, are a subject for a whole other book.

Dams and the sediments they lock have a tremendous impact on the downstream environment. The regions downstream of the dam, which were usually floodplains, used to get a constant or yearly supply of fresh sediments with each year's flooding. This cycle has been interrupted with the dam's construction; now most if not all of the sediments are being locked behind the dam and stored within the reservoir.

This occurrence of silt or sediment accumulation firstly takes away the available water storage space in the reservoir, specifically in the Balkans, most of the dams were constructed during the Tito era and are multipurpose dams, meaning that irrigation or water supply was one of the core, if not core functionality. When this is diminished or parallel to it, the ability for efficient hydropower production is affected, and usually, if nothing is done, the dam loses the core stability of its structure. As the specific mass of sediments or silt is much higher than that of water, the weight of sediments is something that imposes an imminent risk to the dam structure. This impacts dam safety all over the world, especially in regions where torrential floods are a regular occurrence, dam failure due to fast inflow of sediments is very common.

On the other hand, the lack of sediments downstream is also causing problems. First problems are usually visible in environmental degradation of the habitats, for example, the decline of fish species as most small fish procreate and or eat the nutrients carried with sediments that the regular flooding was bringing and, etc. Secondly, once fully established new river bed together with the barges and bridges it holds becomes unstable, as the lack of sediment deposition or flooding increases the instability and the level of erosion along the shores of the river bed. Adding to this the usual occurrence which follows the dam construction of land-use change whereby, the forests or local flora is replaced by agricultural land.

Agricultural land as such is not good support for soil building, and only in rare cases - not documented in the Balkans - regenerative agroforestry can be of use in mitigating torrential floods and climate change. Agricultural land typical of the Balkan region is very loose and easily washable by heavy rains, increasing the likelihood of landslides and nutrient runoff in case of torrential floods.

Another problem connected to dam construction and land use is urbanization, as in many cases in the Balkan, dams were constructed to provide potable water. This then created an opportunity to turn more land into urban land, and unfortunately, with little to no good planning or taking into account Climate change and what it will bring to the region.

As many cities of Tito's era had only grown since, the initially built parks were usually the last available recreational spaces, and rarely do we see new parks being included in the masterplans. Huge green areas are seen disappearing under pressure for fast development and urbanization, whole reserves gone, massive allies cut down and urban green zones paved. Cities are expanding, during this fast development, the green areas are the ones suffering the most, being lost under massive loads of structures and surrounding pavements. The few remaining green surfaces became the only access for floodwaters to enter the soil.

Land Use and Natural Water Retention Measures in Times of Climate Change

The greying effect of cities is not a worldwide trend, quite the opposite in the EU. The trend of re-greening the urban spaces and re-greening the rivers and their supporting infrastructure has been in place by different policies and initiatives more or less since the formation of the European Union and most definitely since the turn of the century.

Natural water retention is explicitly mentioned in the EU Floods Directive (EU, 2007) and the maximization of its use forms part of specific objectives of the Water Blueprint (EC, 2012d). Other restoration measures for natural areas, like remeandering of natural ponds are (indirectly) recommended by a note on better environmental options for flood risk management (EC, 2011d) and are seen as a better environmental option and alternative to hard (grey) infrastructure. 'Flood risk management should work with nature, rather than against it' (EC, 2011d). Where the WFD (EU, 2000) has the water body as a central concept, limited attention is given to riparian zones, which might hinder the implementation of NWRMs to its full potential (EC, 2015g). NWRMs call for integration between the WFD and the Floods Directive and between nature legislation and all policy fields, where water and land planning needs careful coordination (EC, 2015g). (EEA report, 2016)

Reinstating some of the river's basic functions while using the Natural Water Retention Measures (NWRMs), Green Infrastructures and other measures that work with natural processes are integrated into policies that are widely spread across European countries. They need to find their anchorage in the Balkans. Restoration of river basins, opening meandering plains and creating space for natural floodplains is essential for a river to just be a river. But how can we even consider this with our urban structures, deforestation for the purpose of agriculture and also parallel abandonment of agricultural land?

For the Balkans, land use is a huge issue; sustainable land use or planning for sustainable land use that is used for mitigation and adaptation measures is far from reality. Daily reports from any number of local media will have a report on a loss of a tree alley in a city or a loss of a whole green space for a new shopping mall, or a new initiative on building small hydropower plants in national parks, microgrid and similar. Green space and green infrastructure are disappearing fast and, in some instances, irreversibly. At the same time, important world trends are being disregarded or not considered at all in terms of urbanization and planning for a livable second half of the 21st century.

The Balkan is located in the south of Europe, with the predicted growth of heat extremes and the number of consecutive super hot days only expected to increase in the upcoming years as shown in Figure 2., as well as the rise in the extreme flood events and torrential floods, this would mean focusing on quite the opposite of what has been done till today.

Climate change is expected to affect all water-related functions and policies discussed in Chapter 4 of this report (Ciscar et al., 2014; EC, 2013c; IPCC, 2014b). The shifts in the extremes, rather than the trend in the averages, are likely to be the biggest challenge for adaptation (IPCC, 2012) and are also likely to be the cost drivers for the adaptation of infrastructure (OECD, 2013). (EEA report,2016).

Cities like Athens, Thessaloniki, Skopje, Nis, Sofia, Sarajevo, Podgorica and much of the south of the Balkans, are very prone to fast forest fires, which lead to barren lands, which then, during high rainfalls, do not have anything to hold the soil in place, which further leads to areas prone to landslides depending on the terrain and or further degradation of barren soils. The local authorities need to consider this and create an appropriate response in terms of policies and local regulations, which need to be mandatory and in place for active measures reforestation and river basins revitalization, in order to secure the already active sites with landslides from further movement and creation of further topsoil loss and erosion.

Few studies have estimated future damages from inundation in response to an increase in intense rainfall (e.g. Willems et al., 2012). Pluvial flash floods are the result of short-duration extreme rainfall intensi-

Figure 2. Spatiotemporal characteristics of drought in Serbia
Source: Gocic and Trajkovic,2014

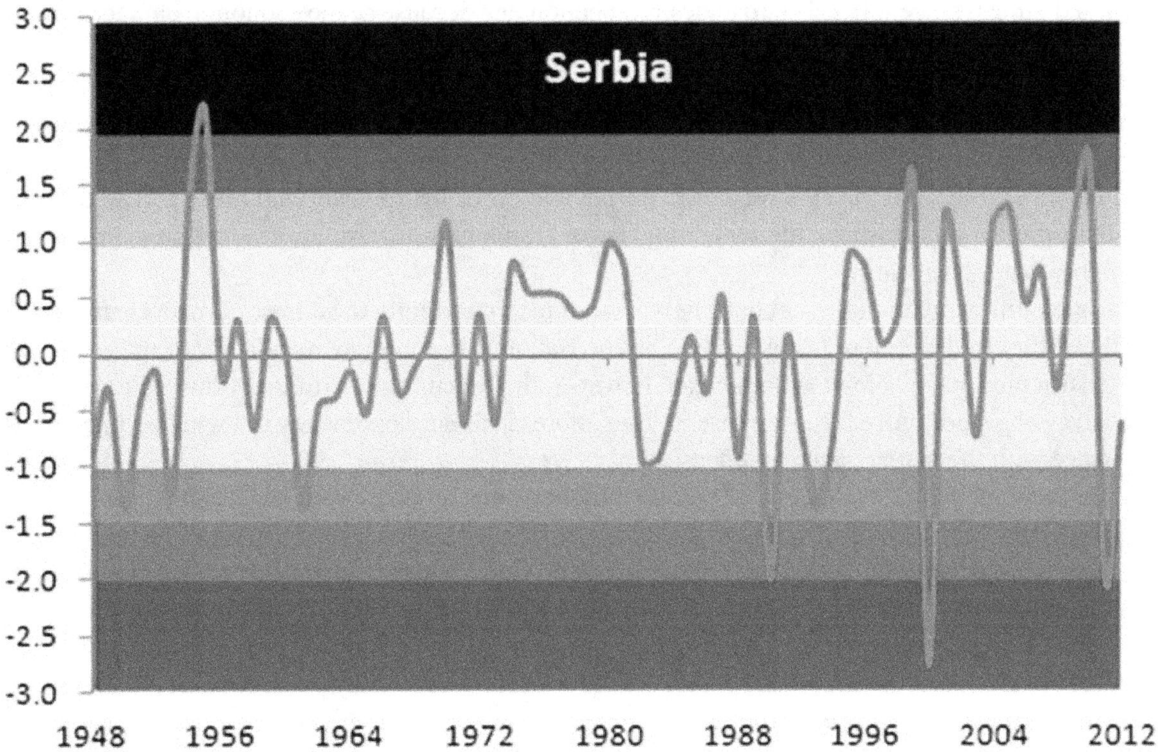

ties, leading to excessive surface run-off and ponding. Processes that influence flash flood risk include increasing exposure from urban expansion and forest fires that lead to erosion and increased surface run-off (Lasda et al., 2010). (EEA report, 2016)

The whole of Balkans suffers from uncontrolled urban expansion, while the south of Balkans is at higher risk from landslides and topsoil loss due to terrain, high heat extremes increase and local fires, as well as soil structures in comparison to the north of Balkans or over the Danube so to speak, predominantly because of its geographic location and outlook, the Pannonian Basin. However, this part of the Balkans has its own problems when thinking about the issue of torrential floods and soil erosion. Very important factors across the Pannonian plain and in the future are groundwaters and their pollution and or rise during floods; ore problems come from the land use change. Big chunks of up to 80 or in some cases 90% of land that used to be occasionally flooded alongside rivers are currently cut off from the river and do not function as active floodplains. Some of these are inherited problems or solutions for flooding from the Austro-Hungarian times, when the digging of first intersected channels started and continued to current times. Dunav-Tisa-Dunav channel is nearly 1000 km long and is the main irrigation supply for the huge part of Vojvodina. Therefore, it is understandable that the main pressures, when thinking about a different look, are economic development, as well as the regulation of water levels, and loss of connectivity as a result of flood-protection measures.

Land use changes that happened in the past 150-200 years in central and southern Europe, mainly from naturally forested areas into agriculture, housing development projects and industries, make irregular and rather frequent floods into undesired phenomena because of the economic damages caused. The whole of Sava river Basin is one of such regions, where forests used to dominate. Now, there is so much development adjacent to the river bed that there is no space even for a minor flood, as it is immediately characterized as moderate to severe risk because major living centers, industries, and agricultural land are surrounding the river. This is why, when possible (and in case of Dunav-Tisa- Dunav entirely regulated flow), the water level is regulated for navigation or hydropower and areas are protected by hard flood-protection measures, the remaining active floodplains, if any, are flooded in extreme cases with higher water levels only.

This grey infrastructure of river control gives a unique opportunity to address climate change and its unpredictability to the region. In some cases, in the Balkans, it is already mandatory to maintain some of these structures navigable or at least clean, however, the execution is a different thing altogether, but it still gives an opportunity for governments to explore different approaches in trying to increase the compliance with the maintenance regimens.

Using the remaining floodplains, as biodiversity hotspots, and may well represent a crucial role in sustainable flood risk management, these would be the locations in which NWRMs can be implemented most methodologically and adequately, serving as an open-door classroom and showplace from one project to another.

Regardless of the names used — NWRMs, building with nature, Room for the River, green measures, etc. — working with natural processes is vital to maximize the common goals and objectives of water management, economic development, nature conservation and ecosystem services. Flood risk management is surrounded by a multitude of uncertainties. Changes in flood regimes (mean annual discharges, maximum discharges, seasonality) show a mixed pattern across Europe. However, even in those cases in which a trend in flow regime is visible, it is difficult to separate a potential climate change signal from other drivers of change (such as land use or infrastructure). There are indications that the increase in reported flood damage should mainly be attributed to economic development as well as to better reporting. Also, an increased flood frequency shall remain uncertain due to climate change. Nevertheless, climate change deserves priority, because the lead-time for the adaptation of measures is often very long. Scenarios and foresight studies are recommended as tools. Sustainable solutions look beyond the protection of flood risk management measures and link it to the overlapping areas of vulnerability, environmental quality and the delivery of ecosystem services. Driving forces and pressures, like socio-economic and political developments on all scales (from local to European and global), can be estimated with only a certain level of detail. This has implications for land use, protection of floodplains, or the availability of funding. (EEA report, 2016)

Left column: Change in the annual maximum number of consecutive dry days (CDD: days with precipitation <1 mm). Right column: Changes in soil moisture (soil moisture anomalies, SMA). Increased dryness is indicated with yellow to red colors; decreased dryness with green to blue. Projected changes are expressed in units of the standard deviation of the interannual variability in the three 20-year periods 1980–1999, 2046–2065, and 2081–2100. The figures show changes for two-time horizons, 2046–2065 and 2081–2100, as compared to late 20th-century values (1980–1999), based on GCM simulations under emissions scenario SRES A2 relative to corresponding simulations for the late 20th century. Results

Figure 3. Projected annual changes in dryness assessed from two indices
Source: IPCC, 2012

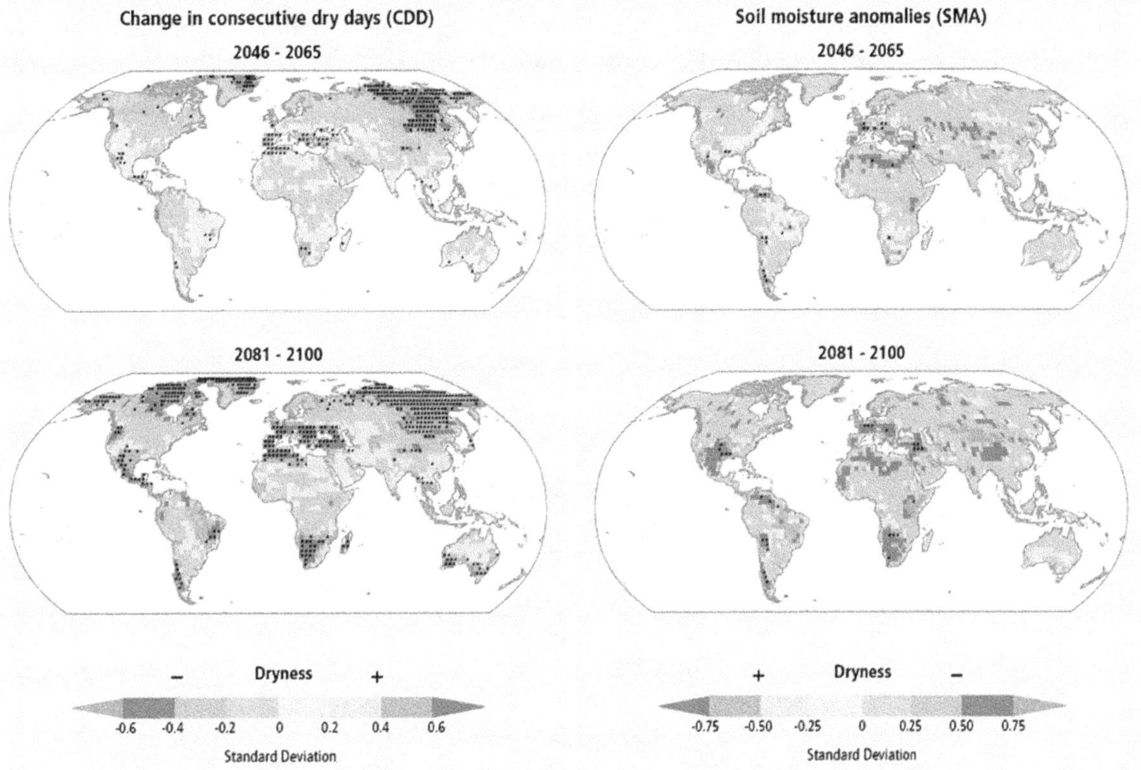

are based on 17 (CDD) and 15 (SMA) GCMs contributing to the CMIP3. Colored shading is applied for areas where at least 66% (12 out of 17 for CDD, 10 out of 15 for SMA) of the models agree on the sign of the change; stippling is added for regions where at least 90% (16 out of 17 for CDD, 14 out of 15 for SMA) of all models agree on the sign of the change. Grey shading indicates where there is an insufficient model agreement (<66%). [3.5.1, Figure 3-9]"

As shown in Figure 3, climate change is expected to affect the Balkans as well as the whole of Southern Europe adversely, but two of the most striking predictions are the increase in summer heat extremes and the predicted increase in the shifts in the extremes of water-related events rather than the continuation of averages, meaning we are to expect that torrential floods enter common vocabulary on a regular basis. In this regard, looking at the current state and the floods we have already experienced and learning from analyzing them and considering actions that can be applied cross basins is crucial to creating fast and flexible mitigation and adaptation measures that can be acted upon and implemented by local governments. Important note from the ISRBC, which the author chooses to stress, is this one as the top priority recommendation not only for the Sava region but the whole of Balkans:

It is very important to designate erosion-prone areas with a set of conditions for their use, to sustain existing forests, and afforest hilly and mountain regions in the Sava River Basin. These may prevent enormous sediment movement and landslides, which induced huge damages in May 2014. (ICPDR & ISRBC, 2015).

The importance of Climate Change influences over flooding events that have happened recently, and that will only continue to increase is perhaps best shown in Figure 4 from the EASCA report (2018).

Figure 4. Trends in different types of natural catastrophe worldwide 1980–2016 (1980 levels set at 100%). MunichRe NatCatSERVICE
Source: EASAC,2018

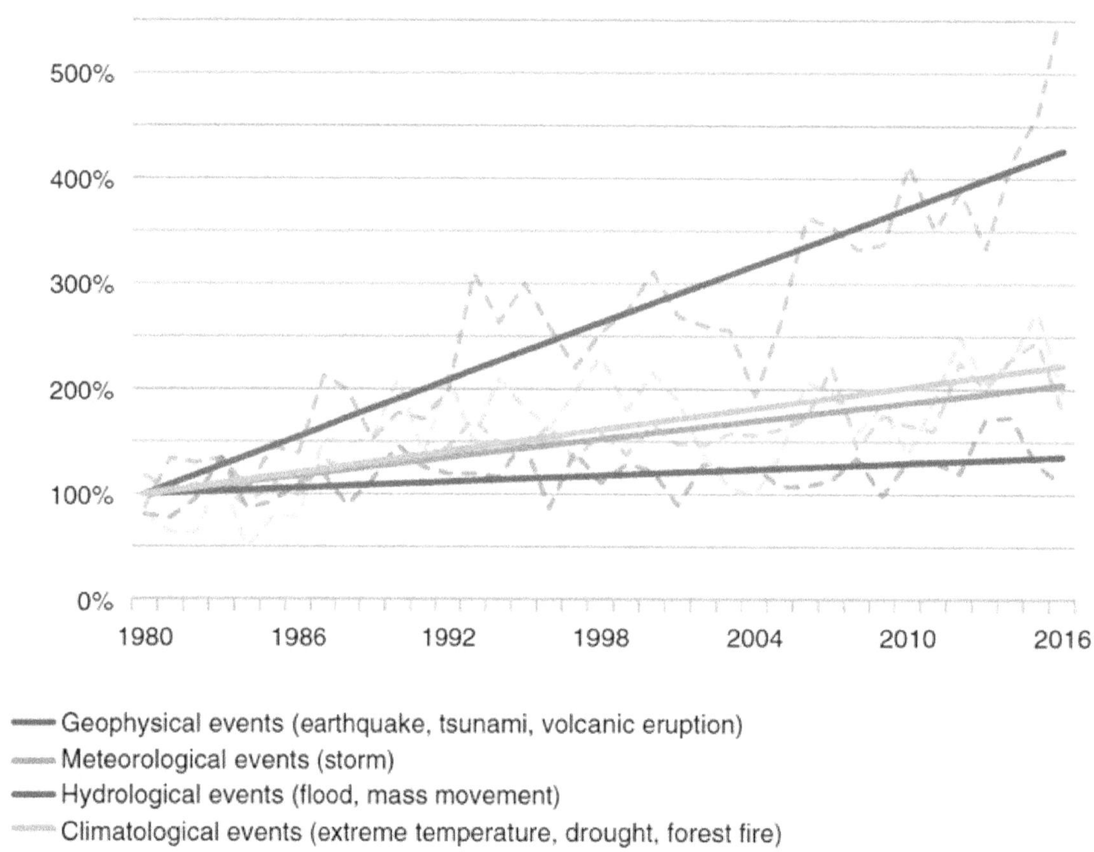

Here one can grasp the severity of the extremes of hydrological events across Europe, Balkans included, where they lead with over 400% increase since the 1980s! With the trend that is expected only to increase in the upcoming decades, it is of crucial importance for the whole region to start addressing the current state of water structures, of which most are legacy of a previous era, outdated and entirely unequipped to deal with the severity of the impending change.

SOLUTIONS AND RECOMMENDATIONS

As mentioned so far in this chapter, stressing the importance of mitigation and adaptation actions needed to be taken from an environmental and engineering standpoint in regards to dams, reservoirs they form as well as the river basins, canals and watercourses they serve and finding a possible solution to this imposing threat of perseverance in difficult times of Climate Change are of vital importance for the Balkans to be able to adapt. Climate change will bring to the region torrential floods, topsoil loss, vegetation loss, land degradation, soil erosion, and increase in summer heat maximums followed closely by desertification of barren soils and then overall economic decline. The social wellbeing of the predominantly agricultural region, that is Balkans, depends on proper land use which cannot be stressed enough, how crucial it will be to focus on nature and to observe its ways rather than trying to work against it both in the cities and in rural areas.

There is a positive side to these problems and the existing dam structures while taking into account all these predictions. It is the easily accessible use of accumulated sediments to build and rebuild riverbeds, meandering and floodplains, as well as fast and efficient afforestation and reforestation of barren soils. Using the channels, barges, dikes, and dams to reforest the barren soils or to increase the existing agricultural production.

In the event of torrential flooding reservoirs and dams get the extra pressure to create the buffer in accumulating the excess water from the flooding. Usually, this is not just the water; it is nearly always a mix of topsoil, vegetation and debris. Many dams had been built with predominantly this servitude in mind, however in the Balkans, as in many other developing regions across the globe dams are not sufficiently maintained, and therefore not able to withstand circumstances that climate change brings. The same goes for navigable and flood channels made across the Pannonian region, proper maintenance is a huge issue.

None of the dams built in the 20th century are, for that matter, made to take climate change into account, even today, we have massive dam construction projects on the top of the Himalayas that are just not suitable for the region and not considering the urgency of Climate Change. One of the main problems identified for the mid-21st century will be the availability of freshwater. Currently, approx. 20% of the world's freshwater is stored in man-made reservoirs. However, these reservoirs sediment over time. This "sediment phenomena" adversely affects the water volume in reservoirs and their sustainable maintenance, potentially jeopardizing the fresh water supply. To answer the "sediment phenomena", here we will explore the possible implementation of a new approach developed at the Technical University of Vienna from 2012-2018 looking to address a no less devastating problem of land degradation and soil erosion caused by torrential floods.

In the Balkan region, sediments are mostly composed of alluvial soil, decomposing organic matter, sands and gravel, making them a perfect soil amendment for degraded lands and barren topsoil terrains destroyed during torrential floods and landslides.

One of the possible solutions to address the existing flood and water retention systems is to make people and authorities realize the urgency and importance of proper maintenance, and do so by making it profitable.

During regular maintenance sediments are transported away from the bottom of the reservoirs in order to use them as a soil kick-starter for surrounding degraded areas. Sediments are dredged, ground in special facilities and further pumped in simple water pipelines to the designated areas or transported via rail or trucks depending on the best carbon option for a specific dam. Sediments can further be used topsoil kick-starter soil amendment, for the already barren lands, or places where landslides have taken off the vegetation cover, or simply along to shores of the reservoir, riverbed or river channel for that matter. For the region, it is crucial to plant trees in these soils, not just any but to focus on successful agroforestry projects to make it profitable.

Why? The Balkan region has always represented a pathway for people's movements or migrations, from south to north from east to west. It is a gate of sorts between Europe and Asia as well as Africa. In the next decades, with the business as usual climate scenario, we must be prepared for a natural heavy inflow of climate refugees. To be prepared, we must welcome them with roads that are shaded and well provided with easily accessible and free food, and these are edible plants and fruit-bearing trees. In the author's opinion it is by far the easiest way to welcome a sea of climate refugees that will cover the roads across the Balkans.

Part of this approach is also the right selection and combination of native and naturalized plants, which on the one hand are able to cope with adversities of the environment and, on the other hand, create a sustainable added value for the ecosystem as a whole. Choosing the right plants is important both from the perspective of rapid implementation as well as the perspective of job creation and retention in the region. For the smaller projects and privately-owned spaces, agroforestry is the best recommendation for being prepared for torrential floods, locking the soil with trees and producing nuts, fruits or similar low maintenance produce that predominantly only need to be harvested if planted in good relations. Using sediments to afforest the soil will increase the seedling survival rate based on up to 500% species and will ensure soil stability. And the very nature of agroforestry, its diversity ensures profitability regardless of the climate scenario or a disaster that hits that year, one of the species in a good agroforestry system will be harvested. Sediments from the river beds and channels in the Pannonian basin can be utilities to secure the banks and shores with agroforestry, and produce soil amendments after phytoremediation, as some sediments in this region may be heavily polluted.

This does not only solve the sedimentation issues, but also the problem of unemployment since a completely new ecosystem is developing around now fertile grounds, where there was only desolate land before. This also goes for the hydropower stations, as the only solution is a clean-up of debris, debris burning thermal power plants must be installed at some locations with a high inflow of debris, these are very easy to calculate cost-benefit analysis, where by installing an environmentally well-equipped incineration plant one saves not only the hydro station from debris, but from future collapse and possible dam collapse, while in the same time creating jobs. Along some rivers in the Balkans, one thermal station may well serve several hydropower stations.

FUTURE RESEARCH DIRECTIONS

It is important to look at all different available climate models, be critical and constantly compare previous predictions with the current reality. Looking specifically at torrential flood events and their configuration, soil erosion and possible outcomes after implementing best practices examples versus status quo. How will existing dams hold up in the Balkans during the next 10, 20 years and until the end of the century? Which are the most vulnerable dams, which regions? Which river basin needs to be addressed first?

In addition, future research could take a global perspective, as climate impacts in the rest of the world will affect Europe, e.g. via global (agriculture) markets or via migration flows. Generally speaking, uncertainties in climate change and impact modelling remain very large. The depth of uncertain analysis largely varies across sectors. These uncertainties need to be reduced in order to allow for robust statements regarding overall climate costs and the importance of costs in one sector relative to another. (Ciscar JC, 2014)

All this takes into account the recent massive torrential floods during the 2021 summer in Austria, Benelux countries and Germany. Countries with impeccable records in river maintenance, dam maintenance and channel maintenance, and countries that set standards for river basin and dam maintenance, where we have reported on the number of dam failures, and a few dams collapses!

Is the Balkan ready for this? And more importantly, what must be done in the next 10 years to be more prepared for such massive torrential floods and soil erosion?

CONCLUSION

There are a number of positive things happening across the Balkans as an answer to the recent torrential flood events. However, most are still at the level of academia and perhaps some have entered the policy level. Still, all are far from actual implementation and integration into current action plans for mitigation and adaptation measures against climate change.

Making actionable plans, policies and actually being able to integrate into the reality of the Balkans will play a crucial role in further dealing with soil erosion and torrential floods but also in all other aspects that Climate Change will bring to this region.

Therefore, first doing proper research on all water bodies, and especially small and under the radar hydropower plants and their impoundments, is crucial for full control over the fresh water system.

Dam maintenance is very specific to each dam, and it depends above all on constant and dependable surveillance and monitoring, but also on the river, its tributaries, its river basin, surrounding terrain, submerged terrain, as well as the frequency of rainfall and flood occurrence. Therefore, it is very difficult to make any type of general recommendations to minimum mandatory dam maintenance in the Balkan region, before knowing all the data. Still, biyearly monitoring performed by a dam safety engineer who would further recommend next steps should be a minimum for medium-impact dams, while for strategic dams like Djerdap, this needs to be daily. Usually, strategic dams also have remote structure monitoring, and many other basin monitoring systems in place. Below, for example, you can see an official recommendation for DSC - dam safety conditions, made by the Queensland government, on the proper operation and maintenance practices for non-strategic dams and their owners.

Proper operation and maintenance are essential for the continued viability and safety of a dam and its associated structures over the life of the dam. Appropriate operation and maintenance reduce dam safety risks through avoidance of:

- activities are not undertaken inappropriately or in an unsafe manner
- 'out of date' procedures being applied
- problems not being addressed because dam safety inspections are not performed or are not
- carried out by appropriate people
- critical equipment not being checked, then not being operational when needed

- poor maintenance that can result in abnormal deterioration of the dam, reduced life expectancy of the dam and increased risk of dam failure
- failure to open gated spillways at the appropriate time, which can cause overtopping of the gates and subsequent failure of the dam
- failure to close gated spillways or outlet works which may empty a reservoir
- poorly trained operations and maintenance personnel.

Queensland Department of Natural Resources Energy and Mine, (2020), Dam Safety Management Guideline.

Creating simple yet easy-to-follow documents like the previously mentioned one is the best recommendation at the moment. However, it is crucial to start with creating a list of all mini, micro, small and medium dams in the region, together with authority responsible for its maintenance and then integrating it into a larger river basin study, should be one of the first next steps. Here it must be clearly stated where one authority stops and the next one begins, further pushing this to the policy level where maintenance of river basins, including small and medium dams, must be prioritized. Abandoned lands and lands adjacent to the riverbed or meandering plans must be made eminent domains, and governments must take action to put in place proper NWRMs and protect national parks from further development, especially from mini and micro hydropower plants.

FUNDING

This research received no specific grant from any funding agency in the public, commercial, or not-for-profit sectors.

REFERENCES

Atkins. (2013). *Impact of Climate Change on Dams & Reservoirs*. Final Guidance Report UK.

Boulange, J., Hanasaki, N., Yamazaki, D., & Pokhrel, Y. (2021). Role of dams in reducing global flood exposure under climate change. *Nature Communications*, *12*(1), 417. doi:10.103841467-020-20704-0 PMID:33462241

Ciscar, Feyen, Soria, Lavalle, Raes, Perry, Nemry, Demirel, Rozsai, Dosio, Donatelli, Srivastava, Fumagalli, Niemeyer, Shrestha, Ciaian, Himics, Van Doorslaer, Barrios, Ibáñez, … Ibarreta. (2014). *Climate Impacts in Europe.* The JRC PESETA II Project. JRC Scientific and Policy Reports, EUR 26586EN.

Dam Safety DNRME. (2020). *Dam Safety Management Guideline.* Author.

European Environment Agency. (2016). *Flood risks and environmental vulnerability - Exploring the synergies between floodplain restoration, water policies and thematic policies.* EEA.

International Commission for the Protection of the Danube River and International Sava River Basin Commission. (2015). *Floods in May 2014 in the Sava River Basin-Brief overview of key events and lessons learned.* Author.

IPCC. (2012). Managing the Risks of Extreme Events and Disasters to Advance Climate Change Adaptation. A Special Report of Working Groups I and II of the Intergovernmental Panel on Climate Change. Cambridge University Press.

Sekularac, I. (2021). *More garbage than water: Serbia promises clean-up of hydro reservoir.* Reuters https://www.reuters.com/article/us-environment-serbia-dam-idUSKBN29B1D7

Tanaskovic, V. (2018). *Making the Sahara Desert Green - Sustainable transport of bottom lake sediments via pipelines to produce new agriculture land from desert* [PhD thesis]. Technical University of Vienna.

ADDITIONAL READING

Abdelsalam, A., Hamid, U., & Ismail, A. E. III. (2008). *Sediment in the Nile River System.* UNESCO-IHP-International Sediment Initiative.

Estrany, J., López-Tarazón, J. A., & Smith, H. G. (2016). Wildfire Effects on Suspended Sediment Delivery Quantified Using Fallout Radionuclide Tracers in a Mediterranean Catchment. *Land Degradation & Development*, *27*(5), 1501–1512. doi:10.1002/ldr.2462

Fearnside, P. (2019). Financiers to discuss hydropower as climate-change mitigation, but dams are not 'clean energy'. https://news.mongabay.com/2019/02/financiers-to-discuss-hydropower-as-climate-change-mitigation-but-dams-are-not-clean-energy-commentary/

Jeuland, M. (2010). Social Discounting of Large Dams with Climate Change Uncertainty. *Water Alternatives*, 3.

Null, S. E., Ligare, S. T., & Viers, J. H. (2013). A Method to Consider Whether Dams Mitigate Climate Change Effects on Stream Temperatures. *Journal of the American Water Resources Association*, *49*(6), 1456–1472. Advance online publication. doi:10.1111/jawr.12102

KEY TERMS AND DEFINITIONS

Adaptation: In human systems, the process of adjustment to actual or expected climate and its effects, in order to moderate harm or exploit beneficial opportunities. In natural systems, the process of adjustment to actual climate and its effects; human intervention may facilitate adjustment to expected climate.

Afforestation: Any new land being developed as a forest that has not been a forest since 31.12.1982 is considered afforestation.

Climate Change: A change in the state of the climate that can be identified (e.g., by using statistical tests) by changes in the mean and/or the variability of its properties and that persists for an extended period, typically decades or longer. Climate change may be due to natural internal processes or external forcings, or to persistent anthropogenic changes in the composition of the atmosphere or in land use.

Drought: A period of abnormally dry weather long enough to cause a serious hydrological imbalance. Drought is a relative term, therefore any discussion in terms of precipitation deficit must refer to the particular precipitation-related activity that is under discussion. For example, shortage of precipitation during the growing season impinges on crop production or ecosystem function in general (due to soil moisture drought, also termed agricultural drought), and during the runoff and percolation season primarily affects water supplies (hydrological drought). Storage changes in soil moisture and groundwater are also affected by increases in actual evapotranspiration in addition to reductions in precipitation. A period with an abnormal precipitation deficit is defined as a meteorological drought. A mega drought is a very lengthy and pervasive drought, lasting much longer than normal, usually a decade or more.

Flood: The overflowing of the normal confines of a stream or other body of water, or the accumulation of water over areas that are not normally submerged. Floods include river (fluvial) floods, flash floods, urban floods, pluvial floods, sewer floods, coastal floods, and glacial lake outburst floods.

Runoff: That part of precipitation that does not evaporate and is not transpired but flows through the ground or over the ground surface and returns to bodies of water.

Vulnerability: The propensity or predisposition to be adversely affected.

Chapter 10
Torrent Monitoring and Early Warning Systems Development:
Application and Lessons Learned

Dejan Vasović
University of Niš, Serbia

Ratko Ristić
University of Belgrade, Serbia

Muhamed Bajrić
University of Sarajevo, Bosnia and Herzegovina

ABSTRACT

The level of sustainability of a modern society is associated with the ability to manage unwanted stressors from the environment, regardless of origin. Torrential floods represent a hydrological hazard whose frequency and intensity have increased in recent years, mainly due to climate changes. In order to effectively manage the risks of torrents, it is necessary to apply early warning systems, since torrential floods are formed very quickly, especially on the watercourses of a small catchment area. The early warning system is part of a comprehensive torrential flood risk management system, seen as a technical entity for the collection, transformation, and rapid distribution of data. Modern early warning systems are the successors of rudimentary methods used in the past, and they are based on ICT and mobile applications developed in relation to the requirements of end users. The chapter presents an analysis of characteristic examples of the use. The main conclusion of the chapter indicates the need to implement early warning systems in national emergency management structures.

INTRODUCTION

Observed through different historical periods, catastrophes and different emergencies have always followed and marked different time-spans in the development of humanity (Stanković, 2019). Such events

DOI: 10.4018/978-1-7998-8459-0.ch010

are an integral part of the human past, present but also future. For centuries, natural and man-made disasters have caused fear and suffering to affected populations and destroyed natural and material goods (Knežević et al., 2018). Investing in disaster prevention can significantly reduce loss of life and human tragedy. In addition to the humanitarian influence realized by investing in prevention, which is rarely recorded in economic analyzes, such investment can also bring significant economic benefits and cost savings to states that would otherwise be burdened by increased post-disaster spending (Ristić, 2014). Global costs caused by disasters are on the rise (Janaćković, 2018). However, most of the international aid is currently consumed to answer to disaster and recovery instead of reducing future risks, which is quite an unsustainable condition. In the context of modern approaches to risk management from various disasters, priority is given to various measures of risk prevention and mitigation (Ristić, 2019). Creating a resilient disaster management system largely depends on the preparatory phase, as well as on learning from previous experience. The resilient disaster management system must be designed so that, in addition to efficient communication between the competent services, it enables a smooth, fast, but also timely flow of relevant information (Vasović, 2018a, b). In that sense, an indispensable part of any modern disaster management system is the early warning system, which has a dual role: timely informing the competent services, but also the average member of the affected population. No less important is the fact that modern early warning systems are a kind of database of actions and activities taken during and after a particular emergency, allowing later data analysis and stimulating build back better processes. In that sense, this chapter is stewarded to early warning systems description and explanation of its role in torrential flood monitoring and management processes. The key objectives of this chapter are a narrative description of early warning systems structure, financial issues related to the torrential flood management and practical experiences with such systems in Bosnia & Hercegovina and the Republic of Serbia.

BACKGROUND

The appearance of the first rudimentary early warning systems dates back to the Middle Ages when miners in brown and hard coal mines used cages with canaries that could warn miners of the presence of methane before an explosive mixture of methane and air appeared in the mine shafts. Since the physiology of canaries is significantly more sensitive to the presence of methane than humans, the weakness or death of canaries was a reliable, early signal that methane had appeared in the shaft and that evacuation and later ventilation were necessary. Such early warning systems were used primarily for those hazards that are not easy to detect, i.e., that are difficult to detect, such as the detection of methane concentrations in the Middle Ages. On the other hand, hazards that create clear signals such as fires did not require special warning systems given the clear propagation of combustion products. In terms of early warning of torrential floods, rudimentary early warning systems involved the creation of warnings in terms of noise, most often caused by drums or shots, less often by light signals. The purpose of each of these systems was to inform that part of the population could be affected by the consequences of torrential floods without even being aware of it. Common to all early, basic systems of early warning of torrential floods is the fact that torrential floods can form very quickly, that their effects are short but destructive and that there is a short time interval for information exchange and possible evacuation of citizens and mobilization of competent services (Dragović, 2017). Later, and especially during the 20th century, information exchange systems were significantly improved but were not used too much in the field of natural disaster risk management until the beginning of the 21st century, and the emergence of the first global initiatives aimed at preventing natural and man-made disaster risks, such as which are the Yokohama Strategy for a Safer World, the Hyogo Framework for Action and especially the Sendai Framework for Disaster Risk Reduction and its global target G on multi-hazard early warning systems. According to the United Nations, the early warning system (EWS) represents: *an adaptive measure for climate change, using integrated communication systems to help communities prepare for hazardous climate-related events. A successful EWS saves lives and jobs, land and infrastructures and supports long-term sustainability. Early warning systems will assist public officials and administrators in planning, saving money in the long run and protecting economies. The UN, working in diverse partnerships, has introduced a number of innovative early warning systems initiatives in vulnerable areas around the world*" (UN, 2021). As stated by (Chaves & De Cola, 2017), an early warning system represents the set of capacities needed to generate and disseminate timely and meaningful warning information that enables at-risk individuals, communities and organizations to prepare and act appropriately and in time to reduce harm or loss. In (Zambrano et al., 2017) modified the definition of an early warning system points that an early warning system allows harm and loss reduction by getting and disseminating warning information about hazards and vulnerabilities in a group of people who are considered to be at risk. Each word has a significant meaning, for example, community involves a network of social interaction, early refers to the prevention of any disaster or reduction of the potential harm or damage, warning means a message that announces danger and system puts all together. According to this author, typical early warning system has four key elements:

- risk knowledge;
- monitoring;
- response capability;
- warning communication.

Furthermore, according to the Serbian National Strategy for Protection and Rescue in Emergency Situations, principles on which the integrated protection and rescue system a based are:

- the right to protection,
- solidarity,
- publicity,
- preventive protection,
- responsibility,
- gradual use of forces and means, as well as active
- equal opportunities policy.

According to this strategy, the existing shortcomings highlight the need for better mapping and capacity building for risk analysis, promoting integrated vulnerability and capacity assessment, as well as improving early warning systems to develop resilient risk reduction strategies and measures that contribute to strengthening resilience at the local level (National Strategy for Protection and Rescue in Emergencies, 2011). Regarding Bosnia & Hercegovina, this area is regulated by Internal law on the protection and rescue of people and material goods from natural or other disasters in B&H, which stresses the importance of early warning system development but also the necessity of improved and prompt cooperation between two main entities (Federation B&H and the Republic of Srpska) and Brčko District, focusing also on cantonal levels (Internal law on the protection and rescue of people and material goods from natural or other disasters in B&H, 2008).

EARLY WARNING SYSTEMS – GLOBAL PERSPECTIVE

At the global scale, probably the most prominent early warning systems are those established by different United Nations divisions focusing on disaster risk reduction (like UNISDR, UNEP, UNDP, etc.), World Meteorological Organization (WMO), World Health Organization (WHO), Food and Agriculture Organization (FAO) and the World Organization for Animal Health (OIE). Particularly interesting are the International Telecommunication Union (ITU) activities, which has developed Multi-hazard Early Warning Systems and the Common Alerting Protocol.

The Food and Agriculture Organization of the United Nations (FAO), the World Organization for Animal Health (OIE), and the World Health Organization (WHO) jointly established the Global Early Warning System for Major Animal Diseases including Zoonosis (GLEWS). GLEWS became one of the mechanisms used by the OIE, FAO, and WHO to monitor data from existing event-based surveillance systems and track and verify relevant animal and zoonotic events. This mechanism has provided a global platform that brought together experts, data, functional networks, operational systems and stakeholders to improve inter-organizational coordination and support to Member countries for detecting, preventing and controlling threats to health and the food chain. GLEWS embodies a cross-sectoral and multidisciplinary collaborative tool in addressing health risks at the human-animal-ecosystems interface (GLEWS, 2021).

The United Nations Office for Disaster Risk Reduction (UNISDR) and the World Meteorological Organization (WMO) Secretariat, along with other international and national agencies, established the International Network for Multi-Hazard Early Warning Systems (IN-MHEWS). This multi-stakeholder partnership system facilitates the share of expertise and good practice on strengthening multi-hazard early warning systems as an integral component of national strategies for disaster risk reduction, climate change adaptation, and building community resilience. IN-MHEWS supports the implementation of the Sendai Framework for Disaster Risk Reduction 2015-2030, especially the achievement of its global target G on multi-hazard early warning systems and the United Nations Plan of Action on Disaster Risk Reduction for Resilience (IN-MHEWS, 2021).

Regarding the ITU activities, Information and Communication Technologies (ICTs) are an essential and integral component of Multi-Hazard Early Warning Systems (MHEWS) that manage and deliver alerting messages to those in affected areas, and broader national or international level which allows them to take action to mitigate the impacts of the hazard. An effective multi-hazard early warning systems should include the participation of different stakeholders and actively involve the people and communities at risk. Their aim is to ensure that the system has an enabling environment, which incorporates the appropriate technology, regulatory and legal frameworks, adequate operational capacities, and clearly defined roles and responsibilities for all participating agencies, including communities. The Common Alerting Protocol (CAP) has been adopted as ITU-T Recommendation X.1303. It is a simple but general format for exchanging all-hazard emergency alerts and public warnings over all kinds of ICT networks, allowing a consistent warning message to be disseminated simultaneously over many different warning systems, thus increasing warning effectiveness while simplifying the warning task. The Common Alerting Protocol - CAP lists technologies to be used: mobile and landline telephones, Internet (e-mail, Google, Facebook, Twitter, WhatsApp, smartphone apps, online advertising, Internet of Things (IoT) devices, in-home smart speakers, etc.), sirens (in-building or outdoor), broadcast radio and television, cable television, emergency radio, amateur radio, direct satellite broadcast, and digital signage networks (highway signs, billboards, automobile and rail traffic control), among others. CAP enables authorities to deliver early warnings and alerts to all people and communities at risk and up to a global scale through different technologies (ITU, 2021).

EFAS – European Flood Awareness System

EFAS (EUROPEAN FLOOD AWARENESS SYSTEM) is the first European flood monitoring and prediction system through hydrological forecasting, used as support to the relevant state institutions across Europe. It is a complex and polyvalent platform operating continuously in real-time (Rimac, 2019). The aim of EFAS is to support preparatory measures before major flood events strike, particularly in the large trans-national river basins and throughout Europe in general. EFAS is the first operational European system for monitoring and forecasting floods across Europe and is a component of the Copernicus Emergency Management Service (EFAS, 2021).

EFAS is a system comprising a consortium of multiple centers, each of which has its own mandate to collect and process hydrological and meteorological data and disseminate relevant hydrological forecasts. EFAS sends its forecast (notifications) about potential floods only to its partners and exclusively as support to their hydrological forecasting activities.

The hydrological model for EFAS is set at 5x5 km2 and provides a daily and hourly flood warning twice a day. EFAS results are based on multiple weather forecasts with different spatial and temporal resolutions, including data from other meteorological services, deterministic and short-, medium-, and long-term products. The majority of EFAS information is based on exceeding critical water levels instead of quantitative predictions. EFAS provides quantitative flow rate predictions, which partners can request to download, only at those stations (river profiles) for which the national hydrometeorological service provides relevant real-time data.

EFAS serves only as support (assistance) to institutions authorized to provide hydrological forecasts in their countries, but not before joining the partner system.

EFAS has flood alert levels for return periods of 1.5, 2, 5, and 20 years, which are interpreted as follows:

- 1.5-year return period – no flooding is expected;
- 2-year return period – bankfull conditions, possible flooding;
- 5-year return period – significant flooding is expected;
- 20-year return period – potentially severe flooding is expected.

EFAS produces the following types of hydrological forecasts:

- medium range;
- flash flood forecast (ERIC and ERICHA) – nowcasting";
- seasonal outlook.

For the flash flood hydrological forecast, EFAS provides different indicators, based on two completely different concepts, termed ERIC and ERICHA.

Figure 1. The ERIC flash flood indicators modelling chain
Source: EFAS official website: https://www.efas.eu/en/flash-flood-indicators

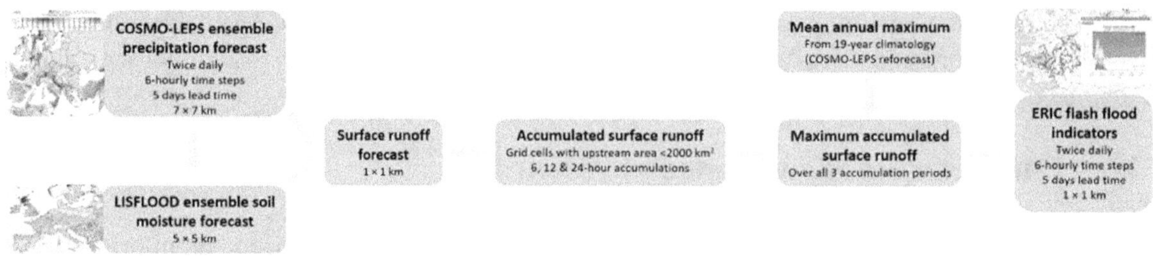

- ERIC flash flood hydrological forecast is generated from high-resolution numerical weather predictions with a lead time of 120 hours (5 days).
- ERICHA flash flood hydrological forecast is generated from radar-based precipitation monitoring and predicts up to 4 hours in advance (the system operates non-stop). Essentially, ERICHA is a case of nowcasting based on the European OPERA radar composite.

As defined on the EFAS official website, the ERIC flash-flood indicator is generated by comparing the forecasted surface runoff accumulated over the upstream catchment with a reference threshold. It is based on the 20-member COSMO-LEPS ensemble precipitation and soil moisture forecasts from the LISFLOOD hydrological model and provides indicators for the next five days for catchments smaller than 2,000km^2. Figure 1 shows the ERIC flash flood indicators modelling chain.

For now, two ERIC products exist:

- *Reporting points* ("ERIC Reporting Points" layers): points in the river network (where the catchment area is <= 2000 km^2) where flash flooding is possible. Enlarged triangles highlight where the flash flood forecast probability over the next 5 days meets specific criteria:
    1. probability of exceeding a 5 year return period magnitude of the surface runoff index is forecasted to be greater than or equal to 30%, and
    2. lead time of the above criterion being satisfied is <= 48 hours in a region where an EFAS partner exists.
- *Affected area* ("ERIC Affected Area"): river network which contributes to each ERIC reporting point, i.e., areas at risk from flash flooding ("ERIC Affected Area")

The ERICHA flash-flood indicator is generated from radar-based precipitation monitoring and nowcasting product, based on the European OPERA radar composite. This aims to capture very localised events difficult to predict from numerical weather prediction systems, but only provide information up to 4 hours. Figure 2 shows the chain of the updated ERICHA system.

Three ERICHA products consists:

- *Hourly precipitation maps*: Hourly precipitation totals from the OPERA radar composite, updated every 15 minutes ("ERICHA hourly accumulation precipitation" layer),
- *Flash flood hazard maps*: Sections of the river network are highlighted because their flash flood forecast probability over the next 4 hours meets certain criteria. The thresholds are based on

regional climatic characteristics and river basin upstream area as published by the MeteoAlarm consortium ("ERICHA - FF hazard levels forecasts" layer), and

- *Daily precipitation maps*: Daily gauge-adjusted radar rainfall accumulation over the last 24 hours ("ERICHA 24-h accumulations" layer).

Figure 2. The chain of the updated ERICHA system producing precipitation and flash flood hazard nowcasts
Source: EFAS official website: https://www.efas.eu/en/flash-flood-indicators

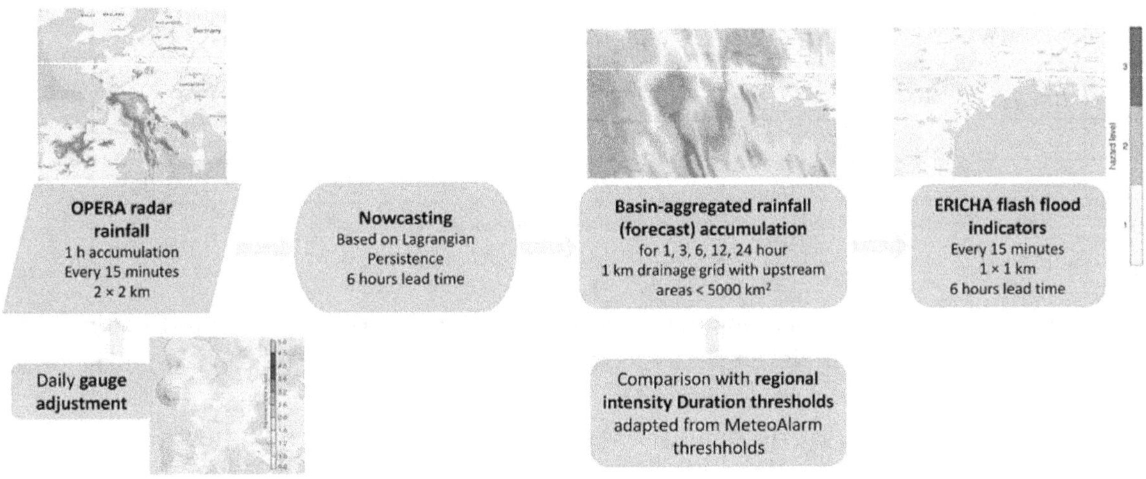

Besides those two flash flood indicators, EFAS provides medium-range flood forecasts as an overview of flood probability over the coming ten days. They are created by comparing the EFAS forecast simulations with EFAS flood threshold levels. These flood threshold levels are calculated for every grid cell, based on a discharge time series simulated by the operational LISFLOOD hydrological model from observed meteorological data (EFAS, 2021).

TORRENT MONITORING AND EARLY WARNING SYSTEMS – EXPERIENCES OF BOSNIA & HERCEGOVINA

Bosnia and Herzegovina (Bosna & Hercegovina – B&H) is predominantly covered by hilly and mountainous terrain, which is home to a large number of torrents with different intensities of destruction as well as to numerous areas affected by erosion processes of all types and intensities, and there is every likelihood that new torrents and erosion hot spots will emerge (Framework Water Management Foundation of B&H, 1994). The possibility primarily refers to the relief energy, geological and pedological foundation, climate condition, and the manner of utilization of the land resources and the vegetation cover, which create favorable conditions for intense erosion processes and torrents that can cause extensive damage (Sokolović et al., 2013). It is difficult to find a branch of the economy in B&H that has not been affected by soil erosion and torrents over the recent period. The damage usually involves siltation in residential areas, commercial and industrial facilities, communication infrastructure, water storage reservoirs, and

Table 1. Torrential basins with torrential flows

No	Primary basin area	Number of torrential flows	Surface area of the primary basin (km^2)	Surface area of the torrential basin (km^2)	%
	1.	2.	3.	4.	5.
1.	Una	96	8.185	1.301	15,8
2.	Vrbas	44	5.400	614	11,4
3.	Bosna	137	10.460	2.909	27,8
4.	Drina	291	7.200	2.546	35,4
5.	Immediate basin of the Sava River	81	5.029	1.536	30,5
6.	Neretva and Trebišnjica	237	11.535	3.782	32,8
7.	Cetina	49	3.320	281	8,8
8.	B&H	935	51.129	12.969	25,4

Source: "Cadaster of Torrential Flows and Erosion Areas in Bosnia and Herzegovina" (1985)

hydro melioration systems, and deterioration of water regimes in rivers, which causes flooding and all the resulting consequences. Special emphasis should be given to the damage due to erosion on forested and agricultural land, which reduces their productivity (Bajrić et al., 2013).

Catastrophic floods that struck B&H in 2014 were greatly responsible for giving this issue much more prominence and a growing sense of urgency. The urgency is only compounded by the effects of climate change, which are becoming increasingly prominent and scheduled to fully influence the occurrence of floods in the near future.

Table 1 shows the data on torrential flows in B&H from 1985 (Cadaster of Torrential Flows and Erosion Areas in Bosnia and Herzegovina, 1985).

The key flood events in B&H were registered in early January 2010 on the rivers Una, Sana, Vrbas, and Bosna. The major Drina basin floods were caused by extreme precipitation in Montenegro and partly from Serbian drainage basins. The flow volume of the Drina at the confluence with the Sava exceeded 4.000 m^3/s, which is the largest flow recorded over the last 50 years.

The heaviest flood in the Sava basin in B&H occurred in May 2014, resulting in severe human casualties and considerable damage to property, land, and businesses, thus leading to high economic loss (Flood Risk Management Plan for the Sava River Basin, 2018). Approximately 25% of average annual precipitation fell in only several days. The B&H floods from May 2014 were the heaviest recorded floods in the previous 120 years. The natural disaster affected one-quarter of the country's territory, with about one million people, which is approximately 27% of the total B&H population of 3.8 million. The total estimated economic damage amounted to over 4 billion convertible marks or 15% of the total B&H's GDP in 2014.

According to the official data from the competent institutions of the entities and the Brčko District, the 2014 floods killed 22 people, while two people were designated as missing. The total damage was estimated at over €2 billion.

The damage and destruction due to floods in B&H in 2014 can be summarized as follows (according to the Ministry of Security of B&H):

- 73 municipalities (50%) in B&H affected by floods;
- 100,000 homes damaged or destroyed;
- 230 schools and hospitals damaged or destroyed;
- 66,080 persons evacuated;
- 7,176 landslides activated;
- €2 billion total estimated damage (Crisis management – the case of 2014. floods, 2015).

Torrential Flood Prevention Measures Undertaken in Bosnia & Hercegovina

Even though numerous torrential basins with torrential flows have been identified in B&H, this issue was almost completely neglected during the previous period. According to the available data on flood prevention investment, some measures to manage erosion areas and torrential basins were indeed implemented, only not to a satisfactory level (Action Plan for Flood Control and River Management in BiH 2014-20, 2014).

Previous activities on erosion and torrent protection in B&H mostly involved works in river beds (about 87% of all works performed), typically mitigation works at the sites most heavily damaged by the torrents (Development of Flood Hazard and Risk Maps at Vrbas River Basin in BiH, 2017). Biological works in the basin, which are particularly beneficial to the country because of landscaping and improved microclimatic conditions, were much more reduced in scope and received only 13% of total investment, treating only about 10.000 ha of erosion areas in the entire country. Such practice could not lead to conservation and biological management of the basins, which would not only resolve the erosion issue but also restore the treated areas to the state where they could be used for manufacturing or commercial purposes.

Over the previous period, an updated erosion map (2012) was released for the Republika Srpska entity, while such a map is currently in development for the Federation of Bosnia and Herzegovina.

In terms of investments required to resolve these issues, estimates have been made for the Republika Srpska (Source: Framework Plan of Water Management Development in the Republika Srpska – Annex 3, 2006). The estimate was performed according to the same methodology developed by professor Gavrilović. The amount of €123,545,461 was allocated for 178 selected basins in the Republika Srpska, whereby the said methodology included the cost estimate for construction work (m^3/km^2) and forestation (ha/km^2) for each basin.

The average annual investments in the proposed program of activities amount to approximately €6 million per year. The sum is divided into €4 million per year for new anti-erosion works (with projects) and €2 million per year for the recovery of damaged anti-erosion facilities and their ongoing maintenance. One-third of the total annual costs are allocated for recovery and maintenance, which, if the plans are realized, will positively affect the optimal performance of anti-erosion works and the already built torrent defense facilities.

The above amount indicates that anti-erosion works require significant investments since almost nothing has been done to build new systems over the last several decades. In addition, the existing systems were not properly maintained and are now neglected, with most of them unusable. However, if the said amount is only placed against the damage and losses due to floods (significant damage also due to torrential flows themselves) from 2014. In that case, it is evident that even such relatively significant investments for the Republika Srpska are acceptable and undoubtedly justifiable.

Existing Flood Early-Warning Systems: Sava FFWS (Flood Forecasting and Warning System in the Sava River Basin)

As part of a joint project for flood management in the Sava river basin, implemented through the support of the World Bank, the project of establishing the Flood Forecasting and Warning System in the Sava River Basin (Sava FFWS) was initiated (Zeljko, 2019).

Sava FFWS is a special regional forecasting system installed and utilized by five countries in the Sava river basin – B&H, Montenegro, Croatia, Serbia, and Slovenia – which takes into account the that each country has its own models, monitoring and data collection systems, forecasting systems, authorized institutions, and interests. The aim of the project is to improve the quality and efficiency of joint flood and drought management in the Sava basin, to increase the capacity for flood and drought forecasting, and to encourage cooperation between the project countries, Slovenia, Croatia, B&H, Serbia, and Montenegro, and the International Sava River Basin Commission - ISRBC.

This system represents a platform for joint management of flood and drought forecasting across the entire Sava basin. The platform includes every existing forecasting model from the member countries. It also allows the production of forecasts for the portions of the Sava basin for which no models were previously available (Sava FFWS, 2021).

To develop a stable and functional system, server platforms were established for the project users in these countries: Slovenia (ARSO – Slovenian Environment Agency) – primary server platform; Serbia (RHMZ – Hydrometeorological Service) – backup server platform; B&H (AVP Sava) – backup server platform; and Croatia (DHMZ) – backup server platform. Forecast users access the system via a client application while the entire system is located on the server platform.

The establishment of Sava FFWS accomplished three main goals:

- flood forecasting and warning operating system for the entire Sava basin was installed in all the basin countries (B&H, Montenegro, Croatia, Serbia, and Slovenia);
- professional personnel in each country were successfully trained;
- recommendations were issued for improving the forecasting system in the future, including the proposal concerning the organizational structure for maintenance, use, and improvement of the system in terms of monitoring, telemetry, and model development and upgrades, with the prioritization of activities and assessment of financial requirements.

Sava FFWS is based on the Delft-FEWS software platform, which has already been used in many river basins throughout the world, but mainly within individual countries. Sava FFWS is implemented as an open structure for data and forecast management, enabling the integration of a broad spectrum of data and models. The concept was of particular importance for the five Sava-sharing countries, which use different models. Sava FFWS integrates the following input components, which are interconnected in the joint forecasting platform:

- observed hydrological and meteorological data in real-time;
- Hydrological Information System of the Sava basin (Sava HIS);
- a variety of Numerical Weather Prediction (NWP) models;
- available radar and satellite weather images;
- outputs and results of existing national forecasting systems;

- different meteorological, hydrological, and hydraulic models.

The modular structure of the Sava FFWS system allows users to use a unique forecasting process to place elements and combine the observed precipitation and air temperature data with different sources of NWP weather forecasts for 3, 5, 7, and 10 days in advance as the input data for the hydrological model and for transforming precipitation into the runoff, followed by runoff propagation through the hydraulic model in order to calculate the water level. Because of the number of available NWP, hydrological, and hydraulic models, Sava FFWS is configured to allow multiple operative processes of forecasting and analysis of result spectrums on predicted water levels and water flow rates. The users can always adjust them according to their own needs.

As the end result, the predicted water level or flow is compared via the automated process against the official reference values when declaring different flood events or flood defense activities. If the reference values have been exceeded, Sava FFWS will generate a warning for the forecaster.

The authorized forecasting service will then use this information to prepare a report in accordance with their official procedures. For the purposes of preparing a report and forwarding the warning to active flood control services and/or emergency services, Sava FFWS provides options that allow the warnings to be forwarded via a separate webpage, online service, SMS, etc., so that appropriate action defined by official procedures could be taken promptly. Through its client application options, Sava FFWS also allows users to conduct various analyses, such as input data assimilation, reliability analysis, what-if scenario analysis, and many others.

Thus far, the use of EFAS elements in the Hydrometeorological Institute of the Federation of B&H has shown that notifications issued to partners were useful for implementing flood control measures.

TORRENT MONITORING AND EARLY WARNING SYSTEMS – EXPERIENCES OF SERBIA

Torrential floods are the most common occurrence out of all so-called 'natural risks' in Serbia (Ristić, 2014), where about 11,500 torrential water courses have been registered, mainly south of the Sava and the Danube but also in the northern Province of Vojvodina (Fruška Gora, Vršac Hill, and Titel Hill). Erosive processes, as one of the contributory factors of torrential floods, are present in 75% of the territory of Serbia (Ristić et al., 2019), with an average annual erosive material production of 30 million m^3, of which approximately 8 million m^3 reaches the river and stream beds (the cause of enormous amounts of mud and rocks in flooded towns and villages). Only within the 1950-2020 period, torrential floods killed over 130 people and resulted in material damage in excess of 12 billion euros. The incidence, intensity, and extent of torrential floods make them a constant threat to the environmental, economic, and social sphere (Ristić et al., 2012a). Figure 3 shows the consequences of torrential floods in the Republic of Serbia.

In May 2014, catastrophic torrential floods occurred in Serbia, causing several dozen human casualties and the total property damage exceeding 2 billion euros. The event was the result of a climatic and meteorological phenomenon that is difficult to predict and highly unlikely to occur, whereby daily rainfall accumulated to 190-218 liters per square meter. Serbian authorities claimed that it was an unexpected natural event (Ristić, 2014), but even a superficial chronological analysis reveals that only in the 1996-2020 period, five devastating torrential floods occurred in the drainage basin of the Kolubara, three in the basin of the Drina, and one in the basins of the South, the Great, and the West Morava,

Figure 3. Torrential floods in Serbia – consequences: • human life losses and material damage; • material damage only; the period from 1950 to 2020
Source: Ristić et al. (2012a)

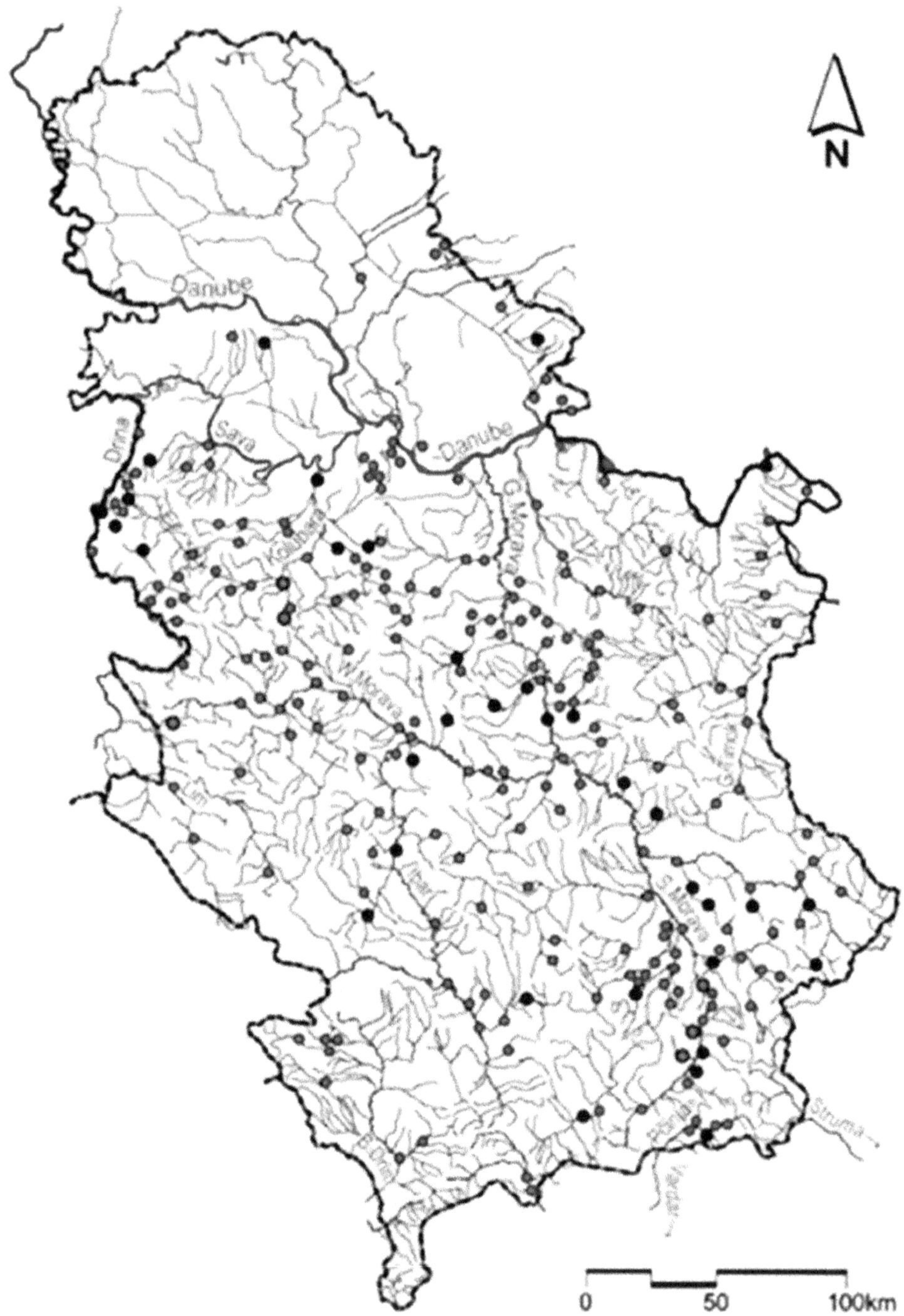

the Timok, and the Pčinja. Over the last 70 years, daily rainfalls exceeding 120 mm were registered 60 times in Serbia, those exceeding 140 mm 25 times, while the highest daily rainfalls were registered at rain gauging stations Rakov Dol (220 mm, South Morava basin) and Planina (218 mm, Drina basin). Properties of the cyclone that formed in May 2014 (static nature, duration, spatial coverage, spatial and temporal rainfall distribution) led to the formation of torrential floods in Serbia over an area of more than 30,000 km^2, which was a precedent compared to the previous meteorological/hydrological events. Yet, there are no guarantees that similar events will not occur in the future, the only uncertainty being when exactly they will occur. It is best exemplified by the ongoing floods in Germany, Austria, Belgium, and Luxembourg, which surpassed the Serbian floods of 2014 in devastation, human casualties (over 190 people), and property damage (over 5 billion euros).

Torrential Flood Prevention Measures Undertaken in the Republic of Serbia

The destructiveness of torrential floods from May 2014 and sporadic springtime flood events from 2016 to 2021 was considerably augmented owing to the absence of preventive action and measures, which has been a noticeable trend since 1990. This primarily refers to the integral management of torrential basins, which involves design and construction of technical (surge barriers, weirs, regulation structures, micro reservoirs, retention basins, bank revetments), biotechnical (gully remediation; slope protection), and biological structures (afforestation of bare land; amelioration of degraded forests, meadows, and pastures; planting of terraced orchards), as well as the implementation of administrative measures (management regulations, land use and protection in vulnerable basins). Likewise, the existing torrent and erosion protection systems (cleaning of regulation structures, natural beds, and barriers from deposits, vegetation, and waste material; repairs of damaged facilities) were not maintained, which significantly reduced their efficiency (Ristić, 2014). In addition, the administrative measures defined in the *Plans for Declaring Erosion-prone Areas* (PPEP) were not implemented in the territories under the jurisdiction of local self-governments (cities/municipalities). The PPEP identify vulnerable areas (according to the methodology prescribed by the Ministry of Agriculture, Forestry and Water Management), after which optimal land use (forested, agricultural, urban) measures in terms of erosion and torrential flood prevention are implemented to the level of cadastral parcels (Ristić, 2014). Numerous local self-governments still possess completely unusable *Operational Plans for Flood Control of Second-order Watercourses* (OPOP_II).

Three levels of responsibility where prompt action for torrential flood prevention was lacking have been observed (Ristić, 2014):

- systemic level (the state);
- local level (cities and municipalities);
- individual level (citizens).

Systemic level of responsibility refers to actions taken by the state in terms of financing flood control works, authority and purview of water management companies, ownership status of regional water management companies, current legislation, and the position of water management in the system of public activities. Local level of responsibility refers to the perception of the problem and the total extent of activities performed by city and municipal authorities, where urban planning (control of legal and illegal building construction in flood zones) and public utility (control of waste disposal along the banks and inside the river and stream beds) regulation take priority. Individual level of responsibility refers to citizens whose actions contribute to the destructiveness of torrential floods and who thus endanger themselves and their environment (illegal construction of residential buildings in flood zones and waste disposal along the banks). In addition, there is also the fact that the media seem to be completely disinterested in these occurrences until they actually take place and that TV stations that broadcast nationwide allocate negligible time in their programming for educational content about torrential floods in Serbia.

Efficient prevention measures for the entirety of Serbia can be prescribed according to the following documents, which need to be created as soon as possible, considering the potential for human casualties and extensive property damage (Ristić, 2014):

- National strategy for soil erosion and torrential flood control;
- Erosion map of Serbia;
- Cadaster of torrential flows in Serbia;
- Cadaster of completed anti-erosion works.

The May 2014 floods were characterized by two observable segments – torrential floods (formed on hilly and mountainous areas of Serbia) and the river flood on the lower stream sections of the Kolubara and the Sava (Ristić, 2014). Characteristics of torrential floods, such as origin, effects and duration, and prevention and control, considerably differ from those of river floods. A torrential flood wave is formed within several hours after heavy rainfall, when it exhibits its devastating effects and then slowly dissipates, as was the case in May 2014 in Krupanj, Ljubovija, Bajina Bašta, Osečina, Paraćin, Svilajnac, and Mali Zvornik. At the same time, a flood wave on larger rivers can be detected several days in advance, which happened in the Sava basin, allowing the authorities to set up flood defenses in Srtemska Mitrovica and Šabac. Therefore, the current *Law on Waters* (Official Gazette of the Republic of Serbia, no. 95/2018) needs to be harmonized with the EU *Floods Directive* (2007/60/EC, Article 2), which recognizes torrential floods in addition to river floods. Likewise, the *Law on Waters* contains the damaging item 5 within Article 23: "Water facilities for watercourse management and flood control on second-order waters and water facilities for erosion and torrent protection, with the exception of water facilities for erosion and torrent protection from item 3 of the current Article, which is publicly owned, are managed by the local self-government unit on whose territory the facilities are located". The 2014 floods showed that local self-governments (cities and municipalities), except for a few exceptions, do not possess either material or professional resources to handle flood control on second-order watercourses (torrential watercourses). Consequently, full jurisdiction of state institutions is warranted for the design, construction, and maintenance of facilities for erosion and torrential flood control (Ristić, 2014). Article 26, item 2 of the *Law on Remediation of Flood Consequences in the Republic of Serbia* (Official Gazette of the Republic of Serbia, no. 68/2015) fails to state that the said facilities also include water structures for watercourse management, flood, erosion, and torrent control, as well as drainage facilities.

Coordination of operations of water management companies, approval of operational programs, and provision of funds should be conducted via a separate organizational unit for erosion and torrent control within the Republic Water Directorate in conjunction with public water management companies Srbijavode, Vode Vojvodine, and Beogradvode. The umbrella Ministry of Agriculture, Forestry and Water Management comprises organizational units that deal with water management, forestry, and agriculture (Water Directorate; Forestry Department; Agricultural Land Department), dominant sectors for spatial treatment in terms of erosion and torrential flood prevention. However, current practice has shown a complete lack of 'horizontal' coordination (Ristić, 2014) during the implementation of projects that promote "sustainable land use practices, improvement of water retention as well as the controlled flooding of certain areas in the case of a flood event" (Directive 2007/60/EC on the assessment and management of flood risks).

Articles 6 and 8 of the current *Law on Forests* (Official Gazette of the Republic of Serbia, no. 95/2018) recognize the general use functions of forests through the following items:

water relations balance and torrent and flood wave prevention; protection of soil, settlements, and infrastructure against erosion and landslides; protective forests are forests whose primary function is to protect the soil against erosion.

However, some forestry activities in hilly and mountainous areas of Serbia actually help intensify the erosion and form torrential floods (choice of locations for forest exploitation; log skidding methods; methods of construction, use, and maintenance of forest roads). Yet, all of this is still not enough to urge PC Srbijašume and PC Vojvodinašume to open new job positions for professionals dealing with erosion and torrent management (Ristić, 2014). In comparison, erosion protection and torrent management are officially included activities within the forestry sector in France and Austria. The *Law on Agricultural Land* (Official Gazette of the Republic of Serbia, no. 95/2018) recognizes the importance of proper treatment of arable areas, intending to create a model of sustainable land use to prevent destructive erosive processes and torrential floods (articles 18 and 19). Unfortunately, the highly useful provisions of the *Law on Forests* and the *Law on Agricultural Land*, pertaining to erosion and torrential flood prevention, are most often not implemented in practice (Ristić, 2014).

Financial Support for Prevention Concept Implementation in the Republic of Serbia

If there is no proper flood control, there is no reason to be optimistic about the safety of the population, the building of infrastructure and new production capacities, and agricultural improvement, which jeopardizes the entire concept of social development. Therefore, the allocation of funds for flood control is a prerequisite for social development planning and necessary action from political and state bodies (Ristić, 2014).

Serbia requires a 90-million euros investment annually (10.58 billion dinars according to the exchange rate 1 euro = 117.5606 dinars, on the day of 26 July 2021), of which 30 million are required for torrential flood control (3.53 billion dinars) and 60 million (7.05 billion dinars) for river flood and internal water flood control over a 10-year period, in order to ensure optimal efficiency of the flood control system and to provide the population and the economy with a higher degree of safety. The sum of 900 million euros is a significant burden for the Serbian national budget, but it should be noted that direct property damage from May 2014 floods alone amounted to over 1.7 billion euros and resulted in negative GDP growth (Ristić, 2014).

After the 2014 floods, a detailed and comprehensive analysis of all causes and manifestations of torrential floods was never performed (Ristić et al., 2017). Mitigation and recovery of the damaged or destroyed water management infrastructure were indeed helpful in reaching a certain level of resilience in the event of flood recurrence. Extensive work was performed to reconstruct and repair the damaged infrastructure, but few new facilities were built. Using different international financial sources (UNDP – United Nations Development Program; UNOPS – United Nations Office for Project Services; governments of Japan, Canada, and Turkey), a total of 44 torrent barriers were built, with the total value of 1.67 million euros. However, it must be noted that 30 to 50 of such structures used to be built during the 1970s and the 1980s in Serbia using only national budget funds (Ristić et al., 2017).

Table 2. Financial support for adverse water-related effects protection

Watercourse management and protection against adverse effects of water	Allocated funds (in billion dinars)	Allocated funds compared to required funds (%)
2018	2.30	21.74
2019	2.88	27.22
2020	2.92	27.57
2021	3.18	30.00

Source: Business plan of PC Srbijavode for 2018, 2019, 2020, and 2021

According to the business plan of Public Company Srbijavode for 2018, 2019, 2020, and 2021, 21.74% to 30% of the necessary funds were allocated for the item entitled "Watercourse management and protection against adverse effects of water" (Table 1).

In 2018, only 7.42 million dinars were allocated for biological and technical works; in 2019 and 2020, the funding was raised to 13 million dinars, while the item was completely removed from the business plan for 2021. The fact is that biological and biotechnical works were almost completely neglected, except for the afforestation of eroded, barren, and burnt surfaces, mainly in hilly and mountainous areas covering from 2,500 to 3,000 hectares. Integrated management of torrential watersheds (Ristić et al., 2012b), which involves not only technical and biotechnical but also biological works, requires at least ten years to reach the full effect of vegetation (water interception and improved infiltration and retention). Stable forest ecosystems substantially modify the process of transforming rainfall into runoff using interception (retaining a portion of the rain on needles and/or leaves) (Ristić & Macan, 2002), infiltration, and retention in the powerful forest soil, thus significantly reducing the probability that fast surface runoff will form (Ristić & Macan, 1997).

The surface area of the entire country of Serbia is 88,361 km^2, of which 29.1% (25,713.05 km^2) is covered by forest vegetation. According to the current Spatial Plan of the Republic of Serbia, the plan is to increase the forested area to 41.4% (36,581.45 km^2), i.e., to increase it by 10,868.40 km^2 (1,086,840 ha) by 2050. The plan is to plant approximately 200,000 hectares of forests as multi-row protective belts, primarily in the autonomous province of Vojvodina and in the plains south of the Sava and the Danube, as well as to plant 886,840 hectares of forests in hilly and mountainous areas. The afforestation process requires an investment of 2.3 billion euros, but in addition to other benefits, it would also significantly increase the level of security regarding erosions and torrential floods.

It is of particular importance to improve the interaction between the complexes of forest and soil covers in order to preserve biodiversity and ecosystem stability (Ristić et al., 2021).

SOLUTIONS AND RECOMMENDATIONS

Efficient torrential flood control is not feasible with the existing financing system, as there are few newly-built facilities and the existing ones are poorly maintained, which reduces the system's functionality and progressively increases the risk from floods (Bajrić & Sokolović, 2015). A stable financing system can be established through appropriate legislative solutions. Operational and funding priorities have to be determined only according to professional criteria, with transparent and controlled processes of asset spending (Ristić et al., 2019). The identified responsible entities on all levels (systemic, local, individual) are obliged to implement optimal preventive and protective measures against erosions and torrential floods in accordance with their level of authority.

FUTURE RESEARCH DIRECTIONS

In addition to protecting people and property, all erosion and torrential flood prevention activities essentially contribute to the improvement of the state of the environment and act as a mitigating factor for the current and forecasted climate fluctuations. The importance of soil protection against erosion and torrential floods has been recognized within the fundamental principles of the Rio conventions pertaining to the climate, biodiversity, and soil protection against desertification (UNFCCC – United Nation Framework Convention on Climate Change; UNCBD – UN Convention on Biological Diversity; UNCCD – United Nation Convention to Combat Desertification).

Furthermore, anti-erosion protection is fully compatible with the concept of Land Degradation Neutrality, as the most important global measure to preserve land resources (UNCCD, 2016). No less important are the principles and guidelines of EU directives ratified both by the Republic of Serbia and Bosnia & Hercegovina, such as the EU Water Framework Directive and Floods Directive, which provides a holistic methodological framework for integrated water resources management (Vasović, 2015).

CONCLUSION

The history of torrential floods in the Republic of Serbia and Bosnia & Hercegovina, current and predicted climate trends, characteristics of hilly and mountainous terrains, as well as the vulnerability of cities and other settlements, require due diligence when preparing and implementing preventive measures. The fact remains that torrential floods resulting in considerable material damage and even human casualties are an increasingly common occurrence both in the Republic of Serbia and the Republic of Bosnia & Hercegovina, which is why it is necessary to strengthen every aspect of the system for torrential flood risk and effects management. Accordingly, it is necessary to further develop early warning systems, both vertically (more applicable to the state in Republic of Serbia) and horizontally (more applicable to the state in Bosnia & Hercegovina). In a technical sense, only early warning systems with "gauges" and water level meters with further readings can be used as a precautionary measure for (relative) early warning on torrential floods. This means that the warning system must be further improved by prescribing a specific response for each category of possible hazards, as well as the measures to be taken on that occasion. It should be reminded that local emergency headquarters are mostly people who do not have special professional qualifications. In order to avoid unnecessary waste of resources, false alarms and panic, with the existing warning system, when it comes to flood waves in torrential streams, it is necessary to invest in the already mentioned early warning systems ("rain gauges," water meters in the upper stream part of rivers …). A particularly difficult challenge is to maintain an effective data flow in such systems and to present the data and information to the average users, who make up the largest part of the population affected by torrential floods. Therefore, it is imperative to develop a user-friendly system interface, with simple visualization and easy-to-understand data, in the future stages of implementation.

The key idea is to educate people who would react preventively in such situations, which will be even more frequent due to climate changes, so professional training in this area is of paramount importance.

ACKNOWLEDGMENT

This research was supported by the Erasmus+ Program of the European Union through the Capacity Building in Higher Education project entitled: Soil Erosion and Torrential Flood prevention: curriculum development at the universities of Western Balkan countries (SETOF), Project No. 598403-EPP-1-2018-1-RS-EPPKA2-CBHE-JP.

REFERENCES

Akcioni plan za zaštitu od poplava i upravljanje rekama u BiH 2014-20 [Action Plan for Flood Control and River Management in BiH 2014-20], Bosnia and Herzegovina, Savet ministara, Sarajevo, 2014.

Bajrić, M., Sokolović, D., Musić, J., & Halilović, V. (2013). Assessment of intensity of erosion and natural coverage of skid roads in post-exploitation period. In *Works of the Faculty of Forestry*. University of Sarajevo.

Bajrić, M., & Sokolović, D. (2015). *Rehabilitation of a secondary network of forest traffic infrastructure*. Glasnik Šumarskog fakulteta. . doi:10.2298/GSF15S1005B

Chaves, J. M., & De Cola, T. (2017). Public Warning Applications: Requirements and Examples. Wireless Public Safety Networks, 3, 1-18. doi:10.1016/B978-1-78548-053-9.50001-9

Dragović, N., Ristić, R., Pülzl, H., & Wolfslehner, B. (2017). *Natural resource management in Southeast Europe: forest, soil and water*. Skopje, GIZ, 270. EFAS - European Flood Awareness System. Available at: https://www.efas.eu/en

EFAS. (n.d.a). *Flash flood indicators*. https://www.efas.eu/en/flash-flood-indicators

EFAS. (n.d.b). *Medium range flood indicators*. Available at: https://www.efas.eu/en/medium-range-flood-forecasts

GLEWS. (n.d.). *The Joint FAO-OIE-WHO Global Early Warning System for health threats and emerging risks at the human-animal-ecosystems interface*. Available at: http://www.glews.net

IN-MHEWS - The International Network for Multi-Hazard Early Warning Systems. (n.d.). Available at: https://mhews.wmo.int/en/partners

Izrada mapa opasnosti i mapa rizika od poplava na slivu reke Vrbas u BiH, Knjiga 3 – "Izdvajanje bujičnih slivova i formiranje modela osjetljivosti na pojavu i razvoj bujičnih poplava, sa kartom erozije u slivu reke Vrbas" [Development of Flood Hazard and Risk Maps at Vrbas River Basin in BiH, Vol. 3 – Separation of Torrential Basins and Creation of Models of Sensitivity to the Occurrence and Development of Torrential Floods with an Erosion Map in the Vrbas River Basin]. (2017). UNDP BIH and "Zavod za vodoprivredu d.o.o." Bijeljina, BiH.

Janaćković, G., Vasović, D., Malenović Nikolić, J., Mušicki, S., & Vranjanac, Ž. (2018). Vulnerability assessment of municipality areas to natural disasters based on group fuzzy analytic hierarchy process. *Journal of Environmental Protection and Ecology, 19*(4).

Katastar bujičnih tokova i erozionih područja u Bosni i Hercegovine [Cadaster of Torrential Flows and Erosion Areas in Bosnia and Herzegovina], Zavod za vodoprivredu, Sarajevo, 1985.

Knežević, J., Gurda, S., Musić, J., Halilović, V., Sokolović, D., & Bajrić, M. (2018). The Impact of Animal Logging on Residual Trees in Mixed Fir and Spruce Stands. *South-East European Forestry, 9*(2), 107–114. doi:10.15177eefor.18-16

Okvirna vodoprivredna osnova BiH [Framework Water Management Foundation of BiH]. JVP Vodoprivreda BiH, Zavod za vodoprivredu Sarajevo, Sarajevo, 1994.

Okvirni plan razvoja vodoprivrede Republike Srpske [Framework Plan for Water Management Development in the Republika Srpska], Ministarstvo poljoprivrede, šumarstva i vodoprivrede, Republička direkcija za vode Bijeljina, 2006.

Okvirni plan razvoja vodoprivrede Republike Srpske [Framework Plan of Water Management Development in the Republika Srpska], Zavod za vodoprivredu, Bijeljina, 2006.

Okvirni zakon o zaštiti i spašavanju ljudi i materijalnih dobara od prirodnih ili drugih nesreća u BiH [Internal law on the protection and rescue of people and material goods from natural or other disasters in B&H], Ministarstvo sigurnosti Bosne i Hercegovine, 2008.

Plan upravljanja rizicima od poplava u slivu reke Save [Flood Risk Management Plan for the Sava River Basin], Investicijski okvir za Zapadni Balkan, Consultant: Eptisa Servicios de Ingeniería S.L. (Španija), 2018.

Rimac, N., Bijedić, A., & Babić, A. (2019). Prvi operativni evropski sistem upozorenja na poplave [First Operative European Flood Warning System]. *Proceedings, Second Water Congress.*

Ristić, R. (2014). Redefinisanje politike prevencije bujičnih poplava u Srbiji [Redefining the Policy of Torrential Flood Prevention in Serbia]. *Polis, 8*, 23–27.

Ristić, R., Kapović Solomun, M., Malušević, I., Ždrale, S., Radić, B., Polovina, S., & Milčanović, V. (2021). Healthy Soils-Healthy People; Soil and Human Health. In The Reality of the Balkan Region (pp. 223-248). CRC Press, Taylor & Francis Group.

Ristić, R., Kostadinov, S., Abolmasov, B., Dragićević, S., Trivan, G., Radić, B., Trifunović, M., & Radosavljević, Z. (2012a). Torrential floods and town and country planning in Serbia. *Natural Hazards and Earth System Sciences, 12*(1), 23-35. doi:10.5194/nhess-12-23-2012

Ristić, R., Kostadinov, S., Radić, B., Trivan, G., & Nikić, Z. (2012b). Torrential Floods in Serbia – Man Made and Natural Hazards. 12th Congress INTERPRAEVENT 2012 Proceedings, 771-779.

Ristić, R., & Macan, G. (1997). The Impact of erosion control measures on runoff processes. IAHS Publ.

Ristić, R., & Macan, G. (2002). Istraživanje procesa intercepcije u bukovo-jelovoj sastojini na planini Goč. *Glasnik Šumarskog fakulteta, 86*, 181-188. doi:10.2298/GSF0286181R

Ristić, R., Malušević, I., Milčanović, V., Polovina, S., & Radić, B. (2019). Prevencija bujičnih poplava u kontekstu tekućih klimatskih promena [Torrent Flood Prevention in the Context of Current Climate Change]. In Scientific conference: "Future of Serbian Agriculture and Forestry", (pp. 113-121). AINS.

Ristić, R., Polovina, S., Malušević, I., & Milčanović, V. (2017). Srbija i bujične poplave: tri godine posle katastrofe u maju 2014. godine [Serbia and Torrent Floods: Three Years After the May 2014 Disaster]. *Vodoprivreda, 288-290*, 245-252. Available at: https://www.voda.ba/sistem-predvidjanja-poplava-i-ranog-upozoravanja-na-slivu-rijeke-save-sava-ffws

Serbian National Strategy for Protection and Rescue in Emergency Situations. Official Gazette of the Republic of Serbia, No 86/2011.

Sokolović, D., Pičman, D., Lojo, A., Gurda, S., Bajrić, M., & Koljić, H. (2013). Određivanje optimalnog prostornog rasporeda mreže sekundarnih šumskih prometnica [Determining the optimal spatial distribution of the network of secondary forest roads]. *Sumarski List, 137*(1-2), 7–22.

Stanković, S., Vasović, D., & Trajković, S. (2019). Model of sustainable water resources management in the conditions of extreme hydrological phenomena. *Journal of Environmental Protection and Ecology, 20*(3), 1393-1401. Available at: https://www.itu.int/en/ITU-D/Emergency-Telecommunications/Pages/Multi-hazard-early-warning-systems.aspx

UNCCD. (2016). *Scaling up Land Degradation Neutrality Target Setting. From Lessons to Actions: 14 Pilot Countries Experiences.* Available at: https://www.unccd.int/publications/scaling-land-degradation-neutrality-target-setting-lessons-actions-14-pilot-countries

United Nations, Climate Actions Division, Early Warning Systems. (n.d.). Available at: https://www.un.org/en/climatechange/climate-solutions/early-warning-systems

Upravljanje krizom – primjer poplava iz 2014. godine [Crisis management – the case of 2014. floods], Parlamentarna skupština Bosne I Hercegovine, Sekretarijat, Istraživački sektor, Sarajevo, 2015.

Vasović, D., Janaćković, G., Malenović Nikolić, J., Milošević, L., & Mušicki, S. (2018b). Promoting reflective practice in resources protection area: a step to forecast outcomes in uncertainty. *Journal of Environmental Protection and Ecology, 19*(3), 1320-1329.

Vasović, D., Janaćković, G., Malenović Nikolić, J., Mušicki, S., & Marković, S. (2018a). Multimodality in the field of resources protection. *Journal of Environmental Protection and Ecology, 19*(4), 1519-1525.

Vasović, D., Malenovic Nikolic, J., & Janackovic, G. (2016). Evaluation and Assessment Model for Environmental Management under the Seveso III, IPPC/IED and Water Framework Directive. *Journal of Environmental Protection and Ecology, 17*(1), 356-365.

Zambrano, A. M., Calderón, X., Jaramillo, S., Zambrano, O. M., Esteve, M., & Palau, C. (2017). Public Warning Applications: Requirements and Examples. Wireless Public Safety Networks, 3, 39-66. doi:10.1016/B978-1-78548-053-9.50003-2

Zeljko, D., Sarač, M., & Mičivoda, H. (2019). Sistem prognoze i upozorenja na poplave u slivu reke Save [Flood Forecasting and Warning System in the Sava River Basin]. In *Proceedings, Second Water Congress.* Udruženje konsultanata inženjera BiH.

ADDITIONAL READING

Alasali, F., Tawalbeh, R., Ghanem, Z., Mohammad, F., & Alghazzawi, M. (2021). A Sustainable Early Warning System Using Rolling Forecasts Based on ANN and Golden Ratio Optimization Methods to Accurately Predict Real-Time Water Levels and Flash Flood. Sensors, 21(13), 4598. doi:10.3390/s21134598

Beden, N., & Keskin, A. Ü. (2020). Flood hazard assessment of a flood-prone intensively urbanized area – A case study from Samsun Province, Turkey. *Geofizika, 37*(1), 1–25. doi:10.15233/gfz.2020.37.2

Costache, R., Barbulescu, A., & Pham, Q. (2021). Integrated Framework for Detecting the Areas Prone to Flooding Generated by Flash-Floods in Small River Catchments. Water, 13(6), 758. doi:10.3390/w13060758

Hofmann, J., & Schüttrumpf, H. (2019). Risk-Based Early Warning System for Pluvial Flash Floods: Approaches and Foundations. Geosciences, 9(3), 127. doi:10.3390/geosciences9030127

Meng, X., Sun, Z., Zhao, H., Ji, X., Wang, H., Xue, L., ... Zhu, Y. (2018). Spring Flood Forecasting Based on the WRF-TSRM Mode. *Tehnicki Vjesnik (Strojarski Fakultet), 25*(1), 141–151. doi:10.17559/TV-20171211112204

Petrić-Stepanović, V. (2008). Zaštita Slavonskog Broda od poplava bujičnih potoka. *Gradevinar, 60*(08), 695–703.

Rapant, P., & Kolejka, J. (2021). Dynamic Pluvial Flash Flooding Hazard Forecast Using Weather Radar Data. Remote Sensing, 13(15), 2943. doi:10.3390/rs13152943

Spada, E., Sinagra, M., Tucciarelli, T., & Biondi, D. (2017). Unsteady State Water Level Analysis for Discharge Hydrograph Estimation in Rivers with Torrential Regime: The Case Study of the February 2016 Flood Event in the Crati River, South Italy. Water, 9(4), 288. doi:10.3390/w9040288

Stanković, D. (2014). Challenges to the Takeover Early Warning System in the EU: The Case of Germany. *Croatian Yearbook of European Law & Policy, 10*(1), 291–324. doi:10.3935/cyelp.10.2014.184

KEY TERMS AND DEFINITIONS

Affected Population: People who are affected, either directly or indirectly, by a hazardous event. Directly affected are those who have suffered an injury, illness or other health effects; who were evacuated, displaced, relocated or have suffered direct damage to their livelihoods, economic, physical, social, cultural and environmental assets. Indirectly affected are people who have suffered consequences, other than or in addition to direct effects, over time, due to disruption or changes in the economy, critical infrastructure, basic services, commerce or work, or social, health and psychological consequences.

Disaster Management: The organization, planning and application of measures preparing for, responding to and recovering from disasters.

Early Warning System: An integrated system of hazard monitoring, forecasting and prediction, disaster risk assessment, communication and preparedness activities systems and processes that enables individuals, communities, governments, businesses and others to take timely action to reduce disaster risks in advance of hazardous events.

Flood: An overflow of water onto normally dry land. The inundation of a normally dry area caused by rising water in an existing waterway, such as a river, stream, or drainage ditch. Ponding of water at or near the point where the rain fell. Flooding is a longer event than flash flooding: it may last days or weeks.

Flood Control: Methods that are used to reduce or prevent the detrimental effects of flood waters.

Flood Relief: Methods that are used to reduce the effects of flood waters or high water levels.

Hazard: A process, phenomenon or human activity that may cause loss of life, injury or other health impacts, property damage, social and economic disruption or environmental degradation.

Hazardous Event: The manifestation of a hazard in a particular place during a particular period of time.

Landslides/Mudslides: Downhill sliding or falling movement of cry soil and rock. Landslides are difficult to estimate as an independent phenomenon. It seems appropriate, therefore, to associate landslides with other hazards such as tropical cyclones, severe local storms and torrential floods. The term landslide is used in its broad sense to include downward and outward movement of slope forming materials (natural rock and soil). It is caused by heavy rain, soil erosion and earth tremors and may also happen in areas under heavy snow.

Management: Process of planning, decision making, organizing, leading, motivation and controlling the human resources, financial, physical, and information resources of an organization to reach its goals efficiently and effectively.

Mitigation: The lessening or minimizing of the adverse impacts of a hazardous event.

Prevention: Activities and measures to avoid existing and new disaster risks.

Safety: Concept that includes all measures and practices taken to preserve the life, health, and bodily integrity of individuals.

System: Group of interacting or interrelated elements that act according to a set of rules to form a unified whole.

Torrential Flood: A flood caused by heavy or excessive rainfall in a short period of time, generally less than 6 hours. Torrential floods are usually characterized by raging torrents after heavy rains that rip through river beds, urban streets, or mountain canyons sweeping everything before them. They can occur within minutes or a few hours of excessive rainfall. They can also occur even if no rain has fallen, for instance, after a levee or dam has failed or after a sudden release of water by a debris or ice jam. It lasts for a shorter time period than a pluvial flood. Because of their rapid nature, torrential floods are difficult to forecast and give people little time to escape or to take food and other essentials with them.

Chapter 11
Response of Erosion to Environmental and Climate Changes During the Anthropocene Within the Endorheic System of Mhabeul, Southeastern Tunisia

Elhoucine Essefi
University of Gabes, Tunisia

Soumaya Hajji
Faculty of Sciences of Sfax, Tunisia

ABSTRACT

This chapter aimed to investigate the record of climatic and environmental change in the sedimentary filling of sebkha Mhabeul and their effect on hydric and eolian erosion within the wetland and its watershed. Along a 37 cm core, the sedimentary, geochemical, and geophysical signals at the Holocene-Anthropocene transition were followed. Sampling was carried out each 1 cm to obtain 37 samples. All studied parameters and clustering techniques indicate that the first 7 cm represent the Anthropocene strata. According to the age model, this upper part of the core records the last 300 yrs. The sedimentary record of the Anthropocene is marked by an increasing rate of sedimentation, grain size fining, heavy metals (Pb, Cu, Ni, Mn, and Fe) enrichment, which is related to increased erosion. Other intrinsic parameters such as CE, pH, Na, K, and CaCO3 enhance sediment erodibility. The measurement of the magnetic susceptibility along a 37 cm core collected from the sebkha Mhabeul shows an obvious upward increase related to a high content of heavy metals for the first 7 cm.

DOI: 10.4018/978-1-7998-8459-0.ch011

INTRODUCTION

The sedimentary record of salt lakes has been used as a proxy to monitor eolian and hydric erosions either from their surfaces (Liu et al., 2011; Han et al., 2014; Young & Evans, 1986; Yang et al., 2021) or from their catchments (Coronato & Del Valle, 1993; Essefi, 2021).

The Anthropocene is a geologic epoch related to the setting of noticeable human direct and indirect impacts on terrestrial environments and subsequent geochemical, geophysical and sedimentological repercussions (e.g., Essefi, 2020; Gharsalli et al., 2020). In spite of the controversy on the Anthropocene starting date and its stratigraphic status, some common geochemical, sedimentary and mineralogical signals, including a variation in the rate of erosion (Bernier et al., 2021) were recorded in inland saline systems related to a polluted environment and changing climate. To set a new chronostratigraphic subdivision at the beginning of the Anthropocene, the signal should be persistent, sharp, global, and synchronous to indicate the required Global Stratotype Section and Point (GSSP). This GSSP would deal on the worldwide signal of erosion as an indicator of the starting date of this epoch since the Anthropocene is marked by an increasing erosion (Marx et al., 2014; Owens, 2020), leading to a noticeable research gap concerning the issue of land erosion (Poesen, 2018)

Actually, it is the first notice in the history of the geological sciences to include the human scale within the geological scale. As a new emerging epoch, the Anthropocene scaled on a few hundreds of years would be integrated on the geological scale extending millions of years. In doing so, even the most efficient traditional dating techniques (e.g., radiocarbon dating) letting errors ranging on hundreds of years are good for nothing to file a case for the setting of the Holocene-Anthropocene transition. In this vein, tephrochronology having a good accuracy seems promising to solve the issue (Wagreich and Draganits, 2017).

Anthropocene geochemical signal includes enrichment with heavy metals, organic pollutants and the modification of other geochemical parameters, including Ce and pH. This enrichment may be related to their amounts increase in the lithosphere, biosphere, hydrosphere and atmosphere as it may be related to their easy dissolution into the eroded materials. Several proxies and methods were developed to evaluate the scale of anthropogenic effect, including the computation of some geochemical parameters such as enrichment and contamination factors, geo-accumulation index and pollution load index (Essefi, 2020). The sedimentation during the Anthropocene is modified in two ways. (1) The geochemical modification of the elemental composition of sediments with industrial and agricultural pollutants due to direct human activities. (2) Modification of sedimentary flux related to eolian activity and/or erosion of the outcropping sediments (Foucher et al., 2021).

In this chapter, the aim was to infer the record of environmental and climatic changes during the Anthropocene strata within the sedimentary filling of sebkha Mhabeul based on the multi-proxies study. These changes would have an effect on eolian and hydric erosion and deposition within this wetland and its hydrological catchment. To make it, a dated core from the sebkha of Mhabeul underwent sedimentary, geochemical, and geophysical analyses to determine the Holocene-Anthropocene transition.

STUDY AREA

Since they are the outlets of the surrounding basins, wetlands are more candidate to record the setting of the Anthropocene more than high lands (Waters et al., 2018). On the other hand, mountains are the last to be influenced by the anthropogenic-induced activities (Gabrieli and Barbante, 2014). Located on the coastal Djeffara plain, the sebkha Mhabeul occupies an area of 4-km^2 (Figure 1). It is plain with a width of 40 Km, covered with marine and lacustrine Quaternary deposits. Bordered to the south by an escarpment of Mesozoic limestones (Gammoudi, 2020), the region is marked by an arid climate imposing desert conditions. The rainfall and temperature considerably vary according to the seasons. The wet period of winter is marked by low temperatures (1 to 2 °C). Oppositely, the dry period of the summer is characterized by excessive temperatures sometimes reaching 40 °C in the shade. The average total rainfall is 100 mm/year with significant variability. The evaporation is at its highest level in July and August. Depending on the season, the wind sector changes and sometimes blows with an instantaneous speed, easily reaching 10 m/s. The winds of the wet period are generally from the west to the east. Those of the dry period come rather from the south to south-south-west. The latter are frequently accompanied with sirocco, especially during June, July and August. The sedimentary filling of sebkha Mhabeul has been considered a good recorder of climatic and environmental changes (Essefi et al., 2014, 2015; Gammoudi, 2020; Gammoudi et al. 2019; Gammoudi et al., 2021). These changes directly affect the hydric and eolian deposition and erosion along the surface of the sebkha and the area of its hydrological watershed.

MATERIALS AND METHODOLOGY

Coring techniques were acknowledged as one the efficient tool to explore shallow subsurface sediment. A 37 core cm was taken in sebkha Mhabeul, (Figure 1). Sampling was carried out at each 1 cm. In the laboratory, grain size distribution was evaluated by FRITSCH laser granulometer. The pH and electric conductivity Ce were measured for all samples. Magnetic properties are an efficient tool for environmental and paleoenvironmental investigations. After each heating at (250 C°, 500 C°, 750 C°, 1000 C°), the low and high frequencies Magnetic Susceptibility (MS) were measured by a Bartington MS2B probe. As for the grain size distribution, computed variables of grain size, including geometric mean, skewness, kurtosis and coefficient of variation, were used to set the Anthropocene strata.

Geochemistry is a more relevant tool to assess the Anthropocene (Gałuszka et al. 2014). Potassium and sodium were measured by Flame Photometer. The carbonate percentage $CaCO_3$ was measured by Bernard Calcimeter. After tri-acids (Nitric, perchloric, chloridric) attack of sediment, the amounts of heavy metals (Pb, Cu, Ni, Mn, and Fe) were measured by Atomic Absorption to calculate the geo-accumulation indices. Indeed, a relevant criterion to evaluate heavy metal pollution is the geo-accumulation index (Geo Index). Müller (1981) determined the degree of metal contamination in sediments by comparing current concentrations with preindustrial levels. It can be calculated by the following equation. Geo Index=$Log_2(C_n/1.5C_{bn})$: where C_n is the measured concentration of the examined metal (n) in the sediment and B_n is the background concentration of the metal (n), and 1.5 is the background matrix correction factor due to lithogenic effects. According to the Müller scale (1981), the calculated geo-accumulation index related to the studied samples showed different levels of contamination for the analyzed metals (Table 1).

Detecting the Anthropocene does not deal only with analytical techniques. Instead, some statistical and computing methods such as Clustering Techniques (CT) may help in individualize anthropogenic samples as a standalone population. Clustering Techniques recently emerged as the most used to highlight pollution and follow its origins, allowing hence the identification of the Anthropocene strata (Essefi, 2020). All computed variables (Na, Ca, K, $CaCO_3$, HFMS25, HFMS250, HFMS500, HFMS750,

Figure 1. The sebkha of Mhabeul within the Mediterranean and Atlantic contexts: (a) Surface Atlantic water circulation in the Mediterranean basin (Bianchi et al., 2012); (b) Geographical location of sebkha Mhabeul (Southeastern Tunisia): (c) Cites of cores of previous and in-progress works on the sebkha of Mhabeul

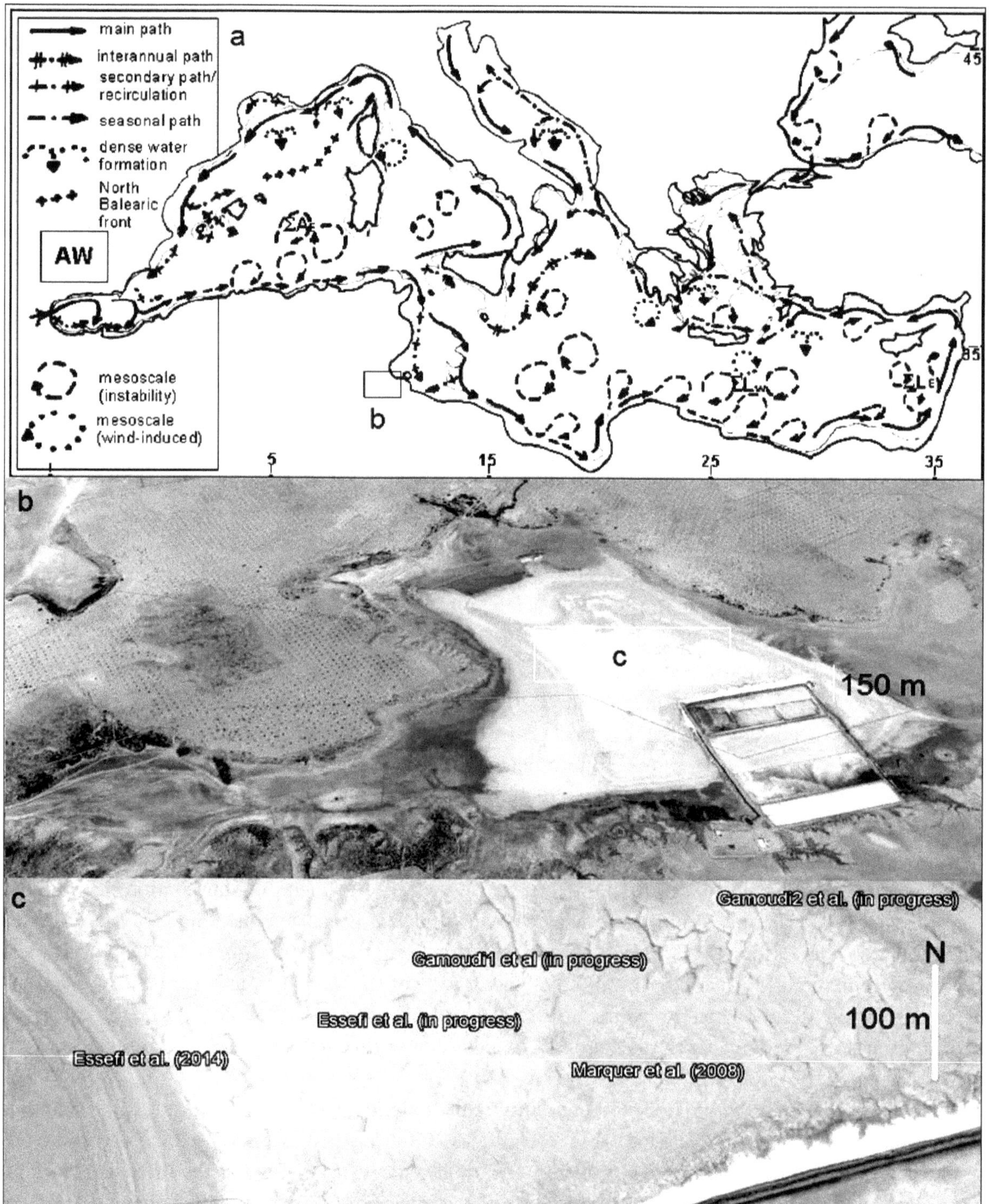

Table 1. Sediment classes according to Geo Index interval Müller scale (1981)

Geo Index interval	Class	Quality of sediment
Less or equal to 0	0	Unpolluted
[0, 1[	1	From unpolluted to moderately polluted
[1, 2[	2	Moderately polluted
[2, 3[	3	From moderately to strongly polluted
[3, 4[	4	Strongly polluted
[4,5[	5	From strongly to extremely polluted
More or equal to 5	6	Extremely polluted

HFMS1000, clay, silt, sand, very fine silt, fine silt, moderate silt, coarse silt, very fine sand, fine sand, moderate sand, coarse sand, heavy metals (Pb, Ni, Cd, Mn, Fe) and their geo-accumulation indices) were included to distinguish between the Anthropocene strata from the remaining part of the Late Holocene. To better identify the Anthropocene strata, three different clustering methods were used: Paired Group, Single Euclidian Linkage and Wards Methods.

RESULTS AND DISCUSSION

Sedimentary Signal of the Anthropocene: Increasing Sedimentary Flux

The setting of the Anthropocene-Holocene transition needs precise dating. Dating in continental sebkha remains a major handicap for a precise age model, which is the principal requirement not only to infer the record of the paleoclimatic variability but also to find out the date of this transition. Within Mhabeul wetland, the first dating attempt was carried out by Marquer et al. (2008). Along the core of Marquer et al. (2008), the age model showed that between 10 cm and 2 cm, the rate of sedimentation is 0.251 mm/yr. For the first upper 2 cm, the rate of sedimentation is equal to 0.238/yr. The rate of sedimentation is higher for the Late Holocene with 0.412 mm/yr. Dating of our core was already carried out by Essefi et al. (2014) based on tephrochronology. The age model (Figure 2) also showed an increasing sedimentary flux from 0.355 mm/yr during the Late Holocene to 0.233 mm/yr during the Anthropocene. As is the case in exoreic saline wetlands (Gharsalli et al., 2020, Essefi et al., 2020), this inland saline system of Mhabeul shows an increase in the rate of sedimentation, which is related to arid climatic conditions of the Anthropocene.

The grain size distribution and its homogeneity directly influence the erosion rate (Dallo, 2021; Li et al., 2020) (Figure 3). Finer and more heterogeneous grains enhance sediment erodibility. Along the first 7 cm, the plot of grain size parameters shows the setting of the Anthropocene strata 300 yr ago. During the Anthropocene, the grain size distribution recorded an upwards fining with a decrease of the geometric mean, probably related to the arid climate. On the other hand, distribution parameters (skewness, kurtosis, coefficient of variation) increase during this period, indicating hence and increasing erosion. The Anthropocene encompasses the Warming Present (WP), and the upper part of the Late Little Ice Age (Late LIA). It is distinguished from the Late Holocene, including the Early Little Ice Age (ELIA), and the Medieval Climate Anomalies (MCA). First, in the uppermost part of the core (3 cm),

Figure 2. Age model of the sebkha Mhabeul: decreasing rate of sedimentation

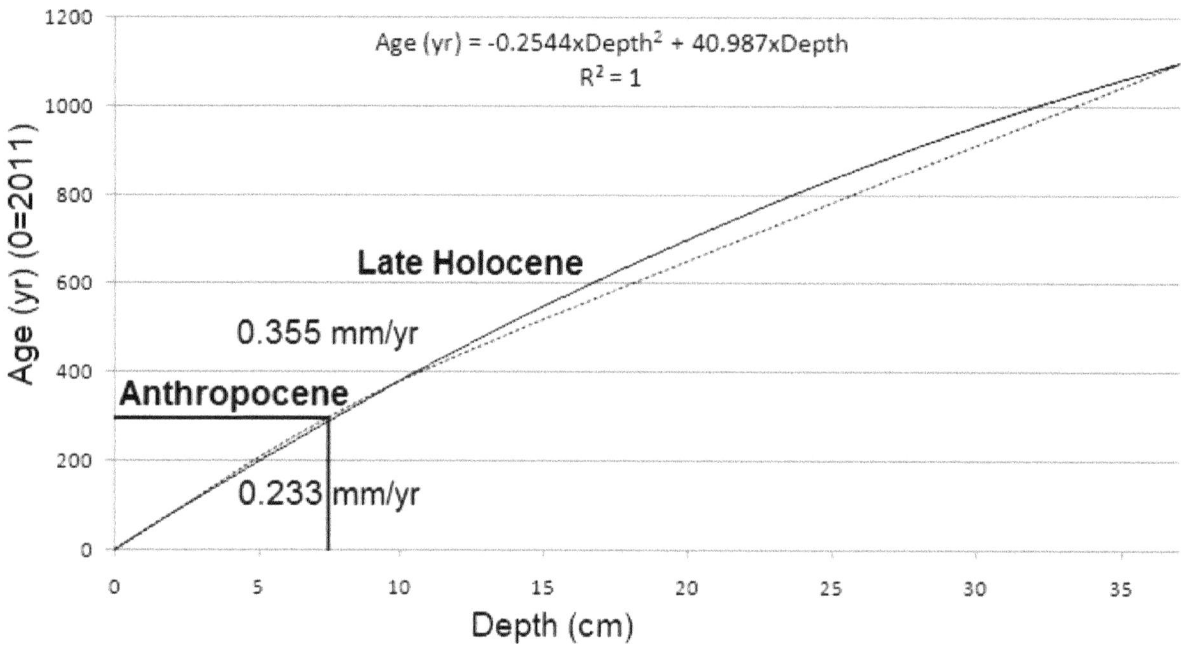

the stage Warming Present (WP) stretches from present to 1918, i.e. ≈95 yr; the establishment of current conditions is characterised by stable conditions. Added to a small salt crust, this period is dominated by clayey sedimentation rich with organic matter. This part may imbed the Great Acceleration (GA≈70 yr).

Coupled with Mn and Fe amounts, the descriptive grain size distribution including clay, fine silt, and fine sand (Figure 4) gives further details to understand the depositional environment and the climatic conditions. The sedimentary filling of sebkha Mhabeul is basically dominated by the very fine sand. The coarse sand is rare; it only appears in some locations along the core. The high amounts of fine fractions coincide with high values of Mn and Fe amounts. This coexistence may be due to the capacity of fine sediment to fix heavy metals related to pollution (Aliff et al., 2020) as it may be related to the enrichment of eolian dust by Mn and Fe (Salama et al., 2019).

More significance has long been attributed to the shapes of the cumulative grain-size curves rather than distribution curves of sediments. However, in the details of their form, grain-size distribution curves are more telling than cumulative curves about constituent sub-populations of particles. At sebkha Mhabeul, wet aeolian sediment is distinguished as a mix of aeolian and aqueous particles based on modes of the grain size distribution (Essefi et al., 2015). According to the genetic approach of the grain size distribution, during the WP, the plot of the secondary mode shows a significant downward increase. During this period, the primary mode does not show such a significant increase. Subsequently, the ratio M/m downward decreases due to wetter climatic conditions. In the Late LIA, the values of secondary mode continue their increase with a significant oscillation. The WP period is dominated by sandy sedimentation. The bottom of the cycle (Figure 5; 2-3 cm) shows an obvious fining; it is characterized by a primary mode (M: ca 103 μm) as an indication of the course eolian sedimentation. The secondary mode (m: 3.8 μm) is, however, an indication of the fine aeolian sedimentation. The upward coarsening

Figure 3. Grain size parameters variation along the Holocene-Anthropocene transition: finer and more heterogeneous sediment grains during the Anthropocene

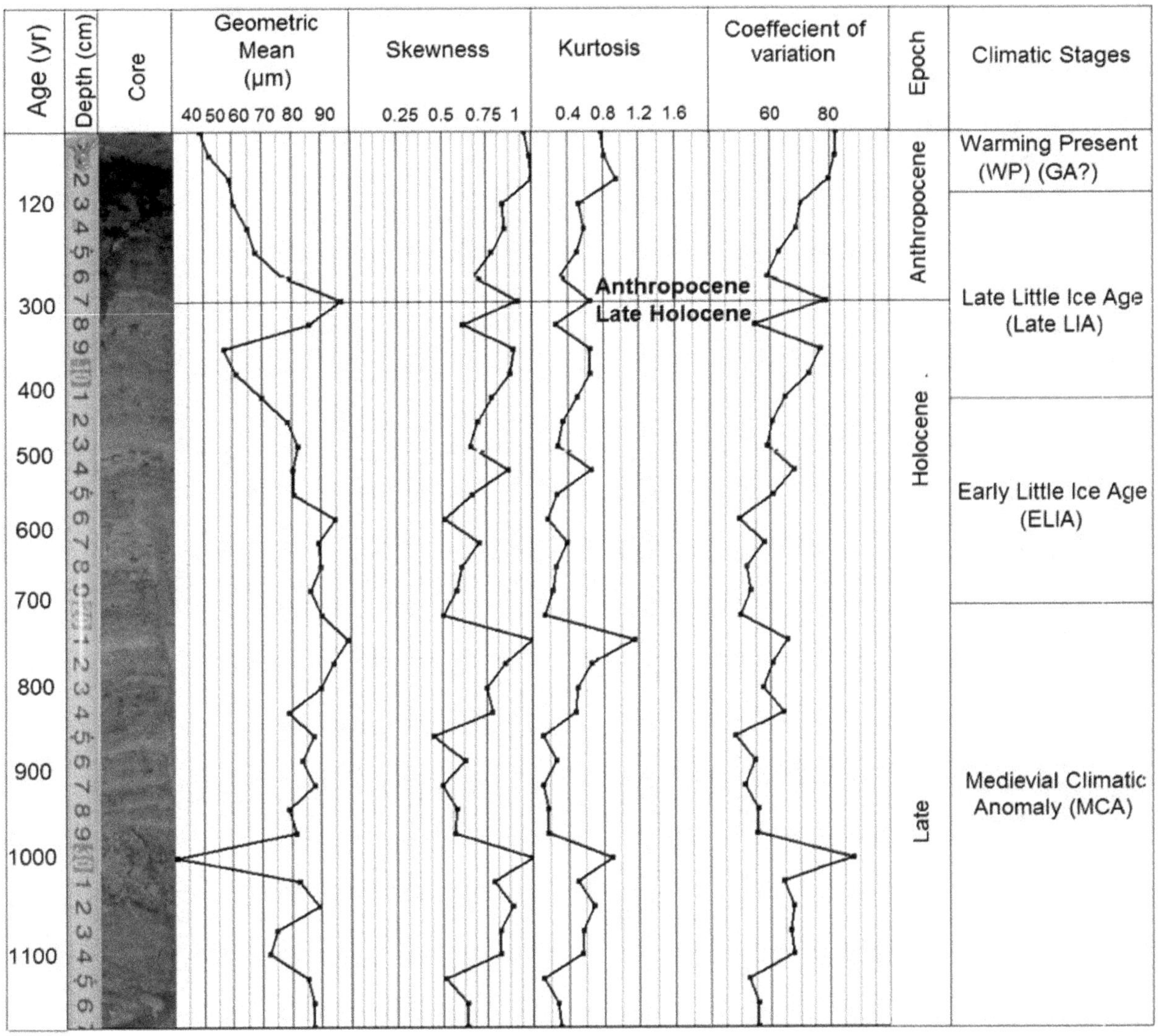

appears due to an increasing dryness during the WP. The second stage is located between 3.5 cm and 14 cm; it is called the Late Little Ice Age (Late LIA). It is characterized by a primary mode (M: ca 114 µm) as an indication of the coarse aeolian sedimentation, a translated secondary mode (m: ca 4.5 µm) as an indication of the fine aeolian sedimentation, and an occluded to absent fraction (O, A: ca 20 µm) (Figure 5). The Late Little Ice Age (Late LIA) is expressed by an upward increase of the fine aeolian fraction (Figure 5). On the other hand, the course aeolian fraction shows a slight decrease. The increase of the fine aeolian fraction is an indicator of decreasing aridity. Weak amounts of the coarse fraction mark wet conditions. This availability of dust in the atmosphere is controlled by aridity.

Figure 4. Variation of clay, fine silt, fine sand, Mn and Fe amounts along Holocene-Anthropocene transition

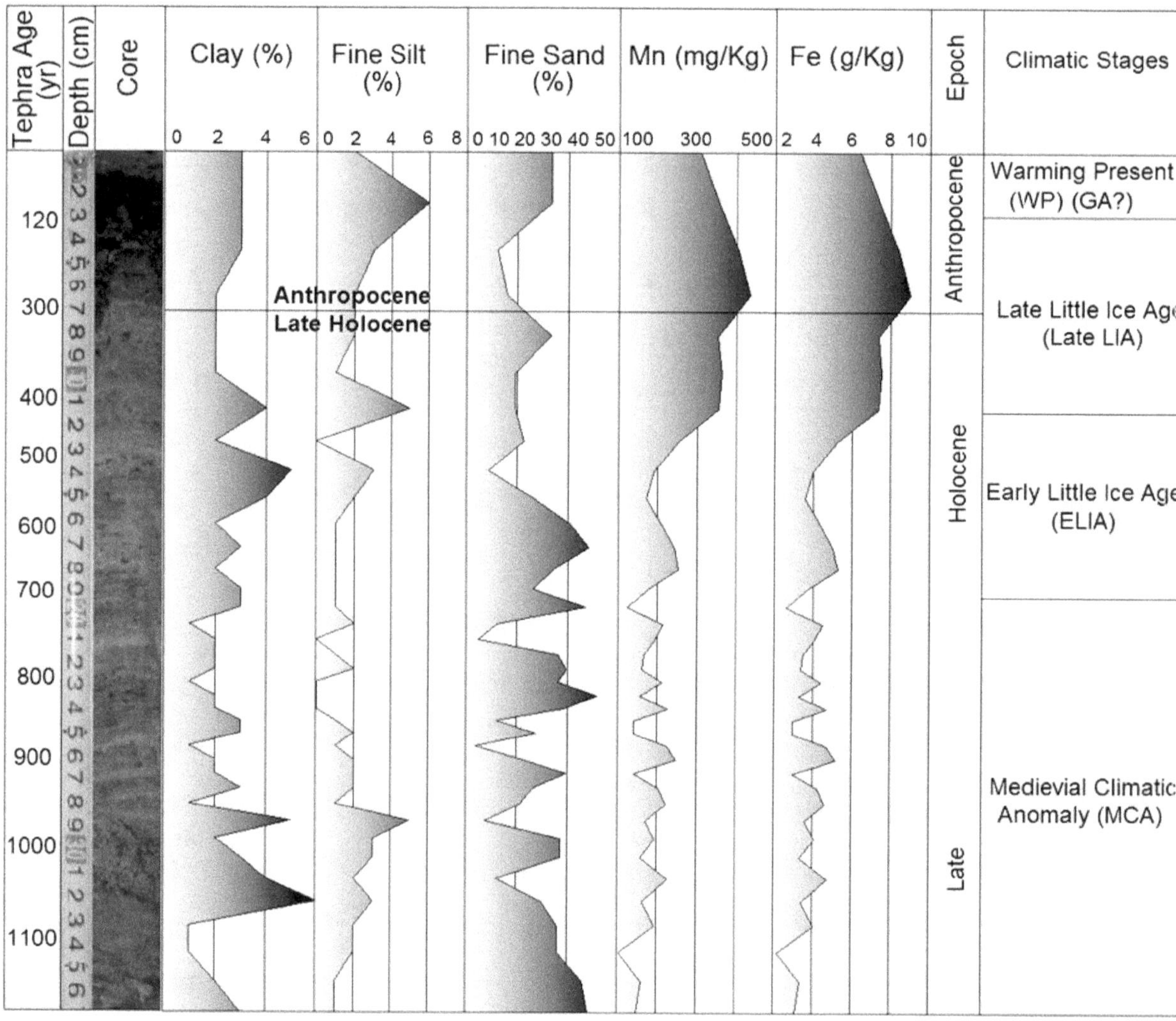

Geochemical Signal of the Anthropocene

Geochemical Parameters of the Anthropocene Strata

Erodibility is the set of intrinsic properties of soil such as CE, K, Na, and $CaCO_3$ that are responsible for resistance to erosion by external agents (Casabella-González et al., 2021). Along the core, values of CE, K, Na, and $CaCO_3$ present fluctuations (Figure 6). Indeed, the first 7 centimeters record a dramatic collective upward increase of all variables related to the setting of the Anthropocene strata. Indeed, the high carbonate contents indicate chemical conditions, which are characterized by high primary productivity and hot temperatures. So, it is an indicator of an arid climate. The proportion of potassium (K) dramatically increases from 4 g/kg to 9 g/kg. This fraction may be due to the precipitation of KCl under a high rate of evaporation (Essefi et al., 2020) related to a very arid climate or increased K-clays (Wasli, 2016). Values of sodium (Na) increase from 2 g/kg to 3 g/kg. This fraction may be due either to

Figure 5. Genetic approach for the setting of the Holocene-Anthropocene transition

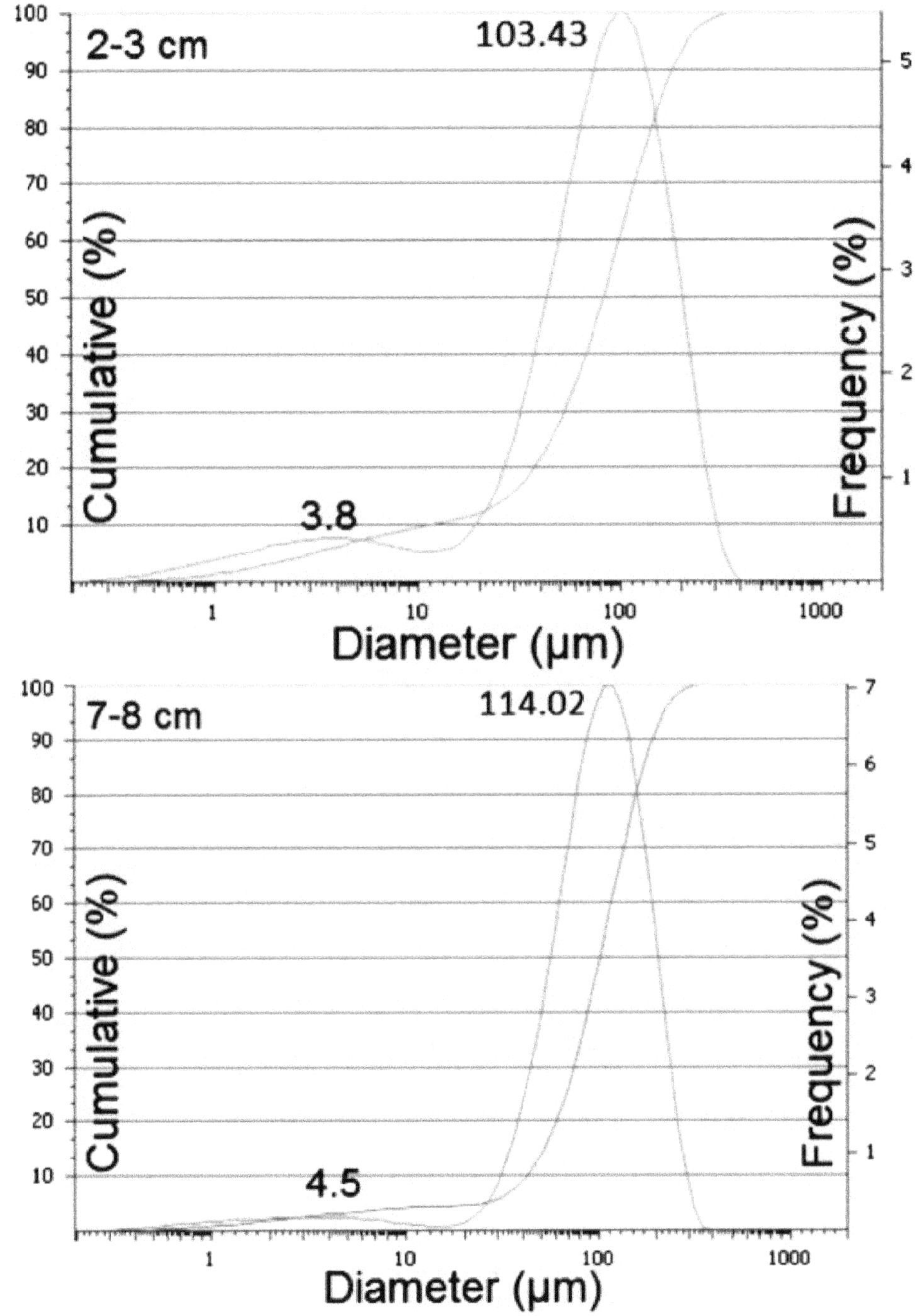

the precipitation of NaCl under a low rate of evaporation (Essefi et al., 2020) related to arid climate or increased Na-clays (Wasli, 2016). Both possibilities are worth consideration. On the one hand, increased CE and $CaCO_3$ are in favor of increasing dryness during the Anthropocene. On the other hand, increasing amounts of the clayey fraction are indicated in the sebkha during the Anthropocene (Figure 4-5).

Figure 6. CE, K, Na, and CaCO3 evolution along the core of sebkha Mhabeul

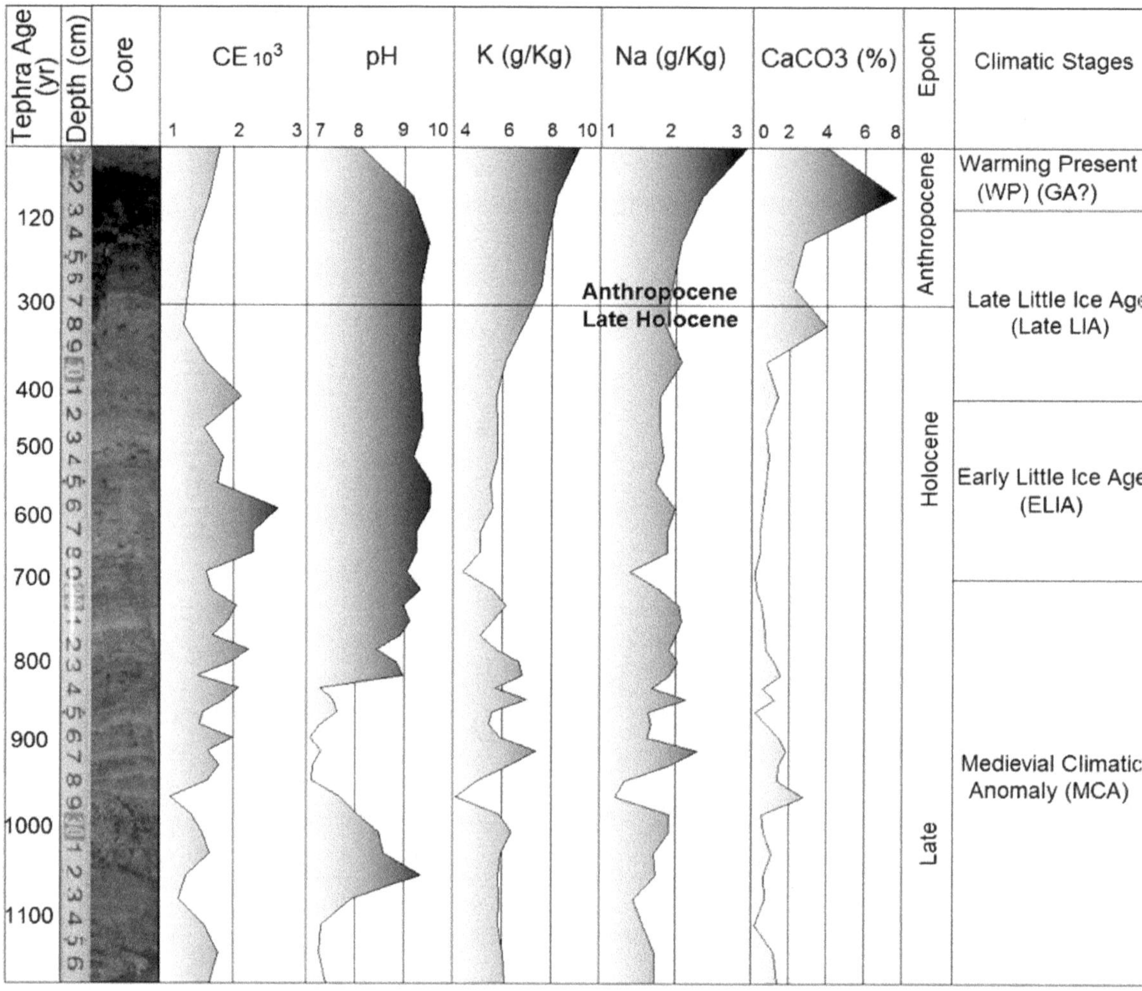

Heavy Metals and the Holocene-Anthropocene Transition

Wetlands are a good recorder of the Anthropocene due to their fine grain size allowing heavy metals fixation and relative preservation of organic pollutants. On the other hand, heavy metals having a natural origin, such as iron, do not show an obvious increase, probably related to an increasing erosion (Angeli, et al., 2021). The plot of the geo-accumulation indices along the core (Figure 7) confirmed the limit of the Anthropocene already identified by sedimentary and other geochemical parameters. According to the results obtained, based on the geo-accumulation index of the metals measured, the values show a remarkable rise from the first seven centimeters of core (Figure 7). The geo-accumulation indexes of Pb, Cu, Ni and Cd are between 1 and 2, they are taken second class, and therefore the soil quality is moderately polluted. Generally, it can be seen that the intensity of contamination is moderately strong, and the limit Anthropocene-Holocene is remarkable in the upper part of the core (from 7 cm). The evolution of the heavy metals content shows upward increasing pollution related to the setting of the Anthropocene. According to the age model, the polluted era may have started 300yr ago. Since this date,

Figure 7. Setting of the Late Holocene-Anthropocene limit based on heavy metals within sebkha Mhabeul, southern Tunisia

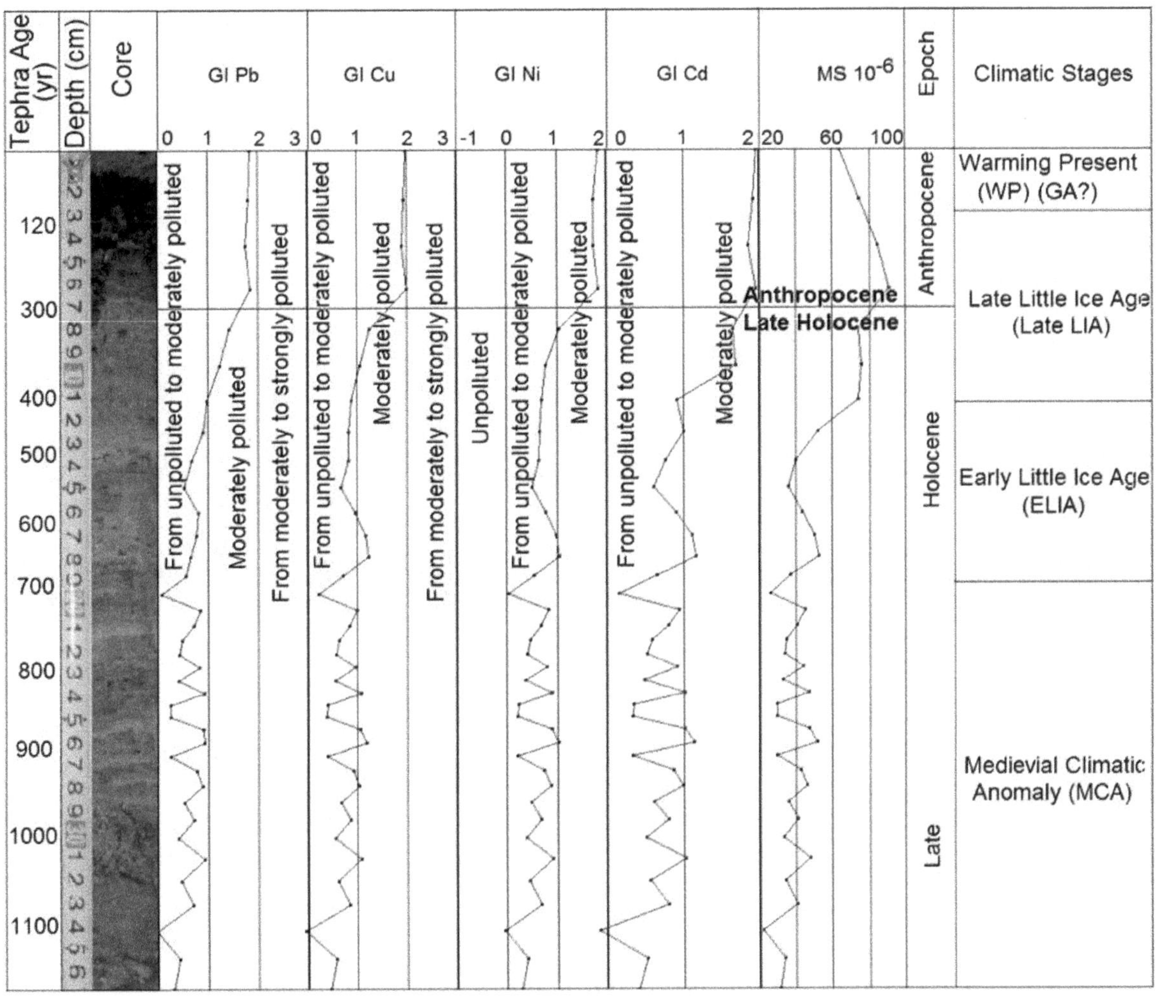

all heavy metals together increased indicating hence values exceeding the background. We notice an increase in low-frequency magnetic susceptibility values that are basically correlated with enrichment with heavy metals and Fe-clays.

Geophysical Signal of the Anthropocene

Worldwide, the magnetic susceptibility increase has been considered as an indicator of an increasing erosion (Peng et al., 2021; Ding et al., 2020; Kruglov et al., 2020; Menshov et al., 2020). The magnetic susceptibility may be due to enrichment with heavy metals. Also, the setting of new microbiology may result in magnetic enhancement (Gillings and Paulsen, 2014). Using the magnetic measurement in wetlands sediments as a marker for the beginning of the Anthropocene revealed relevance. In this vein, special attention is given to the importance of developing and disappearance of magnetic minerals. Actually, the starting of the Anthropocene is worldwide marked by an abrupt increase of the magnetic susceptibility

(Oldfield, 2015). The magnetic signal is obvious due to the Holocene-Anthropocene transition (Figure 8). The heating of sediment does not erase this geophysical signal. Instead, the Anthropocene signal becomes more obvious due to heating and the subsequent geochemical and mineralogical transformation. So, due to mineralogical and geochemical magnetic signals, magnetic susceptibility is a cheap and easy proxy to identify the Anthropocene signal.

Figure 8. Magnetic signals along the Holocene-Anthropocene transition: high-frequency magnetic susceptibility at ambient temperature and after heating at 250 C°, 500 C°, 750 C° and 1000 C°

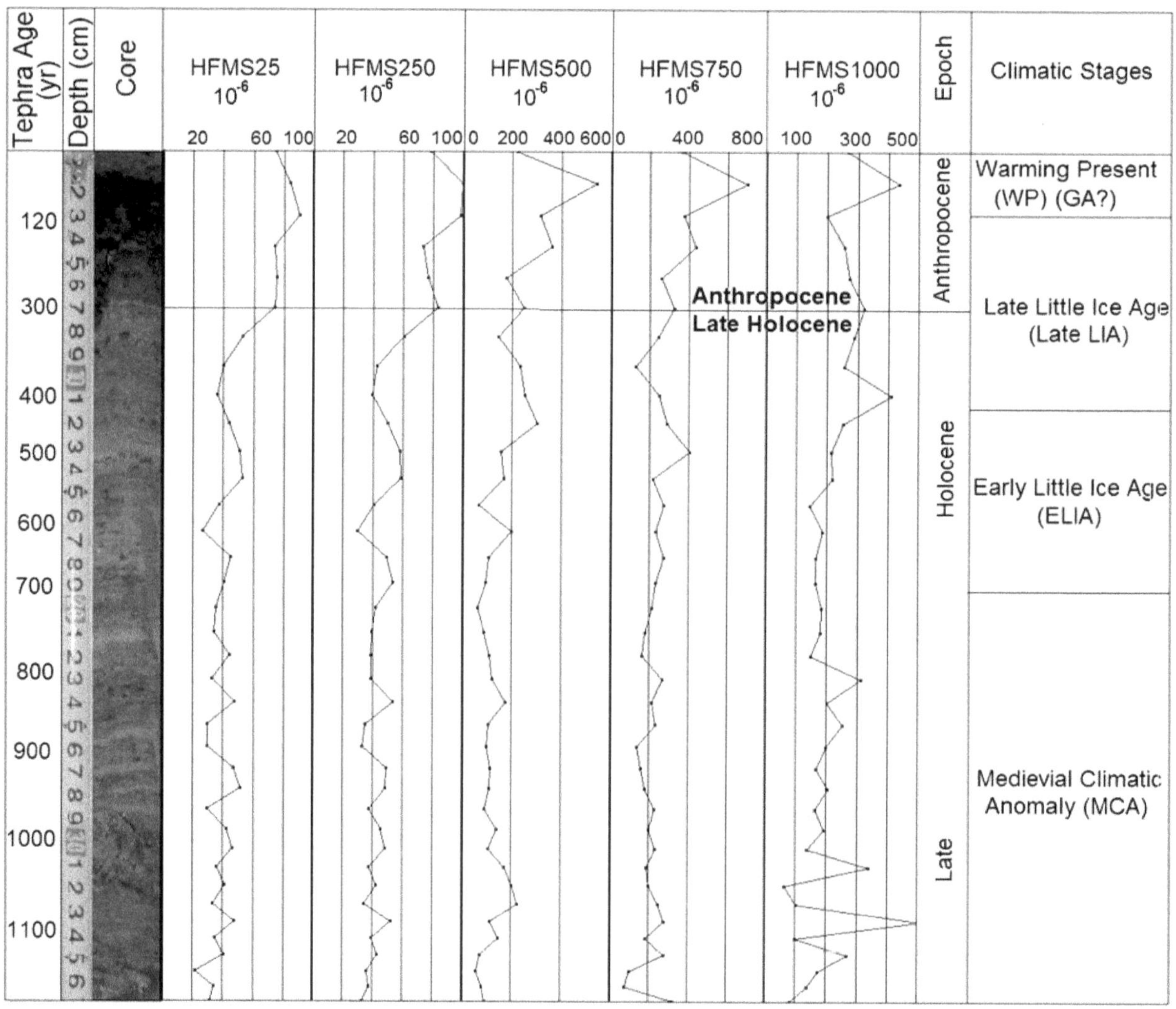

Clustering Technique to Detect Anthropocene

The clustering of all sedimentary, geochemical and geophysical parameters of the 37 samples from the core of the sebkha Mhabeul based on three different techniques (a: Paired Group, b: Single Euclidian Linkage, c: Wards Method) (Figure 9) shows the individualization of the first seven samples (depth 7 cm i.e., 300 yr). The setting of the Anthropocene at the top of this core is due to a radical change of the sedimentary environment-related climate change and/or pollution. Except slight modification of samples order, the change of the clustering technique does not modify the distinction between Late Holocene and Anthropocene samples.

Figure 9. Clustering (a: Paired Group, b: Single Euclidian Linkage, c: Wards Method) of all sedimentary, geochemical and geophysical parameters of the 37 samples from the core of the sebkha Mhabeul, southeastern Tunisia

FUTURE RESEARCH DIRECTIONS

Working on erosion along a drainage basin is not only in the outlet. Instead, the upstream and the surface of the basin may record erosion. Monitoring erosion along the drainage basin and the upstream may be based on remote sensing. Maps of different factors controlling erodibility may be superposed.

CONCLUSION

One of the indicators of the setting of the Anthropocene epoch is the increase of erosion. This increase is related to climatic and environmental changes. A multidisciplinary approach was used to set the limit of the Anthropocene along the sedimentary filling of Mhabeul inland wetland. Based on the previous age model and analyses of the sedimentary, geochemical and geophysical parameters, the limit Anthropocene-Holocene dates back to ca. 300 yr ago. In addition, the clustering technique of all parameters individualized samples of the Anthropocene strata with a distinguished population compared with the remaining part of the Late Holocene. At this age, the geo-accumulations indices increase, indicating the increased pollution. In terms of sedimentary dynamics, the setting of the Anthropocene in sebkha Mhabeul is marked by an increasing sedimentary flux recorded by the dominance of the fine fraction at the expense of the coarse sediment, related probably to an increasing erosion. Based on the age model, the Anthropocene in inland saline systems such as Mhabeul is marked by a decreasing sedimentation rate. From a sedimentary viewpoint, the upward fining during the Anthropocene coincides with high values of Mn and Fe amounts. This is due to the capacity of fine sediment to fix heavy metals related to pollution. Also, it may be connected to the enrichment of eolian dust by Mn and Fe. CE, K, Na, and $CaCO_3$ values record dramatic collective upward increases related to the setting of the Anthropocene strata. Along the core, the magnetic signal of the Anthropocene is obvious. Due to the heating of sediment, the signal becomes more obvious due to heating and the subsequent geochemical and mineralogical transformation. Based on three different clustering techniques of all parameters, the individualization of the first seven samples (depth 7 cm i.e., 300 yr) is recorded in the Anthropocene strata.

REFERENCES

Aliff, M. N., Reavie, E. D., Post, S. P., & Zanko, L. M. (2020). Anthropocene geochemistry of metals in sediment cores from the Laurentian Great Lakes. *PeerJ*, *8*, e9034. doi:10.7717/peerj.9034 PMID:32411522

Angeli, J. L. F., Sartoretto, J. R., Kim, B. S. M., de Lima Ferreira, P. A., de Mahiques, M. M., & Figueira, R. C. L. (2021). Trace element fluxes during the "Anthropocene" in a large South American industrial and port area (Santos and São Vicente estuarine system, SE, Brazil). *Environmental Monitoring and Assessment*, *193*(9), 1–18. doi:10.100710661-021-09378-3 PMID:34426853

Bernier, J. F., Chassiot, L., & Lajeunesse, P. (2021). Assessing bank erosion hazards along large rivers in the Anthropocene: A geospatial framework from the St. Lawrence fluvial system. *Geomatics, Natural Hazards & Risk*, *12*(1), 1584–1615. doi:10.1080/19475705.2021.1935333

Bianchi, C. N., Morri, C., Chiantore, M., Montefalcone, M., Parravicini, V., & Rovere, A. (2012). Mediterranean Sea biodiversity between the legacy from the past and a future of change. *Life in the Mediterranean Sea: A look at habitat changes*, 1-55.

Casabella-González, M. J., Borselli, L., & García-Meza, J. V. (2021). Soil Horizon erodibility assessment in an area of Mexico susceptible to gully erosion. *Journal of South American Earth Sciences*, *111*, 103497. doi:10.1016/j.jsames.2021.103497

Coronato, F. R., & Del Valle, H. F. (1993). Methodological comparison in the estimate of fluvial erosion in an arid closed basin of northeastern Patagonia. *Journal of Arid Environments, 24*(3), 231–239. doi:10.1006/jare.1993.1021

Dallo, Y. A. (2021). Discussion of "Influence of particle-size distribution homogeneity on shearing of soils subjected to internal erosion". *Canadian Geotechnical Journal, 58*(1), 151–153. doi:10.1139/cgj-2020-0146

Ding, Z., Zhang, Z., Li, Y., Zhang, L., & Zhang, K. (2020). Characteristics of magnetic susceptibility on cropland and pastureland slopes in an area influenced by both wind and water erosion and implications for soil redistribution patterns. *Soil & Tillage Research, 199*, 104568. doi:10.1016/j.still.2019.104568

Essefi, E. (2020). On the Geochemistry and Mineralogy of the Anthropocene. *Int J Water Wastewater Treat, 6*(2).

Essefi, E. (2021). *Apport des Approches Géochimique et Sédimentologique au Fonctionnement des Systèmes Endoréiques et Exoréiques de l'Holocène à l'Anthropocène: Intérêts Environnemental et Géoéconomique*. Institut Supérieur des Sciences et Techniques des Eaux de Gabès. Habilitation Universitaire.

Essefi, E., Ben Smida, N., Jandoubi, I., Othmani, M. A., & Tagorti, M. A. (2020). Progressive evaporation of brine of sebkha Mchiguig, central Tunisia: A geo-economical comparative study of salt and brine. *Carbonates and Evaporites, 35*(2), 59. doi:10.100713146-020-00592-7

Essefi, E., Gharsalli, N., & Yaich, C. (2015). Climatic significance of grain size parameters of the wet aeolian sedimentary filling of sebkha mhabeul, southeast tunisia: Contribution of the genetic approach. *Journal of Basic and Applied Research International*, 140-152.

Essefi, E., Mefteh, S., Medhioub, M., & Yaich, C. (2014). Magnetic study of the heated and unheated sedimentary fillings of Sebkha Mhabeul, Southeast Tunisia: A geophysical method for paleoclimatic investigation and tephrochronological dating. *International Journal of Geophysics, 2014*, 2014. doi:10.1155/2014/908395

Foucher, A., Evrard, O., Cerdan, O., Chabert, C., Lefèvre, I., Vandromme, R., & Salvador-Blanes, S. (2021). Deciphering human and climatic controls on soil erosion in intensively cultivated landscapes after 1950 (Loire Valley, France). *Anthropocene, 34*, 100287. doi:10.1016/j.ancene.2021.100287

Gabrieli, J., & Barbante, C. (2014). The Alps in the age of the Anthropocene: The impact of human activities on the cryosphere recorded in the Colle Gnifetti glacier. *Rendiconti Lincei, 25*(1), 71–83. doi:10.100712210-014-0292-2

Gałuszka, A., Migaszewski, Z. M., & Zalasiewicz, J. (2014). Assessing the Anthropocene with geochemical methods. *Geological Society of London, Special Publications, 395*(1), 221–238. doi:10.1144/SP395.5

Gammoudi, A. (2020). *Etude géochimique et cyclostratigraphie holocène des sédiments de la sebkha Mhabeul au Sud-Est de la Tunisie* (PhD Thesis). Higher Institute of Applied Sciences and Technology of Gabes.

Gammoudi, A., Essefi, E., & Rigane, H. (2019). Variability of the grain-sized fractions during the last three millennia within the sedimentary fillings of the saline system of Mhabeul: The Holocene cyclostratigraphy. *Arabian Journal of Geosciences*, *12*(9), 296. doi:10.100712517-019-4472-3

Gammoudi, A., Tlili, A., Essefi, E., El Feki, H., & Rigane, H. (2021). Contribution of the geochemical, physico-chemical, mineralogical, and statistical approaches to the reconstructing of the Holocene depositional environments along South-Eastern Tunisia. *Arabian Journal of Geosciences*, *14*(19), 2029. doi:10.100712517-021-08227-4

Gharsalli, N., Essefi, E., Baydoun, R., & Yaich, C. (2020). The Anthropocene and great acceleration as controversial epoch of human-induced activities: case study of the Halk El Menjel Wetland, Eastern Tunisia. *Applied Ecology and Environmental Research*, *18*(3), 4137–4166. doi:10.15666/aeer/1803_41374166

Gillings, M. R., & Paulsen, I. T. (2014). Microbiology of the Anthropocene. *Anthropocene*, *5*, 1–8. doi:10.1016/j.ancene.2014.06.004

Han, W., Ma, Z., Lai, Z., Appel, E., Fang, X., & Yu, L. (2014). Wind erosion on the north-eastern Tibetan Plateau: Constraints from OSL and U-Th dating of playa salt crust in the Qaidam Basin. *Earth Surface Processes and Landforms*, *39*(6), 779–789. doi:10.1002/esp.3483

Kruglov, O., Menshov, O., Kovalenko, S., Nazarok, P., & Andreeva, O. (2020, November). Magnetic susceptibility and heterogeneities of agronomic indicators of soil as a tool for water erosion assessment. In *XIV International Scientific Conference "Monitoring of Geological Processes and Ecological Condition of the Environment"* (Vol. 2020, No. 1, pp. 1-5). European Association of Geoscientists & Engineers. 10.3997/2214-4609.202056055

Li, S., Russell, A. R., & Muir Wood, D. (2020). Influence of particle-size distribution homogeneity on shearing of soils subjected to internal erosion. *Canadian Geotechnical Journal*, *57*(11), 1684–1694. doi:10.1139/cgj-2019-0273

Liu, D., Abuduwaili, J., Lei, J., Wu, G., & Gui, D. (2011). Wind erosion of saline playa sediments and its ecological effects in Ebinur Lake, Xinjiang, China. *Environmental Earth Sciences*, *63*(2), 241–250. doi:10.100712665-010-0690-4

Marquer, L., Pomel, S., Abichou, A., Schulz, E., Kaniewski, D., & Van Campo, E. (2008). Late Holocene high resolution palaeoclimatic reconstruction inferred from Sebkha Mhabeul, southeast Tunisia. *Quaternary Research*, *70*(2), 240–250. doi:10.1016/j.yqres.2008.06.002

Marx, S. K., McGowan, H. A., Kamber, B. S., Knight, J. M., Denholm, J., & Zawadzki, A. (2014). Unprecedented wind erosion and perturbation of surface geochemistry marks the Anthropocene in Australia. *Journal of Geophysical Research. Earth Surface*, *119*(1), 45–61. doi:10.1002/2013JF002948

Menshov, O., Kruglov, O., Nazarok, P., Andreeva, O., & Kruhlov, B. (2020, September). Soil erosion magnetic measurements as a tool for landslides studies. In *Second EAGE Workshop on Assessment of Landslide Hazards and impact on communities* (Vol. 2020, No. 1, pp. 1-6). European Association of Geoscientists & Engineers. 10.3997/2214-4609.202055019

Oldfield, F. (2015). Can the magnetic signatures from inorganic fly ash be used to mark the onset of the Anthropocene? *The Anthropocene Review*, *2*(1), 3–13. doi:10.1177/2053019614534402

Owens, P. N. (2020). Soil erosion and sediment dynamics in the Anthropocene: A review of human impacts during a period of rapid global environmental change. *Journal of Soils and Sediments*, *20*(12), 1–29. doi:10.100711368-020-02815-9

Peng, J., Bai, X., & Chen, X. (2021). Climate-driven soil erosion processes in alpine environments over the last century: Evidence from the Taibai Mountain (central China). *Catena*, *206*, 105569. doi:10.1016/j.catena.2021.105569

Poesen, J. (2018). Soil erosion in the Anthropocene: Research needs. *Earth Surface Processes and Landforms*, *43*(1), 64–84. doi:10.1002/esp.4250

Salama, W., Anand, R., & Bohu, T. (2019). Biochemical and mechanical dispersion mechanisms of Au and As in areas covered by Permian glacial sediments and aeolian sand. *ASEG Extended Abstracts*, *2019*(1), 1–4. doi:10.1080/22020586.2019.12072948

Vanwalleghem, T., Gómez, J. A., Infante Amate, J., González de Molina, M., Vanderlinden, K., Guzmán, G., Laguna, A., & Giráldez, J. V. (2017). Impact of historical land use and soil management change on soil erosion and agricultural sustainability during the Anthropocene. *Anthropocene*, *17*, 13–29. doi:10.1016/j.ancene.2017.01.002

Wagreich, M., & Draganits, E. (2017). Mining-related geochemical anomalies from 3500 to 2000 BP as potential stratigraphic markers for the base of the Anthropocene. *EGUGA, 5934*.

Wasli, A. (2016). *Holocene cyclostratigraphy within the sedimentary record of exoreic (Boujmel) and endorheic (Mhabeul) saline systems: eustatic and climatic controls* (Master Thesis). Higher Institute of Applied Sciences and Technology of Gabes.

Waters, C. N., Zalasiewicz, J., Summerhayes, C., Fairchild, I. J., & Rose, N. J. … Edgeworth, M. (2018). Global Boundary Stratotype Section and Point (GSSP) for the Anthropocene Series: Where and how to look for potential candidates. *Earth-Science Reviews*, *178*, 379–429. doi:10.1016/j.earscirev.2017.12.016

Yang, R., Wu, D., Li, Z., Yuan, Z., Niu, L., Zhang, H., Chen, J., & Zhou, A. (2021). Holocene-Anthropocene transition in northwestern Yunnan revealed by records of soil erosion and trace metal pollution from the sediments of Lake Jian, southwestern China. *Journal of Paleolimnology*, 1–12. doi:10.100710933-021-00193-x

Young, J. A., & Evans, R. A. (1986). Erosion and deposition of fine sediments from playas. *Journal of Arid Environments*, *10*(2), 103–115. doi:10.1016/S0140-1963(18)31251-5

ADDITIONAL READING

Chafai, A., Brahim, N., & Shimi, N. S. (2020). Mapping of water erosion by GIS/RUSLE approach: Watershed Ayda river—Tunisia study. *Arabian Journal of Geosciences*, *13*(16), 1–14. doi:10.100712517-020-05774-0

Gara, A., Hammami, M., Hammami, S., Aounallah, M. K., Elmouaddab, R., Nahdi, M., & Laajili-Ghezal, L. (2020). Econometric analysis of acceptance of soil and water conservation techniques in the semiarid region of Zaghouan (North-East Tunisia). *Asian Journal of Agriculture and Rural Development, 10*(1), 440–449. doi:10.18488/journal.1005/2020.10.1/1005.1.440.449

Maguire, G. (2020). Human Erosion: Indigenous Peoples and Well-Being in the Anthropocene. *Irish Studies in International Affairs, 31*, 113–130. doi:10.3318/isia.2020.31.3

Mosbahi, M., & Benabdallah, S. (2020). Assessment of land management practices on soil erosion using SWAT model in a Tunisian semi-arid catchment. *Journal of Soils and Sediments, 20*(2), 1129–1139. doi:10.100711368-019-02443-y

Owens, P. N. (2020). Soil erosion and sediment dynamics in the Anthropocene: A review of human impacts during a period of rapid global environmental change. *Journal of Soils and Sediments, 20*(12), 1–29. doi:10.100711368-020-02815-9

Riahi, O. (2021). *Mapping Environmental Risk Degradation Under Climate Stress and Anthropogenic Pressure: Case Study of Abdeladim Watershed.* Environmental Remote Sensing and GIS in Tunisia.

Yang, R., Wu, D., Li, Z., Yuan, Z., Niu, L., Zhang, H., Chen, J., & Zhou, A. (2021). Holocene-Anthropocene transition in northwestern Yunnan revealed by records of soil erosion and trace metal pollution from the sediments of Lake Jian, southwestern China. *Journal of Paleolimnology*, 1–12. doi:10.100710933-021-00193-x

KEY TERMS AND DEFINITIONS

Anthropocene: The Anthropocene Epoch is an unofficial unit of geologic time, used to describe the most recent period in Earth's history when human activity started to have a significant impact on the planet's climate and ecosystems.

Clustering Techniques: Clustering is a Machine Learning technique that involves the grouping of data points.

Endorheic System: An endorheic basin, also called an internal drainage system, is a drainage basin, or watershed, that does not flow to one of the Earth's major oceans.

Erodibility: Soil erodibility reflects the inherent susceptibility of soils to erosion processes and depends therefore on various soil properties. Sedimentary flux: Also known as the deposition rate (DR) is the vertical flux of sediment to the depositional surface. It a good proxy to monitor the increasing erosion.

Heavy Metals: They are generally defined as metals with relatively high densities, atomic weights, or atomic numbers.

Magnetic Susceptibility: It is the degree to which a material can be magnetized in an external magnetic field. Magnetic susceptibility of soil is a fast, cheap, and non-destructive technique that could be used to quantify soil erosion or soil redistribution on a long-term scale.

Chapter 12
Sedimentary Dynamics Within the Sedimentary Filling of Sebkha Sidi El Hani, Eastern Tunisia:
Climatic Variability and Its Effect on Erosion

Elhoucine Essefi
University of Gabes, Tunisia

Soumaya Hajji
Faculty of Sciences of Sfax, Tunisia

Mohamed Ali Tagorti
Univeristy of Sousse, Tunisia

ABSTRACT

The Sidi El Hani Wetland is located in Eastern Tunisia. It represents the natural outlet of an endorheic system, Mechertate-Chrita-Sidi El Hani, and it collects all the eroded sediment from this watershed. In this chapter, the visual core description focused on three reference sandy bands and on the concept of grey scale variability in order to infer the clay pan response to the climatic variability and erosion during the last two millennia. First, in the uppermost part, the stage Warming Present (WP) stretches from (1954-80= 1874) to 1993, i.e. ≈120yrs; the establishment of modern conditions is characterized by stable conditions with high grey scale. Added to a small salt crust, this period is dominated by a clayey sedimentation. Second, the stage C4 is called the Late Little Ice Age (Late LIA); it stretches between the 80yrBP and 400yrBP, i.e., 320yrs. It is characterized by intermediate GS values; the clayey sedimentation makes up the twofold and threefold laminates. Based on laser granulometer, the genetic approach shows the interplay of eolian and hydraulic erosion.

DOI: 10.4018/978-1-7998-8459-0.ch012

INTRODUCTION

Global change of climate would have dramatic repercussions on erosion (Ciampalini et al., 2020). On the one hand, extreme events accentuate hydric erosion (Borrelli, et al., 2020). On the other hand, dry conditions enhance win erosion (Guo et al., 2020). Since wetlands are the natural outlets of all sediment eroded from their drainage basins, they are good recorders of climatic variability and excellent monitors of hydric and eolian erosion (Sun et al., 2020; Sapkota and White, 2021). As it is anticipated by previous tectonic, hydrogeological, hydrological and geochemical studies of the hydrogeological and hydrological basins on the one hand and the sedimentary and the hydric fillings of sebkha Sidi El Hani (as it is first defined englobing the sebkhas of Dkhila, Souassi, and Sidi El Hani sensu stricto) (Essefi, 2009, 2013, 2021; Essefi et al., 2013a, b, c, d, e; Essefi et al., 2014a, b, c) it is rather difficult to find an obvious geochemical signal of the climatic change in the sedimentary filling. This is because Sidi El Hani discharge playa is more governed by its hydrogeological basin rather than its hydrological watershed. Nonetheless, the variation of climatic conditions influences the watershed's erosion from the drainage basin and deposition within Sidi El Hani wetland. This chapter aims to overcome these handicaps and to guess the climatic variability and its effect on erosion during the last two millennia only on the basis of the visual description of cores. Accordingly, before doing any destructive analysis, non-destructive analysis, namely the Visual Cores Description (VCD), could give a lot of information about cores content, texture, and the organization of sedimentary bands. Furthermore, it would be an efficient tool for correlation between different cores from the same playa or different playas in order to compare them and to establish their stratigraphy.

STUDY AREA: VARIATION OF EROSION AND DEPOSITION

The sebkha of Sidi El Hani is part from the Mechertate-Chrita-Sidi El Hani. Previous geomorphologic studies (Ben Jmaa, 2008) investigated erosion along the drainage basin. But this study has neglected an important factor which is groundwater upwelling (Essefi, 2013). In doing so, it considers erosion and deposition related only to climatic and geomorphologic features of the region. Further, the deposition within depressions was considered homogenous, while field investigations advocate a heterogenous deposition even within the same depression such as Sidi El Hani (Figure 1). Obviously, erosion and deposition within Mechertate-Chrita-Sidi El Hani are controlled by several factors such as the topography, the groundwater intervention, and the climate. Figure 1 shows zones of aeolian and hydraulic erosions; these zones are located in the periphery of the system, where an elevated topography and a weak groundwater intervention enhance the hydraulic and aeolian erosions, respectively. On the other hand, zones of accumulation are located within depressions and their vicinities, where a weak topography and groundwater intervention increase accommodation in these zones. Thus, the balance between erosion and deposition, which is the basic concept of sequence stratigraphy, is controlled by groundwater intervention. In this sedimentologic investigation, facies identification allows the identification of the repartition of sediment types along zones of erosion (watershed high lands) and zones of accumulation (low lands of depressions). Subsequently, a profile may collect the interaction between both zones; it may also show the scenario of the evolution of this system due to tectonic subsidence and/or water table fluctuations. Actually, geomorphologists (e.g., Ben Jmaa, 2008) have once given the super voice to this factor in deciding the shape of system landscape; but in this work, we see that this factor competes with the groundwater intervention to determine the sedimentary and geochemical content of Mechertate-Chrita-Sidi El Hani system. It is quite evident that the salinization increases from the periphery toward the basal part of the system. Being the core of the system, Sidi El Hani vicinities are more controlled by the groundwater upwelling effect, which results in high percentages of gypsum. On the other hand, Mechertate and Chrita vicinities seem fed by gypsum through seepage and migration of tectonic faults. Thus, groundwater interferes with aeolian and hydraulic activity to set sediments with different chemical compositions and shapes (Tagorti, 1995; Gaied, 2002; Ben Jmaa, 2008; Essefi, 2009).

Figure 1. Geomorphologic maps showing zones of weak, moderate, high and hard aeolian and hydraulic erosions and deposition within Mechertate-Chrita-Sidi El Hani system
Source: Ben Jmaa (2008); Modified Essefi (2013)

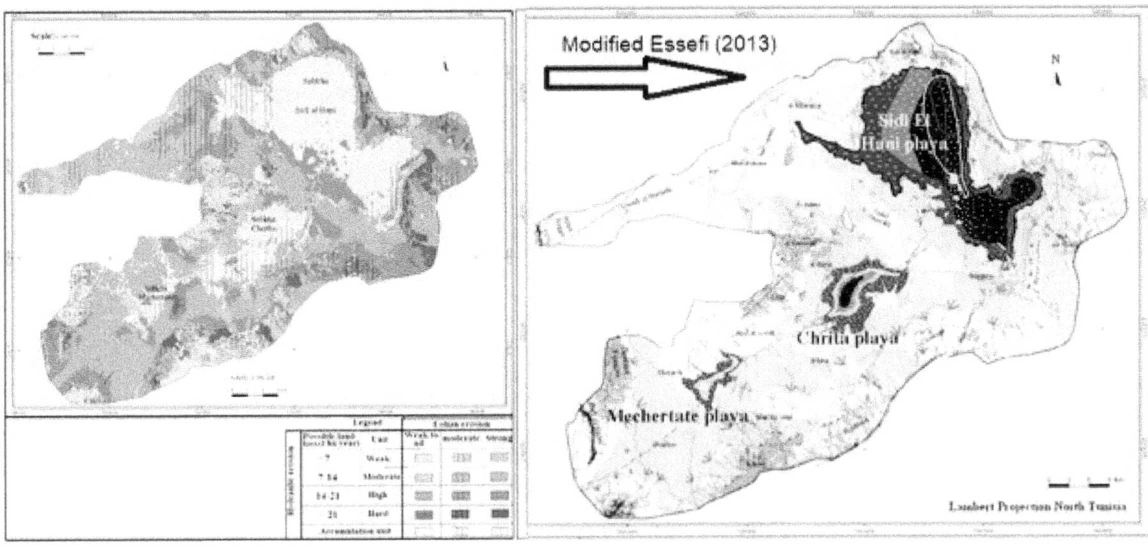

Wind proves vital in the morphological shaping of endorheic systems. In order to infer the aeolian paleo-activity, it is necessary to find out the link between current sedimentation and aeolian activity. Aeolian sediments are candidate to deposit in the Sahel area from different regions in Tunisia. During the year, winds are mainly oriented toward the Sahel area from the regions of Sfax, Kairouan and especially Gafsa (Figure 2). During autumn and summer, the wind coming from Sfax is strong. The wind coming from Kairouan is weak. During winter and spring, winds coming from Kairouan and Gafsa dominate the morphological shaping at the expense of wind coming from Sfax.

METHODS AND MATERIALS

The manual coring of eight cores (Figure 3; Table 1) in Sidi El Hani saline environment was done during three field missions (coordinate was taken during field missions by using a Global Positioning System GPS and localized on Google Earth satellite image). The first mission spanned for one day; it focused on the northwest side of Sidi El Hani saline environment (*stricto sensu*), in the zone of Oued Elmaktaa.

Figure 2. Wind data in Tunisia (1951-1970)

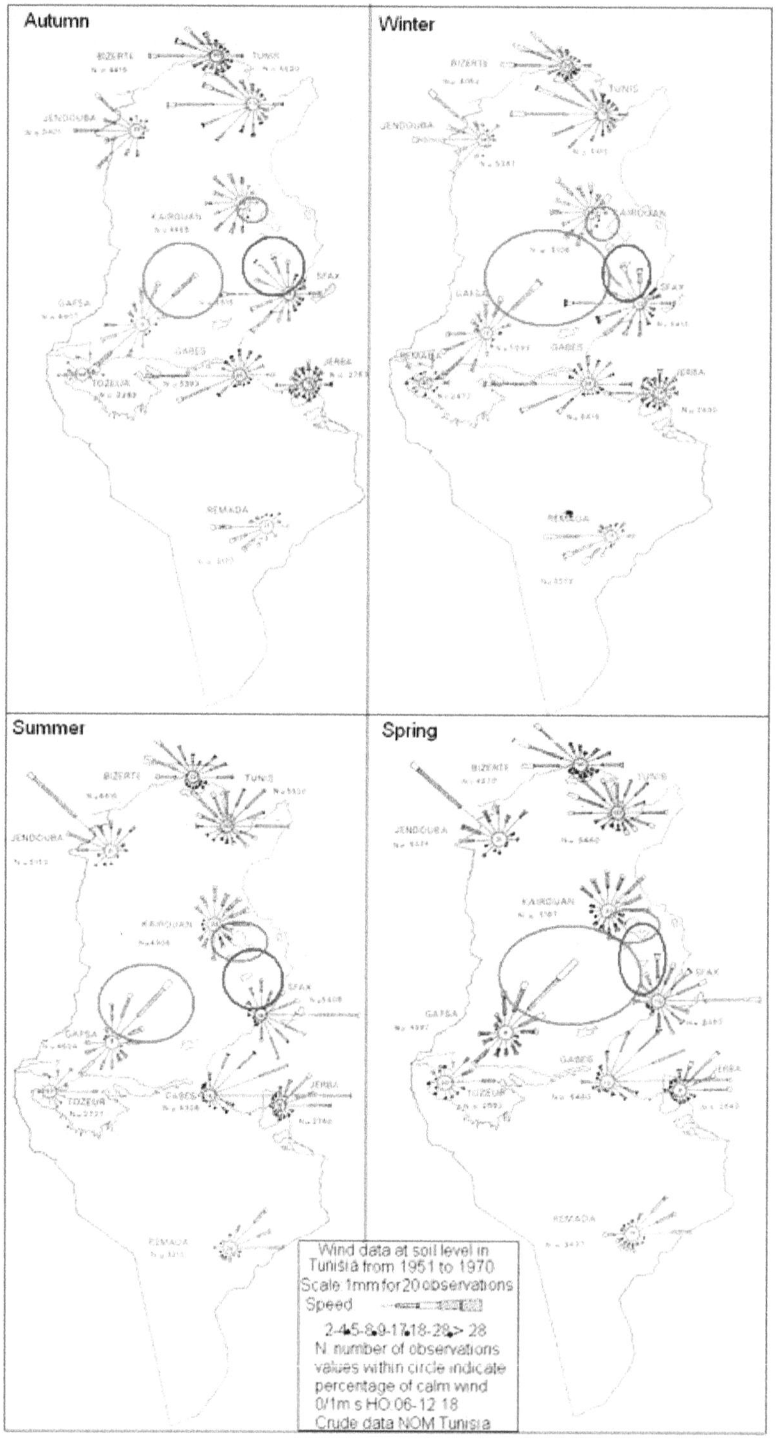

The main purpose of this mission was to have an idea about the manual coring and the used materials. A 310cm core was done (core4). The <u>second mission</u> took place in the field of the margin of the Beta

Table 1. Geological coordinates of Sidi El Hani saline environment cores

Core	Latitude	Longitude
Core 1	35°29'37.90"N	10°34'42.89"E
Core 3	35°33'38.31"N	10°28'47.94"E
Core 4	35°36'47.47"N	10°27'45.04"E
Core 5	35°35'45.23"N	10°29'00.18"E
Core 6	35°29'45.36"N	10°29'07.48"E
Core 7	35°29'45.32"N	10°29'15.60"E
Core 9	35°27'52.97"N	10°31'44.12"E
Core 10	35°35'45.49"N	10°21'39.91"E

Island OuledMoussa-Aalalcha. Two other cores with a length of 55cm (core5) and 72cm (core7) were done in the vicinity of the margin Beta Island-playa. Eventually, I did the third **and most extended mission**, it spanned for 30 days (mission of July 2008). After taking the key concepts of the manual coring, it was judged to do cores with our hands.

In this context, five cores were done in different zones of the saline environment; their lengths are 310cm (core10), 210cm (core1), 193cm (core3), 155cm (core6) and 120cm (core9). It is worth noting that coring was stopped for each core because of serious technical problems resulting from the used materials and/or sediment types. Some cores were aborted at small depths. For an aborted core, another core was restarted approximately in the same location. For instance, core2 and core8 were aborted and replaced respectively by core3 and core9.

Some sophisticated programs have been in progress to program an automatic VCD. For example, the J-CORES-VCD is a program elaborated by J-DESC IODP Information System Working Group (J-DESC-ISWG) with Science Support Group, Center for Deep Earth Exploration (CDEX), in Japan (2006); it is a scientific data management system in order to store and distribute scientific data taken on the Japanese "Chikyu" shipboard. It consists of a relational database on the host computer and an application program on client hosts. It can produce hand-drawn graphic representation anytime on digital images stored in the J-CORES during core-processing onboard (Figure 4).

The aim is not to use such a program, but its method appears quite interesting to follow the transition from **photos** to **logs** of cores. The use of a numeric photographer is sufficient to do the VCD. Many sedimentary structures should by rights appear in Sidi El Hani cores, but the focus will be only on the darkness (the greyscale variability) and the three reference sandy bands.

RESULTS

Visual Core Description of Sidi El Hani Clay Pan Cores

It is worth noting that the VCD of Sidi El Hani cores was not a purpose *per se*. Still, it was meant to select which cores were good to correlate with sebkha Mhabeul core according to the method of significant grey scale variability and the occurrence of the three reference sandy bands. Since the use of the major grey scale necessitates a noticeable variability, only cores with well-elaborated major grey scale variability

Figure 3. Locations of cores in Sidi El Hani clay pan on a satellite photograph (Google Earth, 2008): (a) Setting of Tunisia in North Africa; (b) Panoramic view of cores location; (c) Zoom on core5, core6 and core7 locations

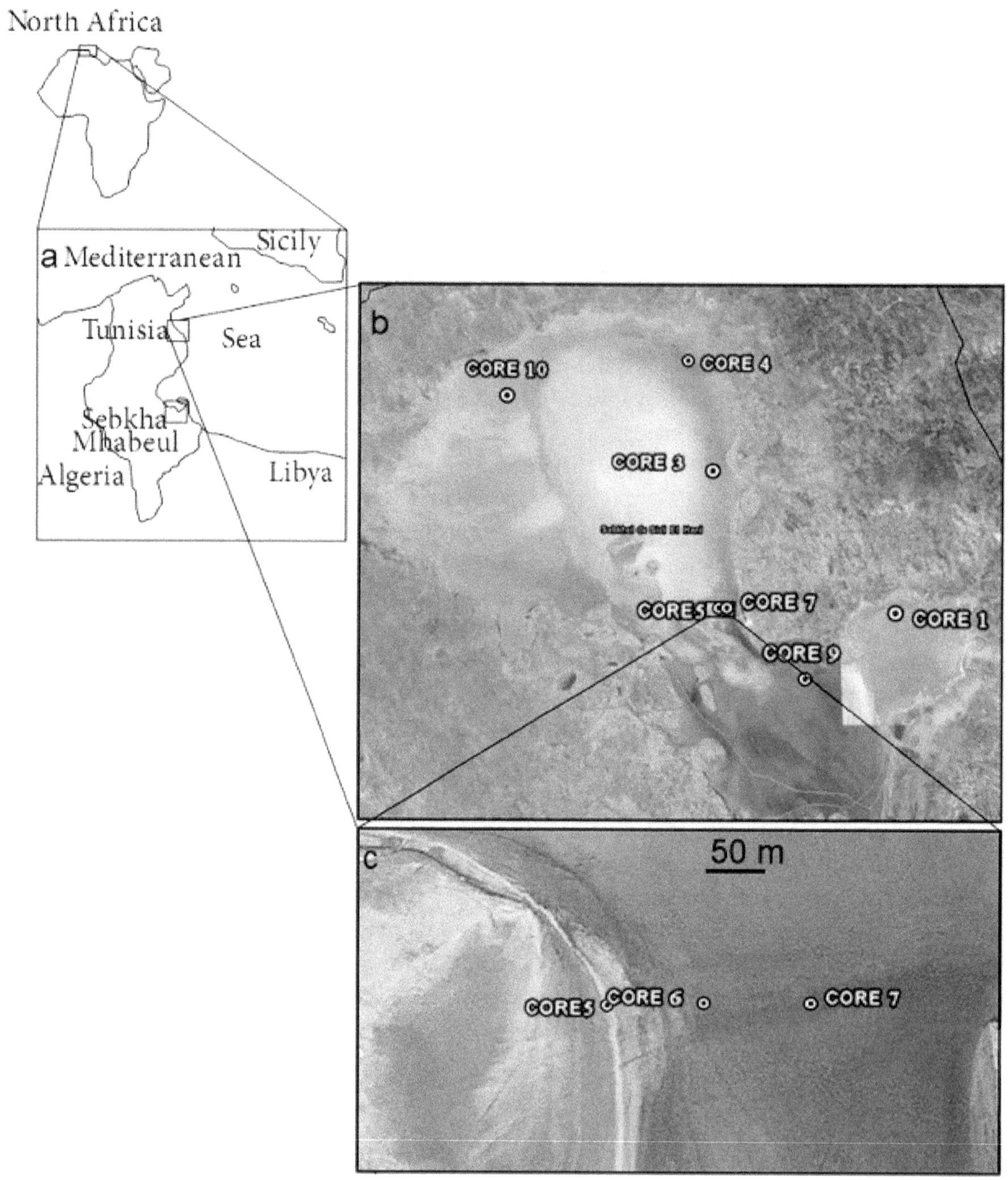

were suitable to be studied by this demarche. Cores had to satisfy this special requirement in order to do correlations between Sidi El Hani cores and a core from sebkha Mhabeul (Marquer et al., 2008); such

Figure 4. J-CORES-VCD program: manipulation (a) and results (b)
Source: J-DESC-ISWG (2006)

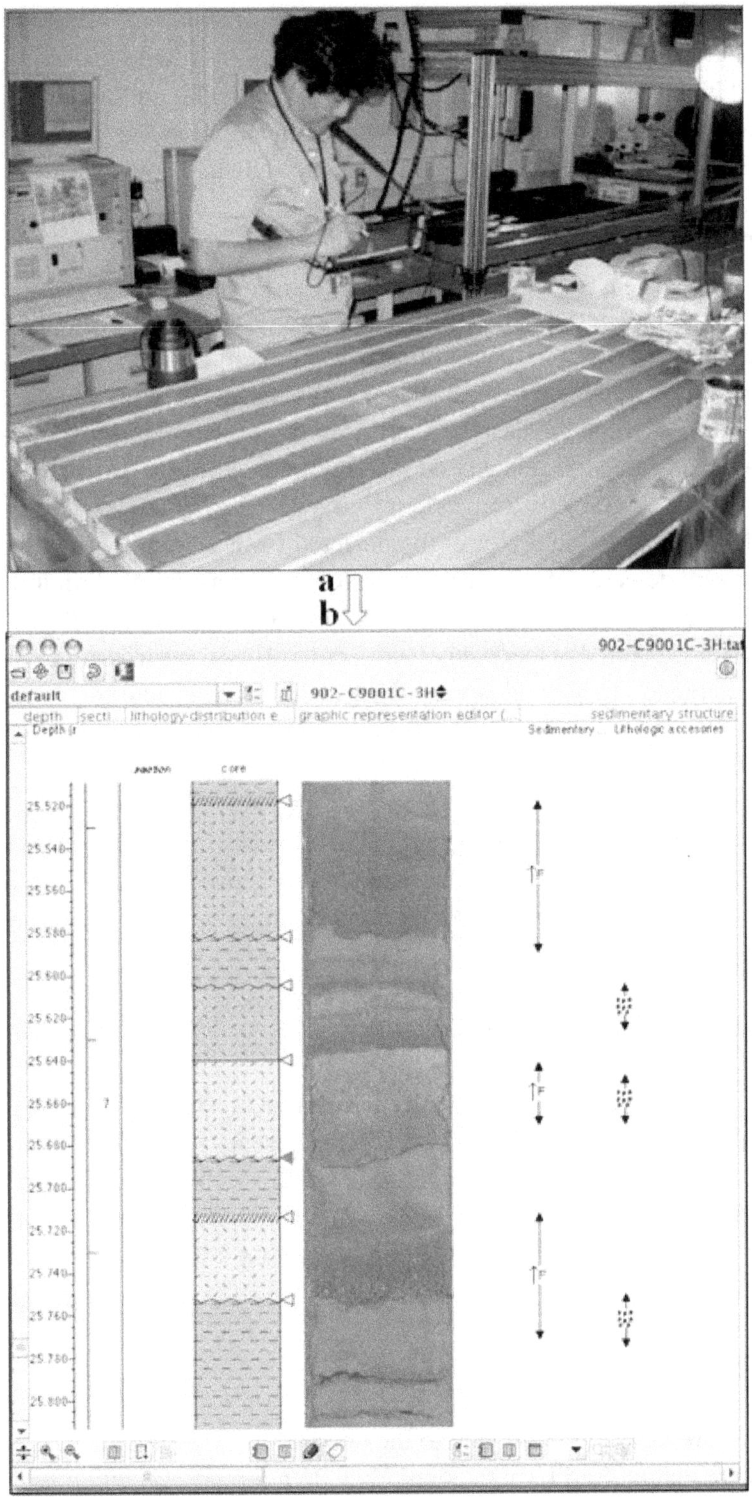

correlations were meant to find in the sedimentary record of Sidi El Hani clay pan the analogous of the climatic signals found in the sedimentary record of sebkha Mhabeul.

Visually, Sidi El Hani cores show different colours ranging from black to yellow and even white. Hence, a distinction could be established on the basis of the grey scale. At a wild guess, Sidi El Hani cores are divided into **three** different types according to their major grey scales and major grey scales variability. Some cores are with high major grey scale variability; others are with low major grey scale variability. The latter may be divided into cores with a high major grey scale and cores with a low major grey scale.

Cores With a High Major Grey Scales Variability

They are made up of bands of sediment that differ sharply in darkness. The alternation of clear and dark sediments gives the possibility to distinguish between them. Core10, core1 and core3, which are represented in (Figure 5), could be the subject of a visual correlation with the reference core of sebkha Mhabeul (Marquer et al., 2008) on the basis of the variability of the major grey scale. Furthermore, the three reference sandy bands are well recognized visually in these cores. **In core10**, the three reference sandy bands are detected respectively from 33cm to 39cm, from 70cm to 77cm and from 117cm to 140cm. A fourth reference sandy band is visually detected from 190cm to 194cm. **In core1**, the three reference sandy bands are detected respectively from 7cm to 10cm, from 54cm to 69cm, and 109cm to 114cm. A fourth reference sandy band is visually detected from 160cm to 180cm. **In core3**, the three reference sandy bands are detected respectively from 7cm to 10cm, from 76cm to 84cm and from 94cm to 104cm. A fourth reference sandy band is visually detected from 155cm to 160cm.

Cores With a Low Major Grey Scales Variability

Two cases are distinguished when the major grey scale variability is low (Figure 6 and Figure 7). **First**, cores with a high major scale along with the core; darkness is not interrupted by any detectable clear band. **Second**, cores with low major grey scale variability along with the core; the clarity is not interrupted by any detectable dark band.

Cores With High Major Grey Scales

These cores are mainly made up of shale; they are characterised by high grey scale due to their confinement and organic matter content. The core9 (Figure 6) represents a concrete example of this category. Image software is needed to detect even the major grey scale variability in this core; hence, it is impossible to correlate it with other cores on the basis of this demarche. As for the three reference sandy bands are also not well distinguished in these cores; the clayey fraction dominates in such a way that no sandy band can appear.

Cores With Low Major Grey Scales

These cores (Figure 7) are clear, probably due to the deposition of sediments in O_2-rich conditions that may destroy the organic matter. Similarly to core9, they do not represent any interest in this demarche; because the grey scale is not well elaborated. As for the three reference sandy bands are also not well

Figure 5. Cores with a high grey scale variability in Sidi El Hani clay pan: the distinction of three reference sandy bands

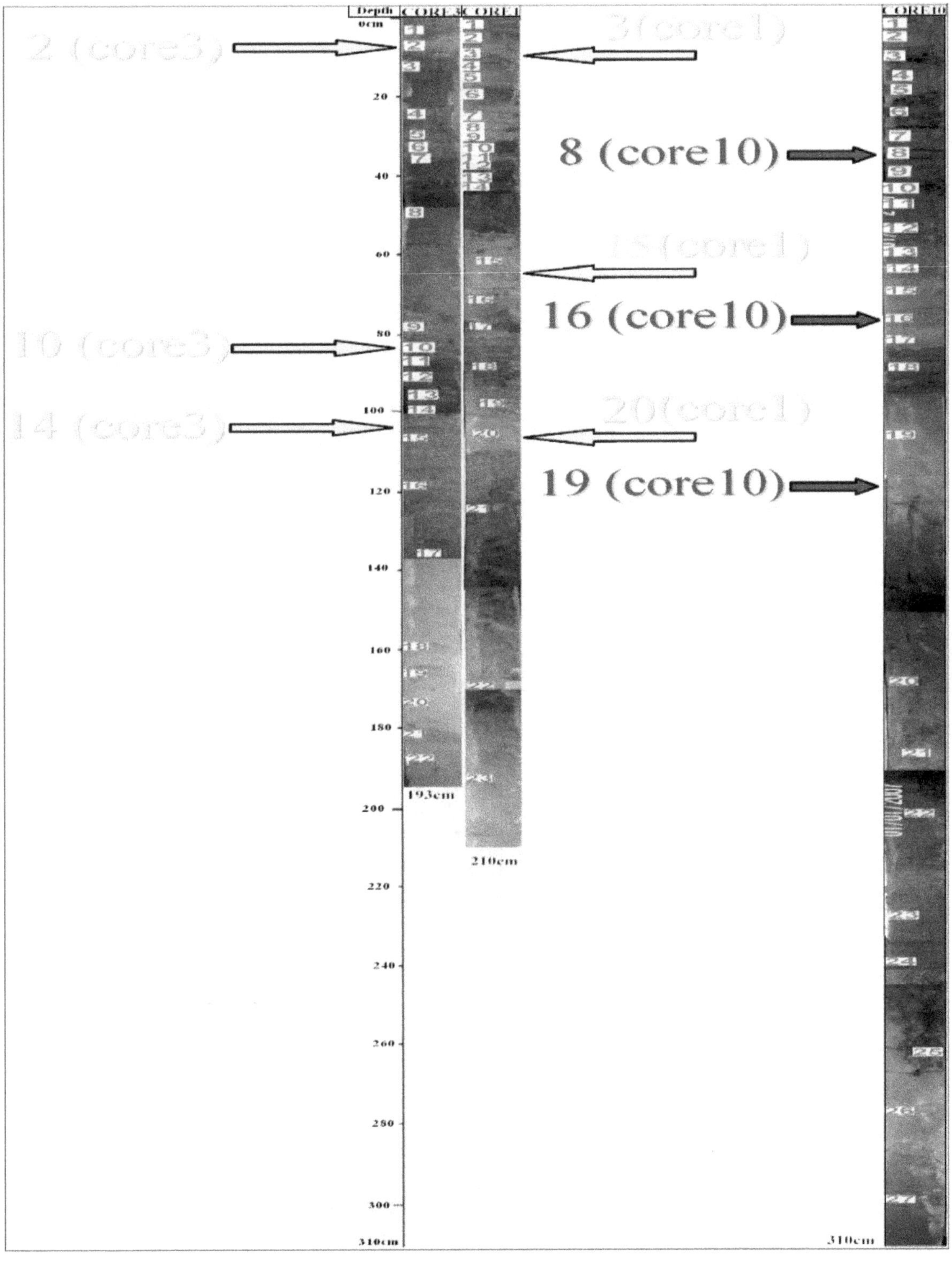

distinguished in these cores; these cores seem filled with a sandy fraction. Thus, even if the sandy bands do exist, they are more likely not to be detected easily.

Figure 6. Core with low major grey scale variability and high grey scales in Sidi El Hani clay pan

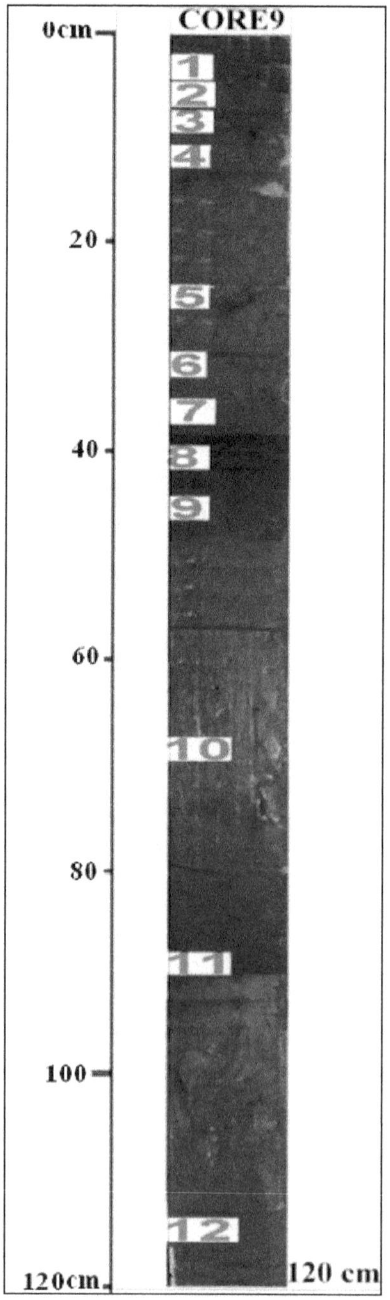

Figure 7. Cores with a low grey scales variability and low grey scales in Sidi El Hani claypan

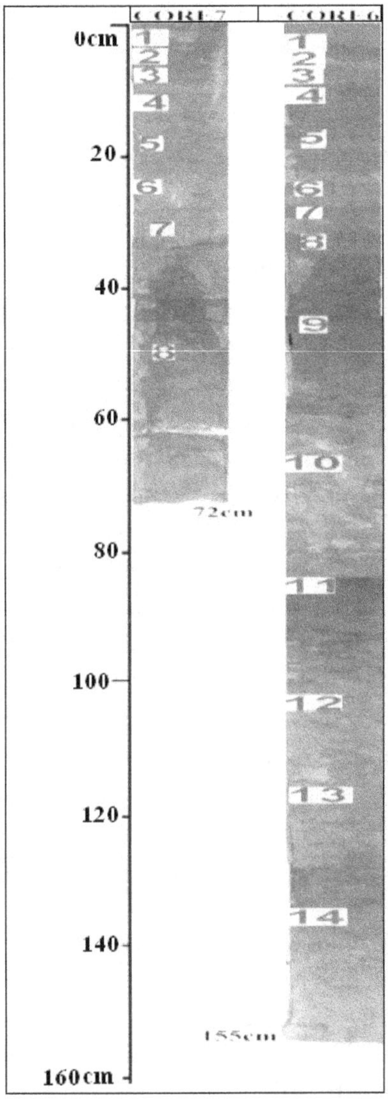

Record of the Climatic Variability During the Last Two Millennia: Correlation Between Sebkha Mhabeul Core and Sidi El Hani Clay Pan Cores

The response of Sidi El Hani claypan to the climatic variability is materialized either by the confinement periods, which are visually expressed by high major grey scales or by oxygenated periods, which are expressed by low major grey scales. Thus, on the basis of the **major grey scale variability** and the **three reference sandy bands**, the correlation was done between the core of sebkha Mhabeul (Southeast Tunisia) recently published by Marquer et al. (2008) and cores from Sidi El Hani clay pan with a high major grey scale variability (core1, core3 and core10) and with well-elaborated three reference sandy bands in order to guess the climatic variability during the last two millennia. Even though it is arbitrary, this method may give satisfactory results. But before doing any correlation, it is strongly recommended to guarantee that any depositional period represented by a band in sebkha Mhabeul has its equivalent band in Sidi El Hani clay pan. That is to say, to ensure that both saline environments behave as ever open depositional environments during the studied period.

Sidi El Hani Clay Pan and Sebkha Mhabeul as Ever Open Depositional Environments

During eustatic cycles of the fourth and fifth orders, the variation of the sea level is not so strong to reverse the sedimentation within continental zones from deposition to erosion. Thus, in addition to the regional sea level of the Mediterranean Sea, sedimentation processes in inland basins such as the Tunisian Sahel area should respond during this short period to local base levels (e.g., Shanley and McCabe, 1994) imposed by the local orography such as local playas levels of Kelbia and Sidi El Hani (Figure 8). The aeolian deflation from a discharge playa is restricted by the surface related to the level of the groundwater table; because the capillarity forbids wind erosion and attracts the dust deposed from the atmosphere. Thus, according to the sequence stratigraphy, these saline environments make the exception by being ever depositional environments. For instance, even though Sidi El Hani claypan is above the regional sea level of the Mediterranean Sea, it seems to be constantly under its **local base level**; erosion is less likely to occur. Furthermore, the near aquifer level, thanks to Kairouan and other aquifers, guarantees at least a sedimentation rate up to nil. In this study, according to the same line of thinking, it is adopted that sebkha Mhabeul represents a local base level, and it may be considered as an ever depositional environment; thus, for any deposition band in Sidi El Hani discharge playa may correspond an analogous band in sebkha Mhabeul. The difference may be in the thickness of bands, which is a function of the rate of sedimentation.

Chronology and Major Climatic Events Seen in Sebkha Mhabeul Core

Sebkha Mhabeul is an inland saline environment located in southeast Tunisia (Figure 9), it was used by Marquer et al. (2008) to infer the climatic variability during the last two millennia on the basis of studying a 65cm core (33°25.02′N, 10°50.52′E, 2m a.s.l.). Our subsequent studies deepened the study of this sebkha based on sedimentary, geochemical, and geophysical tools (Essefi et al., 2014d; Essefi et al., 2015; Gammoudi et al., 2019; Essefi, 2020).

Among other dating and sedimentary tools, Marquer et al. (2008) used the minor grey scale variability as a proxy to guess the climatic variability. Since the purpose is to find out this same climatic variability within Sidi El Hani cores, the minor grey scale of sebkha Mhabeul core, which is detected only by an images software, is transformed towards the major grey scale (Figure 10), which may be detected by the naked eye in sebkha Mhabeul core. The three reference bands are also represented along with the core (Figure 10). Thus, after elaborating the major grey scale variability, an apparent similarity seems to take place between core1, core3 and core10 (Figure 5) on the one hand and the core of sebkha Mhabeul (Figure 10) on the other hand. This core, then, is ready to be used as a tool for detecting the climatic stages already recognized in sebkha Mhabeul within the cores of the Sidi El Hani claypan. As for these climatic stages, according to the tephrochronology, the sebkha Mhabeul core (65cm) covers the last 1700 yr (Figure 10) (Marquer et al., 2008). This time interval encompasses from the top to the bottom **six** different climatic stages.

Figure 8. Playas levels of Kelbia and Sidi El Hani as local base levels during cycles of fourth and fifth orders adapted to Sidi El Hani case after Press and Siever (1986)
Source: Modified Essefi (2009)

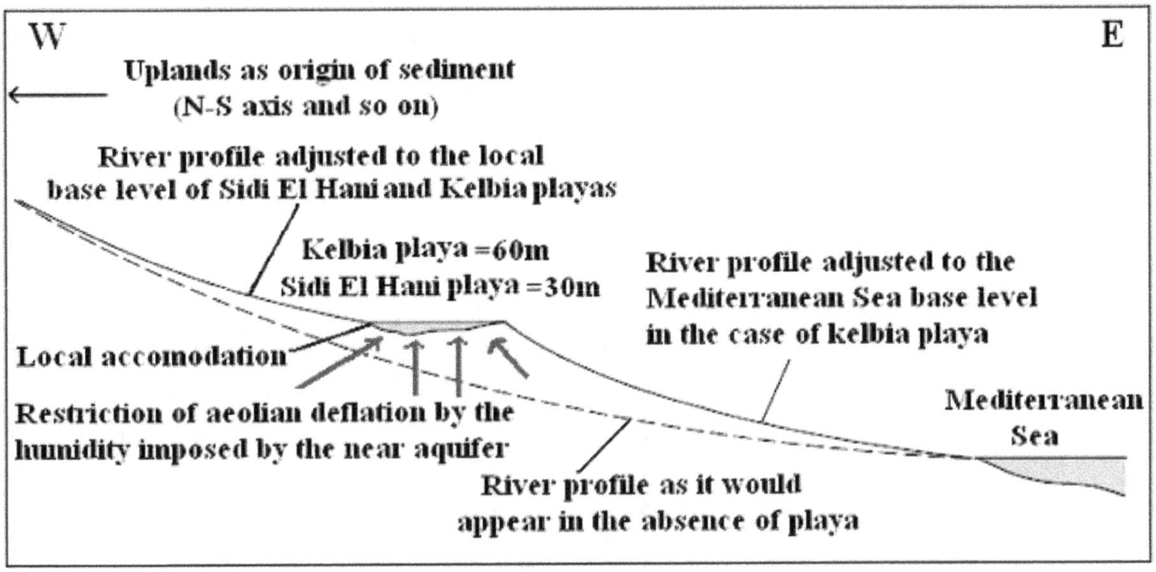

First, in the uppermost part of the core (3cm), the stage **Warming Present** (**WP**) stretches from (1954-80= 1874) to 1993, i.e. **≈120yr**; stable conditions with high grey scale characterise the establishment of modern conditions. Added to a small salt crust, this period is dominated by clayey sedimentation. This period is marked by a worldwide increasing eolian erosion (Ma et al., 2021).

Second, stage **C4** is located between 3cm and 12.5cm; it is called the **Late Little Ice Age** (**Late LIA**) (Wiles et al., 2008); it stretches between the 80yrBP and 400yrBP, i.e., **320yr**. It is characterized by an intermediate GS value; the clayey sedimentation makes up the twofold and threefold laminates. This period is limited at the bottom by the first reference sandy band from 12.5cm to 15cm. This period is marked by a worldwide increasing hydraulic erosion (Li et al., 2021).

Third, from ca. 400yrBP to 600yrBP, during the **200yr** of cycle **C3**, which is called **Early Little Ice Age** (**ELIA**), the GS signal reaches low values. As for the grain size distribution, the sandy sedimentation with a low grey scale value between 12.5cm and 15cm represents the **first reference band** to correlate with cores from Sidi El Hani clay pan; since such sandy bands are rare within such a clay pan. This stage ends at a depth of 23cm. During this period, eolian erosion returns to increase (Castro et al., 2020).

Fourth, the passage to stage **C2**, which is called the **Medieval Climatic Anomaly** **(MCA),** is marked by a sharp increase of the GS. The MCA spanned between ca. 600yrBP to 1000yrBP, i.e. **≈400yr**. The initiation of this cycle C2 is marked by the sharp decrease in the GS values, revealing a return to an intense flood activity. The end of this stage (37cm) is marked by two oscillations of high grey scale. This period is marked by a worldwide increasing hydraulic erosion (Jaramillo et al., 2021)

Fifth, during the stage **(C1)** of the **Dark Age** **(DA)**, which stretches from 37cm to 54cm, a shift from light to dark sediments is recorded between ca. 1000yrBP and 1400yrBP, i.e. **400yr**. As for the grain size distribution, the **second reference sandy band** is detected at the top of the DA. This cold period (Waltgenbach et al., 2021) is marked by decreasing eolian and hydraulic erosions.

Figure 9. Relative position of Sidi El Hani clay pan and sebkha Mhabeul
Source: Marquer et al. (2008)

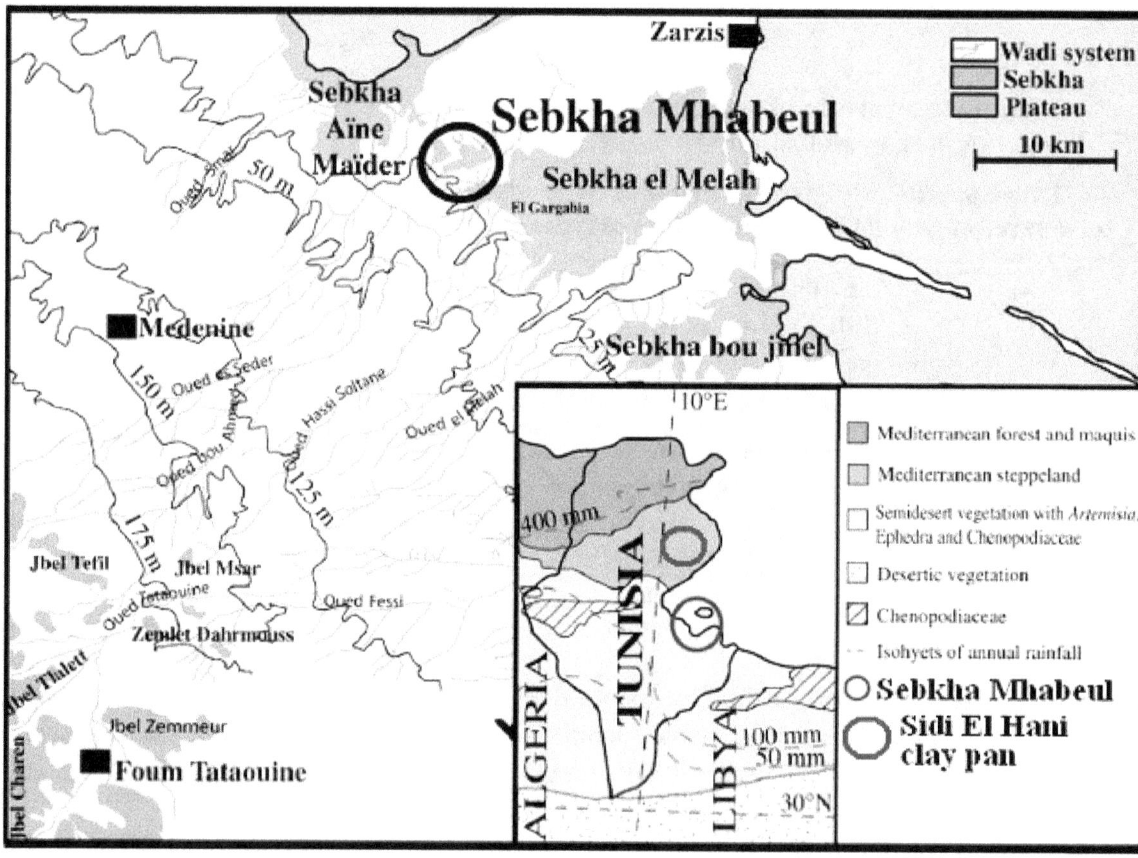

Sixth, the **300yr Roman Warm Period (RWP)** stretches between 1400yrBP and 1750yrBP. During the (RWP), high GS values associated with dark sediments suggest stable climatic conditions with small and/or scarce flood events; this period is recorded in the core of sebkha Mhabeul between 54cm and 65cm. The bottom of this recording is marked by a **third sandy reference band** between 63cm and 65cm. this period is marked by warm conditions leading to increasing eolian erosion (Luse, 2021).

Correlation of Sebkha Mhabeul Core with Core10 from Sidi El Hani Claypan

The correlation between core10 (310cm) from Sidi El Hani and the core from sebkha Mhabeul (65cm) (Figure 11) shows that the climatic variability recorded during the last two millennia in these 65cm seems to be recorded in core10 of Sidi El Hani discharge in the upper 140cm. The sedimentation rate during the last two millennia is higher in this location of Sidi El Hani than in sebkha Mhabeul. Furthermore, the sedimentation in sebkha Mhabeul that follows a linear evolution is marked by a stable rate of sedimentation; in Sidi El Hani discharge playa, it follows a chaotic evolution; sometimes the rate of sedimentation is high and sometimes is low; this phenomenon may be explained by the significant tectonic activity in the Sahel area in comparison with SE Tunisia. Following the major grey scale variability and the three reference bands along with both cores, the **six** climatic stages recognized in sebkha Mhabeul were found

Figure 10. Transformation of the minor grey scale variability of sebkha Mhabeul core toward the major grey scale variability and the representation of the three reference sandy bands
Source: Modified Marquer et al. (2008)

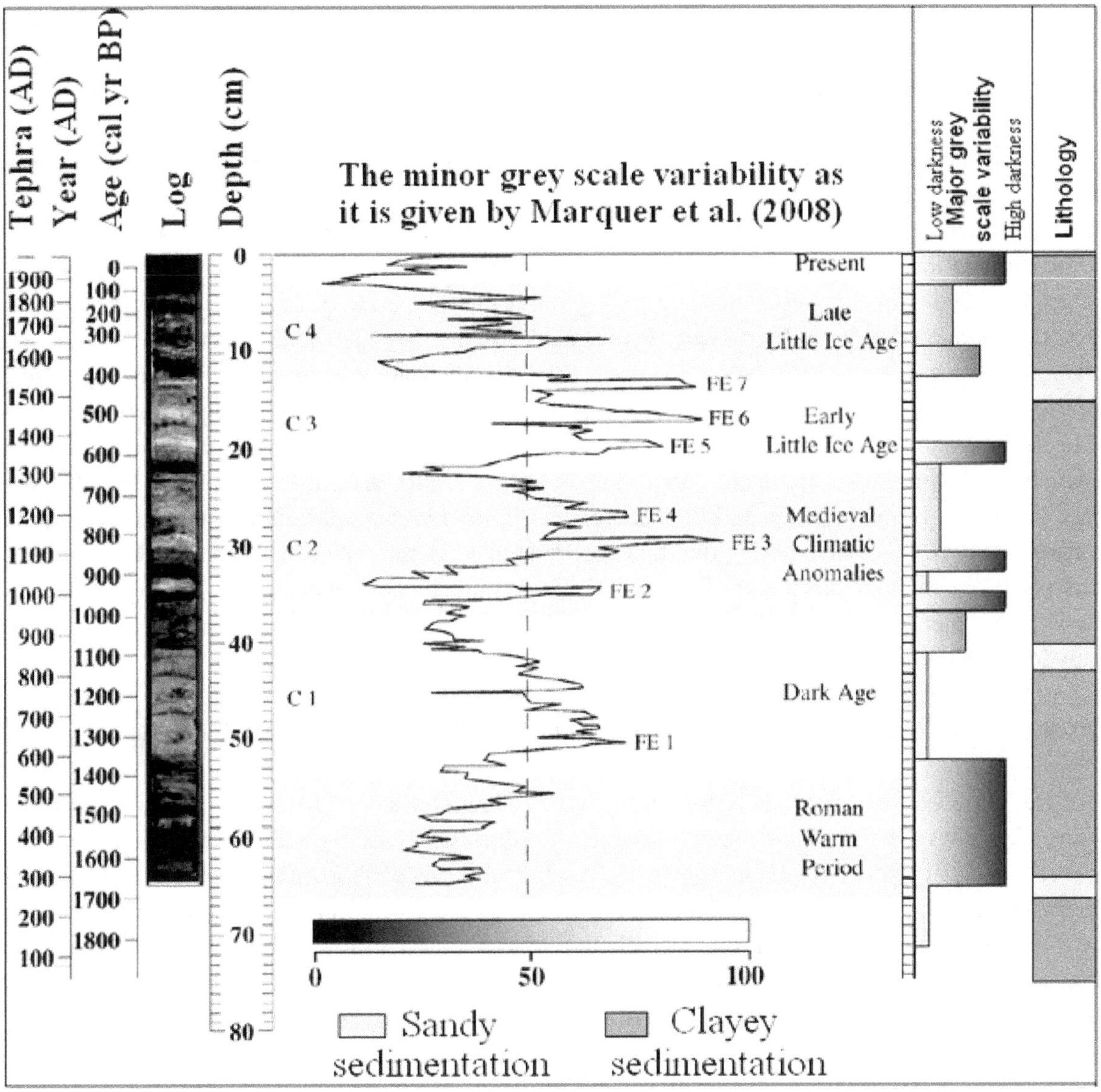

out along core10. **Firstly**, the **Warming Present** (**WP**) stretches during **≈120yr** along the uppermost 5cm of core10; the establishment of modern conditions is characterised by stable conditions with high grey scale. This period is dominated in core10 by silty sedimentation and increasing sedimentary flux in coastal regions in Tunisia (Gharsalli et al., 2020). Climatologically, this stage may be classified as dry. **Secondly**, the **Late Little Ice Age** (**Late LIA**) is located between 5cm and 33cm in core10. Chronologically, it stretches between the 80yrBP and 400yrBP. It starts with an intermediate GS value; thus, it reaches high GS values. Climatologically, this period started quite dry; then, it became wetter. As in the sebkha Mhabeul core, this period is limited in the bottom by the first reference sandy band from 33cm to 37cm.

Thirdly, the part between 33cm and 46cm represents the **Early Little Ice Age** (**ELIA**); chronologically, it stretches from ca. 400yrBP to 600yrBP. Climatologically, the intermediate values of GS indicate that this stage may be classified as moderate. **Fourthly**, the **Medieval Climatic Anomaly (MCA)** spanning from 46cm to 66cm in core10 is marked by a sharp increase of the GS. The (MCA) spanned between ca. 600yrBP to 1000yrBP. Climatologically, the sharp decrease in the GS values reveals a return to an intense flood activity; this stage, hence, may be classified as a wet period. **Fifthly**, **Dark Age (DA)** appears along 66cm to 90cm, a shift from light to dark sediments is recorded between ca. 1000yrBP and 1400yrBP. As for the grain size distribution, the **second reference sandy band** is detected at the top of the DA. **Sixthly**, the **300yr Roman Warm Period (RWP)** appears between 90cm and 140cm; chronologically, it stretches between 1400yrBP and 1750yrBP. According to the major grey scale, this period may be divided into two sub-stages. The first (from 90cm to 109cm in core10) is marked by high GS values associated with dark sediments suggesting stable climatic conditions. The second is recorded in the bottom; it is marked by a **third sandy reference band** and low GS values as signs of wet climate.

The remaining part (170cm) of core10 is made of the alternation of humid (low major grey scale) and arid (high major grey scale) periods. From 140cm to 172cm, the high major grey scale signifies a stable climatic condition. From 172cm to 206cm, the major grey scale is more likely to have intermediate values indicating more moderate conditions. From 206cm to 261cm, the darkness takes the super voice. To the end of the core, the darkness decreases slightly to reach minimum values in the bottom of the core (310cm). It is worth noting that this core is located in the outlet of Wadi Elmaktaa where the occasional rains in the western part of the Sidi El Hani watershed could cause the coming of quantities of detritic sediments mixed with fresh water. Thus, the confinement is destroyed with the humid climate in this location.

Correlation of Sebkha Mhabeul Core with Core1 from Sidi El Hani Claypan

The correlation between core1 (210cm) from Sidi El Hani and a core from sebkha Mhabeul (65cm) (Figure 12) shows that the climatic variability recorded during the last two millennia in 65cm is recorded in core1 of Sidi El Hani discharge in 114cm. As it is proved in core10 location, the rate of sedimentation is higher in this location of Sidi El Hani than in sebkha Mhabeul. Similarly, the sedimentation in this location of Sidi El Hani clay pan follows a chaotic evolution. By following the major grey scale variability and the three reference bands along with both cores, the **six** climatic stages recognized in sebkha Mhabeul and core10 were found out also along core1. **First**, the **Warming Present** (**WP**) stretches along the uppermost 3cm of core1 with its high grey scale as a sign of a dry climate. **Second**, the **Late Little Ice Age** (**Late LIA**) is located between 3cm and 7cm in core1. During this stage, the too low rate of sedimentation in this location of Sidi El Hani clay pan forbids the distinction between intermediate GS values and high GS values. Nevertheless, this period may be classified climatologically dry. As in sebkha Mhabeul core and core10, this period is limited to the bottom by the first reference sandy band. **Third**, the **Early Little Ice Age** (**ELIA**) is located in core1 between 7cm and 28cm. Climatologically, the intermediate values of GS indicate that this stage may be classified as moderate. **Fourth**, the **Medieval Climatic Anomaly** (**MCA**) spanning from 28cm to 40cm in core1 is marked by a sharp increase of the GS revealing a wet period. **Fifth**, **Dark Age** (**DA**) appears along the part between 40cm and 79cm, a shift from light to dark sediments is recorded. As for the grain size distribution, the **second reference sandy band** is detected at the top of the DA. **Sixth**, the **Roman Warm Period** (**RWP**) appears between 79cm and 114cm. According to the major grey scale, this period may be divided into sub-stages. The

Figure 11. Correlation between sebkha Mhabeul core and core10 from Sidi El Hani claypan

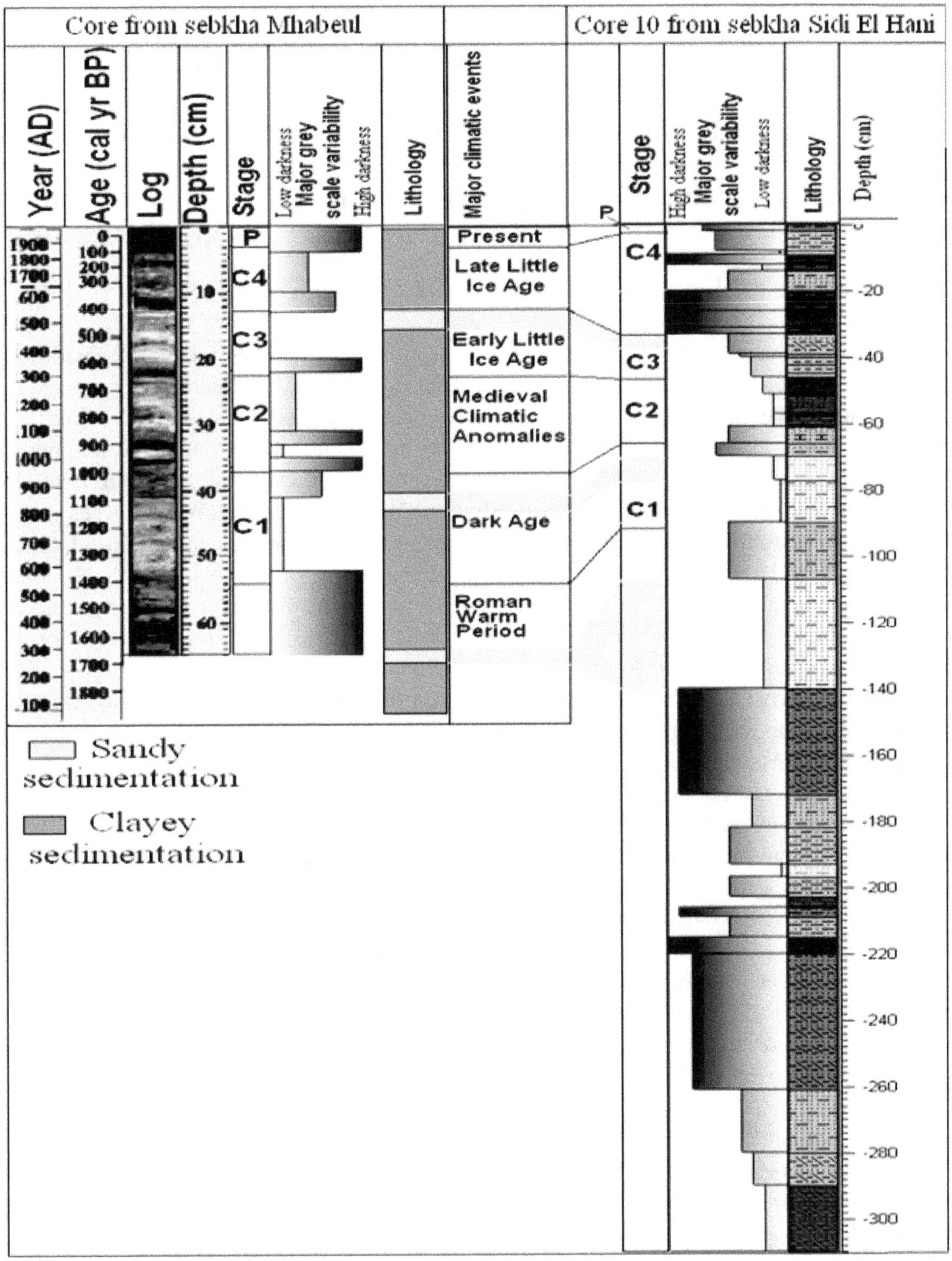

first (from 79cm to 109cm in core1) is marked by high GS values associated with dark sediments suggesting stable climatic conditions. The second (from 109cm to 114cm in core1) is recorded in the bottom is marked by a **third sandy reference band** and low GS values as signs of wet climate.

The remaining part (96cm) of core1 is made of the alternation of humid and arid periods. From 114cm to 180cm, the high major grey scale indicates a stable climatic condition. To the end of the core, the darkness slightly decreases to reach minimum values in the bottom of the core (210cm). It is worth noting that this core is located in the outlet of Wadi Enmira, where the occasional rain in the eastern part of the Sidi El Hani watershed could cause the coming of quantities of detritic sediments mixed with fresh water.

Correlation of sebkha Mhabeul core with core3 from Sidi El Hani claypan

The correlation between core3 (193cm) from Sidi El Hani and a core from sebkha Mhabeul (65cm) (Figure 13) shows that the climatic variability recorded during the last two millennia in 65cm is recorded in core3 of Sidi El Hani discharge in 104cm. Similarly to core10 and core1, the rate of sedimentation, then, is higher in this location of Sidi El Hani than in sebkha Mhabeul. Following the major grey scale variability and the three reference bands along both cores, the **six** climatic stages recognized in sebkha Mhabeul and core10 were found out along core1. **First**, the **Warming Present** (**WP**) stretches along the uppermost first centimetre of core3 with its high grey scale as a sign of a dry climate. **Second**, the **Late Little Ice Age** (**Late LIA**) is located between 1cm and 6cm in core1. During this stage, the too low rate of sedimentation in this location of Sidi El Hani clay pan forbids the distinction between intermediate GS values and high GS values. Nevertheless, this period may be classified climatologically dry. As in sebkha Mhabeul core, core10 and core1, this period is limited the bottom by the first reference sandy band. **Third**, the **Early Little Ice Age** (**ELIA**) is located in core3 between 6cm and 40cm.Climatologically, the intermediate values of GS indicate that this stage may be classified as moderate. **Fourth**, the **Medieval Climatic Anomaly** (**MCA**) spanning from 40cm to 72cm in core3 is marked by an increase of the GS revealing a wet period. **Fifth**, **Dark Age** (**DA**) appears along in the band between 72cm and 84cm, a shift from light to dark sediments is recorded. As for the grain size distribution, the **second reference sandy band** is detected at the top of the DA. **Sixth**, the **Roman Warm Period** (**RWP**) appears between 84cm and 104cm. According to the major grey scale, this period may be divided into sub-stages. The first (from 84cm to 94cm in core3) is marked by high GS values associated with dark sediments suggesting stable climatic conditions. The second (from 94cm to 104cm in core3) is recorded in the bottom is marked by a **third sandy reference band** and low GS values as signs of wet climate.

The remaining part (89cm) of core3 is made of the alternation of humid (low major grey scale) and arid (high major grey scale) periods. From 104cm to 114cm, the high major grey scale is a sign of a stable climatic condition. To the end of the core, the darkness decreases slightly to reach minimum values in the bottom of the core (193cm). It is worth noting that this core is located in the outlet of an unknown temporary Wadi where the occasional rain in the eastern part of the Sidi El Hani watershed could cause the coming of quantities of detritic sediments mixed with fresh water. Thus, the confinement is destroyed with the humid climate in this location. Compared with core10, core3 location seems less subsiding. Sidi El Hani (*stricto sensu*), then, is more subsiding than Souassi. It does not differ much from core1.

The grain size distribution of samples along a core from the sebkha of Sidi El Hani showed the superposition of eolian and hydraulic sedimentation. The fractions centered around 6, 60, 160, 250, 315, 500, 1000, 1600 μm belong to the aeolian component, and the fraction centered around 1 and 280 μm

Figure 12. Correlation between sebkha Mhabeul core and core1 from Sidi El Hani claypan

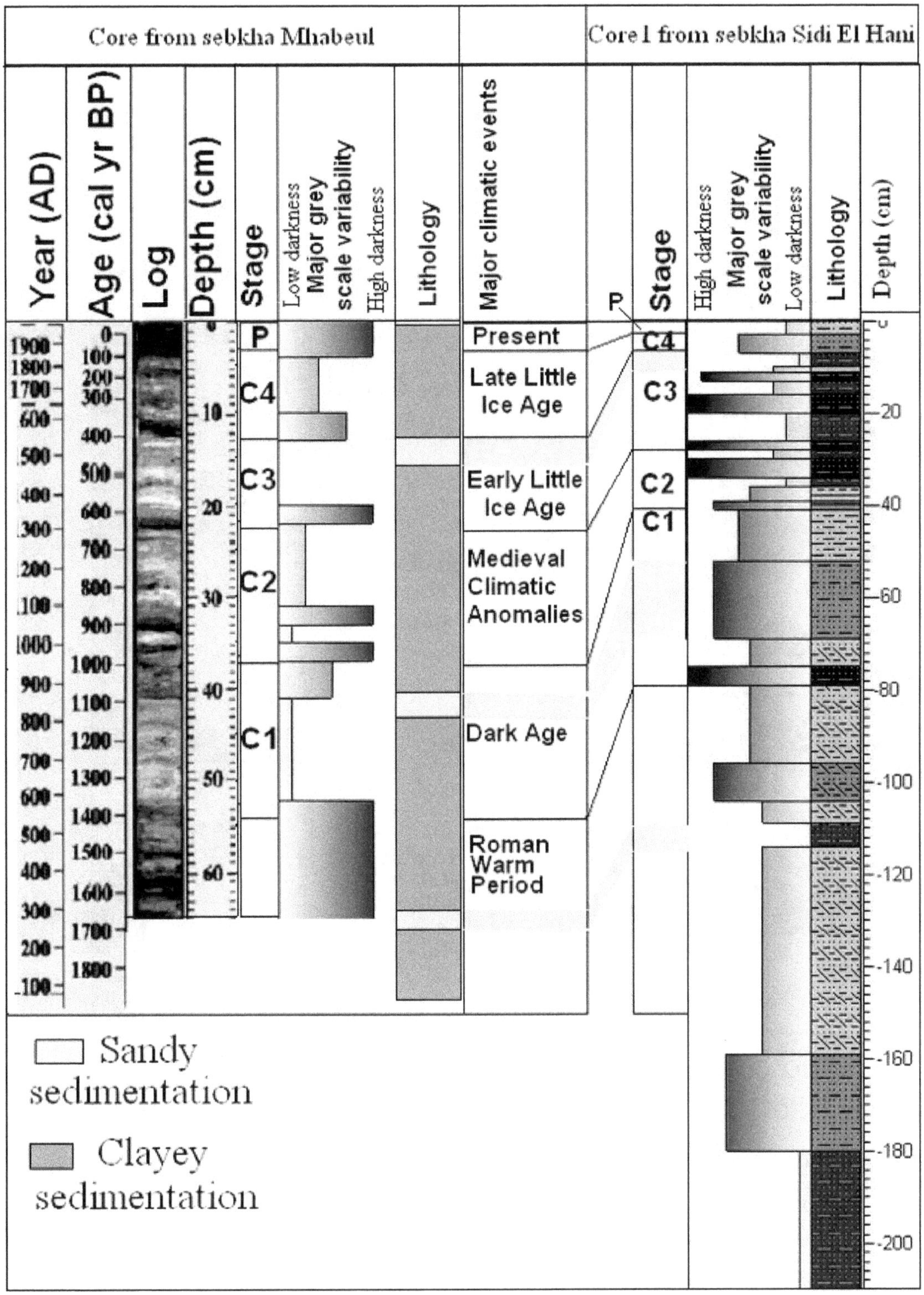

Figure 13. Correlation between sebkha Mhabeul core and core3 from Sidi El Hani claypan

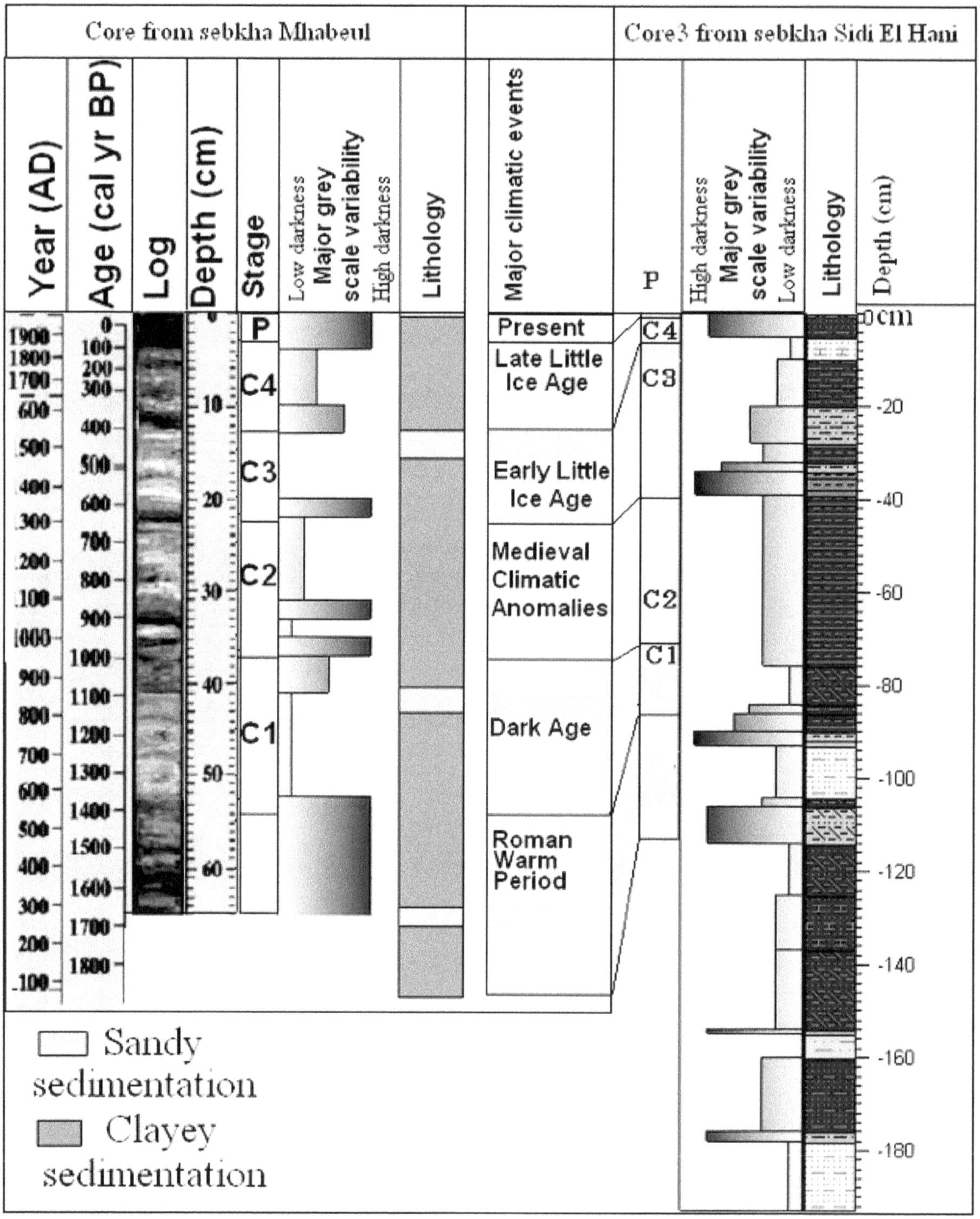

belong to the aqueous component. Sometimes, the fraction is centered around a value next to the primary modes, so we use the pre-modifier translated in order to show the slight modification. For instance, if a fraction is centered at a shoulder of 2 μm instead of 1.5 μm, it is called a translated shoulder.

Figure 14. Superposition of the eolian and hydraulic sedimentations within the sedimentary filling of Sidi El Hani wetland

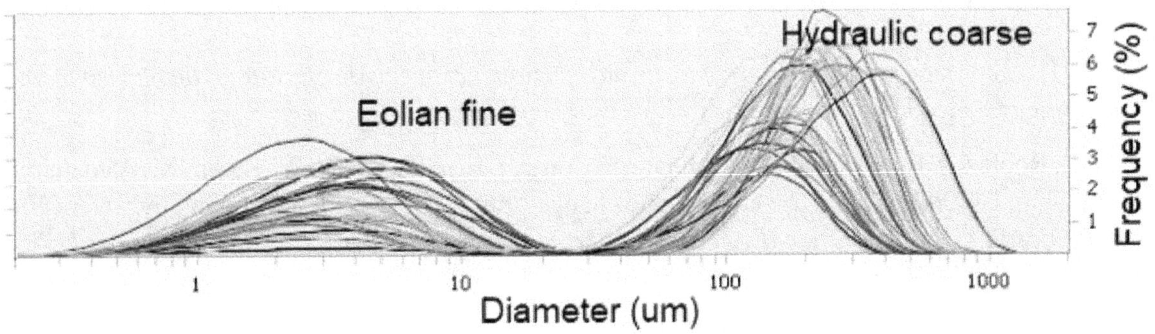

FUTURE RESEARCH DIRECTIONS

Being a serious issue threatening the pedosphere, studying erosion along drainage basins of wetlands proves vital to carry out works of soils and water conservation. During the Anthropocene, erosion has worldwide increased to modify the rate of sedimentation in many watershed outlets. Future studies should give special care to factors increasing the erodibility along the Sidi El Hani basin. Further, remedial solutions to this issue should be proposed in order the maintain the soil budget.

CONCLUSION

The VCD was a useful tool for correlation between Sidi El Hani cores and a core from sebkha Mhabeul in order to guess the climatic variability during the last two millennia in this clay pan. Nevertheless, some cores could not be correlated due to their low major grey scale variability; the use of image software, then, is a must. This method has permitted the guessing of the climatic variability in the Sidi El Hani saline environment during the last two millennia. The six climatic stages recognized in sebkha Mhabeul core were identified in core10, core1 and core3 of Sidi El Hani clay pan. As for the three sandy reference bands, each band belongs to a climatic stage. In all cores, the first sandy reference band is detected in the **Early Little Ice Age** (ELIA); the second sandy reference band is detected in the **Roman Warm Period (RWP);** the third sandy reference band is detected in the **Dark Age (DA)**. Furthermore, all Sidi El Hani claypan cores prove that Sidi El Hani is more subsiding than sebkha Mhabeul.

ACKNOWLEDGMENT AND CONFLICT OF INTEREST

This chapter is inspired by the Thesis of Essefi (2009); Any plagiarism check should exclude the URL of this work.

REFERENCES

Ben Jmaa, H. (2008). *Le système endoréique de Sidi Al Hani, Chrita et Mechertate: Paléoenvironnement et dynamique récente* (Thèse de doctorat). Université de Tunis.

Borrelli, P., Robinson, D. A., Panagos, P., Lugato, E., Yang, J. E., Alewell, C., Wuepper, D., Montanarella, L., & Ballabio, C. (2020). Land use and climate change impacts on global soil erosion by water (2015-2070). *Proceedings of the National Academy of Sciences of the United States of America, 117*(36), 21994–22001. doi:10.1073/pnas.2001403117 PMID:32839306

Castro, D., Souto, M., Fraga, M. I., García-Rodeja, E., Pérez-Díaz, S., Sáez, J. A. L., & Pontevedra-Pombal, X. (2020). High-resolution patterns of palaeoenvironmental changes during the Little Ice Age and the Medieval Climate Anomaly in the northwestern Iberian Peninsula. *Geoscience Frontiers, 11*(5), 1461–1475. doi:10.1016/j.gsf.2020.05.015

Ciampalini, R., Constantine, J. A., Walker-Springett, K. J., Hales, T. C., Ormerod, S. J., & Hall, I. R. (2020). Modelling soil erosion responses to climate change in three catchments of Great Britain. *The Science of the Total Environment, 749*, 141657. doi:10.1016/j.scitotenv.2020.141657 PMID:32841861

Essefi, E. (2021). *Apport des Approches Géochimique et Sédimentologique au Fonctionnement des Systèmes Endoréiques et Exoréiques de l'Holocène à l'Anthropocène: Intérêts Environnemental et Géoéconomique*. Institut Supérieur des Sciences et Techniques des Eaux de Gabès. Habilitation Universitaire.

Essefi, E., Ben Jmaa, H., Touir, J., Tagorti, M. A., & Yaich, C. (2013c). Variability to sedimentary dynamics and climatic conditions during the last two millennia at sebkha Souassi in eastern Tunisia. *Papers on Global Change IGBP, 20*, 29–45. . doi:10.2478/igbp-2013-0001

Essefi, E., Gharsalli, N., & Yaich, C. (2015). Climatic significance aeolian sedimentary filling of sebkha Mhabeul, southeastern Tunisia: Contribution of the genetic approach. *J Basic Appl Res Int, 15*(2–4), 140–152.

Essefi, E., Komatsu, G., Fairén, A. G., Chan, M. A., & Yaich, C. (2014b). *Models of Formation and Activity of Spring Mounds in the Mechertate-ChritaSidi El Hani System, Eastern Tunisia: Implications for the Habitability of Mars*. doi:10.3390/life40x000x

Essefi, E., Komatsu, G., & Fairén, A. G., Chan, M. A., & Yaich, C. (2013a). Groundwater influence on the aeolian sequence stratigraphy of the Mechertate-Chrita- Sidi El Hani system, Tunisian Sahel: Analogies to the wet-dry aeolian sequence stratigraphy at Meridiani Planum, Terby crater, and Gale crater, Mars. *Planetary and Space Science*. Advance online publication. doi:10.1016/j.jafrearsci.2013.05.004

Essefi, E., Mefteh, S., Medhioub, M., & Yaich, C. (2014d). Magnetic Study of the Heated and Unheated Sedimentary Fillings of Sebkha Mhabeul, Southeast Tunisia: A Geophysical Method for Paleoclimatic Investigation and Tephrochronological Dating. *International Journal of Geophysics, 2014*, 1–7. doi:10.1155/2014/908395

Essefi, E., Tagorti, M. A., Touir, J., & Yaich, C. (2014a). *Past human life in the vicinities of saline systems in Tunisia: the geoarcheological approach to link paleoclimatology, paleoepidemiology, and populations dynamics*. Arabian Journal of Earth Sciences.

Essefi, E., Tagorti, M. A., Touir, J., & Yaich, C. (2013e). Hydrocarbons migration through groundwater convergence towardsaline depressions: case study, Sidi El Hani discharge playa, Tunisian Sahel. *ISRN Environmental Chemistry*. . doi:10.1155/2013/709190

Essefi, E., Touir, J., Tagorti, M. A., & Yaich, C. (2013b). Effect of the groundwater contribution, the climatic change, and the human induced activities on the hydrological behavior of discharge playas: A case study Sidi El Hani discharge playa, Tunisian Sahel. *Arabian Journal of Geosciences*, *6*(10), 3997–4009. Advance online publication. doi:10.100712517-012-0659-6

Essefi, E., Touir, J., Tagorti, M.A., & Yaich, C. (2013d). Record of the climatic variability and the sedimentary dynamics during the last two millennia at sebkha Dkhila, eastern Tunisia. *ISRN Geology*, 1-9. . doi:10.1155/2013/936198

Essefi, E., Touir, J., Tagorti, M. A., & Yaich, C. (2014c). *Geodynamic framework of saline systems in eastern Tunisia: saline depressions inherited from the Triassic intrusions and/or the Messinian Salinity Crisis*. ISRN Geology. doi:10.1155/2014/798706

Essefi. (2009). *Multidisciplinary study of Sidi El Hani Saline Environment: The Geological History and the Climatic Variability*. Faculty des Sciences of Sfax.

Essefi. (2013). *Wet Aeolian Sedimentology and Sequence Stratigraphy within the Terrestrial Analogues in Eastern Tunisia: Implications for Wet Aeolian Sedimentology and Sequence Stratigraphy on Mars* (PhD Thesis). National Engineering School of Sfax.

Gaied, M. S., Ghanmi, A., & Ben Youssef. (2002). *Carte géologique simplifiée des ensembles géologiques de la région de Sidi El Hani*. Academic Press.

Gaied, M. S., Ghanmi, A., & Ben Youssef, M. (2002). Carte géologique simplifiée des ensembles géologiques de la région de Sidi El Hani. Ann. Min. Géol.

Gammoudi, A., Essefi, E., & Rigane, H. (2019). Variability of the grain-sized fractions during the last three millennia within the sedimentary fillings of the saline system of Mhabeul: the Holocene cyclostratigraphy. *Arabian Journal of Geosciences, 12*(9).

Gharsalli, N., Essefi, E., Baydoun, R., & Yaich, C. (2020). The Anthropocene and great acceleration as controversial epoch of human-induced activities: case study of the Halk El Menjel Wetland, Eastern Tunisia. *Applied Ecology and Environmental Research, 18*(3), 4137–4166. doi:10.15666/aeer/1803_41374166

Guo, B., Zang, W., Yang, X., Huang, X., Zhang, R., Wu, H., ... Zhang, Y. (2020). Improved evaluation method of the soil wind erosion intensity based on the cloud–AHP model under the stress of global climate change. *The Science of the Total Environment, 746*, 141271.

J-DESC IODP Information System Working Group. (2006). *(J-DESC-ISWG) with Science Support Group, Center for Deep Earth Exploration (CDEX) Guideline of Visual Core Description (VCD) Lithology Classification Scheme on the D/V Chikyu and related on-shore research Version 1.* Author.

Jaramillo, D., Vélez, M. I., Escobar, J., Pardo-Trujillo, A., Vallejo, F., Villegas, J. C., Acevedo, A. L., Curtis, J., Rincón, H., & Trejos-Tamayo, R. (2021). Mid to late holocene dry events in Colombia's super humid Western Cordillera reveal changes in regional atmospheric circulation. *Quaternary Science Reviews, 261*, 106937. doi:10.1016/j.quascirev.2021.106937

Li, T. Y., Xiao, S. Y., Shen, C. C., Zhang, J., Chen, C. J., Cheng, H., ... Yu, T. L. (2021). Little Ice Age climate changes in Southwest China from a stalagmite δ18O record. *Palaeogeography, Palaeoclimatology, Palaeoecology, 562*, 110167.

Luse, M. (2021). *Comparison of Modern and Roman Warm Period Seasonal Climate in Coastal North Carolina Using Isotope Sclerochronology of Archaeological Shells.* Academic Press.

Ma, X., Zhao, C., & Zhu, J. (2021). Aggravated risk of soil erosion with global warming–A global meta-analysis. *Catena, 200*, 105129.

Marquer, L., Pomel, S., Abichou, A., Schulz, E., Kaniewski, D., & Van Campo, E. (2008). Late Holocene high resolution palaeoclimatic reconstruction inferred from Sebkha Mhabeul, southeast Tunisia. *Quaternary Research, 70*(2), 240–250. doi:10.1016/j.yqres.2008.06.002

Press, F., & Siever, R. (1986). *Earth* (4th ed.). W.H. Freeman and Company.

Rodbell, D. T., Seltzer, G. O., Anderson, D. M., Abbott, M. B., Enfield, D. B., & Newman, J. H. (1999). An 15,000-year record of El Nino-driven alluviation in southwestern Ecuador. *Science, 283*(5401), 516–520. doi:10.1126cience.283.5401.516 PMID:9915694

Sapkota, Y., & White, J. R. (2021). Long-term fate of rapidly eroding carbon stock soil profiles in coastal wetlands. *The Science of the Total Environment, 753*, 141913. doi:10.1016/j.scitotenv.2020.141913 PMID:32906042

Sun, R., Wang, D., Wang, Y., Zhang, L., & Gu, Y. (2020). Study on durability against dry-wet cycles and chloride ion erosion of concrete revetment materials at the water-level-fluctuations zone in yellow river delta wetlands. *Wetlands, 40*(6), 2713–2727. doi:10.100713157-020-01326-0

Tagorti, M. A. (1995). *Rapport d'étude bibliographique sur le secteur de la sebkha de Sidi El Hani.* Academic Press.

Toimil, A., Camus, P., Losada, I. J., Le Cozannet, G., Nicholls, R. J., Idier, D., & Maspataud, A. (2020). Climate change-driven coastal erosion modelling in temperate sandy beaches: Methods and uncertainty treatment. *Earth-Science Reviews, 202*, 103110. doi:10.1016/j.earscirev.2020.103110

Waltgenbach, S., Riechelmann, D. F., Spötl, C., Jochum, K. P., Fohlmeister, J., Schröder-Ritzrau, A., & Scholz, D. (2021). Climate Variability in Central Europe during the Last 2500 Years Reconstructed from Four High-Resolution Multi-Proxy Speleothem Records. *Geosciences*, *11*(4), 166. doi:10.3390/geosciences11040166

Wiles, G. C., Barclay, D. J., Calkin, P. E., & Lowell, T. V. (2008). Century to millennial-scale temperature variations for the last two thousand years indicated from glacial geologic records of Southern Alaska. *Global and Planetary Change*, *60*(1-2), 115–125. doi:10.1016/j.gloplacha.2006.07.036

ADDITIONAL READING

Ben Zaied, M., Jomaa, S., & Ouessar, M. (2021). Soil Erosion Estimates in Arid Region: A Case Study of the Koutine Catchment, Southeastern Tunisia. *Applied Sciences (Basel, Switzerland)*, *11*(15), 6763. doi:10.3390/app11156763

Chafai, A., Brahim, N., & Shimi, N. S. (2020). Mapping of water erosion by GIS/RUSLE approach: Watershed Ayda river—Tunisia study. *Arabian Journal of Geosciences*, *13*(16), 1–14. doi:10.100712517-020-05774-0

Jemai, S., Kallel, A., Agoubi, B., & Abida, H. (2021). Soil Erosion Estimation in Arid Area by USLE Model Applying GIS and RS: Case of Oued El Hamma Catchment, South-Eastern Tunisia. *Photonirvachak (Dehra Dun)*, *49*(6), 1–13. doi:10.100712524-021-01320-x

Jouquet, P., Henry-Des-Tureaux, T., Bouet, C., Labiadh, M., Caquineau, S., Boukbida, H. A., ... Podwojewski, P. (2021). Bioturbation and soil resistance to wind erosion in Southern Tunisia. *Geoderma*, *403*, 115198. doi:10.1016/j.geoderma.2021.115198

KEY TERMS AND DEFINITIONS

Climate Variability: Climate variability includes all the variations in the climate that last longer than individual weather events.

Drainage Basin: A drainage basin is any area of land where precipitation collects and drains off into a common outlet, such as into a river, bay, or other body of water.

Eolian: Eolian Erosion. Wind erodes the Earth's surface by deflation, the removal of loose, fine-grained particles by the turbulent eddy action.

Erosion: Erosion is the geological process in which earthen materials are worn away and transported by natural forces such as wind or water.

Hydraulic: Hydraulic erosion is the process by which water transforms terrain over time. This is mostly caused by rainfall, but also by ocean waves hitting the shore and the flow of rivers.

Outlet: The point where a stream flows out of a lake or pond.

Sedimentary Record: The geologic record in stratigraphy, paleontology and other natural sciences refers to the entirety of the layers of rock strata.

Chapter 13
Torrential Floods and Soil Erosion:
Review on Characteristic Examples and Prevention Praxis in the Republic of Serbia

Slađan Đikić
Leskovac City Administration, Serbia

ABSTRACT

Floods on large rivers and torrential floods are the most common natural disasters in the Republic of Serbia. Floods on rivers are natural phenomena that go far beyond the framework of water management and hydro-technical measures. Given the distribution of hilly and mountainous areas in the Republic of Serbia and the developed hydrographic network, torrential floods occur very often, almost every year. Torrential floods and soil erosion are inseparable natural phenomena that shaped the relief long before the appearance of living beings on Earth. Erosion processes are difficult to notice and slow and are most often noticed only when large areas are exposed, and then the problem of erosion becomes a difficult-to-solve or unsolvable problem. For the classification of erosion processes in the Republic of Serbia, the EPM method (erosion potential method) is used, which classifies erosion into five categories that have their own quantitative characteristics.

INTRODUCTION

The floods that occur, due to their unpredictability and speed of happening, as well as due to the great lack of time for the organized reaction of people, cause great damage, not only material but also those irreparable, expressed in lost human lives. Defense plans with exact data, computer models of potential events, and experiences from the past must be elaborated in detail and based on real requirements, but also on real possibilities, in order to be applicable in a given situation. In order to reduce the risk of loss of human lives and the occurrence of flood damage, the term must be precisely defined, and the question of what a flood is must be answered. The professional definition is as follows: A flood is a temporary

DOI: 10.4018/978-1-7998-8459-0.ch013

water coverage of land that is not usually covered with water. These include floods caused by rivers, mountain streams, torrent watercourses, as well as floods caused by the sea in coastal areas. The previous definition is given in the EU Floods Directive - directive on assessment and management of flood risks (hereinafter: DAR), and it is noticeable that floods caused by water spills from sewage systems, whether atmospheric, sanitary or industrial sewage systems, are not treated as floods. As the definition of flood given in the DAR roughly states, floods can be divided according to their origin into floods caused by:

- rivers,
- streams,
- torrents,
- coastal waters or sea (Directive 2007/60/EC, 2007).

There are many causes of floods in river valleys, and they can generally be divided into three basic groups:

- those that are a consequence of natural phenomena,
- those caused by anthropogenic influence and
- those that are the result of a combination of natural and anthropogenic influences.

BACKGROUND

The anthropogenic impact is mostly related to activities in the riverbed itself, but also in the basin. Deforestation, construction of buildings and roads, channels and other activities increase the flow rate and the amount of runoff water from the basin, and shorten the time of water concentration in the main riverbed, i.e. increase the coefficient of runoff from the basin area. Also, by regulating the bed of watercourses, "cutting off dead branches", dead or still waters, and building facilities on the banks, or even in the watercourse bed itself, reduces the flow time through the riverbed, but also reduces the flow profile, which inevitably reflects on increasing water level in the riverbed. Due to certain anthropogenic influences, such as, for example, deforestation of slopes, floods caused by sliding slopes are not uncommon, and even entire hills slide down, ending up in the waterbed as a kind of a dam. The anthropogenic impact is also expressed by partitioning or narrowing the riverbed (stops, dams, bridges), which causes the formation of a flood lake upstream of the dam, and uncontrolled construction of buildings in inundations reduces the flow capacity of high waters through waterbeds. According to the statistical data, the periods of floods during the year in the Republic of Serbia differ mainly due to the size and character of the basin, i.e. due to the main cause of floods. On large plain rivers, floods occur most often in the period of early spring and spring, depending on the temperature characteristics in the upper courses of the rivers, i.e. on the melting of the snow cover in the higher basin zones. On watercourses with smaller basin areas, floods occur in periods of long-lasting and intensive rains, mainly in spring and autumn, depending on the weather conditions in the basin area. A large number of studies dealing with natural disasters begin with devastating statistics which show that their intensity and frequency are increasing, and that the consequences are increasing both in terms of damage and the number of victims. The reason for this, on the one hand, lies in natural processes that have their own dynamics and trends of intensity and frequency. Although they argue about the causes that lead to changes in the timing of hydrometeorological processes and phenomena, the direction in which they take place and the significance of these changes, most scientists agree that changes at the global level are evident, especially in the atmosphere and hydrosphere (Gosling & Arnell, 2013). On the other hand, the spatial distribution and properties of the elements that are exposed to natural disasters (primarily the population and their material goods), as well as the natural processes, influence the natural process to acquire the characteristics of a disaster. Population growth and their migration to areas where natural conditions are less suitable for life and economic activities, lead to more frequent events that we call natural disasters, more precisely to events due to which those affected need the help of other communities. Also, the increased population density in certain locations and the concentration of economic activities with accompanying infrastructure has led to such areas (cities) becoming the most sensitive to various types of disasters caused by natural and other types of hazards (Gencer, 2013). In such an atmosphere and environment, world conferences (Yokohama, 1994; Kobe, 2005; Sendai, 2015) began on the topic of "fighting" natural disasters, resulting in a number of accompanying documents, frameworks, strategies, proposed measures, which over time have been turned into legal acts across the planet to a greater or lesser extent. In this sense, disaster risk management systems have been developed within which the preparation of risk assessment studies is one of the preventive measures in that management (Wisner et al, 2004) and is legally binding for many state and local communities (Directive 2007/60/EC, 2007). What has not yet been fully defined is the methodology for risk assessment. Everything is at the level of recommendations with the aim of standardizing both the methodology in the broadest sense and the collection of data for certain levels of detail.

Reducing the risk of disasters is the goal set at the already mentioned world conferences in Yokohama, Kobe and Sendai. They are dedicated to disaster risk reduction, and the following documents have emerged from them: Hyogo Framework 2005-2015 and Sendai Framework 2015-2025 (UN, 2009). The set priorities, which are highlighted in the latter document, relate to understanding disaster risk; strengthening the organizational system for disaster risk management (prevention, mitigation, preparedness, response, recovery, etc.) in the function of disaster risk management; investing in disaster risk reduction with the aim to increase/strengthen resilience; improving preparedness for effective disaster response and for "Build-Back-Better" in the disaster recovery, remediation and reconstruction phases. Tasks set in European documents such as the Water Framework Directive (Directive 2000/60/EC) and the Flood Risk Assessment and Management Directive (Directive 2007/60/EC), which are binding on the EU Member States, have also been incorporated into laws in the Republic of Serbia, and are contained in the Law on Waters (Official Gazette of RS 30/10, 93/12, 101/2016, 95/2018, 95/2018), as well as in the Law on Emergency Situations (Official Gazette of RS 111/09, 92/11, 93/12) stating that flood risk and threat assessments are part of water risk management and protection, i.e., that are part of preventive protection and rescue measures. Today, it is necessary to keep in mind that global and climate changes are some of the biggest challenges of today. Floods caused by heavy regional rainfall are becoming more frequent, more intense and can overcome the catastrophic floods recorded. At present, floods occur as a consequence of a combination of natural and anthropogenic factors, i.e. relations in the human-environment system. Anthropogenic impact has caused the phenomenon of climate change, which increases the probability of floods and other natural disasters. In this sense, climate change scenarios are extremely important as well as flood forecasting based on them. As a consequence of the change in the precipitation regime, there was a large imbalance between useful - small and medium waters and, on the other hand large, flooding and harmful waters, in favor of the latter (Brajković & Gavrilović, 1989). Earlier it was thought that climate change would be a problem of the distant future, and even if it did happen, it would have no effect on the man or the world around us. However, over the past years, extreme events such as large fires, floods, soil erosion, storms and tropical heat waves have made us all think and ask ourselves whether these phenomena are really normal or whether climate change is happening. According to the data of the Environmental Protection Agency of the Republic of Serbia, in the last 15 years, temperatures in our country have obviously been going up. It has also been proven that global warming does not happen by itself and that it is de facto a consequence of human influence. Climate change is affecting our lives in every way. No system is not affected by them. They affect the ecosystem, people, urban, economic, and social systems (Gavrilović & Brajković, 1989). With the draft Law on Climate Change, the Republic of Serbia should establish a system for monitoring, reporting and verification (MRV) of data and information relevant to climate change, as a key precondition for sustainable planning and reduction of greenhouse gas (GHG) emissions and adaptation to the changed climatic conditions (Stefanović et al., 2014).

FLOODS

Floods on large rivers and torrential floods are the most common natural disasters in the Republic of Serbia, conditioned by the position and relief of the Republic of Serbia. From the 19th century until today, the defense against these natural disasters has been institutionally organized. Through its specialized services and public enterprises, the State has organized flood defenses on major rivers and the protection of international and other major roads. Due to the increase of unprotected facilities, the existing protection system has weakened, along with a lack of funds for the construction of new and non-maintenance of existing protection systems. But despite that, the existing protection system for protection against floods and torrents withstood the catastrophic flood of the Danube during 2006. The question is where the limits of the protective ability of the existing system are. The examples of catastrophic floods and torrents that hit European countries in the period from 2000 to the present testify to the negative impact of deception on the protective ability of the existing system. The damage from these floods conditioned the revision of the existing protection system, as well as the change of criteria and the creation of a new concept of defense. In the Republic of Serbia, the Law on Waters and accompanying bylaws prescribe the obligation of local self-government in the field of flood defense. The implementation of this obligation has pointed to a number of problems due to the lack of legislation that has ruled out the existence of a torrential flood while transferring too many responsibilities to local communities. Due to that, the training for protection against floods and torrents has been reduced to a negligible level. The occurrence of torrential floods or short torrents requires full readiness and organization of this complex activity at the level of local self-government. It can be said that there is no local self-government in the Republic of Serbia, or individuals who have not faced floods or torrents because these types of natural disasters were very common in the past. The role of local self-government in flood defense is of paramount importance, and a well-designed organization and clear division of responsibilities in the preparation and implementation of defense can significantly reduce the risk of catastrophic damage and human casualties (Stefanović et al., 2014). The Law on Emergency Situations of the Republic of Serbia ("Official Gazette of the RS", No. 111/2009, 92/2011 and 93/2012), defined the role of all subjects of state administration in emergency situations more precisely. This Law clearly defines the role of local self-government in acting in emergency situations, among which floods are a regular occurrence.

Surface waters on the territory of the Republic of Serbia, according to their importance for water management and based on established criteria: position of watercourses in relation to the state border, size and characteristics of basins, regimes and characteristics of watercourses from the aspect of water use, water protection and protection from harmful effects of the water are divided into:

- waters of the first order,
- waters of the second order.

The government determines the list of waters of the first order. Decision on determining the Census of waters of the first order (Official Gazette of RS 83/2010). Waters of the first order are interstate waters, other waters (natural and artificial watercourses). All surface waters that are not determined as waters of the first order are considered as waters of the second order. The first ones are large and medium rivers, and they are taken care of by the Republic Water Directorate and the company Srbijavode (Water Management Basis of the Republic of Serbia, 2002). Others are smaller rivers, streams, or torrents and are under the jurisdiction of local governments. Waters of the second order are critical because of their nature. They react violently, they also withdraw the same way and make terrible damage. Thus, for example, a stream water level can rise up to several meters. There are between twelve and twelve and a half thousand torrent watercourses in our country, and even much more developed countries cannot cope with that. The forecasting of torrential floods is the most difficult type of hydrological forecasting. Meteorologist Sparavalo explains that local torrents are limited to short-term floods caused by heavy local showers and a large amount of precipitation in a short time in a relatively small area. He states that such torrents, as a rule, last for a short time, several hours or at most one day, and are characterized by a high speed at which the water moves. Unlike waters of the second order, Sparavalo states that the overflow of larger rivers can take days. "It is a consequence of longer and more abundant precipitation in a wide area of river basins, sometimes in combination with sudden melting of snow, depending on the season." "These floods can be predicted in time and thus reduce the damage with timely protection," (BBC News, 2021).

Floods on rivers are natural phenomena that go far beyond the framework of water management and hydrotechnical measures. It is known that rivers and floods had a significant impact on the development of human society through history. The outpouring of large waters from riverbeds and the flooding of river valleys are among the oldest human experiences, as is their antipode - drought. Floods on large rivers bring fertility to the land even now. Today, it is necessary to keep in mind that global and climate changes are among the biggest challenges. Floods caused by heavy regional rainfall are becoming more frequent, more intense and can overcome the catastrophic floods recorded. As a special type of flood on watercourses, torrential floods stand out. Their appearance is related to torrential watercourses, whose basic characteristic is a specific hydrological and psammological regime (sediment regime). As a result of soil erosion in the basin, sediment production occurs and it is moved from the watershed to the watercourse and is further transported by the hydrographic network. The main characteristic of torrents is a small amount of water during most of the year, but large flows after intense precipitation. According to research on experimental basins in the Republic of Serbia, about 80% of the total runoff water, as well as 90% of sediment transport from torrent basins, runs off during 3-4 flood waves, and during the rest of the year, we have very little flow or torrent flow dries up in the warm season (Land Degradation and Protection - Thematic collection, 2016). Torrential floods occur in basins with intensive erosion processes, due to intense precipitation or sudden melting of the snow cover and are characterized by the rapid formation of torrent waves. The river Vrbas and its tributaries, as a consequence of the natural characteristics of the basin and the most often inadequate action of the human factor in terms of land management, cause frequent floods that are different in intensity, area and the damage they cause. Floods of larger areas are caused by the river Vrbas and the main tributary, while local floods caused by torrents are more frequent, with a very destructive effect. Flood defense in the Vrbas basin with such characteristics is a very complex problem and requires complex solutions. Although the risk of torrential floods also exists in those parts of the basin that are outside the flood zone, only those parts of the basin that are directly exposed to the risk, i.e., the areas where the torrential flood is realized, were selected for risk assessment (Kostadinov et al., 2019).

Three basins (Skrapež, Belica and Lužnica) were selected for the analysis at the macro level, while a more detailed analysis at the meso level was shown on the example of the river Skrapež. The results showed that the hazard indices are the highest in the areas with the largest area of the geomorphological flood zone and the highest erosion, and that the Skrapež basin is the most endangered from the occurrence of torrents of the three selected basins. Exposure and vulnerability indices are the highest in urban settlements, and as the Belica basin is the most populated, these indices are the highest in it and the lowest in Lužnica. The risk index is highest in the Skrapež basin, followed by Belica, and lowest in the Lužnica basin.

Torrential flooding is a natural process of increased intensity, which is an outpouring of water with a high concentration of sediment from the riverbed. Unlike the so-called river floods, which are characteristic of large plain rivers and which are realized in the lower parts of the basin, torrential floods are associated with smaller occasional, periodic and permanent watercourses whose basins are affected by erosion processes. Also, river floods appear suddenly after intense rains and the sudden melting of snow and last for a shorter time. Since they endanger all the elements of space (population, its activities, material goods and nature), they can have the characteristics of a natural disaster. For these reasons, their research and risk assessment require an interdisciplinary approach, i.e. they are based on the study of natural and social components of space (Smith & Petley, 2009).

Floods within Europe in the period 2000-2011 affected 3.4 million people, taking more than 1,000 lives. It is estimated that due to climate change, the frequency and number of flood events in Europe will increase. Without the adoption of special adaptation strategies, the number of people who will be exposed to floods annually will increase from today's 775,000 inhabitants to the value of 5 million inhabitants per year by 2085.

Pluvial floods - sudden rain floods that primarily cause great damage in urban areas, depending on local conditions and population density, as well as population structure (especially vulnerable elderly population, poor people, sick people and people who do not own a car) primarily in old urban cores along the coast where there is no space and possibility for drainage of heavy rains of large inflow, so the water level rises sharply, flooding the streets and the lowest buildings, causing great damage, including possible human casualties.

Urban floods can be of fluvial or pluvial origin. In the case of fluvial floods, they are most often the consequence of increased inflows from the wider basin area. Due to the increase in water levels and outflow of water out of the riverbed, they can cause very great damage to buildings, human property and potentially endanger human lives. Torrential floods in urban areas are especially dangerous, primarily due to the speed of formation and propagation of flood waves, frequent saturation of water with suspended sediments and mud, and high flow velocities (torrential floods in the Mediterranean and mountainous areas). In the case of fluvial floods, the causes should most often be sought in uncontrolled urbanization, which occupies the space of natural watercourses, narrows the inundation areas and increases flow rates by regulatory works (Prohaska, 2003).

With the pluvial origin of urban floods, the main cause lies in extreme rainfall, when extremely large amounts of rain fall in a relatively short time. At such times no existing system of receiving these precipitations is able to accept the amounts of water that appear as an underground or aboveground flow. The streets of the cities are turning into torrents, and the local depressions are being filled with water. Problems of urban drainage of extreme rains occur all over the world. The first to face them were large cities (Tokyo, Sao Paulo, London....) where due to large urbanization, excessive concreting and large asphalt surfaces, large floods occurred which were impossible to prevent and stop (Blagojević, 2019).

Torrents are caused by heavy rains falling from clouds known as cumulonimbus clouds, which also cause hail. The hail affects a relatively narrow area of 100 m to 300 m, while the rain from these clouds covers an area of 10 km^2 to 30 km^2. The worrying fact is the appearance of cumulonimbus systems, which cover an area of more than 600 km2, with extremely high rainfall in a short time (1 - 5 hours), which is a characteristic of the tropical climate. In the shortest possible time, such rains turn relatively large rivers into destructive torrents, which, in addition to destroying everything in their way, also take human victims. Great floods in the past, which used to be recorded once in a hundred years, are now recurring more and more often. Mild winters with little snow but heavy rain falls for hours or heavy snow with extreme temperature oscillations increase the risk of flooding. On the other hand, people tried to limit the space that used to be occupied by rivers and streams. This has been done since the dawn of civilization by building embankments and deepening riverbeds. It is commonly understood that this was done in order to make the land for agriculture, which is partly true. Numerous methods and types of facilities have been developed for river regulation and torrent regulation, which have increased the efficiency of protection over time. However, that does not mean that the danger of torrents and floods has been removed, it has only been reduced because there is always the probability of some phenomenon that will exceed the defense capability of the built protection system (Prohaska, 2003).

Given the distribution of hilly and mountainous areas in the Republic of Serbia and the developed hydrographic network, torrential floods occur very often, almost every year. As it is known, the genesis of large waters takes place in the upper part of the basin, while floods occur in river valleys, in the lower course. In this regard, it should be noted that the lower courses of most torrent watercourses in the Republic of Serbia, with developed river valleys, are of great social and economic importance. In these valleys. there is a large number of urban and rural settlements, as well as significant traffic infrastructure. Most river valleys have an agricultural purpose, while industrial zones are often next to the settlements. This means that floods threaten very valuable goods in the coastal area - settlements, roads, agriculture and industry. When we talk about torrential floods, we should keep in mind that this term is much broader than in the case of floods on larger rivers. That is why it is more correct to talk about "torrential processes" than about floods, because it is really a set of phenomena that take place in a torrential watercourse and the coastal area, when large waves of water occur. In addition to the classic manifestation of floods (due to the outpouring of large waters from the riverbed), the phenomena of torrential lava, landfalls and landslides also occur simultaneously. Torrential waves are associated with another phenomenon, which has a great impact on the scale of the phenomenon, which can only be conditionally called a flood. Namely, due to the sudden inflow of large waters, torrent waves have a very exposed steep forehead ("torrent fist"), which has great destructive power. The head of the wave destroys trees, undermines shores, creates landslides and landfalls. All that affected material mixed with water destroys everything in its way. Only well-designed and constructed torrent management facilities can withstand it and perform a protective function against the flow of a mass called "torrent lava" because it really looks like that phenomenon. On the other hand, the duration of swollen waters is very short, several hours. Hydrograms of torrent waves have a short time base, with a particularly short rise time (ascending branch) due to the rapid formation and sudden arrival of large waters. During the year, the flow duration curves of torrential watercourses also have a characteristic shape, with a very short duration of large waters (several days a year) and a long duration of small waters and medium waters. In the category of smaller watercourses, the torrential character of the hydrological regime is manifested, above all, by the specific genesis, fast concentration and short duration of large waters. Large water waves on smaller streams have the typical characteristics of torrent waves, with a sudden onset and a short time base. The two-phase character of the flow of large waters is especially visible due to the large mass of suspended and drawn sediment transmitted in waves, which prevents psammological types of observation and analysis, common on large rivers. Unlike classic torrents, the torrent character of large waters is not always equally visible in the category of smaller watercourses. Depending on the distribution and intensity of precipitation in the basin, the genesis of large waters can be different, in spatial and temporal terms. Hence, there are waves of large waters with a longer time base and a smaller maximum flow, but also typical torrent waves, with a characteristic shape of the hydrogram. Every torrential flood causes damage to the river bank. In the case of small outpourings of large waters of torrent watercourses, which occur more often, the damage is not so great. However, occasionally, massive torrential floods occur in torrential areas of Serbia, with human casualties and great material consequences (Prohaska, 2006).

FLOOD PRONE AREAS

Flood zones are areas of land that flood during floods. In relation to the character of the flood, flood zones are divided into the field and torrential. Field floodplains are on the shores of large watercourses and occupy relatively large areas. The fields are natural wetlands and field forests, but over the past millennia, they have been conquered and turned into arable land and settlements. In that way, the natural flood zone was significantly reduced. The danger of flooding the protected area has not disappeared, but it has been mitigated by effective defense measures. Recent floods on major rivers in Europe have shown that nature punishes any defense failure by flooding the former fields again. Field floods are characterized by a relatively slow rise, long duration and slow decline of the flood wave. Field floods leave behind damage on all flooded facilities that do not have the projected resistance to floods. Damaged crops and forests in accordance with the resistance of a particular species to flooding also suffer. Heavy floods cause the greatest damage to settlements and roads. It is not uncommon for less well-founded buildings to be demolished. When it comes to urban environments, great damage occurs to infrastructure facilities (water supply, sewerage, energy, telecommunications, etc.). Torrential flood zones are set in the coastal area of torrential watercourses: torrential floods are characteristic of smaller watercourses. The occurrence and duration of a torrential flood, even on a relatively large torrential flow, rarely exceeds a few hours from the beginning of a heavy downpour to the end of a torrential flood. Torrential floods are always devastating. Although field floods and torrential floods have completely different characteristics, similar principles are applied to determine flood zones, with field flood zones being larger and easier to describe in square kilometers, while torrential flood zones are described in hectares. The calculation of the flood zone depends on the accuracy of the data on the riverbed and the potential flood zone and the flow conditions. Such data are obtained by geodetic surveying, and the required accuracy is determined based on the needs of a specific task, in accordance with the regulations in the field of geodesy, surveying and land cadaster. Geodetic surveys with satisfactory accuracy can be performed with classical surveying, which is the most expensive and slowest type when it comes to large areas. Classical surveying is only cost-effective in the case of small areas. For recording large areas, various methods of remote detection are used, recording from various aircraft and satellites. Imaging can be photographic, multispectral, radar, laser or combined.

Today, the best results of recording flood zones of torrents and small rivers have been achieved with the LIDAR system. It is a laser and photographic image system that provides data in several levels applicable for purposes other than those for determining flood zones.

In short, the LIDAR system provides the following data classes:

- terrain data - DTM (digital terrain model),
- data on vegetation and its height,
- data on constructed facilities,
- orthophoto of the terrain,
- infrared imaging of the terrain (used to detect nearby underground water).

Due to this set of data and high precision, which is about ten times higher than other remote detection methods and is close to the accuracy of classical recording, the LIDAR system is today the most common recording method in the EU and on other continents.

The investment in the LIDAR system has shown high profitability. In addition to precisely defining the flood zone and flood risk zone, this method of surveying is used to develop urban and infrastructure plans, as well as to create a complete geodetic base necessary for projecting, especially for working with numerous software design/project programs (Rosier et al., 2021).

EROSION, TORRENTS AND EROSION AREAS

Torrential floods and soil erosion are inseparable natural phenomena that shaped the relief long before the appearance of living beings on Earth. The significant effects of climatic and meteorological factors, such as torrential rains and storms, cause torrential floods, which are manifested by devastation, high speed of occurrence and short duration. On the other hand, there is an imperceptible effect of climatic factors manifested as soil erosion. Erosion is slow and less noticeable until it strikes entire provinces, states and parts of continents. Traces of civilizations that have disappeared due to the effects of erosion are present all over the world. Erosion processes are difficult to notice and slow. They are most often seen only when large areas are exposed, and then the problem of erosion becomes a difficult or unsolvable problem. Due to the change in the microclimatic characteristics of the area, a new precipitation regime is being established, and it is characterized by reduced total precipitation in the area, but also more frequent high-intensity rains and long dry periods. Thus, deserts are gradually created. Eroding (crushing) of the soil is achieved first by thermal stress, because moisture, frost and heat gradually crush the soil and prepare it for rain and wind, which move the crushed material. Of course, not every rain or wind has enough energy to start the prepared material. The people gave each type of rain and wind a name that characterizes them. The pouring rain sprinkles the ground and sticks to the clothes, but it does not give any runoff. Mild rain is something that every farmer wants because almost every drop of rain ends up in the soil and there is hardly any surface runoff. A downpour is a type of heavy rain of high intensity, greater than 30 mm per hour. These rains have great impact energy. Raindrops of this type of rain have a diameter of 1.5 mm to 7 mm and each drop is a projectile that falls to the ground at a speed of 3 to 6 m/s. The average diameter of drops in such rains is about 3 mm in the conditions of our climate. During the downpour, 10 to 30 tons of such projectiles per km^2 fall to the ground every second. As the surface of a typical rain cloud that delivers rainfall is usually around 10-30 km^2, it is not difficult to calculate the total amount of water that fell to the ground. Showers also erode flat soil, and on the steep ground, almost all rain water flows away along with sediment, even on completely dry and permeable soil. Although it seems illogical, it should be understood that regardless of the water permeability and dryness of the soil, everything that exceeds the power of absorption flows away. Thus, in our region, it is not uncommon for torrents to appear on sandy terrains that have the highest power and speed of absorption. Wind also has various names, from breeze to storm. Strong, stormy winds blow in our plains, there are even tornadoes. The strength of the winds in the plains is great, it often happens that they blow the seeds out of the soil, so for such winds, the removal of small soil particles is not a problem at all. Both forms of erosion are present in the Republic of Serbia, but when it comes to the transport of moving particles, the wind has a transit role because every wind-driven particle is captured by water at some point and taken to the watercourse (Climate Change Vulnerability Assessment, 2012).

From ancient times until today, people have been trying not to be in the way of a devastating torrential flood, if at all possible, and where that was not possible, more or less successful buildings were built for protection from torrents. Bare surfaces are ideal for unobstructed drainage during heavy rain. As there are no obstacles that would reduce the flow rate, the water erodes the soil easier, creates ravines and forms a torrential flood wave. In Vojvodina, oak forests have disappeared on all higher terrains and only those in the flood zone of large rivers have survived. Erosion did its job and took away the fertile layer of soil and sand reached the surface. That is how the Deliblatska, Ramsko-Golubačka, and Subotička sandstones were created in the today's Republic of Serbia. Similar sandstones exist near Szeged (Hungary) and Đurđevac (Republic of Croatia).

In summary, the intensity of erosion depends on four main factors. The three factors represent the natural characteristics of the area: geological-pedological background, relief and climate, while the way the land is used is a factor that is mostly under the control of people, which is why it is subject to dynamic and rapid changes. For the classification of erosion processes in the Republic of Serbia, the EPM method (erosion potential method) is used, classifying erosion into five categories with their own quantitative characteristics. This current situation can be easily changed by some planned or irresponsible change in the way the land is used (Prohaska, 2006).

Comparing the state of erosion, sediment production and planning the necessary works requires regionalization, some kind of division of the territory. As natural units in the Republic of Serbia, basins are the basis for review, analysis and comparison.

The production of erosion deposits means the initiated and removed volume of material created by climatic factors on the surface layer of the soil. The abstracted relation of erosion categories and sediment production is a double increase or decrease of sediment volume within one erosion category. The produced quantities of erosion deposits are usually expressed in specific and total annual values for defined basin units. Specific production is expressed in m^3/km^2 per year, i.e. the volume of eroded and removed soil that occurs on 1 square kilometer of surface during one year. The total production for the entire basin is expressed in m^3/year, i.e. as the total eroded and removed soil from the basin during one year. The specific production of erosion deposits is a practical quantity because it indicates the intensity of erosion processes regardless of the size of the basin and is most often used to compare the situation before and after the works performed, as well as the expected results of the planned work. The initiated sediment can be expressed in the form of the thickness of the layer of removed soil. When the intensity of erosion reaches 1000 m^3/km^2 per year, it corresponds to the reduction of one millimeter in the thickness of the fertile soil layer. In our conditions, the intensity of erosion can vary from 100 m^3/km^2 to 4000 m^3/km^2 per year (Gavrilović & Brajković, 1989).

Erosion areas and zones are most often equalized. I In fact, erosion zones are areas affected by different classes and categories of erosion classified according to appropriate methods of mapping erosion processes, while erosion areas are areas where a strong erosion process does not have to be developed but which can become foci of erosion if any of the factors important for the development of erosion changes. Simply put, it is a natural predisposition of a specific terrain for the development of erosion processes, and that characteristic is unchangeable and cannot be influenced by any works and measures. In the past, when entire basins were affected by strong erosion processes, all of them were declared an erosion area not because they were such as a whole, but in order to facilitate the implementation of extensive administrative measures against erosion land management. Another reason was that no methods for identifying erosion areas had been developed at the time. Although the provision on declaring erosion areas has been active for a long time, the terms erosion processes and erosion areas are often equated or confused. This creates difficulties for local governments, because they have to implement this obligation. Also, a problem for local governments is very expensive works on the sanation of land destroyed by erosion due to the complexity, difficult accessibility of the terrain on which the works are performed and the necessary high expertise in the design and execution of works. The development of erosion processes in erosion areas is mitigated by the application of non-investment anti-erosion measures prescribed to owners and users of land in the erosion area (Stefanović et al., 2014).

These measures are simple and consist of prohibitions and obligations, as follows:

I Prohibitions

- Prohibition of pruning branches (for animal food)
- Prohibition of growing root crops on steep fields
- Prohibition of plowing on the slope of the land
- Prohibition of clear cutting of forests on sloping terrains
- Prohibition of grazing on degraded pastures

II Obligations

- Obligation to plow following isohypsis (contour)
- Obligation to convert degraded fields into meadows
- Obligation to ameliorate degraded pastures
- Obligation to afforest bare slopes
- Obligation to convert annual crops into perennial crops on degraded areas
- Obligation of anti-erosion land management
- Obligation of anti-erosion forest management

These measures are prescribed for each individual parcel located in the erosion area and are entered in the part of the parcel data table. The measure that will best suit the anti-erosion method of land management is prescribed, without reducing the income of users from that area. Only in extreme cases can it be ordered to take some measure that drastically changes the conditions and culture of land use. These are only severely eroded arable lands that must be excluded from agricultural production for a longer period of time and protected by perennial forest vegetation. In places where erosion has reached large proportions, administrative measures are useless until expensive technical, biotechnical and biological works are performed. Let the detail from Stara Planina and Trgoviški Timok serve as an example, where erosion processes have reached extreme proportions. Wild forest exploitation on typical eroded soil influenced the formation of a range of deep ravines. This area is afforested with a combination of biotechnical works for ravine consolidation and afforestation.

The problem that characterizes the erosion areas is not the current intensity of erosion that is the result of the applied anti-erosion measures, but complying with the restrictions in land management in the erosion area. Namely, a significant volume of anti-erosion works is afforestation of erosion areas with the application of a set of technical and biotechnical works. The natural susceptibility of the terrain to erosion is the basis for the determination of the erosion area. If this is not done and erosion areas are not declared, there is no possibility to impose the obligation to implement administrative measures, but that does not mean that the erosion area has ceased to exist. The biggest problems are in the field of forestry. Intensive logging on steep slopes in the erosion area deepens the ravines that were created during the period of excessive exploitation in the first half of the twentieth century.

In some places in the Republic of Serbia, the soil is such that a high leveling of the terrain is not needed for the development of erosion. Let us use the example from the vicinity of Veliko Gradište, where a torrent ravine cut the road, and the total denivelation from the top of the basin to the mouth is about 40 m. Agriculture in the erosion area requires complying with anti-erosion land management. Plowing on the slope of such land destroys the field and quickly turns it into a deep ravine, although the land is not so steep (Land Degradation and Protection - Thematic collection, 2016).

Erosion areas are identified and displayed on maps. Detailed displays are on maps of scale 1:25000, while overview maps of small dimensions show the percentage of the occurrence of erosion areas. The map of erosion areas is an important document because it clearly identifies areas where a small change in the purpose of the surface changes the intensity of erosion processes. Such surfaces must be managed only in a way that mitigates the intensity of erosion, often with the additional application of protective anti-erosion works. Such a way of managing the purpose of the land is called "integral" and the law prescribes it in the area that has been declared erosive. The plain part is not so affected by erosion.

The Republic of Serbia is naturally destined for torrents and torrential floods. Relief, geological background subject to erosion, climate characterized by heavy rains and parts of the country where strong winds occur - re prerequisites for the appearance of torrents not only in mountainous areas but also in plains. Torrents were not classified as a problem as long as the endangered values were small. The railway and road network development conditioned the "conquest" of river valleys in which entire settlements were built. The torrents also conditioned the construction of facilities for protection against torrential floods, which was later extended to works for protection against erosion. People in the known torrent areas where the protection system was built were convinced that they were protected from torrents. This was somewhat true, but only until the occurrence of floods that exceeded the level of the protection system, projected for the probability of a once-in-a-hundred-year flood (1%).

The torrents caused damage in parts of the Republic of Serbia that were covered by a smaller scope of protection systems. It is clear that torrential floods are more common in areas that are insufficiently covered by works for torrent management and erosion protection, but that there are also catastrophic torrential floods on watercourses that are relatively well covered by protective works. The current state of erosion is the result of work on anti-erosion and torrent management in the period of several decades, which has made great progress compared to the period of fifty years ago. The average intensity of erosion has been reduced by one, and in some areas by two categories. As erosion is a phenomenon and a process that always exists, maintaining the current condition requires maintaining the existing protection system, and improving the condition requires additional work. These works have been performed for decades and are necessary because significant areas are still under processes from excessive to medium erosion category. Simply put, works on torrent management (barriers, thresholds, regulations, etc.) serve as a "skeleton" of the system and prevent changes in torrent waterbeds, and biological and biotechnical works (afforestation, grassing, raising orchards and the application of administrative anti-erosion measures, etc.) represent "suit" of integral protection against torrents and erosion. After each major torrential flood that caused great damage, analyzes were done with the goal to determine the existence of some responsibility. The most common problem was that there were no precise data on rains, nothing was left but a rough estimate. A retrospective of torrential floods that have caused great damage during the last decades has been made. Countless minor torrential floods destroyed one or two houses, destroyed a less important bridge, destroyed several fields and, unfortunately, only the affected knew about them, who mostly did not have insurance, so there is no record of the damage.

TORRENTIAL FLOODS EXAMPLES

Torrential Floods in Europe

Torrential floods have been characterized as one of the most significant natural disasters in the world (Smith & Petley, 2009) and in Europe (Gencer, 2013), which pose serious risks to human lives, facilities and infrastructure. Given that torrential floods often occur in smaller basins where water levels and flows are not measured, a project (Hydrate) of their recording (550 documented torrential floods) has been started in Europe and the spatial and temporal regularity of their occurrence has been analyzed. They have been found to have a greater magnitude in Mediterranean countries, as well as to have a clear seasonality in their occurrence in each region. Major torrential floods with significant casualties occurred in 1952 in the United Kingdom (34 victims), in 1962 in Spain (more than 400 victims), and in 1968 in Italy (72 victims), in 1994 in Italy (69 victims), in 1999 in France (35 victims) (Gaume et al, 2009).

Torrential Floods in the Republic of Serbia

The Korbevačka river is right tributary of the river Južna Morava, joining it near Vranjski Priboj. The torrent wave on this river is formed on the slopes of Besna kobila and descends almost a thousand meters to the mouth of the river Južna Morava. The basin area of this torrent is about 90 km^2. This river has been a nightmare for decades for both the population and the economy, as well as those working in the anti-torrent profession. It often interrupted traffic on the Belgrade-Skopje railway. Torrential barriers proved to be insufficient and were demolished several times along with the railway bridge. Two torrential floods on this river had severe consequences. On July 23, 1963, two of the three bridges over the Korbevačka river, which often changed direction, were destroyed. On that occasion, the only anti-flood barrier was destroyed. More of them were planned; however, only one was built. The torrential wave cannot occur without heavy rain, but none of the pluviometer stations registered rain either that or the previous days because the pluviometers were at a great distance from the basin. The collapse of the bridge on the Korbevačka river happened on May 15, 1975, as a consequence of low throughput for such a torrent wave. The bridge was demolished just before the train reached it, which led to a record number of human casualties caused by a torrential flood. The tragic balance of this accident was 13 human casualties, with enormous material damage.

Another example is the railway accident caused by the torrential flood of the Korbevačka river in 1975. The Korbevačka river belongs to the hydrologically unexplored basins where no observations and measurements were performed, which is still valid today, so in both cases, only assessments were made on the basis of recording the situation. The only change is that the torrent bed is regulated in the zone of the railway and the road and that afforestation of larger scope has been carried out in the basin, but the issue of determining and defining the authoritative phenomenon and the degree of system safety still remains unresolved.

The Sejanička river is a small torrent tributary of the Južna Morava, with a basin area of about 12 km^2. On July 2, 1983, it caused great damage to the urban part of the town of Grdelica with a torrential flood. The Textile factory was completely destroyed. According to meteorological observations, the total precipitation that preceded the torrential wave was about 90 mm. The rain lasted a relatively short time (about 3 hours) and was very strong. The analysis determined that the probability of the occurrence of such a rain was once in two hundred years (0.5%). Analyzes also confirmed that the rank of high waters was at the level of occurrence once in a hundred years, which is a direct consequence of the performed anti-erosion works, which were not sufficient.

The river Vlasina is a left tributary of the Južna Morava, with a basin area of about 900 km^2. The upper part of the basin, on the slopes of the mountain Čemernik, has a characteristic mountain relief. There are two municipal centers (Babušnica and Crna Trava) in the basin of the river Vlasina, and at the mouth is a third municipal center and the largest populated place - Vlasotince.

The torrential flood of the river Vlasina on June 26, 1988 is one of the largest torrential floods in the territory of Serbia recorded in history and water management annals. The causes and consequences of that flood are of extreme character, both in terms of the probability of their occurrence and in terms of catastrophic proportions. Precipitation in the Vlasina basin that caused the flood wave is one of the meteorological phenomena with an extremely rare probability of occurrence. On the other hand, the great waters of Vlasina caused torrential floods of enormous proportions, with catastrophic damage. The torrents were caused by extremely heavy rain from 150 mm to 220 mm, which fell in four hours. The probability of the occurrence of that torrential flood was once in 300 years (0.33%). The catastrophic scale of the flood of the river Vlasina can be illustrated by data on 4 human victims, several thousand destroyed houses and other facilities, 32 destroyed bridges, as well as tens of kilometers of damaged regional roads. The water intake plant "Ljuberađa", which supplies water to Babušnica, Bela Palanka and a part of Niš, was severely damaged. In Vlasotince itself, the right side of the dam was broken, from where a torrent wave broke through the settlement and wiped out the shopping center and the bus station. That saved the settlement behind the left bank embankment, where the houses were flooded. Only a few bridges were damaged, and the rest were completely wiped out by the force of a torrent.

The torrential flood also occurred in November 2007, which, in addition to the Vlasina basin, also affected parts of the Nišava basin. Bridges, parts of roads, as well as the water supply plant "Ljuberađa" with a pipeline for the supply of Niš, Babušnica, Bela Palanka and other places were destroyed again. The interruption of traffic and water supply lasted for weeks. The official conclusion was that this phenomenon was a consequence of the penetration of the heat wave from the southeastern and southern parts of Europe and the sudden melting of the snow, which was accompanied by moderate precipitation - rain. The damage from that flood has not yet been statistically processed. What can be stated is the fact that the scale of that flood is not comparable to the catastrophic flood from June 1988. However, so many times, the fact has been confirmed that the damages from torrential floods, always and without exception, are proportional to the values of the goods that are hit by the torrential wave. The flood on the river Vlasina in 2007 indicates the need to improve the system of protection against torrents and erosion. The flood of November 2007 damaged the water supply system for the supply of water to Niš, which runs along the bed of the Koritnica river.

During July 1999, a large part of Central Europe and the Mediterranean was hit by above-average rainfall. Precipitation over the Republic of Serbia occurred in several isolated but relatively close in time waves. Under the influence of a strong altitude cyclone, the first rain wave appeared between July 8 and 11. The strongest precipitation was in the valley of the river Velika Morava. After a short-term improvement in the weather conditions, there was a worsening with showers and thunder. The precipitation was high in places on July 13 and 15 and was higher in the Zapadna Morava basin than in the Velika Morava basin itself. The third wave occurred in the second part of the last decade of July, and especially high precipitation was in the Zapadna Morava basin and in the Sava River basin. Unusually high successive precipitations led to the outpouring of a large number of smaller watercourses and characteristic torrents - immediate tributaries of the Velika Morava, Zapadna Morava and the immediate basin of the Sava river (the Kolubara, the Topčiderska river, etc.). Floods and torrents have caused enormous damage to settlements, industrial plants, infrastructure and agriculture. The most intense flooding of the Velika Morava and the left tributaries of the Zapadna Morava, then the Kolubara basin and the immediate tributaries of the Sava river. Torrential floods in the immediate basin area of the Velika Morava occurred on the following rivers: Lugomir, Belica, Lepenica, Jasenica, Kubršnica, Ravanica, Crnica and Resava. In the Zapadna Morava basin, large water outflows and torrents have been registered on the following rivers: Skrapež, Bjelica, Ribnica, Gruža, Despotovica, Dičina and Rasina. The most severe consequences of the flood were in Smederevska Palanka. Due to the insufficient capacity of the riverbed and the burst of the embankment in several places, the rivers Jasenica and Kubršnica overflowed. The intensity of the floods and the amount of damage were very high in the area of Jagodina due to the overflow of the rivers Lugomir and Belica. In addition, the river Lepenica flooded Batočina and caused very great damage. The floods on the right bank of the Morava river were of slightly lower intensity, but still caused a lot of damage. The river Resava in the area of Svilajnac overflowed, then the river Ravanica in Ćuprija, Crnica in Paraćin and the river Jovanovačka in the vicinity of Ćićevac.

In the immediate basin area of the Zapadna Morava, the largest floods occurred on its left tributaries: the Čemernica with tributaries the Dičina and the Despotovica and on the river Gruža. Floods of somewhat lower intensity occurred in the basins of Skrapež, Bjelica, Ribnica and Rasina. In some of the designated areas, extremely high waters have occurred several times over the recent years and devastating torrential floods have occurred (Jasenica, Kolubara, Topčiderska river, Jadar), which have been characterized as once-in-a-hundred-years phenomena, and even as phenomena of a longer return period. The flood on the river Vlasina from 1988 is said to have a much longer return period (over three hundred years).

The municipality of Topola suffered great damage from torrents during July 1999. Torrential floods were not concentrated only in the hilly and mountainous areas of the Republic of Serbia. They were also found in typical lowland areas. From Požarevac to Golubac, a series of torrential floods occurred on extremely small torrential watercourses. The regional road Veliko Gradište - Ram suffered special damage, on which a torrent cut a canal through the asphalt in 1999. That road is important because there is a ferry from Ram to Banatska Palanka, and that is the fastest way to cross from Stig to Banat and Romania.

Intense precipitation during June 2001, which affected the central and western part of the Republic of Serbia, caused extremely high waters on several watercourses in the Sava and Drina basins, which resulted in catastrophic floods and torrents accompanied by great damage. The biggest floods occurred in the basins of the Kolubara, Jadra and Ljubovidja rivers. Heavy rains fell in the entire area from June 19 to 21, which caused a sudden rise in water levels and the appearance of extremely high water levels in all smaller watercourses. The floods were especially severe in six municipalities: Šabac, Loznica, Ljubovija, Valjevo, Pecka (municipality of Osečina) and Osečina, and by their nature, they can be classified as torrential. Settlements, industrial facilities, roads and communal infrastructure, as well as agricultural land, were flooded at several locations. The outflow occurred not only on parts of unregulated streams but also on sections of regulated riverbeds with built defensive lines on the following watercourses:

- the Obnica, Jablanica, Kolubara, Ljig (upstream from Ljig), Lepenica and Ribnica rivers overflowed in the vicinity of Valjevo,
- the area of Loznica and Lešnica was flooded by the Jadar and Krivaja rivers, and large flows occurred in other watercourses in the city area of Loznica (Žeravija and Štira), although there was no large-scale flooding on these watercourses,
- the area of Šabac, more precisely its southwestern part, was flooded due to the outpouring of water from the Cer circumferential canal on a stretch of about 32 km, i.e. from the confluence of the Cer circumferential canal and the Sava near Prnjavor,
- in the area of Osečina, the river Jadar and all its tributaries overflowed (Pecka, Ostružanka, Lovačka river); downstream from the settlement of Pecka, the river Pecka overflowed and caused great damage to the traffic infrastructure,
- in the town of Ljubovija, the river Ljuboviđa overflowed in the city area itself.

Based on meteorological data, it can be concluded that in the entire area, precipitation on June 20, 2001 (recorded on June 21 at 07 AM) was very abundant and ranged from 35 mm to 40 mm in the lowlands (Šabac and Sremska Mitrovica) and from 80 mm to 100 mm in the upper parts of the considered basins (Petkovica, Osečina, Ljubovija, Donja Orovica, Poćute). The maximum daily precipitation, which was registered mainly on June 21, 2001, at some stations, had values close to the averages for June (Osečina, Donja Orovica and Poćute). If we observe the precipitation that caused the flood wave in June 2001, it can be noticed that the most abundant precipitation occurred in the following areas:

- the upper part of the basin of the Cer circumferential canal (Petkovica and Badovinci);
- the entire Jadra basin, and especially its upper part, where very intense precipitation was registered at the stations Dvorska, Krupanj, Osečina, Stave and Pecka;
- in the Ljuboviđa river basin (Ljubovija, Donja Orovica);
- in the basins of the watercourses that make up the Kolubara river, with high precipitation at the stations Bogatić, Donje Leskovice, Majinovići, Mratičići and Brežđe; maximum precipitation in the Kolubara basin occurred at some stations on June 20, and at others on June 21;
- in the Tamnava basin, precipitation was quite high, although they did not have significant extreme values.

The Jadar river and its tributaries had a series of torrential floods almost every other year and all were accompanied by great damage.

On May 15, 2010, Trgovište, a small municipality in the southern part of the Republic of Serbia, was hit by a catastrophic torrential flood. The torrent appeared really unexpectedly and destroyed several bridges and damaged roads and several houses and business facilities. Meteorological data from stations in Serbia did not indicate the reason for the occurrence of torrents. The analysis of the radar image established that 110 mm of rain fell on the basin in Macedonia in 26 minutes, which is an unprecedented intensity in our region. That torrential flood showed that the border could not stop it. Two people were killed and great material damage was caused.

2014 is characterized by a series of floods and torrents of catastrophic proportions. In the period from May 14 to 17, large amounts of precipitation fell on the entire territory of the Republic of Serbia. The highest precipitation was in the western parts of Serbia, which caused torrential floods of all tributaries of the Drina and the Kolubara. Rains and torrents moved east towards the tributaries of the Velika Morava, the Mlava and the Pek, so the municipal centers of Krupanj, Loznica, Ljubovija, Mali Zvornik, Valjevo, Ub, and especially catastrophically Obrenovac, were destroyed. As soon as in September 2014, an unprecedented storm hit eastern Serbia and torrents destroyed Tekija, Podvrška, Velika Kamenica, Grabovica and Rečica, which are located in the basins of the eponymous tributaries of the Danube. Bor and Majdanpek were also damaged on that occasion. The analysis of all those floods and torrents has not been finished yet, so no one can come forward with a precise statement of what really happened. The fact is that during May 2014, large floods and torrents hit Croatia and Bosnia and Herzegovina and caused great damage to the whole area.

SOLUTIONS AND RECOMMENDATIONS

The basic feature of floods on torrential watercourses arises from the specific dynamics of torrential phenomena. The characteristic genesis and fast concentration of large waters in torrential basins prevent the application of classical hydrotechnical principles and methods of flood protection. The sudden arrival and short duration of high waters often do not leave enough time to declare flood defense (which can never be regular, but only emergency). Unfortunately, in most torrential floods, no active defense measures are taken in our country, but everything comes down to passive remediation of the consequences of the flood. Considering the character and dynamics of torrential processes, preventive protection measures are of the greatest importance. The main goal of preventive flood protection measures on torrent watercourses is, on the one hand, to reduce the probability of high water spills, and on the other hand, to reduce potential flood damage. Having in mind the fundamental feature of small watercourses - the inseparable connection of phenomena in the basin and the river course, preventive measures of flood protection on torrent watercourses must include the spatial whole of watercourses, coastal areas and basins. The approach to flood defense must be adapted to the presented characteristics of torrent processes, in order to rationally and effectively protect against water. This means that the dynamics of torrential phenomena and a very short time for the organization and implementation of flood protection measures must be taken into account. On the other hand, the greatest attention must be paid to preventive activities before the appearance of torrent waves. In other words, all potential flood protection measures must be prepared in advance and elaborated in detail so that in the event of a wave of high waters, the activation of all planned measures can be started immediately. Defense against torrential floods cannot be successful without adequate organization, adapted to the character of the phenomenon. This means that all flood defense activities must be coordinated and synchronized and all defense actors involved at the right time. In that sense, a clear and solid structure of the flood defense headquarters is necessary.

FUTURE RESEARCH DIRECTIONS

Integral defense against the harmful effects of water implies undertaking certain activities, works and measures in parts of the municipality that are potentially endangered by the harmful effects of water, in conditions that are treated as natural disasters, such as torrential rains and torrential phenomena. Defense against these phenomena is treated as flood defense with special measures taken in potentially endangered area by the influx of torrential sediments (torrential lava), landslides and landfalls. In addition to the mentioned torrential phenomena, the accompanying phenomena of the harmful effect of water also include erosion processes. These processes are the subject of special plans - erosion protection plans (which are a special obligation of municipalities). Municipal torrential flood defense plans must also address these phenomena, given that their consequences (deposition in drains, in the zone of bridges and culverts) can be a potential cause of floods (Plan for the proclamation of erosion areas on the territory of the municipality of Leskovac, 2007; Flood protection plan for the area of the municipality of Leskovac, 2007). Erosion protection has the character of preventive activities in order to reduce the intensity of erosion processes. Control of erosion processes includes the following activities:

- identification of erosion hotspots locations,
- identification of areas that may be endangered by erosion processes,
- systematic implementation of remedial measures and works on these sites,
- preventive planning of the use of space in the endangered area.

These measures are the subject of erosion protection plans for the area, while the plans for flood protection envisage measures of timely provision of conditions for water runoff, by intervention in known critical places. The realization of these activities is under the jurisdiction of the Municipal Assembly, the Directorate for Construction, the competent utility company, i.e. the competent Section for Roads.

CONCLUSION

Contemporary European trends in the field of flood protection can be summarized in several fundamental principles:

- It is important to solve the problems of flood protection in accordance with the sustainable development and preservation of natural resources and the environment as a whole,
- It is necessary to "give" more space to rivers. Floods of 1998 and 2002 showed a general lack of inundation spaces in Europe,
- All plants, which may be sources of water pollution, must be dislocated outside the flood zones. This is true for both existing and new facilities, the construction of which is planned.

Hydrometeorological services must play a much larger and more responsible role in forecasting catastrophic flood hazards. Flood risk maps are very important instruments in determining the degree of flood risk. In many European countries, mapping of floodplains is forthcoming, and GIS is recommended for this purpose.

Spatial planning is also an important instrument, which must be used in order to reduce flood damage. Commercial and residential buildings are prohibited from building in flood zones. Flood protection financing can take the form of public-private partnerships. Informing and involving the public in all phases of flood protection is invaluable for successful flood defense.

ACKNOWLEDGMENT

This research was realized with a support of the Erasmus+ Programme of the European Union throughout the Capacity Building in Higher Education project entitled: Soil Erosion and Torrential Flood prevention: curriculum development at the universities of Western Balkan countries (SETOF), Project No. 598403-EPP-1-2018-1-RS-EPPKA2-CBHE-JP.

REFERENCES

Blagojević, I. (2019). *Flood risk assessment with improvement measures* (Graduate thesis). University of Belgrade, Faculty of Security.

Brajković, M., & Gavrilović, Z. (1989). Causes and consequences of some recorded "sudden" torrential floods. *Yugoslav Conference on Torrential Floods, Kopaonik*, 433-440.

Climate Change Vulnerability Assessment. (2012). WWF (World Wide Fund for Nature) & Center for Environmental Improvement.

Decision on determining the Census of waters of the first order (Official Gazette of RS 83/2010)

Directive 2000/60/EC of the European Parliament and of the Council establishing a framework for the Community action in the field of water policy - EU Water Framework Directive (WFD). OJL 327, 2000.

Directive 2007/60/EC on the assessment and management of flood risks. OJL 288, 2007.

Flood protection plan for the area of the municipality of Leskovac. (2007). Institute of Water Management "Jaroslav Cerni" AD Belgrade.

Floods, Serbia and Europe: What is the danger and how to protect yourself from the water element. (2021, July). BBC News.

Gaume, E., Bain, V., Bernardara, P., Newinger, O., Barbuc, M., Bateman, A., Blaškovičová, L., Blöschl, G., Borga, M., Dumitrescu, A., Daliakopoulos, I., Garcia, J., Irimescu, A., Kohnova, S., Koutroulis, A., Marchi, L., Matreata, S., Medina, V., Preciso, E., ... Viglione, A. (2009). A Collation of Data on European Flash Floods. *Journal of Hydrology (Amsterdam)*, *367*(1), 70–78. doi:10.1016/j.jhydrol.2008.12.028

Gavrilović, Z., & Brajković, M. (1989). Natural and anthropogenic factors important for the formation of "Sudden" floods. *Yugoslav Conference on Torrential Floods, Kopaonik*, 46-53.

Gencer, E. A. (2013). Natural Disasters, Urban Vulnerability, and Risk Management: A Theoretical Overview. In E. A. Gencer (Ed.), *The Interplay between Urban Development, Vulnerability, and Risk Management, Mediterranean Studies* (Vol. 7, pp. 7–43). Springer-Verlag. doi:10.1007/978-3-642-29470-9_2

Gosling, S. N., & Arnell, N. W. (2016). A global assessment of the impact of climate change on water scarcity. *Climatic Change*, *134*(3), 371–385. doi:10.100710584-013-0853-x

Kostadinov, S., Tošić, R., Hrkalović, D., Nikolić, S., Sudar, N., Kapović Solomun, M., Dragičević, S., Momirović, N., Cupać, R., Bosankić, G., & Bundalo, S. (2019). Soil erosion control in the function of reducing the risk of floods in the Vrbas River Basin. *Vodoprivreda*, *51*(300-302), 211–224.

Land Degradation and Protection - Thematic collection. (2016). University of Belgrade, Faculty of Forestry.

Law on Emergency Situations ("Official Gazette of the RS", No. 111/2009, 92/2011 and 93/2012)

Law on Waters ("Official Gazette of the RS", No. 30/2010, 93/2012, 101/2016, 95/2018 and 95/2018 - other law)

Plan for the proclamation of erosion areas on the territory of the municipality of Leskovac. (2007). Institute of Water Management "Jaroslav Cerni" AD Belgrade.

Prohaska, S. (2003). *Hydrology: Hydrometeorology, hydrometry and water regime. Part 1*. Faculty of Mining and Geology.

Prohaska, S. (2006). *Hydrology. Part II, Institute of Water Management "Jaroslav Cerni", Faculty of Mining and Geology, Belgrade*. Republic Hydrometeorological Institute.

Rosier, I., Diels, J., Somers, B., & Van Orshoven, J. (2021). A Workflow to Extract the Geometry and Type of Vegetated Landscape Elements from Airborne LiDAR Point Clouds. *Remote Sensing*, *13*(20), 4031. doi:10.3390/rs13204031

Smith, K., & Petley, D. N. (2009). *Environmental hazards: Assessing risk and reducing disaster*. Routledge. doi:10.4324/9780203884805

Stefanović, M., Gavrilović, Z., & Bajčetić, R. (2014). Local community and the issue of torrential floods. Organization for European Security and Co-operation, Mission to Serbia.

Water Management Basis of the Republic of Serbia. (2002). Institute for Water Management "Jaroslav Cerni", AD, Belgrade.

Wisner, B., Blaikie, P., Cannon, T., & Davis, I. (2004). *At Risk: Natural Hazards, Peoples' Vulnerabilities and Disasters*. Routledge.

ADDITIONAL READING

Jackovics, P. (2019). Evaluation a City Emergency Management Exercise for Organizational Learning. *Interdisciplinary Description of Complex Systems*, *17*(1, 1-B), 177–186. doi:10.7906/indecs.17.1.17

Janev Holcer, N., & Jeličić, P. (2015). Health protection and risks for rescuers in cases of floods. *Arhiv za Higijenu Rada i Toksikologiju*, *66*(1), 9–12. doi:10.1515/aiht-2015-66-2559 PMID:25741935

Juárez, A., Alfredsen, K., Stickler, M., Adeva-Bustos, A., Suárez, R., Seguín-García, S., & Hansen, B. (2021). A Conflict between Traditional Flood Measures and Maintaining River Ecosystems? A Case Study Based upon the River Lærdal, Norway. *Water, 13*(14), 1884. doi:10.3390/w13141884

Korolov, V., Kurowska, K., Korolova, O., Zaiets, Y., Milkovich, I., & Kryszk, H. (2021). Methodology for Determining the Nearest Destinations for the Evacuation of People and Equipment from a Disaster Area to a Safe Area. *Remote Sensing, 13*(11), 2170. doi:10.3390/rs13112170

Krogulec, E., Zabłocki, S., & Sawicka, K. (2020). Temporal and Spatial Diversity of Renewable Groundwater Resources in the River Valley. *Applied Sciences, 10*(14), 4827. doi:10.3390/app10144827

Lorenzoni, N., Stühlinger, V., Stummer, H., & Raich, M. (2020). Long-Term Impact of Disasters on the Public Health System: A Multi-Case Analysis. *International Journal of Environmental Research and Public Health, 17*(17), 6251. doi:10.3390/ijerph17176251

Prohaska, S., & Zlatanović, N. (2016). Rekonstrukcija katastrofalne svibanjske poplave 2014. godine u slivu rijeke Kolubare (Srbija). *Hrvatske Vode*, *24*(98), 261–274.

Raffellini, M., Martina, F., Silvestro, F., Giannoni, F., & Rebora, N. (2021). Performance Evaluation of a Nowcasting Modelling Chain Operatively Employed in Very Small Catchments in the Mediterranean Environment for Civil Protection Purposes. *Atmosphere, 12*(6), 783. doi:10.3390/atmos12060783

Šimić, Z. & Veira, M.P. (2016). Lessons Learned from Missing Flooding Barriers Operating Experience. *Journal of Energy, 65*(3-4).

KEY TERMS AND DEFINITIONS

Capacity: The combination of all the strengths, attributes, and resources available within an organization, community or society to manage and reduce disaster risks and strengthen resilience. Capacity may include infrastructure, institutions, human knowledge and skills, and collective attributes such as social relationships, leadership, and management.

Contingency Plan: A plan developed to assist with managing a gap in capability to ensure services are maintained. This plan describes organized and coordinated courses of action with clearly identified institutional roles and resources, information processes and operational arrangements for specific actions at times of need. Contingency planning can be done as deliberate planning or immediate planning as it seeks to address gaps on an as needs basis.

Disaster: An event, either man-made or natural, sudden, or progressive, the impact of which is such that the affected community must respond through exceptional measures.

Disaster Management: There could not be a single organization solely responsible for all aspects of disaster management. The management task is to bring together, in an integrated organizational structure, the resources of many organizations that can take appropriate action in times of disasters.

Disaster Risk Reduction: Disaster risk reduction is aimed at preventing new and reducing existing disaster risk and managing residual risk, all of which contribute to strengthening resilience and therefore to the achievement of sustainable development.

Hazard Analysis: Part of the overall planning process which identifies and describes hazards and their effects on the community.

Recovery: The taking of appropriate measures to recover from an event, including the action taken to support disaster-affected communities in the reconstruction of infrastructure, the restoration of emotional, social, economic, and physical wellbeing, and the restoration of the environment.

Resources: Any asset, physical, human, economic or environmental which can be used to assist in achieving the objectives of the plan (people, equipment, relief supplies, water, roads, warehouses, and money).

Compilation of References

Abramov, V., Popov, N., & Shilin, M. (2021). Geo-information Support Tools for Natural Risks Management within Northern Sea Route. *Transportation Research Procedia, 54*, 144–149. doi:10.1016/j.trpro.2021.02.058

Adegbeye, M. J., Reddy, P. R. K., Obaisi, A. I., Elghandour, M. M. M. Y., Oyebamiji, K. J., & Salem, A. Z. M., Morakinyo-Fasipe, O. T., Cipriano-Salazar, MCamacho-Díaz, L. M. (2020). Sustainable agriculture options for production, greenhouse gasses and pollution alleviation, and nutrient recycling in emerging and transitional nations - An overview. *Journal of Cleaner Production, 242*, 118319. doi:10.1016/j.jclepro.2019.118319

Aghahowa, E. N. (2008). *Developing Key Performance Indicators for Airport Safety and Security: A Study of Three Scottish Airports* (Doctoral thesis). School of the Built Environment, Heriot-Watt University, Edinburgh, UK. Retrieved from https://www.ros-test.hw.ac.uk/handle/10399/2147

Agovino, M., Casaccia, M., Ciommi, M., Ferrara, M., & Marchesano, K. (2019). Agriculture, climate change and sustainability: The case of EU-28. *Ecological Indicators, 105*, 525–543. doi:10.1016/j.ecolind.2018.04.064

Agriculture and Agri-Food Canada. (2010). *Design Guidelines for Farmyard, Field, Roadside, Livestock, Wildlife, and Riparian Buffer Plantings on the Prairies*. Author.

Ahmad, I., Verma, V., & Verma, M. K. (2015). Application of curve number method for estimation of runoff potential in GIS environment. *2nd International Conference on Geological and Civil Engineering, 80*(4), 16-20.

Ahmad, N., & Yamano, N. (2011). *Carbon dioxide emissions embodied in goods and services: Domestic consumption versus production*. OECD Science, Technology and Industry Working Papers, No. 2003/15. Paris: OECD Publishing.

Ahmad, N. S. B. N., Mustafa, F. B., & Didams, G. (2020). A systematic review of soil erosion control practices on the agricultural land in Asia. *International Soil and Water Conservation Research, 8*(2), 103–115. doi:10.1016/j.iswcr.2020.04.001

Aiello, A., Adamo, M., & Canora, F. (2015). Remote sensing and GIS to assess soil erosion with RUSLE3D and USPED at river basin scale in southern Italy. *Catena, 131*, 174–185. doi:10.1016/j.catena.2015.04.003

Akcioni plan za zaštitu od poplava i upravljanje rekama u BiH 2014-20 [Action Plan for Flood Control and River Management in BiH 2014-20], Bosnia and Herzegovina, Savet ministara, Sarajevo, 2014.

Al-Abadi, A. M. A., Ghalib, H. B., & Al-Qurnawi, W. S. (2016). Estimation of soil erosion in northern Kirkuk governorate, Iraq using rusle, remote sensing and gis. *Carpathian Journal of Earth and Environmental Sciences, 11*(1), 153–166.

Alcon, F., Marín-Miñano, C., Zabala, J. A., de-Miguel, M. D., & Martínez-Paz, J. M. (2020). Valuing diversification benefits through intercropping in Mediterranean agroecosystems: A choice experiment approach. *Ecological Economics, 171*, 106593. doi:10.1016/j.ecolecon.2020.106593

Alderfer, C. P. (1969). An empirical test of a new theory of human needs. *Organizational Behavior and Human Performance, 4*(2), 142–175. doi:10.1016/0030-5073(69)90004-X

Alemu, M. M. (2016). Ecological Benefits of Trees as Windbreaks and Shelterbelts. *International Journal of Ecosystem, 6*(1), 10–13. doi:10.5923/j.ije.20160601.02

Alexander, T. J., Vonlanthen, P., & Seehausen, O. (2017). Does eutrophication-driven evolution change aquatic ecosystems? *Philosophical Transactions of The Royal Society B: Biological Sciences, 372*(1712), 20160041. doi:10.1098/rstb.2016.0041

Alfieri, L., Bisselink, B., Dottori, F., Naumann, G., de Roo, A., Salamon, P., Wyser, K., & Feyen, L. (2017). Global projections of river flood risk in a warmer world. *Earth's Future, 5*(2), 171–182. doi:10.1002/2016EF000485

Aliff, M. N., Reavie, E. D., Post, S. P., & Zanko, L. M. (2020). Anthropocene geochemistry of metals in sediment cores from the Laurentian Great Lakes. *PeerJ, 8*, e9034. doi:10.7717/peerj.9034 PMID:32411522

Allardt, E. (1977). On the relationship between objective and subjective predicaments (No. 16). Helsinki: Research Group for Comparative Sociology, University of Helsinki.

Alletto, L., & Coquet, Y. (2009). Temporal and spatial variability of soil bulk density and near-saturated hydraulic conductivity under two contrasted tillage management systems. *Geoderma, 152*(1-2), 85–94. doi:10.1016/j.geoderma.2009.05.023

Altaf, S., Meraj, G., & Romshoo, S. A. (2014). Morphometry and land cover based multi-criteria analysis for assessing the soil erosion susceptibility of the western Himalayan watershed. *Environmental Monitoring and Assessment, 186*(12), 8391–8412. doi:10.100710661-014-4012-2 PMID:25154685

Al-Wadaey, A., & Ziadat, F. (2014). A participatory GIS approach to identify critical land degradation areas and prioritize soil conservation for mountainous olive groves (case study). *Journal of Mountain Science, 11*(3), 782–791. doi:10.100711629-013-2827-x

Ameri, A. A., Pourghasemi, H. R., & Cerda, A. (2018). Erodibility prioritization of sub-watersheds using morphometric parameters analysis and its mapping: A comparison among TOPSIS, VIKOR, SAW, and CF multi-criteria decision making models. *The Science of the Total Environment, 613–614*, 1385–1400. doi:10.1016/j.scitotenv.2017.09.210 PMID:29898506

Amichev, B., Murray, B., Kulshreshtha, S., Laroque, C. P., Piwowar, J. M., & Van Rees, K. (2016a). Carbon sequestration and growth of six common tree and shrub shelterbelts in Saskatchewan, Canada. *Canadian Journal of Soil Science, 97*(3), 368–381. doi:10.1139/CJSS-2016-0107

Amichev, B., Murray, B., Kurz, W., Laroque, C. P., Kulshreshtha, S., Piwowar, J. M., & Van Rees, K. (2016b). Carbon sequestration by white spruce shelterbelts in Saskatchewan, Canada: 3PG and CBM-CFS3 model simulations. *Ecological Modelling, 325*, 35–46. doi:10.1016/j.ecolmodel.2016.01.003

Angeli, J. L. F., Sartoretto, J. R., Kim, B. S. M., de Lima Ferreira, P. A., de Mahiques, M. M., & Figueira, R. C. L. (2021). Trace element fluxes during the "Anthropocene" in a large South American industrial and port area (Santos and São Vicente estuarine system, SE, Brazil). *Environmental Monitoring and Assessment, 193*(9), 1–18. doi:10.100710661-021-09378-3 PMID:34426853

Arruti, A., Fernández-Olmo, I., & Irabien, Á. (2010). Evaluation of the contribution of local sources to trace metals levels in urban PM2. 5 and PM10 in the Cantabria region (Northern Spain). *Journal of Environmental Monitoring, 12*(7), 1451–1458. doi:10.1039/b926740a PMID:20517581

Arsovski, S. (2005). Put od kvaliteta do kvaliteta života. *Zbornik radova Festival kvaliteta, 32*, 3-16.

Arundati, R., Sutiono, H. T., & Suryono, I. A. (2020). Effect of ecological awareness, personal norms and ecological attitude to conservation behavior. *Proceedings on Engineering*, *2*(2), 187–196. doi:10.24874/PES02.02.009

Asadzadeh, A., Kötter, T., Salehi, P., & Birkmann, J. (2017). Operationalizing a concept: The systematic review of composite indicator building for measuring community disaster resilience. *International Journal of Disaster Risk Reduction*, *25*, 147–162. doi:10.1016/j.ijdrr.2017.09.015

Asfaw, D., & Workineh, G. (2019). Quantitative analysis of morphometry on Ribb and Gumara watersheds: Implications for soil and water conservation. *International Soil and Water Conservation Research*, *7*(2), 150–157. doi:10.1016/j.iswcr.2019.02.003

Ashiagbor, G., Forkuo, E. K., Laari, P., & Aabeyir, R. (2013). Modeling soil erosion using RUSLE and GIS tools. *Int J Remote Sens Geosci*, *2*(4), 1–17.

Ashiagbor, G., Forkuo, E. K., Laari, P., & Aabeyir, R. (2013). Modeling soil erosion using RUSLE and GIS tools. *International Journal of Remote Sensing and Geosciences*, *2*(4), 1–17.

Ashmore, P. (1993). Canadian Landscape Charter. *Progress in Physical Geography*, *17*(2), 190–204. doi:10.1177/030913339301700206

Asian Disaster Preparedness Centre. (2000, October). *Workshop on Urban Disaster Mitigation*. University of Karlsruhe.

Atkins. (2013). *Impact of Climate Change on Dams & Reservoirs*. Final Guidance Report UK.

Atoma, H., Suryabhagavan, K. V., & Balakrishnan, M. (2020). Soil erosion assessment using RUSLE model and GIS in Huluka watershed, Central Ethiopia. *Sustainable Water Resources Management*, *6*(1), 1–17. doi:10.100740899-020-00365-z

Baartman, J. E., van Lynden, G. W., Reed, M. S., Ritsema, C. J., & Hessel, R. (2007). *Desertification and land degradation: origins, processes and solutions. DESIRE Report series: Scientific Report*. DESIRE.

Babu, S. (2017). *Hazard vs Disaster: The principle behind disaster management*. https://eco-intelligent.com/2017/01/21/hazard-vs-disaster-the-principle-behind-disaster-management/

Bai, S., Gao, J., Sun, D., & Tian, M. (2020). Monitoring Water Transparency in Shallow and Eutrophic Lake Waters Based on GOCI Observations. *Remote Sensing*, *12*(1), 163. doi:10.3390/rs12010163

Bajocco, S., De Angelis, A., Perini, L., Ferrara, A., & Salvati, L. (2012). The impact of land use/land cover changes on land degradation dynamics: A Mediterranean case study. *Environmental Management*, *49*(5), 980–989. doi:10.100700267-012-9831-8 PMID:22419398

Bajrić, M., & Sokolović, D. (2015). *Rehabilitation of a secondary network of forest traffic infrastructure*. Glasnik Šumarskog fakulteta. . doi:10.2298/GSF15S1005B

Bajrić, M., Sokolović, D., Musić, J., & Halilović, V. (2013). Assessment of intensity of erosion and natural coverage of skid roads in post-exploitation period. In *Works of the Faculty of Forestry*. University of Sarajevo.

Bakker, M. M., Govers, G., Kosmas, C., Vanacker, V., Van Oost, K., & Rounsevell, M. (2005). Soil erosion as a driver of land-use change. *Agriculture, Ecosystems & Environment*, *105*(3), 467–481. doi:10.1016/j.agee.2004.07.009

Balestra, C., & Sultan, J. (2013). *Home sweet home: The determinants of residential satisfaction and its relation with well-being*. OECD Statistics Working Papers 2013/05. Paris: OECD Publishing.

Balestra, C., & Dottori, D. (2012). Aging society, health and the environment. *Journal of Population Economics*, *25*(3), 1045–1076. doi:10.100700148-011-0380-x

Banerjee, S., Black, R., Mishra, A., & Kniveton, D. (2019). Assessing the vulnerability of remittance-recipient and nonrecipient households in rural communities affected by extreme weather events: Case studies from South-West China and North-East India. *Population Space and Place*, *25*(2), e2157. doi:10.1002/psp.2157

Barbano, A., Braca, G., Bussettini, M., Dessì, B., Inghilesi, R., Lastoria, B., Monacelli, G., Morucci, S., Piva, F., Sinapi, L., & Spizzichino, D. (2012). Proposta metodologica per l'aggiornamento delle mappe di pericolosità e di rischio - Attuazione della Direttiva 2007/60/CE relativa alla valutazione e alla gestione dei rischi da alluvioni (Decreto Legislativo n.49/2010). Manuali e Linee Guida ISPRA, 82/2012 (in Italian).

Bardgett, R. D., & Van Der Putten, W. H. (2014). Belowground biodiversity and ecosystem functioning. *Nature*, *515*(7528), 505–511. doi:10.1038/nature13855 PMID:25428498

Barman, B. K., Rao, K. S., Sonowal, K., Prasad, N. S. R., & Sahoo, U. K. (2020). Soil erosion assessment using revised universal soil loss equation model and geo-spatial technology: A case study of upper Tuirial river basin, Mizoram, India. *AIMS Geosciences*, *6*(4), 525–544. doi:10.3934/geosci.2020030

Barredo, J. I. (2007). Major flood disasters in Europe: 1950–2005. *Natural Hazards*, *42*(1), 125–148. doi:10.100711069-006-9065-2

Baumgertel, A., Lukić, S., Belanović Simić, S., & Kadović, R. (2019). Identifying Areas Sensitive to Wind Erosion—A Case Study of the AP Vojvodina (Serbia). *Applied Sciences (Basel, Switzerland)*, *9*(23), 5106. doi:10.3390/app9235106

Baumhardt, R. L., & Blanco-Canqui, H. (2014). Soil: Conservation Practices. In Encyclopedia of Agriculture and Food Systems. Academic Press. doi:10.1016/B978-0-444-52512-3.00091-7

Baveye, P. C., Baveye, J., & Gowdy, J. (2016). Soil "ecosystem" services and natural capital: Critical appraisal of research on uncertain ground. *Frontiers in Environmental Science*, *4*, 41. doi:10.3389/fenvs.2016.00041

Beillouin, D., Ben-Ari, T., & Makowski, D. (2019). Evidence map of crop diversification strategies at the global scale. *Environmental Research Letters, 14*(12).

Belasri, A., & Lakhouili, A. (2016). Estimation of soil erosion risk using the universal soil loss equation (USLE) and geo-information technology in Oued El Makhazine Watershed, Morocco. *Journal of Geographic Information System*, *8*(01), 98–107. doi:10.4236/jgis.2016.81010

Ben Jmaa, H. (2008). *Le système endoréique de Sidi Al Hani, Chrita et Mechertate: Paléoenvironnement et dynamique récente* (Thèse de doctorat). Université de Tunis.

Berg, A., & Pattillo, C. (1999). Predicting currency crises: The indicators approach and an alternative. *Journal of International Money and Finance*, *18*(4), 561–586. doi:10.1016/S0261-5606(99)00024-8

Berke, P. R., Kartez, J., & Wenger, D. (1993). Recovery after disaster: Achieving sustainable development, mitigation and equity. *Disasters*, *17*(2), 93–109. doi:10.1111/j.1467-7717.1993.tb01137.x PMID:20958760

Bernier, J. F., Chassiot, L., & Lajeunesse, P. (2021). Assessing bank erosion hazards along large rivers in the Anthropocene: A geospatial framework from the St. Lawrence fluvial system. *Geomatics, Natural Hazards & Risk*, *12*(1), 1584–1615. doi:10.1080/19475705.2021.1935333

Bertoldo, R. (2021). A comparative frame for a contextualized analysis of local natural risk management. *International Journal of Disaster Risk Reduction*, *52*, 101945. doi:10.1016/j.ijdrr.2020.101945

Bianchi, C. N., Morri, C., Chiantore, M., Montefalcone, M., Parravicini, V., & Rovere, A. (2012). Mediterranean Sea biodiversity between the legacy from the past and a future of change. *Life in the Mediterranean Sea: A look at habitat changes*, 1-55.

Biswas, H., Raizada, A., Mandal, D., Kumar, S., Srinivas, S., & Mishra, P. K. (2015). Identification of areas vulnerable to soil erosion risk in India using GIS methods. *Solid Earth, 6*(4), 1247–1257. doi:10.5194e-6-1247-2015

Blagojević, I. (2019). *Flood risk assessment with improvement measures* (Graduate thesis). University of Belgrade, Faculty of Security.

Blanco-Canqui, H., & Lal, R. (2008). Principles of Soil Conservation and Management. Springer Science+Business Media B.V.

Blinkov, I. (2015). The Balkans - The most erosive part of Europe? *Bulletin of the Faculty of Forestry, 111*, 9-20.

Boardman, J., & Poesen, J. (2006). Soil Erosion in Europe. John Wiley & Sons Ltd. doi:10.1002/0470859202

Boardman, J. (2006). Soil erosion science: Reflections on the limitations of current approaches. *Catena, 68*(2-3), 73–86. doi:10.1016/j.catena.2006.03.007

Boardman, J. (2010). A short history of muddy floods. *Land Degradation & Development, 21*(4), 303–309. doi:10.1002/ldr.1007

Boardman, J., & Vandaele, K. (2020). Managing muddy floods: Balancing engineered and alternative approaches. *Journal of Flood Risk Management, 13*(1), e12578. doi:10.1111/jfr3.12578

Bockstaller, C., Guichard, L., Makowski, D., Aveline, A., Girardin, P., & Plantureux, S. (2008). Agri-environmental indicators to assess cropping and farming systems. A review. *Agronomy for Sustainable Development, 28*(1), 139–149. doi:10.1051/agro:2007052

Boer, M., & Puigdefábregas, J. (2005). Effects of spatially structured vegetation patterns on hillslope erosion in a semiarid Mediterranean environment: A simulation study. *Earth Surface Processes and Landforms: The Journal of the British Geomorphological Research Group, 30*(2), 149–167. doi:10.1002/esp.1180

Boeuf, B., & Fritsch, O. (2016). Studying the implementation of the Water Framework Directive in Europe: A meta-analysis of 89 journal articles. *Ecology and Society, 21*(2), 19. doi:10.5751/ES-08411-210219

Bogunovic, I., Pereira, P., Kisic, I., Sajko, K., & Sraka, M. (2018). Tillage management impacts on soil compaction, erosion and crop yield in Stagnosols (Croatia). *Catena, 160*, 376–384. doi:10.1016/j.catena.2017.10.009

Bohlen, P. J., Parmelee, R. W., McCartney, D. A., & Edwards, C. A. (1997). Earthworm effects on carbon and nitrogen dynamics of surface litter in corn agroecosystems. *Ecological Applications, 7*(4), 1341–1349. doi:10.1890/1051-0761(1997)007[1341:EEOCAN]2.0.CO;2

Boin, A., Hart, P., Stern, E., & Sundelius, B. (2005). *The Politics of Crisis Management: Public Leadership under Pressure*. Cambridge University Press. doi:10.1017/CBO9780511490880

Bombino, G., Denisi, P., Gómez, J. A., & Zema, D. A. (2021). Mulching as best management practice to reduce surface runoff and erosion in steep clayey olive groves. *International Soil and Water Conservation Research, 9*(1), 26–36. doi:10.1016/j.iswcr.2020.10.002

Bongiorno, G., Bünemann, E. K., Oguejiofor, C. U., Meier, J., Gort, G., Comans, R., Mader, P., Brussaard, L., & de Goede, R. (2019). Sensitivity of labile carbon fractions to tillage and organic matter management and their potential as comprehensive soil quality indicators across pedoclimatic conditions in Europe. *Ecological Indicators, 99*, 38–50. doi:10.1016/j.ecolind.2018.12.008

Borrelli, P., Panagos, P., Hessel, R., Riksen, M., & Stolte, J. (2015). *Soil erosion by wind in Europe*. EU RECARE Project. doi:10.2788/488050

Borrelli, P., Robinson, D. A., Fleischer, L. R., Lugato, E., Ballabio, C., Alewell, C., & Panagos, P. (2017). An assessment of the global impact of 21st century land use change on soil erosion. *Nature Communications*, *8*(1), 1–13. doi:10.103841467-017-02142-7 PMID:29222506

Borrelli, P., Robinson, D. A., Panagos, P., Lugato, E., Yang, J. E., Alewell, C., Wuepper, D., Montanarella, L., & Ballabio, C. (2020). Land use and climate change impacts on global soil erosion by water (2015-2070). *Proceedings of the National Academy of Sciences of the United States of America*, *117*(36), 21994–22001. doi:10.1073/pnas.2001403117 PMID:32839306

Bosco, G., & Simeoni, L. (2018). Environmental impact of intensive farming in sloping areas. In *MATEC Web of Conferences* (Vol. 174, p. 01001). EDP Sciences. 10.1051/matecconf/201817401001

Bou-Imajjane, L., & Belfoul, M. A. (2020). Soil loss assessment in Western High Atlas of Morocco: Beni Mohand watershed study case. *Applied and Environmental Soil Science*, 2020.

Boulange, J., Hanasaki, N., Yamazaki, D., & Pokhrel, Y. (2021). Role of dams in reducing global flood exposure under climate change. *Nature Communications*, *12*(1), 417. doi:10.103841467-020-20704-0 PMID:33462241

Bowyer, C., Withana, S., Fenn, I., Bassi, S., Lewis, M., Cooper, T., Benito, P., & Mudgal, S. (2009). *Land Degradation and Desertification*. European Parliament. https://www.europarl.europa.eu/RegData/etudes/etudes/join/2009/416203/IPOL-ENVI_ET(2009)416203_EN.pdf

Brajković, M., & Gavrilović, Z. (1989). Causes and consequences of some recorded "sudden" torrential floods. *Yugoslav Conference on Torrential Floods, Kopaonik*, 433-440.

Brajša-Žganec, A., Merkaš, M., & Šverko, I. (2011). Quality of life and leisure activities: How do leisure activities contribute to subjective well-being? *Social Indicators Research*, *102*(1), 81–91. doi:10.100711205-010-9724-2

Brandle, J., Hodges, L., Tyndall, J., & Sudmeyer, R. (2009). Windbreak Practice. In North American Agroforestry: An Integrated Science and Practice (2nd ed.). American Society of Agronomy. doi:10.2134/2009.northamericanagroforestry.2ed.c5

Brevik, E. C., & Burgess, L. C. (2014). The Influence of Soils on Human Health. *Nature Education Knowledge*, *5*(12), 1.

Bronick, C. J., & Lal, R. (2005). Soil structure and management: A review. *Geoderma*, *124*(1–2), 3–22. doi:10.1016/j.geoderma.2004.03.005

Brönnimann, S., Pfister, C., & White, S. (2018). In S. White, C. Pfister, & F. Mauelshagen (Eds.), *The Palgrave Handbook of Climate History* (pp. 27–36). Palgrave Macmillan. doi:10.1057/978-1-137-43020-5_3

Brown, C., & Grant, M. (2007). Natural medicine for planners. *Town and Country Planning-Special Issue: Planning for Healthy Communities*, *76*(2), 67–69.

Buschiazzo, D., & Funk, R. (2015). Wind Erosion of Agriculutral Soils and the Carbon Cycle 2015. In S. A. Banwart, E. Noellemeyer, & E. Milne (Eds.), *Soil Carbon: Science, Management and Policy for Multiple Benefits*. CAB International. doi:10.1079/9781780645322.0161

Byers, J. E. (2002). Impact of non-indigenous species on natives enhanced by anthropogenic alteration of selection regimes. *Oikos*, *97*(3), 449–458. doi:10.1034/j.1600-0706.2002.970316.x

Cai, W. J., Hu, X., Huang, W.-J., Murrell, M. C., Lehrter, J. C., Lohrenz, S. E., Chou, W.-C., Zhai, W., Hollibaugh, J. T., Wang, Y., Zhao, P., Guo, X., Gundersen, K., Dai, M., & Gong, G.-C. (2011). Acidification of subsurface coastal waters enhanced by eutrophication. *Nature Geoscience*, *4*(11), 766–770. doi:10.1038/ngeo1297

Canada European Environmental Agency (JRC). (2010). *The State of Soil in Europe*. European Commission.

Canyon, D. (2013). Pre-crisis damage containment and leadership policy in health services. *Leadership in Health Services*, *26*(4), 283–293. doi:10.1108/LHS-04-2012-0010

Casabella-González, M. J., Borselli, L., & García-Meza, J. V. (2021). Soil Horizon erodibility assessment in an area of Mexico susceptible to gully erosion. *Journal of South American Earth Sciences*, *111*, 103497. doi:10.1016/j.jsames.2021.103497

Castro, D., Souto, M., Fraga, M. I., García-Rodeja, E., Pérez-Díaz, S., Sáez, J. A. L., & Pontevedra-Pombal, X. (2020). High-resolution patterns of palaeoenvironmental changes during the Little Ice Age and the Medieval Climate Anomaly in the northwestern Iberian Peninsula. *Geoscience Frontiers*, *11*(5), 1461–1475. doi:10.1016/j.gsf.2020.05.015

Centre for Research on the Epidemiology of Disasters. (2004). *Thirty years of natural disasters 1974-2003: The numbers*. https://www.unisdr.org/files/1078_8761.pdf

Chaves, J. M., & De Cola, T. (2017). Public Warning Applications: Requirements and Examples. Wireless Public Safety Networks, 3, 1-18. doi:10.1016/B978-1-78548-053-9.50001-9

Chendev, Y., Sauer, T., Ramirez, G., & Burras, C. (2015). History of East European Chernozem Soil Degradation; Protection and Restoration by Tree Windbreaks in the Russian Steppe. *Sustainability*, *7*(1), 705–724. doi:10.3390u7010705

Cheng, T., Xu, Z., Hong, S., & Song, S. (2017). Flood risk zoning by using 2D hydrodynamic modeling: A case study in Jinan City. *Mathematical Problems in Engineering*, *2017*, 2017. doi:10.1155/2017/5659197

Chen, J., Yu, X., Sun, F., Lun, X., Fu, Y., Jia, G., Zhang, Z., Liu, X., Mo, L., & Bi, H. (2015). The Concentrations and Reduction of Airborne Particulate Matter (PM_{10}, $PM_{2.5}$, PM_1) at Shelterbelt Site in Beijing. *Atmosphere*, *6*(5), 650–676. doi:10.3390/atmos6050650

Chen, X. A., Cai, Q., Zheng, M. G., & Li, J. (2011). Empirical soil erosion model for single rainstorm in Chabagou drainage basin. *Progress in Geography*, *30*, 325–329.

Chow, V. T. (1964). *Handbook of applied hydrology: A compendium of water-resources technology*. McGraw-Hill.

Christiaensen, L., Demery, L., & Kuhl, J. (2011). The (evolving) role of agriculture in poverty reduction-An empirical perspective. *Journal of Development Economics*, *96*(2), 239–254. doi:10.1016/j.jdeveco.2010.10.006

Chu, C.-R., Chang, C.-Y., Huang, C.-J., Wu, T.-R., Wang, C.-Y., & Liu, M.-Y. (2013). Windbreak protection for road vehicles against crosswind. *Journal of Wind Engineering and Industrial Aerodynamics*, *116*, 61–69. doi:10.1016/j.jweia.2013.02.001

Chuenchum, P., Xu, M., & Tang, W. (2020). Estimation of soil erosion and sediment yield in the Lancang–Mekong river using the modified revised universal soil loss equation and GIS techniques. *Water (Basel)*, *12*(1), 135. doi:10.3390/w12010135

Chu, X., Zhan, J., Li, Z., Zhang, F., & Qi, W. (2019). Assessment on Forest Carbon Sequestration in the Three-North Shelterbelt Program Region, China. *Journal of Cleaner Production*, *215*, 382–389. Advance online publication. doi:10.1016/j.jclepro.2018.12.296

Ciampalini, R., Constantine, J. A., Walker-Springett, K. J., Hales, T. C., Ormerod, S. J., & Hall, I. R. (2020). Modelling soil erosion responses to climate change in three catchments of Great Britain. *The Science of the Total Environment*, *749*, 141657. doi:10.1016/j.scitotenv.2020.141657 PMID:32841861

Ciscar, Feyen, Soria, Lavalle, Raes, Perry, Nemry, Demirel, Rozsai, Dosio, Donatelli, Srivastava, Fumagalli, Niemeyer, Shrestha, Ciaian, Himics, Van Doorslaer, Barrios, Ibáñez, … Ibarreta. (2014). *Climate Impacts in Europe*. The JRC PESETA II Project. JRC Scientific and Policy Reports, EUR 26586EN.

Climate Change Vulnerability Assessment. (2012). WWF (World Wide Fund for Nature) & Center for Environmental Improvement.

Coan, T. G., & Holman, M. R. (2008). Voting green. *Social Science Quarterly*, *89*(5), 1121–1135. doi:10.1111/j.1540-6237.2008.00564.x

Cohen, M. J., Brown, M. T., & Shepherd, K. D. (2006). Estimating the environmental costs of soil erosion at multiple scales in Kenya using energy synthesis. *Agriculture, Ecosystems & Environment*, *114*(2-4), 249–269. doi:10.1016/j.agee.2005.10.021

Colazo, J. C., & Buschuazzo, D. E. (2010). Soil dry aggregate stability and wind erodible fraction in a semiarid environment of Argentina. *Geoderma, Elsevier*, *159*(1–2), 228–236. doi:10.1016/j.geoderma.2010.07.016

Coombs, W. T. (2007). *Ongoing crisis communication: Planning, managing, and responding*. Sage Publication, Thousand Oaks.

Coronato, F. R., & Del Valle, H. F. (1993). Methodological comparison in the estimate of fluvial erosion in an arid closed basin of northeastern Patagonia. *Journal of Arid Environments*, *24*(3), 231–239. doi:10.1006/jare.1993.1021

Costanza, R., De Groot, R., Braat, L., Kubiszewski, I., Fioramonti, L., Sutton, P., Farber, S., & Grasso, M. (2017). Twenty years of ecosystem services: How far have we come and how far do we still need to go? *Ecosystem Services*, *28*, 1–16. doi:10.1016/j.ecoser.2017.09.008

Costanza, R., Fisher, B., Ali, S., Beer, C., Bond, L., Boumans, R., ... Gayer, D. E. (2008). An integrative approach to quality of life measurement, research, and policy. *Sapiens*, *1*(1), 17–21. doi:10.5194apiens-1-11-2008

Council of Europe Parliamentary Assembly. (2009). *Drafting an additional protocol to the European Convention on Human Rights concerning the right to a healthy environment*. Recommendation 1885. Strasbourg Cedex, France: Council of Europe.

Crawford, M. H., Crowley, K., Potter, S. H., Saunders, W. S. A., & Johnston, D. M. (2018). Risk modelling as a tool to support natural hazard risk management in New Zealand local government. *International Journal of Disaster Risk Reduction*, *28*, 610–619. doi:10.1016/j.ijdrr.2018.01.011

Crossley, M. S., Meier, A. R., Baldwin, E. M., Berry, L. L., Crenshaw, L. C., Hartman, G. L., Lagos-Kutz, D., Nichols, D. H., Patel, K., Varriano, S., Snyder, W. E., & Moran, M. D. (2020). No net insect abundance and diversity declines across US Long Term Ecological Research sites. *Nature Ecology & Evolution*, *4*(10), 1368–1376. doi:10.103841559-020-1269-4 PMID:32778751

Ćuzović, Đ. (2019). Rizik i ekonomske štete od poplava–osvrt na međunarodna iskustva i prilike u Srbiji. *Revija*, *2*, 22–30.

Czerepowicz, L., Case, B. S., & Doscher, C. (2012). Using satellite image data to estimate aboveground shelterbelt carbon stocks across an agricultural landscape. *Agriculture, Ecosystems & Environment*, *156*, 142–150. doi:10.1016/j.agee.2012.05.014

Dallo, Y. A. (2021). Discussion of "Influence of particle-size distribution homogeneity on shearing of soils subjected to internal erosion". *Canadian Geotechnical Journal*, *58*(1), 151–153. doi:10.1139/cgj-2020-0146

Dam Safety DNRME. (2020). *Dam Safety Management Guideline*. Author.

Daniel, A., & Michaela, C. (2021). Mental health and health-related quality of life in victims of the 2013 flood disaster in Germany–A longitudinal study of health-related flood consequences and evaluation of institutionalized low-threshold psycho-social support. *International Journal of Disaster Risk Reduction*, *58*, 102179. doi:10.1016/j.ijdrr.2021.102179

De Almeida, A. T., Cavalcante, C. A. V., Alencar, M. H., Ferreira, R. J. P., de Almeida-Filho, A. T., & Garcez, T. V. (2015). *Multicriteria and Multiobjective Models for Risk, Reliability and Maintenance Decision Analysis*. International Series in Operations Research & Management Science., doi:10.1007/978-3-319-17969-8

de Almeida, W. S., Seitz, S., de Oliveira, L. F. C., & de Carvalho, D. F. (2021). Duration and intensity of rainfall events with the same erosivity change sediment yield and runoff rates. *International Soil and Water Conservation Research, 9*(1), 69–75. doi:10.1016/j.iswcr.2020.10.004

De Baets, S., Poesen, J., Gyssels, G., & Knapen, A. (2006). Effects of grass roots on the erodibility of topsoils during concentrated flow. *Geomorphology, 76*(1-2), 54–67. doi:10.1016/j.geomorph.2005.10.002

De Baets, S., Poesen, J., Reubens, B., Wemans, K., De Baerdemaeker, J., & Muys, B. (2008). Root tensile strength and root distribution of typical Mediterranean plant species and their contribution to soil shear strength. *Plant and Soil, 305*(1), 207–226. doi:10.100711104-008-9553-0

De Groot, J., & Steg, L. (2006). Impact of transport pricing on quality of life, acceptability, and intentions to reduce car use: An exploratory study in five European countries. *Journal of Transport Geography, 14*(6), 463–470. doi:10.1016/j.jtrangeo.2006.02.011

de Mello, C. R., Alves, G. J., Beskow, S., & Norton, L. D. (2020). Daily rainfall erosivity as an indicator for natural disasters: Assessment in mountainous regions of southeastern Brazil. *Natural Hazards, 103*(1), 947–966. doi:10.100711069-020-04020-w

De Vente, J., & Poesen, J. (2005). Predicting soil erosion and sediment yield at the basin scale: Scale issues and semi-quantitative models. *Earth-Science Reviews, 71*(1-2), 95–125. doi:10.1016/j.earscirev.2005.02.002

Decision on determining the Census of waters of the first order (Official Gazette of RS 83/2010)

Đekić, S. (2010). *Agrarni menadžment (Agrarian management)*. University of Niš, Faculty of Economics.

Deutsche Gesellschaft für Technische Zusammenarbeit. (2002). *Disaster risk management*. https://scarp.ubc.ca/sites/scarp.ubc.ca/files/Disaster%20Risk%20Management.pdf

Dhillon, G. S., & Van Rees, K. C. J. (2017). Soil organic carbon sequestration by shelterbelt agroforestry systems in Saskatchewan. *Canadian Journal of Soil Science*, 1–16. doi:10.1139/cjss-2016-0094

Diaz-Balteiro, L., González-Pachón, J., & Romero, C. (2017). Measuring systems sustainability with multi-criteria methods: A critical review. *European Journal of Operational Research, 258*(2), 607–616. doi:10.1016/j.ejor.2016.08.075

Diener, E. (2000). Subjective well-being: The science of happiness and a proposal for a national index. *The American Psychologist, 55*(1), 34–43. doi:10.1037/0003-066X.55.1.34 PMID:11392863

Diener, E., & Biswas-Diener, R. (2011). *Happiness: Unlocking the mysteries of psychological wealth*. John Wiley & Sons.

Diener, E., Lucas, R., Helliwell, J. F., Schimmack, U., & Helliwell, J. (2009). *Well-being for public policy*. Oxford University Press. doi:10.1093/acprof:oso/9780195334074.001.0001

Diener, E., & Suh, E. (1997). Measuring quality of life: Economic, social, and subjective indicators. *Social Indicators Research, 40*(1-2), 189–216. doi:10.1023/A:1006859511756

Ding, Z., Zhang, Z., Li, Y., Zhang, L., & Zhang, K. (2020). Characteristics of magnetic susceptibility on cropland and pastureland slopes in an area influenced by both wind and water erosion and implications for soil redistribution patterns. *Soil & Tillage Research, 199*, 104568. doi:10.1016/j.still.2019.104568

Diodato, N., Ljungqvist, F. C., & Bellocchi, G. (2020). Historical predictability of rainfall erosivity: a reconstruction for monitoring extremes over Northern Italy (1500–2019). *NPJ Climate and Atmospheric Science, 3*(1), 1-10.

Diodato, N., Borrelli, P., Fiener, P., Bellocchi, G., & Romano, N. (2017). Discovering historical rainfall erosivity with a parsimonious approach: A case study in Western Germany. *Journal of Hydrology (Amsterdam), 544*, 1–9. doi:10.1016/j.jhydrol.2016.11.023

Diodato, N., Borrelli, P., Panagos, P., Bellocchi, G., & Bertolin, C. (2019). Communicating hydrological hazard-prone areas in Italy with geospatial probability maps. *Frontiers in Environmental Science, 7*, 193. doi:10.3389/fenvs.2019.00193

Diodato, N., Ljungqvist, F. C., & Bellocchi, G. (2021). Climate Patterns in the World's Longest History of Storm-Erosivity: The Arno River Basin, Italy, 1000–2019 CE. *Frontiers of Earth Science, 9*, 108.

Directive 2000/60/EC of the European Parliament and of the Council establishing a framework for the Community action in the field of water policy - EU Water Framework Directive (WFD). OJL 327, 2000.

Directive 2007/60/EC on the assessment and management of flood risks. OJL 288, 2007.

Donahue, A., & Joyce, P. (2001). A Framework for Analysing Emergency Management with an Application to Federal Budgeting. *Public Administration Review, 61*(6), 728–740. doi:10.1111/0033-3352.00143

Đorđević, A., Jovanović, M., & Rađenović, Ž. (2020). Marketing plan kao deo biznis plana preduzeća. In D. Živković & I. Simić (Eds.), *Inovativno preduzetništvo* (pp. 125–161). Innovation Centre Universty of Niš.

Dorinet, E., Jouvet, P. A., & Wolfersberger, J. (2021). Is the agricultural sector cursed too? Evidence from Sub-Saharan Africa. *World Development, 140*, 105250. doi:10.1016/j.worlddev.2020.105250

Dragović, N., & Vulević, T. (2020). Soil Degradation Processes, Causes, and Assessment Approaches. In Life on Land, Encyclopedia of the UN Sustainable Development Goals. Springer Nature Switzerland AG 2020 W. doi:10.1007/978-3-319-71065-5_86-1

Dragović, N., Ristić, R., Pülzl, H., & Wolfslehner, B. (2017). *Natural resource management in Southeast Europe: forest, soil and water*. Skopje, GIZ, 270. EFAS - European Flood Awareness System. Available at: https://www.efas.eu/en

Dragović, N., Vulević, T., Zlatić, M., Ristić, R., & Todosijević, M. (2018). Compliance of the Legal Framework for Soil and Water Resources Management in Serbia and Some Regional Countries with the EU Legislation. *Advances in Geoecology, 45*.

Dragović, N., & Ristić, R. (2013). Institutional Organisation and Financing of Works for the Protection of Soil from Erosion and Torrent Control in Serbia: Current Problems and Possible Solutions. *Wildbach- und Lawinenverbau, Journal of Torrent, Avalanche, Landslide and Rock Fall Engineering*, (171), 256–266.

Dragović, N., Zlatić, M., & Petković, S. (2007). Comparative Analysis of Institutional Organization of Erosion and Torrent Control in Serbia and Same EU Countries. In *Conference Proceedings Erosion and Torrent Control as a Factor in Sustainable River Basin Management*. University of Belgrade.

Drennan, L., & McConnell, A. (2007). *Risk and Crisis Management in the Public Sector*. Routledge.

Duarte, C. M., Losada, I. J., Hendriks, I. E., Mazarrasa, I., & Marbà, N. (2013). The role of coastal plant communities for climate change mitigation and adaptation. *Nature Climate Change, 3*(11), 961–968. doi:10.1038/nclimate1970

Duhita, A. D. P., Rahardjo, A. P., & Hairani, A. (2021). The Effect of Slope on the Infiltration Capacity and Erosion of Mount Merapi Slope Materials. *Journal of the Civil Engineering, 7*(1), 71–84. doi:10.22146/jcef.58350

Duile, J. (1826). *Ueber Verbauung der Wildbäche in Gebirgs-Ländern, vorzüglich in der Provinz Tirol, und Vorarlberg*. Rauchische Schriften.

Du, M., Mu, X., Zhao, G., Gao, P., & Sun, W. (2021). Changes in Runoff and Sediment Load and Potential Causes in the Malian River Basin on the Loess Plateau. *Sustainability*, *13*(2), 443. doi:10.3390u13020443

Dunlop, A. (2000). *Spatial and Temporal Aspects of Saskatchewan Field Shelterbelts* (Ph.D. Thesis). University of Saskatchewan.

Duulatov, E., Chen, X., Issanova, G., Orozbaev, R., Mukanov, Y., & Amanambu, A. C. (2021). Introduction and background of rainfall erosivity processes and soil erosion. In *Current and future trends of rainfall erosivity and soil erosion in Central Asia* (pp. 1–7). Springer. doi:10.1007/978-3-030-63509-1_1

Dzator, J., & Dzator, M. (2021). Natural Disasters: Macroeconomic Implications and Measurement Issues. In Economic Effects of Natural Disasters (pp. 317-333). Academic Press. doi:10.1016/B978-0-12-817465-4.00020-0

EC. (2021a). *Political agreement on new Common Agricultural Policy: fairer, greener, more flexible*. https://ec.europa.eu/commission/presscorner/detail/en/IP_21_2711

EC. (2021b). *Working with Parliament and Council to make the CAP reform fit for the European Green Deal*. https://ec.europa.eu/info/sites/default/files/food-farming-fisheries/key_policies/documents/factsheet-cap-reform-to-fit-european-green-deal_en.pdf

EC. (2021c). *Soil and Land*. https://ec.europa.eu/environment/soil/index_en.htm

Edwards, W., Shipitalo, M., Owens, L., & Norton, L. (1990). Effect of *Lumbricus terrestris* L. burrows on hydrology of continuous no-till corn fields. *Geoderma*, *46*(1-3), 73–84. doi:10.1016/0016-7061(90)90008-W

EFAS. (n.d.a). *Flash flood indicators*. https://www.efas.eu/en/flash-flood-indicators

EFAS. (n.d.b). *Medium range flood indicators*. Available at: https://www.efas.eu/en/medium-range-flood-forecasts

Elliot, D., Harris, K., & Baron, S. (2005). Crisis Management and Services Marketing. *Journal of Services Marketing*, *19*(5), 336–345. doi:10.1108/08876040510609943

Elnashar, A., Zeng, H., Wu, B., Fenta, A. A., Nabil, M., & Duerler, R. (2021). Soil erosion assessment in the Blue Nile Basin driven by a novel RUSLE-GEE framework. *The Science of the Total Environment*, *793*, 148–466. doi:10.1016/j.scitotenv.2021.148466 PMID:34175609

Emergency Events Database. (2017). *Annual Disaster Statistical Review 2016: The numbers and trends*. https://reliefweb.int/report/world/annual-disaster-statistical-review-2016-numbers-and-trends

Essefi, E., Ben Jmaa, H., Touir, J., Tagorti, M. A., & Yaich, C. (2013c). Variability to sedimentary dynamics and climatic conditions during the last two millennia at sebkha Souassi in eastern Tunisia. *Papers on Global Change IGBP*, *20*, 29–45. . doi:10.2478/igbp-2013-0001

Essefi, E., Gharsalli, N., & Yaich, C. (2015). Climatic significance of grain size parameters of the wet aeolian sedimentary filling of sebkha mhabeul, southeast tunisia: Contribution of the genetic approach. *Journal of Basic and Applied Research International*, 140-152.

Essefi, E., Komatsu, G., Fairén, A. G., Chan, M. A., & Yaich, C. (2014b). *Models of Formation and Activity of Spring Mounds in the Mechertate-ChritaSidi El Hani System, Eastern Tunisia: Implications for the Habitability of Mars*. doi:10.3390/life40x000x

Essefi, E., Tagorti, M. A., Touir, J., & Yaich, C. (2013e). Hydrocarbons migration through groundwater convergence towardsaline depressions: case study, Sidi El Hani discharge playa, Tunisian Sahel. *ISRN Environmental Chemistry*. . doi:10.1155/2013/709190

Essefi, E., Touir, J., Tagorti, M.A., & Yaich, C. (2013d). Record of the climatic variability and the sedimentary dynamics during the last two millennia at sebkha Dkhila, eastern Tunisia. *ISRN Geology*, 1-9. . doi:10.1155/2013/936198

Essefi. (2009). *Multidisciplinary study of Sidi El Hani Saline Environment: The Geological History and the Climatic Variability*. Faculty des Sciences of Sfax.

Essefi. (2013). *Wet Aeolian Sedimentology and Sequence Stratigraphy within the Terrestrial Analogues in Eastern Tunisia: Implications for Wet Aeolian Sedimentology and Sequence Stratigraphy on Mars* (PhD Thesis). National Engineering School of Sfax.

Essefi, E. (2020). On the Geochemistry and Mineralogy of the Anthropocene. *Int J Water Wastewater Treat*, *6*(2).

Essefi, E. (2021). *Apport des Approches Géochimique et Sédimentologique au Fonctionnement des Systèmes Endoréiques et Exoréiques de l'Holocène à l'Anthropocène: Intérêts Environnemental et Géoéconomique*. Institut Supérieur des Sciences et Techniques des Eaux de Gabès. Habilitation Universitaire.

Essefi, E., Ben Smida, N., Jandoubi, I., Othmani, M. A., & Tagorti, M. A. (2020). Progressive evaporation of brine of sebkha Mchiguig, central Tunisia: A geo-economical comparative study of salt and brine. *Carbonates and Evaporites*, *35*(2), 59. doi:10.100713146-020-00592-7

Essefi, E., Gharsalli, N., & Yaich, C. (2015). Climatic significance aeolian sedimentary filling of sebkha Mhabeul, southeastern Tunisia: Contribution of the genetic approach. *J Basic Appl Res Int*, *15*(2–4), 140–152.

Essefi, E., Komatsu, G., & Fairén, A. G., Chan, M. A., & Yaich, C. (2013a). Groundwater influence on the aeolian sequence stratigraphy of the Mechertate-Chrita- Sidi El Hani system, Tunisian Sahel: Analogies to the wet-dry aeolian sequence stratigraphy at Meridiani Planum, Terby crater, and Gale crater, Mars. *Planetary and Space Science*. Advance online publication. doi:10.1016/j.jafrearsci.2013.05.004

Essefi, E., Mefteh, S., Medhioub, M., & Yaich, C. (2014). Magnetic study of the heated and unheated sedimentary fillings of Sebkha Mhabeul, Southeast Tunisia: A geophysical method for paleoclimatic investigation and tephrochronological dating. *International Journal of Geophysics*, *2014*, 2014. doi:10.1155/2014/908395

Essefi, E., Tagorti, M. A., Touir, J., & Yaich, C. (2014a). *Past human life in the vicinities of saline systems in Tunisia: the geoarcheological approach to link paleoclimatology, paleoepidemiology, and populations dynamics*. Arabian Journal of Earth Sciences.

Essefi, E., Touir, J., Tagorti, M. A., & Yaich, C. (2013b). Effect of the groundwater contribution, the climatic change, and the human induced activities on the hydrological behavior of discharge playas: A case study Sidi El Hani discharge playa, Tunisian Sahel. *Arabian Journal of Geosciences*, *6*(10), 3997–4009. Advance online publication. doi:10.100712517-012-0659-6

Essefi, E., Touir, J., Tagorti, M. A., & Yaich, C. (2014c). *Geodynamic framework of saline systems in eastern Tunisia: saline depressions inherited from the Triassic intrusions and/or the Messinian Salinity Crisis*. ISRN Geology. doi:10.1155/2014/798706

European Environment Agency. (2016). *Flood risks and environmental vulnerability - Exploring the synergies between floodplain restoration, water policies and thematic policies*. EEA.

Eurostat. (2015). *Quality of life: Facts and views* (2015 edition). Luxembourg: Publications Office of the European Union.

Eurostat. (2021). [Data set]. https://ec.europa.eu/eurostat/data/database

Expósito, A., & Velasco, F. (2020). Exploring environmental efficiency of the European agricultural sector in the use of mineral fertilizers. *Journal of Cleaner Production, 253*, 119971. doi:10.1016/j.jclepro.2020.119971

Fahey, T., Whelan, C. T., & Maître, B. (2005). *First European quality of life survey: Income inequalities and deprivation*. Office for Official Publications in the European Communities.

Fan, Y., Liu, J., & Cai, Q. (2008). The effects of wetting rate on aggregate stability in three soils. In *15th ISCO Congress "Soil and Water Conservation, Climate Change and Environmental Sensitivity", 18– 23 May 2008 Budapest* (p. 4). Geographical Research Institute, Hungarian Academy of Science.

Fan, F., Deng, Y., Hu, X., & Weng, Q. (2013). Estimating composite curve number using an improved SCS-CN method with remotely sensed variables in Guangzhou, China. *Remote Sensing*, *5*(3), 1425–1438. doi:10.3390/rs5031425

Faniran, A. (1968). The Index of Drainage Intensity - A Provisional New Drainage Factor. *Australian Journal of Science*, *31*, 328–330.

FAO. (2005). Organic matter decomposition and the soil food web. In *The importance of soil organic matter: Key to drought-resistant soil and sustained food and production*. Food and Agriculture Organization of the United Nation.

FAO. (2015). *Status of the world's soil resources (SWSR)–main report*. Food and agriculture organization of the United Nations and intergovernmental technical panel on soils, Rome, Italy, 650. Retrieved from https://www.fao.org/3/i5199e/I5199E.pdf

FAO. (2015). Status of the world's soil resources (SWSR)–main report. Food and agriculture organization of the United Nations and intergovernmental technical panel on soils, Rome, Italy.

Fawcett, W., Ellingham, I., & Platt, S. (2008). Reconciling the architectural preferences of architects and the public: The ordered preference model. *Environment and Behavior*, *40*(5), 599–618. doi:10.1177/0013916507304695

Fayas, C. M., Abeysingha, N. S., Nirmanee, K. G. S., Samaratunga, D., & Mallawatantri, A. (2019). Soil loss estimation using rusle model to prioritize erosion control in KELANI river basin in Sri Lanka. *International Soil and Water Conservation Research*, *7*(2), 130–137. doi:10.1016/j.iswcr.2019.01.003

Fearn-Banks, K. (2002). *Crisis Communications: A Casebook Approach*. Associates Lawrence Erlbaum, Inc.

Federal Ministry of Agriculture. Forestry, Environment, and Water Management. (2016). Austrian Service for Torrent and Avalanche Control. Author.

Federal Ministry of Agriculture. Water Management, and Forestry Water Management. (2012). Strategy of the Federation of Bosnia and Herzegovina 2010 – 2022. Author.

Feely, R., Alin, S., Newton, J., Sabine, C., Warner, M., Devol, A., Krembs, C., & Maloy, C. (2010). The combined effects of ocean acidification, mixing, and respiration on pH and carbonate saturation in an urbanized estuary. *Estuarine, Coastal and Shelf Science*, *88*(4), 442–449. doi:10.1016/j.ecss.2010.05.004

Felix, R., & Garcia-Vega, J. (2012). Quality of life in Mexico: A formative measurement approach. *Applied Research in Quality of Life*, *7*(3), 223–238. doi:10.100711482-011-9164-4

Ferriss, A. L. (2006). A theory of social structure and the quality of life. *Applied Research in Quality of Life*, *1*(1), 117–123. doi:10.100711482-006-9003-1

Fierera, N., Schimela, J. P., & Holden, P. A. (2003). Variations in microbial community composition through two soil depth profiles. *Soil Biology & Biochemistry*, *35*(1), 167–176. doi:10.1016/S0038-0717(02)00251-1

Flood Defense Plan, R. S. Official Gazette of RS, No. 9/19, (2019), https://ruczrs.org/wp-content/uploads/2019/09/Plan-odbrane-od-poplava-u-Republici-Srpskoj-u-2019.-godini.pdf

Flood protection plan for the area of the municipality of Leskovac. (2007). Institute of Water Management "Jaroslav Cerni" AD Belgrade.

Floods, Serbia and Europe: What is the danger and how to protect yourself from the water element. (2021, July). BBC News.

Foley, J. A., Ramankutty, N., Brauman, K. A., Cassidy, E. S., Gerber, J. S., Johnston, M., Mueller, N. D., O'Connell, C., Ray, D. K., West, P. C., Balzer, C., Bennett, E. M., Carpenter, S. R., Hill, J., Monfreda, C., Polasky, S., Rockström, J., Sheehan, J., Siebert, S., ... Zaks, D. P. (2011). Solutions for a cultivated planet. *Nature*, *478*(7369), 337–342. doi:10.1038/nature10452 PMID:21993620

Foster, D. L., Bardos, C. P., & Southgate, D. D. (1987). Soil erosion and water treatment costs. *Journal of Soil and Water Conservation*, *42*, 349–352.

Foucher, A., Evrard, O., Cerdan, O., Chabert, C., Lefèvre, I., Vandromme, R., & Salvador-Blanes, S. (2021). Deciphering human and climatic controls on soil erosion in intensively cultivated landscapes after 1950 (Loire Valley, France). *Anthropocene*, *34*, 100287. doi:10.1016/j.ancene.2021.100287

Foulds, S. A., Griffiths, H. M., Macklin, M. G., & Brewer, P. A. (2014). Geomorphological records of extreme floods and their relationship to decadal-scale climate change. *Geomorphology*, *216*, 193–207. doi:10.1016/j.geomorph.2014.04.003

Freeman, P. K., Martin, L. A., Linnerooth-Bayer, J., Mechler, R., Pflug, G., & Warner, K. (2003). *National Systems for Comprehensive Disaster Management: Financing Reconstruction.* https://publications.iadb.org/publications/english/document/Disaster-Risk-Management-National-Systems-for-the-Comprehensive-Management-of-Disaster-Risk-and-Financial-Strategies-for-Natural-Disaster-Reconstruction.pdf

Frisch, D., Morton, P. K., Chowdhury, P. R., Culver, B. W., Colbourne, J. K., Weider, L. J., & Jeyasingh, P. D. (2014). A millennial-scale chronicle of evolutionary responses to cultural eutrophication in Daphnia. *Ecology Letters*, *17*(3), 360–368. doi:10.1111/ele.12237 PMID:24400978

Fritsch, O., & Benson, D. (2020). Mutual Learning and Policy Transfer in Integrated Water Resources Management: A Research Agenda. *Water (Basel)*, *12*(1), 72. Advance online publication. doi:10.3390/w12010072

Fryrear, D. W., Sutherland, P. L., Davis, G., Hardee, G., & Dollar, M. (2001). Wind erosion estimates with RWEQ and WEQ. In *Sustaining the Global Farm.* Pardue University and USDA-ARS National Soil Erosion Research Laboratory.

Funk, R., & Reuter, H. I. (2006). Wind Erosion. In J. Boardman & J. Poesen (Eds.), *Soil Erosion in Europe.* John Wiley & Sons. doi:10.1002/0470859202.ch41

Furlan, C., & Mortarino, C. (2018). Forecasting the impact of renewable energies in competition with non-renewable sources. *Renewable & Sustainable Energy Reviews*, *81*, 1879–1886. doi:10.1016/j.rser.2017.05.284

Gabrieli, J., & Barbante, C. (2014). The Alps in the age of the Anthropocene: The impact of human activities on the cryosphere recorded in the Colle Gnifetti glacier. *Rendiconti Lincei*, *25*(1), 71–83. doi:10.100712210-014-0292-2

Gaied, M. S., Ghanmi, A., & Ben Youssef, M. (2002). Carte géologique simplifiée des ensembles géologiques de la région de Sidi El Hani. Ann. Min. Géol.

Gaied, M. S., Ghanmi, A., & Ben Youssef. (2002). *Carte géologique simplifiée des ensembles géologiques de la région de Sidi El Hani.* Academic Press.

Galinha, I. C. C., & Pais-Ribeiro, J. L. (2008). The structure and stability of subjective well-being: A structure equation modelling analysis. *Applied Research in Quality of Life*, *3*(4), 293–314. doi:10.100711482-009-9063-0

Galloway, G., Seminara, G., Blöschl, G., Garcia, M., Montanari, A., & Solari, L. (2017). Saving a World Treasure: Protecting Florence from Flooding. Firenze University Press.

Gałuszka, A., Migaszewski, Z. M., & Zalasiewicz, J. (2014). Assessing the Anthropocene with geochemical methods. *Geological Society of London, Special Publications*, *395*(1), 221–238. doi:10.1144/SP395.5

Gammoudi, A. (2020). *Etude géochimique et cyclostratigraphie holocène des sédiments de la sebkha Mhabeul au Sud-Est de la Tunisie* (PhD Thesis). Higher Institute of Applied Sciences and Technology of Gabes.

Gammoudi, A., Essefi, E., & Rigane, H. (2019). Variability of the grain-sized fractions during the last three millennia within the sedimentary fillings of the saline system of Mhabeul: the Holocene cyclostratigraphy. *Arabian Journal of Geosciences, 12*(9).

Gammoudi, A., Essefi, E., & Rigane, H. (2019). Variability of the grain-sized fractions during the last three millennia within the sedimentary fillings of the saline system of Mhabeul: The Holocene cyclostratigraphy. *Arabian Journal of Geosciences*, *12*(9), 296. doi:10.100712517-019-4472-3

Gammoudi, A., Tlili, A., Essefi, E., El Feki, H., & Rigane, H. (2021). Contribution of the geochemical, physico-chemical, mineralogical, and statistical approaches to the reconstructing of the Holocene depositional environments along South-Eastern Tunisia. *Arabian Journal of Geosciences*, *14*(19), 2029. doi:10.100712517-021-08227-4

Gao, Q., Ci, L., & Yu, M. (2002). Modeling wind and water erosion in northern China under climate and land use changes. *Journal of Soil and Water Conservation*, *57*, 46–55.

Garambois, P. A., Larnier, K., Roux, H., Labat, D., & Dartus, D. (2014). Analysis of flash flood-triggering rainfall for a process-oriented hydrological model. *Atmospheric Research*, *137*, 14–24. doi:10.1016/j.atmosres.2013.09.016

García-Ruiz, J. M., Beguería, S., Nadal-Romero, E., González-Hidalgo, J. C., Lana-Renault, N., & Sanjuán, Y. (2015). A meta-analysis of soil erosion rates across the world. *Geomorphology*, *239*, 160–173. doi:10.1016/j.geomorph.2015.03.008

Gardiner, B., Berry, P., & Moulia, B. (2016). Review: Wind impacts on plant growth, mechanics and damage. *Plant Science*, *245*, 94–118. doi:10.1016/j.plantsci.2016.01.006 PMID:26940495

Gasper, D. (2010). Understanding the diversity of conceptions of well-being and quality of life. *Journal of Socio-Economics*, *39*(3), 351–360. doi:10.1016/j.socec.2009.11.006

Gaume, E., Bain, V., Bernardara, P., Newinger, O., Barbuc, M., Bateman, A., Blaškovičová, L., Blöschl, G., Borga, M., Dumitrescu, A., Daliakopoulos, I., Garcia, J., Irimescu, A., Kohnova, S., Koutroulis, A., Marchi, L., Matreata, S., Medina, V., Preciso, E., ... Viglione, A. (2009). A Collation of Data on European Flash Floods. *Journal of Hydrology (Amsterdam)*, *367*(1), 70–78. doi:10.1016/j.jhydrol.2008.12.028

Gaunt, P., & Ollenburger, J. (1995). Issue management revisited: A tool that deserves another look. *Public Relations Review*, *21*(3), 199–210. doi:10.1016/0363-8111(95)90021-7

Gavrilović, Z., & Brajković, M. (1989). Natural and anthropogenic factors important for the formation of "Sudden" floods. *Yugoslav Conference on Torrential Floods, Kopaonik*, 46-53.

Gelagay, H. S., & Minale, A. S. (2016). Soil loss estimation using GIS and Remote sensing techniques: A case of Koga watershed, Northwestern Ethiopia. *International Soil and Water Conservation Research*, *4*(2), 126–136. doi:10.1016/j.iswcr.2016.01.002

Gencer, E. A. (2013). Natural Disasters, Urban Vulnerability, and Risk Management: A Theoretical Overview. In E. A. Gencer (Ed.), *The Interplay between Urban Development, Vulnerability, and Risk Management, Mediterranean Studies* (Vol. 7, pp. 7–43). Springer-Verlag. doi:10.1007/978-3-642-29470-9_2

Gencer, E. A. (2013). The Interplay Between Urban Development, Vulnerability, and Risk Management. *Mediterranean Studies (Warrensburg, Mo.)*, *7*, 7–43.

Gharsalli, N., Essefi, E., Baydoun, R., & Yaich, C. (2020). The Anthropocene and great acceleration as controversial epoch of human-induced activities: case study of the Halk El Menjel Wetland, Eastern Tunisia. *Applied Ecology and Environmental Research*, *18*(3), 4137–4166. doi:10.15666/aeer/1803_41374166

Gholami, L., Darvishan, A. K., Spalevic, V., Cerdà, A., & Kavian, A. (2021). Effect of storm pattern on soil erosion in damaged rangeland; field rainfall simulation approach. *Journal of Mountain Science*, *18*(3), 706–715. doi:10.100711629-019-5633-2

Ghosh, K., Kumar De, S., Bandyopadhyay, S., & Saha, S. (2013). Assessment of soil loss of the Dhalai river basin, Tripura, India using USLE. *International Journal of Geosciences*, *04*(01), 11–23. doi:10.4236/ijg.2013.41002

Gilbert, N. (2012). (2012). One-third of our greenhouse gas emissions come from agriculture. *Nature*. Advance online publication. doi:10.1038/nature.2012.11708

Gillings, M. R., & Paulsen, I. T. (2014). Microbiology of the Anthropocene. *Anthropocene*, *5*, 1–8. doi:10.1016/j.ancene.2014.06.004

Gioti, E., Riga, C., Kalogeropoulos, K., & Chalkias, C. (2013). A GIS-based flash flood runoff model using high resolution DEM and meteorological data. *EARSeL eProceedings*, *12*(1), 33–43.

Glaesser, D. (2006). *Crisis Management in the Tourism Industry* (2nd ed.). Butterworth Heinemann. doi:10.4324/9780080464596

GLEWS. (n.d.). *The Joint FAO-OIE-WHO Global Early Warning System for health threats and emerging risks at the human-animal-ecosystems interface*. Available at: http://www.glews.net

Gobin, A., Govers, G., Jones, R., Kirkby, M., & Kosmas, C. (2003). *Assessment and Reporting on Soil Erosion*. Technical Report 94. European Environment Agency.

Goldberg, M. S., Burnett, R. T., Bailar, J. C. III, Brook, J., Bonvalot, Y., Tamblyn, R., Singh, R., & Valois, M. F. (2001). The association between daily mortality and ambient air particle pollution in Montreal, Quebec: 1. Nonaccidental mortality. *Environmental Research*, *86*(1), 12–25. doi:10.1006/enrs.2001.4242 PMID:11386737

Gomes, L., Arrúe, J. L., López, M. V., Sterk, G., Richard, D., Gracia, R., Sabre, M., Gaudichet, A., & Frangi, J. P. (2003). Wind erosion in a semiarid agricultural area of Spain: The WELSONS project. *Catena*, *52*(3-4), 235–256. doi:10.1016/S0341-8162(03)00016-X

Gomiero, T. (2016). Soil degradation, land scarcity and food security: Reviewing a complex challenge. *Sustainability*, *8*(3), 281. doi:10.3390u8030281

Gonzales, H.B., Casada, M.E., Hagen, L.J., Tatarko, J., & Maghirang, R.G. (2017). Sand Transprt and Abrasion within Simulated Standing Vegetation. *The American Society of Agricultural and Biological Engineers*, 791-802.

Gosling, S. N., & Arnell, N. W. (2016). A global assessment of the impact of climate change on water scarcity. *Climatic Change*, *134*(3), 371–385. doi:10.100710584-013-0853-x

Guerra, C. A., Maes, J., Geijzendorffer, I., & Metzger, M. J. (2016). An assessment of soil erosion prevention by vegetation in Mediterranean Europe: Current trends of ecosystem service provision. *Ecological Indicators*, *60*, 213–222. doi:10.1016/j.ecolind.2015.06.043

Guerra, C. A., Rosa, I. M., Valentini, E., Wolf, F., Filipponi, F., Karger, D. N., & Eisenhauer, N. (2020). Global vulnerability of soil ecosystems to erosion. *Landscape Ecology*, *35*(4), 823–842. doi:10.100710980-020-00984-z PMID:32587435

Guo, B., Zang, W., Yang, X., Huang, X., Zhang, R., Wu, H., ... Zhang, Y. (2020). Improved evaluation method of the soil wind erosion intensity based on the cloud–AHP model under the stress of global climate change. *The Science of the Total Environment*, *746*, 141271.

Gutiérrez, F., Parise, M., De Waele, J., & Jourde, H. (2014). A review on natural and human-induced geohazards and impacts in karst. *Earth-Science Reviews*, *138*, 61–88. doi:10.1016/j.earscirev.2014.08.002

Guzzetti, F., Cardinali, M., & Reichenbach, P. (1994). The AVI Project: A Bibliographical and Archive Inventory of Landslides and Floods in Italy. *Environmental Management*, *18*(4), 623–633. doi:10.1007/BF02400865

Guzzetti, F., Crosta, G., Marchetti, M., & Reichenbach, P. (1992). Debris flows triggered by the July, 17-19, 1987 storm in the Valtellina area Northern Italy. *Proceedings Interpraevent*, *1992*(2), 193–204.

Gwapedza, D., Nyamela, N., Hughes, D. A., Slaughter, A. R., Mantel, S. K., & van der Waal, B. (2021). Prediction of sediment yield of the Inxu River catchment (South Africa) using the MUSLE. *International Soil and Water Conservation Research*, *9*(1), 37–48. doi:10.1016/j.iswcr.2020.10.003

Gyssels, G., Poesen, J., Bochet, E., & Li, Y. (2005). Impact of plant roots on the resistance of soils to erosion by water: A review. *Progress in Physical Geography*, *29*(2), 189–217. doi:10.1191/0309133305pp443ra

Hajabbasi, M., Jalalian, A. A., & Karimzadeh, H. R. (1997). Deforestation effects on soil physical and chemical properties, Lordegan, Iran. *Plant and Soil*, *190*(2), 301–308. doi:10.1023/A:1004243702208

Hansen, J. W. (1996). Is agricultural sustainability a useful concept? *Agricultural Systems*, *50*(2), 117–143. doi:10.1016/0308-521X(95)00011-S

Hansen, L. T., Breneman, V. E., Davison, C. W., & Dicken, C. W. (2002). The Cost of Soil Erosion to Downstream Navigation. *Journal of Soil and Water Conservation*, *57*(4), 205–212.

Han, W., Ma, Z., Lai, Z., Appel, E., Fang, X., & Yu, L. (2014). Wind erosion on the north-eastern Tibetan Plateau: Constraints from OSL and U-Th dating of playa salt crust in the Qaidam Basin. *Earth Surface Processes and Landforms*, *39*(6), 779–789. doi:10.1002/esp.3483

Harris, R. M., Beaumont, L. J., Vance, T. R., Tozer, C. R., Remenyi, T. A., Perkins-Kirkpatrick, S. E., & Bowman, D. M. J. S. (2018). Biological responses to the press and pulse of climate trends and extreme events. *Nature Climate Change*, *8*(7), 579–587. doi:10.103841558-018-0187-9

Hedelin, B., Evers, M., Alkan-Olsson, J., & Jonsson, A. (2017). Participatory modelling for sustainable development: Key issues derived from five cases of natural resource and disaster risk management. *Environmental Science & Policy*, *76*, 185-196. doi:10.1016/j.envsci.2017.07.001

Heisler, G., & Dewalle, D. (1988). Effects of Windbreaks Structure on Wind Flow. *Agriculture, Ecosystems & Environment*, *22/23*, 41–69. doi:10.1016/0167-8809(88)90007-2

Henderson, H. (2002). Quality of life: Issues and questions. In D. Lamberton (Ed.), *Managing the global: Globalization, employment and quality of life (Humanity and security and global governance)* (pp. 51–95). I. B. Tauris.

Henry, P. C. (1995). Soil as a factor in sugarcane ratoon yield decline on an irrigated estate in Swaziland. *Proc S Afr Sug Tech Ass*, *69*, 122-126.

Herslund, L. B., Jalayer, F., Jean-Baptiste, N., Jørgensen, G., Kabisch, S., Kombe, W., & Vedeld, T. (2016). A multi-dimensional assessment of urban vulnerability to climate change in Sub-Saharan Africa. *Natural Hazards*, *82*(2), 149–172. doi:10.100711069-015-1856-x

Hicks, D. H., & Anthony, T. (2001). *Soil conservation technical handbook.* The Ministry for the Environment. http://www.coastalwiki.org

Holling, C. S. (2001). Understanding the complexity of economic, ecological, and social systems. *Ecosystems (New York, N.Y.)*, *4*(5), 390–405. doi:10.100710021-001-0101-5

Holy, M. (1980). *Erosion and environment- Environmental sciences and application* (Vol. 9). Pergamon Press Ltd.

Horton, R. E. (1932). Drainage Basin Characteristics. *Transactions - American Geophysical Union*, *13*(1), 350–361. doi:10.1029/TR013i001p00350

Horton, R. E. (1945). Erosional Development of Streams and Their Drainage Basins: Hydrophysical Approach to Quantitative Morphology. *Geological Society of America Bulletin*, *56*(3), 275–370. doi:10.1130/0016-7606(1945)56[275:EDOSAT]2.0.CO;2

Huang, J., Mendoza, B., Daniel, J. S., Nielsen, C. J., Rotstayn, L., & Wild, O. (2013). Anthropogenic and natural radiative forcing. *Climate Change 2013 the Physical Science Basis: Working Group I Contribution to the Fifth Assessment Report of the Intergovernmental Panel on Climate Change*, 659–740.

Huang, J., Yu, H., Guan, X., Wang, G., & Guo, R. (2015). Accelerated dryland expansion under climate change. *Nature Climate Change*, *6*(2), 166–171. doi:10.1038/nclimate2837

Inglehart, R., & Klingemann, H. D. (2000). Genes, culture, democracy, and happiness. Culture and subjective well-being. In E. Diener & E. M. Suh (Eds.), *Culture and subjective well-being* (pp. 165–183). MIT Press.

IN-MHEWS - The International Network for Multi-Hazard Early Warning Systems. (n.d.). Available at: https://mhews.wmo.int/en/partners

Institute for Water Management. (2012). *The Strategy of integrated water management of the Republic of Srpska until 2024, Framework plan for the development of water management of the Republic of Srpska - Annex 3 Protection against erosion and torrents.* Author.

International Commission for the Protection of the Danube River and International Sava River Basin Commission. (2015). *Floods in May 2014 in the Sava River Basin-Brief overview of key events and lessons learned.* Author.

Iosub, M., Minea, I., Chelariu, O. E., & Ursu, A. (2020). Assessment of flash flood susceptibility potential in Moldavian Plain (Romania). *Journal of Flood Risk Management*, *13*(4), e12588. doi:10.1111/jfr3.12588

IPBES. (2018). The IPBES assessment report on land degradation and restoration. In L. Montanarella, R. Scholes, & A. Brainich (Eds.), *Secretariat of the Intergovernmental Science-Policy Platform on Biodiversity and Ecosystem Services.* doi:10.5281/zenodo.3237392

IPCC. (2012). Managing the Risks of Extreme Events and Disasters to Advance Climate Change Adaptation. A Special Report of Working Groups I and II of the Intergovernmental Panel on Climate Change. Cambridge University Press.

IPCC. (2013). Summary for Policymakers. In Climate Change 2013: The Physical Science Basis. Cambridge University Press.

ISPRA. (2018). Landslides and floods in Italy: hazards and risk indicators. Summary Report 2018. ISPRA, Rapporti 287 bis/2018.

Issaka, S., & Ashrafa, M. A. (2017). Impact of soil erosion and degradation on water quality: a review. *Geology, Ecology, and Landscapes, 1*(1), 1–11. doi:10.1080/24749508.2017.1301053

Ivčević, A., Mazurek, H., Siame, L., Moussa, A. B., & Bellier, O. (2019). Indicators in risk management: Are they a user-friendly interface between natural hazards and societal responses? Challenges and opportunities after UN Sendai conference in 2015. *International Journal of Disaster Risk Reduction, 41*, 101301. doi:10.1016/j.ijdrr.2019.101301

Ivica, K., Igor, B., Zeljka, Z., & Darija, B. (2018). Effects of soil erosion by water under different tillage treatments on distribution of soil chemical parameters. *Soil and Water Research, 13*(1), 36–43. doi:10.17221/25/2017-SWR

Izrada mapa opasnosti i mapa rizika od poplava na slivu reke Vrbas u BiH, Knjiga 3 – "Izdvajanje bujičnih slivova i formiranje modela osjetljivosti na pojavu i razvoj bujičnih poplava, sa kartom erozije u slivu reke Vrbas" [Development of Flood Hazard and Risk Maps at Vrbas River Basin in BiH, Vol. 3 – Separation of Torrential Basins and Creation of Models of Sensitivity to the Occurrence and Development of Torrential Floods with an Erosion Map in the Vrbas River Basin]. (2017). UNDP BIH and "Zavod za vodoprivredu d.o.o." Bijeljina, BiH.

Jacinthea, P. A., Lal, R., & Kimble, J. M. (2002). Carbon dioxide evolution in runoff from simulated rainfall on long-term no-till and plowed soils in southwestern Ohio. *Soil & Tillage Research, 66*(1), 23–33. doi:10.1016/S0167-1987(02)00010-7

Janaćković, G., Vasović, D., Malenović Nikolić, J., Mušicki, S., & Vranjanac, Ž. (2018). Vulnerability assessment of municipality areas to natural disasters based on group fuzzy analytic hierarchy process. *Journal of Environmental Protection and Ecology, 19*(4).

Janjić, I., Đorđević, A., & Đokić, M. (2020). Internacionalizacija poslovanja. In D. Živković & I. Simić (Eds.), *Inovativno preduzetništvo* (pp. 333–370). Innovation Centre Universty of Niš.

Janjić, I., & Rađenović, T. (2019). The importance of managing innovation in modern enterprises. *Ekonomika (Nis), 65*(3), 45–54. doi:10.5937/ekonomika1903045J

Jaramillo, D., Vélez, M. I., Escobar, J., Pardo-Trujillo, A., Vallejo, F., Villegas, J. C., Acevedo, A. L., Curtis, J., Rincón, H., & Trejos-Tamayo, R. (2021). Mid to late holocene dry events in Colombia's super humid Western Cordillera reveal changes in regional atmospheric circulation. *Quaternary Science Reviews, 261*, 106937. doi:10.1016/j.quascirev.2021.106937

Javed, A., Khanday, M. Y., & Ahmed, R. (2009). Prioritisation of sub watersheds based on Morphometric and Land use analysis using remote sensing and GIS techniques. *Photonirvachak (Dehra Dun), 37*(2), 261–274. doi:10.100712524-009-0016-8

J-DESC IODP Information System Working Group. (2006). *(J-DESC-ISWG) with Science Support Group, Center for Deep Earth Exploration (CDEX) Guideline of Visual Core Description (VCD) Lithology Classification Scheme on the D/V Chikyu and related on-shore research Version 1*. Author.

Jiang, C., Xiong, L., Wang, D., Liu, P., Guo, S., & Xu, C. Y. (2015). Separating the impacts of climate change and human activities on runoff using the Budyko-type equations with time-varying parameters. *Journal of Hydrology (Amsterdam), 522*, 326–338. doi:10.1016/j.jhydrol.2014.12.060

Jia, P., Liu, R., Ma, M., Liu, Q., Wang, Y., Zhai, X., & Wang, D. (2019). Flash flood simulation for ungauged catchments based on the distributed hydrological model. *Water (Basel), 11*(1), 76. doi:10.3390/w11010076

Jónsson, J. Ö. G., & Davíðsdóttir, B. (2016). Classification and valuation of soil ecosystem services. *Agricultural Systems, 145*, 24–38. doi:10.1016/j.agsy.2016.02.010

Jovanović, M. (2018). Innovation of Serbia in relation to neighboring countries as a determinant of competitiveness. *Ekonomika (Nis)*, *64*(1), 57–78. doi:10.5937/ekonomika1801065J

Jovanović, M., & Đokić, M. (2019). National competitiveness and goods market efficiency in Serbia and neighboring countries. In B. Krstić (Ed.), *Enhancing competitiveness of national economies and enterprises* (pp. 109–142). Faculty of Economics University of Niš.

Jovanović, M., & Đokić, M. (2020). Innovativeness of Serbian Enterprises as a Factor of Competitiveness Improvement at the International Market. In M. Janowicz-Lomott, K. Lyskawa, P. Polychronidou, & A. Karasavvoglou (Eds.), *Economic and Financial Challenges for Balkan and Eastern European Countries* (pp. 101–115). Springer Proceedings in Business and Economics. doi:10.1007/978-3-030-39927-6_7

Jovanović, M., Đokić, M., & Janjić, I. (2020). Plan organizacije. In D. Živković & I. Simić (Eds.), *Inovativno preduzetništvo* (pp. 209–250). Innovation Centre Universty of Niš.

Jozaghi, A., Alizadeh, B., Hatami, M., Flood, I., Khorrami, M., Khodaei, N., & Ghasemi Tousi, E. (2018). A comparative study of the AHP and TOPSIS techniques for dam site selection using GIS: A case study of Sistan and Baluchestan Province, Iran. *Geosciences*, *8*(12), 494. doi:10.3390/geosciences8120494

Kaboli, S., Hekmatzadeh, A. A., Darabi, H., & Haghighi, A. T. (2021). Variation in physical characteristics of rainfall in Iran, determined using daily rainfall concentration index and monthly rainfall percentage index. *Theoretical and Applied Climatology*, *144*(1), 507–520. doi:10.100700704-021-03553-9

Kairis, O., Karavitis, C., Salvati, L., Kounalaki, A., & Kosmas, K. (2015). Exploring the Impact of Overgrazing on Soil Erosion and Land Degradation in a Dry Mediterranean Agro-Forest Landscape (Crete, Greece). *Arid Land Research and Management*, *29*(3), 360–374. doi:10.1080/15324982.2014.968691

Kallis, G. (2005). Beyond limits and efficiency, what? Assessing developments in EU water policy. *International Journal of Water*, *3*(2), 121–145.

Katastar bujičnih tokova i erozionih područja u Bosni i Hercegovine [Cadaster of Torrential Flows and Erosion Areas in Bosnia and Herzegovina], Zavod za vodoprivredu, Sarajevo, 1985.

Katra, I. (2020). Soil Erosion by Wind and Dust Emission in Semi-Arid Soils Due to Agricultural Activities. *Agronomy (Basel)*, *10*(1), 89. doi:10.3390/agronomy10010089

Katra, I., Gross, A., Swet, N., Tanner, S., Krasnov, H., & Angert, A. (2016). Substantial dust loss of bioavailable phosphorus from agricultural soils. *Scientific Reports*, *6*(1), 24736. doi:10.1038rep24736 PMID:27095629

Kattsov, V., Federation, R., Reason, C., Africa, S., Uk, A. A., Uk, T. A., Baehr, J., Uk, A. B., Catto, J., Canada, J. S., & Uk, A. S. (2013). Evaluation of climate models. *Climate Change 2013 the Physical Science Basis: Working Group I Contribution to the Fifth Assessment Report of the Intergovernmental Panel on Climate Change*, 741–866.

Keesstra, S. D., Bouma, J., Wallinga, J., Tittonell, P., Smith, P., Cerdà, A., Montanarella, L., Quinton, J. N., Pachepsky, Y., van der Putten, W. H., Bardgett, R. D., Moolenaar, S., Mol, G., Jansen, B., & Fresco, L. O. (2016). The significance of soils and soil science towards realization of the United Nations Sustainable Development Goals. *Soil (Göttingen)*, *2*(2), 111–128. doi:10.5194oil-2-111-2016

Keesstra, S., Mol, G., De Leeuw, J., Okx, J., De Cleen, M., & Visser, S. (2018). Soil-related sustainable development goals: Four concepts to make land degradation neutrality and restoration work. *Land (Basel)*, *7*(4), 133. doi:10.3390/land7040133

Keković, Z., Marić, P., & Komazec, N. (2011). Republic of Serbia natural and other disaster risk assessment – methodology. *Journal of Criminalistics and Law, 16*(2), 1-17.

Kelam, I. (2016). Odgovorno upravljanje poljoprivrednim zemljištem (Responsible management of agricultural land). *JAHR: Europski časopis za bioetiku, 7*(2), 203-215.

Keles, R. (2010). Globalization and Ecological Changes, *20th Congress of Spiritual Health and Diseases*, Bodrum.

Keles, R. (2012). The quality of life and the environment. *Procedia: Social and Behavioral Sciences*, *35*, 23–32. doi:10.1016/j.sbspro.2012.02.059

Kešetović, Ž. (2008). *Krizni menadžment.* Beograd: Fakultet bezbednosti, Univerzitet u Beogradu.

Kešetović, Ž., & Toth, I. (2012). *Problemi kriznog menadžmenta.* Velika Gorica: Veleučilište Velika Gorica.

Kisić, I. (2017). Erozija vjetrom. *Hrvatske Vode*, *25*(99), 1–12.

Klasen, S., Meyer, K. M., Dislich, C., Euler, M., Faust, H., Gatto, M., Hettig, E., Melati, D. N., Jaya, I. N. S., Otten, F., Pérez-Cruzado, C., Steinebach, S., Tarigan, S., & Wiegand, K. (2016). Economic and ecological trade-offs of agricultural specialization at different spatial scales. *Ecological Economics*, *122*, 111–120. doi:10.1016/j.ecolecon.2016.01.001

Knežević, J., Gurda, S., Musić, J., Halilović, V., Sokolović, D., & Bajrić, M. (2018). The Impact of Animal Logging on Residual Trees in Mixed Fir and Spruce Stands. *South-East European Forestry*, *9*(2), 107–114. doi:10.15177eefor.18-16

Koković, D. (2010). Ekologija kao način života. *Svarog: Časopis za društvene i prirodne nauke, 1*(1), 75-84.

Kolasa-Więcek, A. (2018). Neural modeling of greenhouse gas emission from agricultural sector in European Union member countries. *Water, Air, and Soil Pollution*, *229*(6), 1–12. doi:10.100711270-018-3861-7

Kort, J., & Turnock, R. (1998). Carbon reservoir and biomass in Canadian prairie shelterbelts. *Agroforestry Systems*, *44*(2/3), 175–186. doi:10.1023/A:1006226006785

Kostadinov, S., Borisavljević, A., & Mlađan, D. (2014). Torrents and torrential floods in Serbia: Characteristics and possibilities of its control. *Advances in GeoEcology*, *41*, 278–285.

Kostadinov, S., Tošić, R., Hrkalović, D., Nikolić, S., Sudar, N., Kapović Solomun, M., Dragičević, S., Momirović, N., Cupać, R., Bosankić, G., & Bundalo, S. (2019). Soil erosion control in the function of reducing the risk of floods in the Vrbas River Basin. *Vodoprivreda*, *51*(300-302), 211–224.

Kouzmin, A. (2008). Crisis Management in Crisis? *Administrative Theory & Praxis*, *30*(2), 155–183. doi:10.1080/10841806.2008.11029631

Kovoor-Misrca, S., Zammuto, R. F., & Mitroff, I. I. (2000). Crisis preparation in organizations: Prescription versus reality. *Technological Forecasting and Social Change*, *63*(1), 43–62. doi:10.1016/S0040-1625(99)00049-9

Ko, Y. (2018). Trees and vegetation for residential energy conservation: A critical review for evidence-based urban greening in North America. *Urban Forestry & Urban Greening*, *34*, 318–335. doi:10.1016/j.ufug.2018.07.021

Krausmann, E., Girgin, S., & Necci, A. (2019). Natural hazard impacts on industry and critical infrastructure: Natech risk drivers and risk management performance indicators. *International Journal of Disaster Risk Reduction*, *40*, 101163. doi:10.1016/j.ijdrr.2019.101163

Křeček, J., Nováková, J., Palán, L., Pažourková, E., & Stuchlík, E. (2021). Role of forests in headwater control with changing environment and society. *International Soil and Water Conservation Research*, *9*(1), 143–157. doi:10.1016/j.iswcr.2020.11.002

Kröbel, R., Moore, J., Ni, Y. Z., McPherson, A., Poppy, L., Soolanayakanahally, R. Y., Amichev, B. Y., Ward, T., Laroque, C., Van Rees, K. C. J., & Akhter, F. (2020). Demonstration and Testing of the Improved Shelterbelt Component in the Holos Model. *Frontiers in Environmental Science*, *8*, 149. Advance online publication. doi:10.3389/fenvs.2020.00149

Krstić, B., & Janjić, I. (2019). Institutions as a factor of competitiveness: The case of South-eastern countries. In B. Krstić (Ed.), *Building national competitiveness* (pp. 1–20). Faculty of Economics, University of Niš.

Kruglov, O., Menshov, O., Kovalenko, S., Nazarok, P., & Andreeva, O. (2020, November). Magnetic susceptibility and heterogeneities of agronomic indicators of soil as a tool for water erosion assessment. In *XIV International Scientific Conference "Monitoring of Geological Processes and Ecological Condition of the Environment"* (Vol. 2020, No. 1, pp. 1-5). European Association of Geoscientists & Engineers. 10.3997/2214-4609.202056055

Kučera, J., & Podhrázská, J. (2016). *Analysis of Soil Aggregate Degradation in Heavy Soils Situated in Localities at Risk of Wind Erosion*. Conference MendelNet 2016, Brno, Chez Republic.

Kulshreshtha, S., & Rempel, J. (2014). Shelterbelts on Saskatchewan farms: an asset or nuisance. In S. Lac & M. McHenry (Ed.) Climate change and forest ecosystems (pp. 37–54). Nova Publishers.

Kumar, A., Devi, M., & Deshmukh, B. (2014). Integrated remote sensing and geographic information system based RUSLE Modelling for estimation of soil loss in Western Himalaya, India. *Water Resources Management*, *28*(10), 3307–3317. doi:10.100711269-014-0680-5

Kumar, P., Garg, V., Mittal, S., & Murthy, Y. K. (2021). GIS-based hazard and vulnerability assessment of a torrential watershed. *Environment, Development and Sustainability*, 1–31. doi:10.100710668-021-01476-z

Kumar, R., Kumar, S., Lohani, A. K., Nema, R. K., & Singh, R. D. (2000). Evaluation of Geomorphological Characteristics of a Catchment Using GIS. *GIS India*, *9*(3), 13–17.

Kusky, M. T. (2020). Encyclopedia of Earth and Space Science. Academic Press.

Kusumastuti, R. D., Husodo, Z. A., Suardi, L., & Danarsari, D. N. (2014). Developing a resilience index towards natural disasters in Indonesia. *International Journal of Disaster Risk Reduction*, *10*, 327–340. doi:10.1016/j.ijdrr.2014.10.007

Kwak, J., St-Hilaire, A., Chebana, F., & Kim, G. (2017). Summer season water temperature modeling under the climate change: Case study for fourchue river, Quebec, Canada. *Water (Basel)*, *9*(5), 346. doi:10.3390/w9050346

Kyhn, H. S. (2008). *Situational Crisis Communication Theory: Its Use in a Complex Crisis with Scandinavian Airlines Grounding of Dash 8-Q400 Airplane* (Master thesis). Airbus School of Business, Handelshojskolen. Retrieved from https://dl.icdst.org/pdfs/files/0e8d769c306ae51d9075fa87693ede99.pdf

Laity, J. E. (2016). Eolian Erosional Processes and Landforms. In The International Encyclopedia of Geography. John Wiley & Sons, Ltd.

Lal, R. (1994). *Soil Erosion Research Methods* (2nd ed.). Soil and Water Conservation Society. doi:10.1201/9780203739358

Lal, R. (2001). World cropland soils as a source or sink for atmospheric carbon. *Advances in Agronomy*, *71*, 145–191. doi:10.1016/S0065-2113(01)71014-0

Lal, R. (2003). Soil erosion and the global carbon budget. *Environment International*, *29*(4), 437–450. doi:10.1016/S0160-4120(02)00192-7 PMID:12705941

Lal, R. (2004a). Soil carbon sequestration impacts on global climate change and food security. *Science*, *304*(5677), 1623–1627. doi:10.1126cience.1097396 PMID:15192216

Lal, R. (2004b). Soil carbon sequestration to mitigate climate change. *Geoderma, 123*(1-2), 1–22. doi:10.1016/j.geoderma.2004.01.032

Lal, R. A. T. T. A. N. (2001). Soil degradation by erosion. *Land Degradation & Development, 12*(6), 519–539.

Lal, R., & Stewart, B. A. (1990). *Soil Degradation.* Springer-Verlag.

Land Degradation and Protection - Thematic collection. (2016). University of Belgrade, Faculty of Forestry.

Langdale, G. W., West, L. T., Bruce, R. R., Miller, W. P., & Thomas, A. W. (1992). Restoration of eroded soil with conservation tillage. *Soil Technology, 5*(1), 81–90.

Lastoria, B., Simonetti, M. R., Casaioli, M., Mariani, S., & Monacelli, G. (2006). Socio-economic impacts of major floods in Italy from 1951 to 2003. *Advances in Geosciences, 7*, 223–229.

Law of Water Protection, Official Gazette of the Brčko District of B&H, No. 1/05, 19/07 (2007), www.podaci.net/_gBiH/propis/Zakon_o_zastiti/Z-zvoda_05v0425-0719. html

Law on Crisis Management, Official Gazette of the Republic of Macedonia No. 29/05 and Amendments Official Gazette 36/11, 41/14, 104/15, (2015), https://www.refworld. org/cgi-in/texis/vtx/rwmain/opendocpdf.pdf?reldoc=y&docid=5d31a4344

Law on Disaster Risk Reduction and Emergency Management. Official Gazette of RS, No. 87/2018 (2018), https://www.paragraf.rs/propisi/zakon-o-smanjenju-rizika-od-katastrofa-i-upravljanju-vanrednim-situacijama.html

Law on Emergency Situations ("Official Gazette of the RS", No. 111/2009, 92/2011 and 93/2012)

Law on Forests, Official Gazette of the Republic of Macedonia No. 64/09, and amendments 24/11, 53/11, 25/13, 79/13, 147/13 and 43/14, (2014), http://extwprlegs1.fao.org/docs/pdf/mac167487.pdf

Law on Hydro meteorological Activity, Official Gazette of the Republic of Macedonia No. 103/08; 53/11; 51/15, (2015) https://dejure.mk/zakon/zakon-za-hidrometeo roloshka-dejnost

Law on Local Self-Government, Official Gazette of the Republic of Macedonia No. 5/2002, (2002), https://aa.mk/WBStorage/Files/Zakon_lokalnata_samouprava.pdf

Law on Meteorological and Hydrological Activities. Official Gazette of RS" No. 88/2010, (20110), https://www.pravno-informacioni-sistem.rs/SlGlasnikPortal/eli/rep/sgrs/ skupstina/zakon/2010/88/9/reg

Law on Protection and Rescue, Official Gazette of the Republic of Macedonia No. 36/04) and amendments 49/04, 86/08, 124/1018/11, 41/14, editorially revised texts in 2019, https://www.ifrc.org/Global/Publications/IDRL/DM%20acts/Macedonia.pdf

Law on Public Administration of the BD of B&H. Official Gazette of the Brčko District of B&H", No. 36/09, 60/10, 37/14, 13/16 i 14/19 (2019), https://skupstinabd.ba/3-zakon/ba/Zakon%20o%20Javnoj%20upravi%20Brc-ko%20Distrikta%20BiH/07B 25-20%20Zakon%20o%20javnoj%20upravi%20-%20prec--is--c-eni%20tekst.pdf

Law on the Government of the BD of B&H, Official Gazette of the Brčko District of B&H, No. 49/18, 8/19, 10/19 i 32/19 (2019), https://www.paragraf.ba/propisi/ brcko/zakon-o-vladi-brcko-distrikta-bosne-i-hercegovine.html

Law on the organization and work of the bodies of the state administration, Official Gazette of the Republic of Macedonia No. 58/2000) and amendments 44/02, 82/08, 167/10, 51/11, 96/19, 110/19) (2019), https://mioa.gov.mk/sites/default / files/pbl_files/documents/legislation/Zakon_za_organizacija_i_rabota_na_organite_na_drzavnata_uprava_konsolidiran_15102015.pdf

Law on water economy - (Official Gazette of the Republic of Macedonia No. 51/2015) and 193/15 (2015) https://vodostopanstvo.mk/Zakoni/MK/.51-15.pdf

Law on Water. Official Gazette of RS No. 95/2018 (2018), https://www.paragraf.rs/ propisi/zakon_o_vodama.html

Law on Waters ("Official Gazette of the RS", No. 30/2010, 93/2012, 101/2016, 95/2018 and 95/2018 - other law)

Law on Waters, Official Gazette of the Republic of Macedonia No. 87/08; and amendments 6/2009; 161/2009; 83/2010; 51/2011 44/2012, 23/2013, 163/2013, 180/2014, 52/16, (2016), https://www.moepp.gov.mk/wp-content/uploads/2014/10/LAW-ON-WATERS.pdf

Law on Waters. Official Gazette of FB&H, No. 70/06, (2006) https://regeneracija.ba/wp-content/uploads/2016/12/Zakon-o-vodama-22Slu% C5%BEbene-novine-FBiH22-br-7006.pdf

Law on Waters. Official Gazette of RS, number: 50/06, 92/09 and 121/12, (2012) http://www.voders.org/images/PDF/zakoni/Zakon121_12b.pdf

Lee, J. J., Phillips, D. L., & Benson, V. W. (1999). Soil erosion and climate change: Assessing potential impacts and adaptation practices. *Journal of Soil and Water Conservation*, *54*, 529–536.

Leitmann, J. (1999). *Sustaining cities: Environmental planning and management in urban design*. McGraw-Hill.

Letić, Lj., Savić, R., & Nikolić, V. (2009). *Wind Erosion Hazard in Western Bačka, Šumarstvo*. Academic Press.

Liddicoat, C., Hayman, P., Alexander, B., Rowland, J., Maschmedt, D., Young, M.A., Hall, J., Herrmann., T, & Sweeney, S. (2012). *Climate change, wheat production and erosion risk in South Australia's cropping zone: Linking crop simulation modelling to soil landscape mapping*. Government of South Australia, through Department of Environment, Water and Natural Resources.

Liere, K. D. V., & Dunlap, R. E. (1980). The social bases of environmental concern: A review of hypotheses, explanations and empirical evidence. *Public Opinion Quarterly*, *44*(2), 181–197. doi:10.1086/268583

Li, H., Tatarko, J., Kucharski, M., & Dong, Z. (2015). PM2.5 and PM10 emissions from agricultural soils by wind erosion. *Aeolian Research*, *19*, 171–182. doi:10.1016/j.aeolia.2015.02.003

Li, J., Ma, X., & Zhang, C. (2019). Predicting the spatiotemporal variation in soil wind erosion across Central Asia in response to climate change in the 21st century. *The Science of the Total Environment*, *709*, 136060. doi:10.1016/j.scitotenv.2019.136060 PMID:31905572

Li, M. M., Liu, A. T., Zou, C. J., Xu, W. D., Shimizu, H., & Wang, K. Y. (2012). An overview of the "Three-North" Shelterbelt project in China. *Forestry Studies in China*, *14*(1), 70–79. doi:10.100711632-012-0108-3

Li, S., Russell, A. R., & Muir Wood, D. (2020). Influence of particle-size distribution homogeneity on shearing of soils subjected to internal erosion. *Canadian Geotechnical Journal*, *57*(11), 1684–1694. doi:10.1139/cgj-2019-0273

Liste, M., Grifoll, M., & Monbaliu, J. (2014). River plume dispersion in response to flash flood events. Application to the Catalan shelf. *Continental Shelf Research*, *87*, 96–108.

Li, T. Y., Xiao, S. Y., Shen, C. C., Zhang, J., Chen, C. J., Cheng, H., ... Yu, T. L. (2021). Little Ice Age climate changes in Southwest China from a stalagmite δ18O record. *Palaeogeography, Palaeoclimatology, Palaeoecology*, *562*, 110167.

Li, T., Zhang, H., Wang, X., Cheng, S., Fang, H., Liu, G., & Yuan, W. (2019). Soil erosion affects variations of soil organic carbon and soil respiration along a slope in Northeast China. *Ecological Processes*, *8*(28), 1–10. doi:10.118613717-019-0184-6

Littlejohn, C. P., Hofmann, R. W., & Wratten, S. D. (2019). Delivery of multiple ecosystem services in pasture by shelter created from the hybrid sterile bioenergy grass Miscanthus x giganteus. *Scientific Reports*, *9*(1), 5575. doi:10.103841598-019-40696-2 PMID:30944349

Liu, D., Abuduwaili, J., Lei, J., Wu, G., & Gui, D. (2011). Wind erosion of saline playa sediments and its ecological effects in Ebinur Lake, Xinjiang, China. *Environmental Earth Sciences*, *63*(2), 241–250. doi:10.100712665-010-0690-4

Liu, Y., Amin, A., Rasool, S. F., & Zaman, Q. U. (2020). The role of agriculture and foreign remittances in mitigating rural poverty: Empirical evidence from Pakistan. *Risk Management and Healthcare Policy*, *13*, 13–26. doi:10.2147/RMHP.S235580 PMID:32021516

Li, W., Wang, F., & Bell, S. (2007). Simulating the sheltering effects of windbreaks in urban outdoor open space. *Journal of Wind Engineering and Industrial Aerodynamics*, *95*(7), 533–549. doi:10.1016/j.jweia.2006.11.001

Luino, F., Turconi, L., Paliaga, G., Faccini, F., & Marincioni, F. (2018). Torrential floods in the upper Soana Valley (NW Italian Alps): Geomorphological processes and risk-reduction strategies. *International Journal of Disaster Risk Reduction*, *27*, 343–354.

Lukić, S. (2006). *The effect of standard windbreaks and usual agricultural crops on wind velocity* (Master's thesis). University of Belgrade, Faculty of Forestry.

Lukić, S., & Dožić, S. (2006). Influence of black walnut belt along the road Pančevo - Kovin on the change of wind speed. Thematic collection of works. *Melioration - Land and Water - Novi Sad.*

Lukic, S., Belanovic Simic, S., Pantić, D., Beloica, J., Baumgertel, A., Miljkovic, P., Borota, D., & Kadović, R. (2018). Carbon storage in shelterbelts in the agroforestry systems of the Bačka Palanka area (Serbia). *AGROFOR*, *3*(2). Advance online publication. doi:10.7251/AGRENG1802080L

Luse, M. (2021). *Comparison of Modern and Roman Warm Period Seasonal Climate in Coastal North Carolina Using Isotope Sclerochronology of Archaeological Shells*. Academic Press.

Mabuhay, J. A., Nakagoshi, N., & Isagi, Y. (2004). Influence of erosion on soil microbial biomass, abundance and community diversity. *Land Degradation & Development*, *15*(2), 183–195. doi:10.1002/ldr.607

Mace, B. L., Bell, P. A., & Loomis, R. J. (1999). Aesthetic, affective, and cognitive effects of noise on natural landscape assessment. *Society & Natural Resources*, *12*(3), 225–242. doi:10.1080/089419299279713

Maleki, M., Ebrahimi, S., Asadzadeh, F., & Tabrizi, M. E. (2016). Performance of microbial-induced carbonate precipitation on wind erosion control of sandy soil. *International Journal of Environmental Science and Technology*, *13*(3), 937–944. doi:10.100713762-015-0921-z

Mandal, A., Sarkar, B., Mandal, S., Vithanage, M., Patra, A. K., & Manna, M. C. (2020). Impact of agrochemicals on soil health. In Agrochemicals Detection, Treatment and Remediation: Pesticides and Chemical Fertilizers (pp. 161-187). Elsevier.

Marin, G., Modica, M., Paleari, S., & Zoboli, R. (2021). Assessing disaster risk by integrating natural and socio-economic dimensions: A decision-support tool. *Socio-Economic Planning Sciences, 101032*. Advance online publication. doi:10.1016/j.seps.2021.101032

Markose, V. J., & Jayappa, K. S. (2016). Soil loss estimation and prioritization of sub-watersheds of Kali River basin, Karnataka, India, using RUSLE and GIS. *Environmental Monitoring and Assessment*, *188*(4), 225.

Marković, M. (2015). Adjustment of EU agricultural policy to the new environment. [Economics of Agriculture]. *Ekonomika Poljoprivrede*, *62*(4), 1031–1044. doi:10.5937/ekoPolj1504031M

Marković, M. (2018a). The economic dimension of sustainable agricultural development of Serbia. In D. Karabašević, S. Vukotić, & M. Maksimović (Eds.), *International Thematic Monograph - Thematic proceedings: Innovation as an initiator of the development "Innovations – Basis for development"* (pp. 410–425). Faculty of Applied Management, Economy and Finance.

Marković, M. (2018b). Organic agriculture as a form of sustainable agricultural development in Serbia. *Economics of Sustainable Development*, *2*(2), 29–36.

Marquer, L., Pomel, S., Abichou, A., Schulz, E., Kaniewski, D., & Van Campo, E. (2008). Late Holocene high resolution palaeoclimatic reconstruction inferred from Sebkha Mhabeul, southeast Tunisia. *Quaternary Research*, *70*(2), 240–250. doi:10.1016/j.yqres.2008.06.002

Marques, J. F. (1998). Costs of soil erosion due to their on-site and off-site effects in the area of agricultural production. *Revista Brasileira de Economia e Sociologia Rural*, *36*, 61–79.

Marshall, J. K. (1967). The effect of shelter on the productivity of grasslands and field crops. *Field Crop Abstracts*, *20*, 1–14.

Marx, S. K., McGowan, H. A., Kamber, B. S., Knight, J. M., Denholm, J., & Zawadzki, A. (2014). Unprecedented wind erosion and perturbation of surface geochemistry marks the Anthropocene in Australia. *Journal of Geophysical Research. Earth Surface*, *119*(1), 45–61. doi:10.1002/2013JF002948

Ma, X., Zhao, C., & Zhu, J. (2021). Aggravated risk of soil erosion with global warming–A global meta-analysis. *Catena*, *200*, 105129.

Mayrinck, C. M., Laroque, C. P., Amichev, B., & Van Rees, K. (2019). Above- and Below- Ground Carbon Sequestation in Shelterbelts Trees in Canada: A Review. *Forests*, *10*(10), 922. Advance online publication. doi:10.3390/f10100922

McRae, T., Smith, C. A. S., & Gregorich, L. J. (2000). *Environmental sustainability of Canadian agriculture. Report of the Agri-Environmental Indicators Project*. Agriculture and Agri-Food Canada.

Mechler, R. (2003). Natural Disaster Risk and Cost-Benefit Analysis. In A. Kreimer, M. Arnold, & A. Carlin (Eds.), *Building Safer Cities: The Future of Disaster Risk* (pp. 45–55). The World Bank Disaster Management Facility.

Meena, N. K., Gautam, R., Tiwari, P., & Sharma, P. (2017). Nutrient losses in soil due to erosion. *Journal of Pharmacognosy and Phytochemistry*, *SP1*, 1009–1011.

Melzner, F., Thomsen, J., Koeve, W., Oschlies, A., Gutowska, M. A., Bange, H. W., Hansen, H. P., & Kortzinger, A. (2013). Future ocean acidification will be amplified by hypoxia in coastal habitats. *Marine Biology*, *160*(8), 1875–1888. doi:10.100700227-012-1954-1

Menshov, O., Kruglov, O., Nazarok, P., Andreeva, O., & Kruhlov, B. (2020, September). Soil erosion magnetic measurements as a tool for landslides studies. In *Second EAGE Workshop on Assessment of Landslide Hazards and impact on communities* (Vol. 2020, No. 1, pp. 1-6). European Association of Geoscientists & Engineers. 10.3997/2214-4609.202055019

Merritt, D., & Wohl, E. E. (2002). Processes governing hydrochory along rivers: Hydraulics, hydrology, and dispersal phenology. *Ecological Applications*, *12*(4), 1071–1087. doi:10.1890/1051-0761(2002)012[1071:PGHARH]2.0.CO;2

Mezősi, G., Blanka, V., Bata, T., Ladányi, Z., Kemény, G., & Meyer, B. C. (2016). Assessment of future scenarios for wind erosion sensitivity changes based on ALADIN and REMO regional climate model simulation data. *Open Geosciences, 8*(1), 465-477.

Mezösi, G., Blanka, V., Bata, T., Kovacs, F., & Meyer, B. (2015). Estimation of regional differences in wind erosion-sensitivity in Hungary. *Natural Hazards and Earth System Sciences*, *15*(1), 97–107. doi:10.5194/nhess-15-97-2015

Milanović, S. (2017). Human Resource Management in the Conditions of Globalization. *TEME*, *XLI*(1), 297–299. doi:10.22190/TEME1701297M

Milanović, S., & Marjanović, I. (2019). Labour Market Efficiency - Case of EU Members and Candidate Countries. In B. Krstić (Ed.), *Enhancing Competitiveness of National Economies and Enterprises* (pp. 65–86). Faculty of Economics University of Niš.

Milanović, S., & Marjanović, I. (2020). Obezbeđenje potrebnih resursa. In D. Živković & I. Simić (Eds.), *Inovativno preduzetništvo* (pp. 251–298). Innovation Centre Universty of Niš.

Mileti, D. S. (1999). *Disaster by Design: A Reassessment of Natural Hazards in United States*. Joseph Henry Press.

Milligan, C., Gatrell, A., & Bingley, A. (2004). Cultivating health. Therapeutic landscapes and older people in Northern England. *Social Science & Medicine*, *58*(9), 1781–1793. doi:10.1016/S0277-9536(03)00397-6 PMID:14990378

Ministry of Agriculture, Forestry, and Water Management. (2014). *Workshop on Water-Food-Energy Ecosystems Nexus Assessment in the Sava, River Basin, Zagreb, Croatia*. https://unece.org/fileadmin/DAM/env/Water_Convention_Sava_River_Workshop/8.section_3_merita_borota.pdf

Ministry of Agriculture. (2006). *Framework Plan for the Development of Water Management of the Republic of Srpska*. Republic Water Directorate Bijeljina.

Mitasova, H., Barton, C. M., Ullah, I., Hofierka, J., & Harmon, R. S. (2013). GIS-based soil erosion modeling. In *Remote Sensing and GIScience in Geomorphology* (pp. 228–258). Elsevier Inc.

Mitroff, I. I. (1994). Crisis Management and environmentalism: A natural fit. *California Management Review*, *36*(2), 101–113. doi:10.2307/41165747

Mitroff, I., & Pearson, C. M. (1993). *Crisis Management: a Diagnostic Guide for Improving your organizations crisis-preparedness*. Jossey-Bass.

Mize, C. W., Brandle, J. R., Schoeneberger, M. M., & Bentrup, G. (2008). Ecological Development and function of Shelterbelts in Temperate North America. *Toward Agroforestry Design*, 27–54. doi:10.1007/978-1-4020-6572-9_3

Moe, T. L., Gehbauer, F., Sentz, S., & Mueller, M. (2007). Balanced Scorecard for Natural Disaster Management Projects. *Disaster Prevention and Management*, *16*(5), 785–806. doi:10.1108/09653560710837073

Mohammadi, M., Khaledi Darvishan, A. K., Spalevic, V., Dudic, B., & Billi, P. (2021). Analysis of the Impact of Land Use Changes on Soil Erosion Intensity and Sediment Yield Using the IntErO Model in the Talar Watershed of Iran. *Water (Basel)*, *13*(6), 881.

Monstad, T. H. (2003). *Issues and crisis management – same, same but different* (Master thesis). Uppsala University, Sweden. Retrieved from www.dis.uu.se/mkv/education/essays/Helauppsatser/2003/up2003 31.pdf

Montagnini, F., & Nair, P. K. R. (2004). Carbon sequestration: An underexploited environmental benefit of agroforestry systems. In New Vistas in Agroforestry. Springer. doi:10.1007/978-94-017-2424-1_20

Montanarella, L. (2007). Climate and land degradation. In M. V. K. Sivakumar & N. Ndiang'ui (Eds.), *Trends in land degradation in Europe* (pp. 83–104). Springer.

Montanarella, L., Pennock, D. J., McKenzie, N., Badraoui, M., Chude, V., Baptista, I., Mamo, T., Yemefack, M., Singh Aulakh, M., Yagi, K., Young Hong, S., Vijarnsorn, P., Zhang, G.-L., Arrouays, D., Black, H., Krasilnikov, P., Sobocká, J., Alegre, J., Henriquez, C. R., ... Vargas, R. (2016). World's soils are under threat. *Soil (Göttingen)*, *2*(1), 79–82. doi:10.5194oil-2-79-2016

Moynihan, K. P., & Vasconcelos, J. G. (2014). SWMM modeling of a rural watershed in the lower coastal plains of the United States. *J. Water Manag. Model*, 1-12.

Mullan, D. (2013). Managing soil erosion in Northern Ireland: A review of past and present approaches. *Agriculture*, *3*(4), 684–699.

Munir, B. A., Ahmad, S. R., & Hafeez, S. (2020). Integrated hazard modeling for simulating torrential stream response to flash flood events. *ISPRS International Journal of Geo-Information*, *9*(1), 1.

Munir, B. A., Ahmad, S. R., & Rehan, R. (2021). Torrential Flood Water Management: Rainwater Harvesting through Relation Based Dam Suitability Analysis and Quantification of Erosion Potential. *ISPRS International Journal of Geo-Information*, *10*(1), 27.

Mysiak, J., Testella, F., Bonaiuto, M., Carrus, G., De Dominicis, S., Ganucci Cancellieri, U., Firus, K., & Grifoni, P. (2013). Flood risk management in Italy: Challenges and opportunities for the implementation of the EU Floods Directive (2007/60/EC). *Natural Hazards and Earth System Sciences*, *13*, 2883–2890.

Nascimento, K. R. D. S., & Alencar, M. H. (2016). Management of risks in natural disasters: A systematic review of the literature on NATECH events. *Journal of Loss Prevention in the Process Industries*, *44*, 347–359. doi:10.1016/j.jlp.2016.10.003

National Action Plan to Combat Desertification, version *prepared for adopting by the Government.*

National Water Strategy, 2012-2042, Official Gazette of the Republic of Macedonia no.122/2012, (2012)

Nearing, M. A., Xie, Y., Liu, B., & Ye, Y. (2017). Natural and anthropogenic rates of soil erosion. *International Soil and Water Conservation Research*, *5*(2), 77–84.

Négyesi, G., Lóki, J., Buró, B., Bertalan-Balázs, B., & Pásztor, L. (2019). Wind erosion research in Hungary – past, present and future possibilities. Hungarian Geographical Bulletin, 68(3), 223-240.

Nenadović, S., Kljajević, L., Nenadović, M., Milanović, M., & Markovi, S. (2013). Physico-chemical soil analysis of Rudovci region. *Geonauka*, *1*(2), 1–8. doi:10.14438/gn.2013.07

Nielsen, U. N., Wall, D. H., & Six, J. (2015). Soil biodiversity and the environment. *Annual Review of Environment and Resources*, *40*, 63–90.

Nikolić Popadić, S. (2020). Legal aspects of flood risk management. *Proceedings of the Faculty of Law in Nis, 59*(86), 201-221

Norbiato, D., Borga, M., Esposti, S. D., Gaume, E., & Anquetin, S. (2008). Flash flood warning based on rainfall thresholds and soil moisture conditions: An assessment for gauged and ungauged basins. *Journal of Hydrology (Amsterdam)*, *362*(3-4), 274–290. doi:10.1016/j.jhydrol.2008.08.023

Nowak, D. J., Appleton, N., Ellis, A., & Greenfield, R. (2017). Residential building energy conservation and avoided power plant emissions by urban and community trees in the United States. *Urban Forestry & Urban Greening*, *21*, 158–165. doi:10.1016/j.ufug.2016.12.004

Oktay, D., Rüstemli, A., & Marans, R. W. (2009). Neighborhood satisfaction, sense of community, and attachment: Initial findings from Famagusta quality of urban life study. *ITU A/Z Journal, 6*(1), 6-20.

Okvirna vodoprivredna osnova BiH [Framework Water Management Foundation of BiH]. JVP Vodoprivreda BiH, Zavod za vodoprivredu Sarajevo, Sarajevo, 1994.

Okvirni plan razvoja vodoprivrede Republike Srpske [Framework Plan for Water Management Development in the Republika Srpska], Ministarstvo poljoprivrede, šumarstva i vodoprivrede, Republička direkcija za vode Bijeljina, 2006.

Okvirni plan razvoja vodoprivrede Republike Srpske [Framework Plan of Water Management Development in the Republika Srpska], Zavod za vodoprivredu, Bijeljina, 2006.

Okvirni zakon o zaštiti i spašavanju ljudi i materijalnih dobara od prirodnih ili drugih nesreća u BiH [Internal law on the protection and rescue of people and material goods from natural or other disasters in B&H], Ministarstvo sigurnosti Bosne i Hercegovine, 2008.

Ola, A., Dodd, I. C., & Quinton, J. N. (2015). Can we manipulate root system architecture to control soil erosion? *Soil (Göttingen)*, *1*(2), 603–612.

Oldeman, L. R., Hakkeling, R. A., & Sombroek, W. G. (1991). *Global Assessment of Soil Degradation (GLASOD): World Map of the Status of Human-Induced Soil Degradation.* United Nations Environment Programme, International Soil Reference and Information Centre.

Oldemann, L. R., Hakkeling, R. T. A., & Sombroek, W. G. (1990). World Map of Human-induced Soil Degradation: an Explanatory Note. Global Assessment of Soil Degradation (GLASOD), ISRIC and United Nations Environment Program (UNEP), FAO-ITC. International Soil Reference and Information Centre (ISRIC), Wageningen. United Nations Environment Programme (UNEP), in cooperation with Winand Staring Centre, Food and Agriculture Organization of the United Nations (FAO). International Institute for Aerospace Survey and Earth Science (ITC).

Oldfield, F. (2015). Can the magnetic signatures from inorganic fly ash be used to mark the onset of the Anthropocene? *The Anthropocene Review*, *2*(1), 3–13. doi:10.1177/2053019614534402

Olson, K. R., Al-Kaisi, M., Lal, R., & Cihacek, L. (2016). Impact of soil erosion on soil organic carbon stocks. *Journal of Soil and Water Conservation*, *71*(3), 61A–67A. doi:10.2489/jswc.71.3.61A

Orgiazzi, A., & Panagos, P. (2018). Soil biodiversity and soil erosion: It is time to get married. *Global Ecology and Biogeography*, *27*(10), 1155–1167. doi:10.1111/geb.12782

Osbaldiston, R., & Sheldon, K. M. (2003). Promoting internalized motivation for environmentally responsible behavior: A prospective study of environmental goals. *Journal of Environmental Psychology*, *23*(4), 349–357. doi:10.1016/S0272-4944(03)00035-5

Osman, K. T. (2014). *Soil Degradation, Conversation and Remediation.* Springer Science+Business Media.

Osorio, R., Barden, C., & Ciampitti, A. (2018). GIS approach to estimate windbreak crop effect in Kansas-Nebraska. *Agroforestry Systems*, *93*(4), 1567–1576. doi:10.100710457-018-0270-2

Owens, P. N. (2020). Soil erosion and sediment dynamics in the Anthropocene: A review of human impacts during a period of rapid global environmental change. *Journal of Soils and Sediments*, *20*(12), 1–29. doi:10.100711368-020-02815-9

Panagos, P., Borrelli, P., Poesen, J., Ballabio, C., Lugato, E., Meusburger, K., Montanarella, L., & Alewell, C. (2015). The new assessment of soil loss by water erosion in Europe Environ. *Environmental Science & Policy*, *54*, 438–447. doi:10.1016/j.envsci.2015.08.012

Panagos, P., Standardi, G., Borrelli, P., Lugato, E., Montanarella, L., & Bosello, F. (2018). Cost of agricultural productivity loss due to soil erosion in the European Union: From direct cost evaluation approaches to the use of macroeconomic models. *Land Degradation & Development*, *29*(3), 471–484. doi:10.1002/ldr.2879

Pandey, S., Kumar, P., Zlatic, M., Nautiyal, R., & Panwar, V. P. (2021). Recent advances in assessment of soil erosion vulnerability in a watershed. *International Soil and Water Conservation Research*, 1–14.

Pant, N., Dubey, R. K., Bhatt, A., Rai, S. P., Semwal, P., & Mishra, S. (2020). Soil erosion and flood hazard zonation using morphometric and morphotectonic parameters in Upper Alaknanda river basin. *Natural Hazards*, *103*(3), 3263–3301.

Parsons, A. J. (2019). How reliable are our methods for estimating soil erosion by water? *The Science of the Total Environment*, *676*, 215–221.

Parveen, R., Kumar, U., & Singh, V. K. (2012). Geomorphometric Characterization of Upper South Koel Basin, Jharkhand: A Remote Sensing & GIS Approach. *Journal of Water Resource and Protection*, *2012*(4), 1042–1050. doi:10.4236/jwarp.2012.412120

Pastuović, N., Kolesarić, V., & Križmanić, M. (1995). Psychological variables as predictors of quality of life. *Review of Psychology*, *2*(1-2), 49–62.

Pe'er, G., Bonn, A., Bruelheide, H., Dieker, P., Eisenhauer, N., Feindt, P. H., Hagedorn, G., Hansjürgens, B., Herzon, I., Lomba, Â., Marquard, E., Moreira, F., Nitsch, H., Oppermann, R., Perino, A., Röder, N., Schleyer, C., Schindler, S., Wolf, C., ... Lakner, S. (2020). Action needed for the EU Common Agricultural Policy to address sustainability challenges. *People and Nature*, *2*(2), 305–316. doi:10.1002/pan3.10080 PMID:32626843

Pearl, H., & Otten, T. (2013). Harmful Cyanobacterial Blooms: Causes, Consequences, and Controls. *Microbial Ecology*, *65*(4), 995–1010. Advance online publication. doi:10.100700248-012-0159-y PMID:23314096

Peltonen-Sainio, P., Jauhiainen, L., & Latukka, A. (2020). Interactive tool for farmers to diversify high-latitude cereal-dominated crop rotations. *International Journal of Agricultural Sustainability*, *18*(4), 319–333. doi:10.1080/14735903.2020.1775931

Pena, S. B., Abreu, M. M., Magalhães, M. R., & Cortez, N. (2020). Water erosion aspects of land degradation neutrality to landscape planning tools at national scale. *Geoderma*, *363*, 114093. doi:10.1016/j.geoderma.2019.114093

Peng, J., Bai, X., & Chen, X. (2021). Climate-driven soil erosion processes in alpine environments over the last century: Evidence from the Taibai Mountain (central China). *Catena*, *206*, 105569. doi:10.1016/j.catena.2021.105569

Perović, M. (2018). *Model procene ekonomskih i finansijskih efekata uticaja prirodnih katastrofa i mogućnosti njihovog ublažavanja u Srbiji*. Univerzitet Privredna akademija u Novom Sadu, Fakultet za ekonomiju i inženjerski menadžment.

Perović, V., Kadović, R., Đurđević, V., Pavlović, D., Pavlović, M., Čakmak, D., Mitrović, M., & Pavlović, P. (2021). Major drivers of land degradation risk in Western Serbia: Current trends and future scenarios. *Ecological Indicators*, *123*, 107377. doi:10.1016/j.ecolind.2021.107377

Petrović, G., & Grujović, M. (2015). The economic damage from natural disasters in Serbia and Sumadija district. *Ekonomski signali: poslovni magazin*, *10*(2), 99-107.

Petrović, A. M., Novković, I., & Kostadinov, S. (2021). Hydrological analysis of the September 2014 torrential floods of the Danube tributaries in the Eastern Serbia. *Natural Hazards*, 1–15.

Pheng, L., Ho, D., & Ann, Y. (1999). Crisis management: A survey of property development firms. *Property Management*, *17*(3), 231–251. doi:10.1108/02637479910286916

Pickett, S. T., Cadenasso, M. L., Grove, J. M., Nilon, C. H., Pouyat, R. V., Zipperer, W. C., & Costanza, R. (2001). Urban ecological systems: Linking terrestrial ecological, physical, and socioeconomic components of metropolitan areas. *Annual Review of Ecology and Systematics*, *32*(1), 127–157. doi:10.1146/annurev.ecolsys.32.081501.114012

Pimentel, D. (2006). Soil erosion: A food and environmental threat. *Environment, Development and Sustainability*, *8*(1), 119–137. doi:10.100710668-005-1262-8

Pimentel, D., Harvey, C., Resosudarmo, P., Sinclair, K., Kurz, D., McNair, M., ... Blair, R. (1995). Environmental and economic costs of soil erosion and conservation benefits. *Science*, *267*(5201), 1117–1123.

Plan for the proclamation of erosion areas on the territory of the municipality of Leskovac. (2007). Institute of Water Management "Jaroslav Cerni" AD Belgrade.

Plan upravljanja rizicima od poplava u slivu reke Save [Flood Risk Management Plan for the Sava River Basin], Investicijski okvir za Zapadni Balkan, Consultant: Eptisa Servicios de Ingeniería S.L. (Španija), 2018.

Podhrázská, J., Kučera, J., Doubrava, D., & Doležal, P. (2021). Functions of Windbreaks in the Landscape Ecological Network and Methods of Their Evaluation. *Forests*, *12*(1), 67. doi:10.3390/f12010067

Poesen, J. (2018). Soil erosion in the Anthropocene: Research needs. *Earth Surface Processes and Landforms*, *43*(1), 64–84. doi:10.1002/esp.4250

Poleto, C., & Beier, E. V. (2012). Siltation and erosion processes on a tributary of Lake Itaipu due a dam reservoir. *Lakes, Reservoirs and Ponds*, *6*(2), 108–119.

Polovina, S., Radic, B., Ristic, R., & Milcanovic, V. (2016). Spatial and temporal analysis of natural resources degradation in the Likodra river watershed. *Glasnik Sumarskog fakulteta*, *144*, 169-188.

Polovina, S., Radić, B., Ristić, R., Kovačević, J., Milčanović, V., & Živanović, N. (2021). Soil Erosion Assessment and Prediction in Urban Landscapes: A New G2 Model Approach. *Applied Sciences (Basel, Switzerland)*, *11*(9), 4154.

Popović, T., Đurđević, V., Živković, M., Jović, B., & Jovanović, M. (2009). Promena klime u Srbiji i očekivani uticaji. In Peta regionalna konferencija EnE09-Životna sredina ka Evropi, Ambasadori životne sredine i PKS (pp. 6-11). Agencija za zaštitu životne sredine.

Possu, B. (2015). *Carbon storage potential of windbreaks on agricultural lands of the continental United States* (Doctoral Dissertation). University of Nebraska-Lincoln.

Possu, W. B., Brandle, J. R., Domke, G. M., Schoenberger, M., & Blankenship, E. (2017). Estimating carbon storage in windbreak trees on U.S. agricultural lands. *Agroforestry Systems*, *90*(5), 889–904. doi:10.100710457-016-9896-0

Prăvălie, R., Patriche, C., Borrelli, P., Panagos, P., Roşca, B., Dumitraşcu, M., Nita, I.-A., Săvulescu, I., Birsan, M.-V., & Bandoc, G. (2021). Arable lands under the pressure of multiple land degradation processes. A global perspective. *Environmental Research*, *194*, 110697. doi:10.1016/j.envres.2020.110697 PMID:33428912

Press, F., & Siever, R. (1986). *Earth* (4th ed.). W.H. Freeman and Company.

Pribadi, D. O., Vollmer, D., & Pauleit, S. (2018). Impact of peri-urban agriculture on runoff and soil erosion in the rapidly developing metropolitan area of Jakarta, Indonesia. *Regional Environmental Change*, *18*(7), 2129–2143.

Priest, S. J., Suykens, C., Van Rijswick, H. F. M. W., Schellenberger, T., Goytia, S. B., Kundzewicz, Z. W., Van Doorn-Hoekveld, W. J., Beyers, J.-C., & Homewood, S. (2016). The European Union approach to flood risk management and improving societal resilience: Lessons from the implementation of the Floods Directive in six European countries. *Ecology and Society*, *21*(4), 50. https://doi. org/10.5751/ES-08913-210450

Prohaska, S. (2003). *Hydrology: Hydrometeorology, hydrometry and water regime. Part 1*. Faculty of Mining and Geology.

Prohaska, S. (2006). *Hydrology. Part II, Institute of Water Management "Jaroslav Cerni", Faculty of Mining and Geology, Belgrade*. Republic Hydrometeorological Institute.

Puigdefábregas, J. (2005). The role of vegetation patterns in structuring runoff and sediment fluxes in drylands. *Earth Surface Processes and Landforms: The Journal of the British Geomorphological Research Group*, *30*(2), 133–147.

Qin, B. Q., Gao, G., Zhu, G. W., Zhang, Y. L., Song, Y. Z., Tang, X. M., Xu, H., & Deng, J. M. (2013). Lake eutrophication and its ecosystem response. *Chinese Science Bulletin*, *58*(9), 961–970. doi:10.100711434-012-5560-x

Quinton, J. N., Edwards, G. M., & Morgan, R. P. C. (1997). The influence of vegetation species and plant properties on runoff and soil erosion: Results from a rainfall simulation study in south east Spain. *Soil Use and Management*, *13*(3), 143–148.

Quinton, J. N., Govers, G., Van Oost, K., & Bardgett, R. D. (2010). The impact of agricultural soil erosion on biogeochemical cycling. *Nature Geoscience*, *3*(5), 311–314.

Rađenović, Ž., Krstić, B., & Marković, M. (2020). Smart farming in agricultural industry: Mobile technology perspective. *Ekonomika Poljoprivrede*, *67*(3), 925–938. doi:10.5937/ekoPolj2003925R

Radukić, S., & Marković, M. (2015). Limitation of trade margins as a measure of food price controls: Experience of Serbia. *Ekonomika Poljoprivrede*, *62*(1), 193–205. doi:10.5937/ekoPolj1501193R

Ramachandran Nair, P. K., Mohan Kumar, B., & Nair, V. D. (2009). Agroforestry as a strategy for carbon sequestration. *Journal of Plant Nutrition and Soil Science*, *172*(1), 10–23. doi:10.1002/jpln.200800030

Ranke, U. (2016). *Natural Disaster Risk Management*. Springer. doi:10.1007/978-3-319-20675-2

Rao, C. S., Kareemulla, K., Krishnan, P., Murthy, G. R. K., Ramesh, P., Ananthan, P. S., & Joshi, P. K. (2019). Agro-ecosystem based sustainability indicators for climate resilient agriculture in India: A conceptual framework. *Ecological Indicators*, *105*, 621–633. doi:10.1016/j.ecolind.2018.06.038

Rastogi, K., Surekha, C., Nayar, S., & Gupta, T. (2021). Impact of Urban Environmental Quality, Residential Satisfaction, Personality on Quality of Life among Delhi Residents. *SPAST Abstracts, 1*(1).

Ratnam, N. K., Srivastava, Y. K., Rao, V. V., Amminedu, E., & Murthy, K. S. R. (2005). Checkdam positioning by prioritization micro-watersheds using SYI model and morphometric analysis – remote sensing and GIS perspective. *Photonirvachak (Dehra Dun), 33*(1), 25–38. doi:10.1007/BF02989988

Ray, S. J. (1999). *Strategic Communications in Crisis Management: Lessons from the Airline Industry*. Quorum Books.

Rehacek, D., Khel, T., Kucera, J., Vopravil, J., & Petera, M. (2017). Effect of Windbreaks on Wind Speed Reduction and Soil Protection against Wind Erosion. *Soil and Water Research*, *12*(2), 128–135. doi:10.17221/45/2016-SWR

Reid, W. V. (2005). Ecosystems and human well-being: a report on the conceptual framework working group of the Millenium Ecosystem Assessment. Academic Press.

Reimann, L., Vafeidis, A. T., Brown, S., Hinkel, J., & Tol, R. S. (2018). Mediterranean UNESCO World Heritage at risk from coastal flooding and erosion due to sea-level rise. *Nature Communications*, *9*(1), 1–11.

Renner, A., Cadillo-Benalcazar, J. J., Benini, L., & Giampietro, M. (2020). Environmental pressure of the European agricultural system: Anticipating the biophysical consequences of internalization. *Ecosystem Services*, *46*, 101195. doi:10.1016/j.ecoser.2020.101195

Renschler, C. S. (2003). Designing geo-spatial interfaces to scale process models: The GeoWEPP approach. *Hydrological Processes*, *17*(5), 1005–1017.

Rhoades, J. L., Demyan, M. S., & Orr, B. (2011). Impacts of Deforestation and Land Cover Change on Mountain Soils in Hrazdan, Armenia. *Journal of Sustainable Forestry*, *30*(7), 677–696. doi:10.1080/10549811.2011.567965

Riksen, M. J. P. M., & De Graaff, J. (2001). On-site and off-site effects of wind erosion on European light soils. *Land Degradation & Development*, *12*(1), 1–11. doi:10.1002/ldr.423

Rimac, N., Bijedić, A., & Babić, A. (2019). Prvi operativni evropski sistem upozorenja na poplave [First Operative European Flood Warning System]. *Proceedings, Second Water Congress.*

Rineau, F., Malina, R., Beenaerts, N., Arnauts, N., Bardgett, R. D., Berg, M. P., & Vangronsveld, J. (2019). Towards more predictive and interdisciplinary climate change ecosystem experiments. *Nature Climate Change*, *9*(11), 809–816.

Ristić, R., & Macan, G. (1997). The Impact of erosion control measures on runoff processes. IAHS Publ.

Ristić, R., & Macan, G. (2002). Istraživanje procesa intercepcije u bukovo-jelovoj sastojini na planini Goč. *Glasnik Šumarskog fakulteta, 86*, 181-188. doi:10.2298/GSF0286181R

Ristić, R., Kapović Solomun, M., Malušević, I., Ždrale, S., Radić, B., Polovina, S., & Milčanović, V. (2021). Healthy Soils-Healthy People; Soil and Human Health. In The Reality of the Balkan Region (pp. 223-248). CRC Press, Taylor & Francis Group.

Ristić, R., Kostadinov, S., Abolmasov, B., Dragićević, S., Trivan, G., Radić, B., Trifunović, M., & Radosavljević, Z. (2012a). Torrential floods and town and country planning in Serbia. *Natural Hazards and Earth System Sciences, 12*(1), 23-35. doi:10.5194/nhess-12-23-2012

Ristić, R., Kostadinov, S., Radić, B., Trivan, G., & Nikić, Z. (2012b). Torrential Floods in Serbia – Man Made and Natural Hazards. 12th Congress INTERPRAEVENT 2012 Proceedings, 771-779.

Ristić, R., Malušević, I., Milčanović, V., Polovina, S., & Radić, B. (2019). Prevencija bujičnih poplava u kontekstu tekućih klimatskih promena [Torrent Flood Prevention in the Context of Current Climate Change]. In Scientific conference: "Future of Serbian Agriculture and Forestry", (pp. 113-121). AINS.

Ristić, R., Polovina, S., Malušević, I., & Milčanović, V. (2017). Srbija i bujične poplave: tri godine posle katastrofe u maju 2014. godine [Serbia and Torrent Floods: Three Years After the May 2014 Disaster]. *Vodoprivreda, 288-290*, 245-252. Available at: https://www.voda.ba/sistem-predvidjanja-poplava-i-ranog-upozoravanja-na-slivu-rijeke-save-sava-ffws

Ristić, R., Radivojević, S., Radić, B., Vasiljević, N., & Bjedov, I. (2010). Restoration of eroded surfaces in ski resorts of Serbia. *Global change: Challenges for soil management*, 165-174.

Ristić, R. (2014). Redefinisanje politike prevencije bujičnih poplava u Srbiji [Redefining the Policy of Torrential Flood Prevention in Serbia]. *Polis*, *8*, 23–27.

Ristić, R., Kostadinov, S., Abolmasov, B., Dragićević, S., Trivan, G., Radić, B., ... Radosavljević, Z. (2012). Torrential floods and town and country planning in Serbia. *Natural Hazards and Earth System Sciences*, *12*(1), 23–35.

Ristić, R., Kostadinov, S., Radić, B., Trivan, G., & Nikić, Z. (2012). *Torrential floods in Serbia–man made and natural hazards. 12th congress interpraevent.*

Ritchie, H., & Roser, M. (2014). *Natural Disasters.* https://ourworldindata.org/natural-disasters

Rodbell, D. T., Seltzer, G. O., Anderson, D. M., Abbott, M. B., Enfield, D. B., & Newman, J. H. (1999). An 15,000-year record of El Nino-driven alluviation in southwestern Ecuador. *Science*, *283*(5401), 516–520. doi:10.1126cience.283.5401.516 PMID:9915694

Rosier, I., Diels, J., Somers, B., & Van Orshoven, J. (2021). A Workflow to Extract the Geometry and Type of Vegetated Landscape Elements from Airborne LiDAR Point Clouds. *Remote Sensing*, *13*(20), 4031. doi:10.3390/rs13204031

Rudd, L. (2020). *Carbon cycle assessment of shelterbelts in Saskatchewan* (Master's dissertation). University of Saskatchewan.

Russo Lopes, G., Bastos Lima, M. G., & dos Reis, T. N. (2021). Maldevelopment revisited: Inclusiveness and social impacts of soy expansion over Brazil's Cerrado in Matopiba. *World Development, 139*(C), 105316. doi:10.1016/j.worlddev.2020.105316

Sabiha, N. E., Salim, R., Rahman, S., & Rola-Rubzen, M. F. (2016). Measuring environmental sustainability in agriculture: A composite environmental impact index approach. *Journal of Environmental Management, 166*, 84–93. doi:10.1016/j.jenvman.2015.10.003 PMID:26492465

Salama, W., Anand, R., & Bohu, T. (2019). Biochemical and mechanical dispersion mechanisms of Au and As in areas covered by Permian glacial sediments and aeolian sand. *ASEG Extended Abstracts, 2019*(1), 1–4. doi:10.1080/22020586.2019.12072948

Salvati, L., & Zitti, M. (2009). The environmental "risky" region: Identifying land degradation processes through integration of socio-economic and ecological indicators in a multivariate regionalization model. *Environmental Management, 44*(5), 888–898. doi:10.100700267-009-9378-5 PMID:19787396

Salvati, P., Bianchi, C., Rossi, M., & Guzzetti, F. (2010). Societal landslide and flood risk in Italy. *Natural Hazards and Earth System Sciences, 10*, 465–483.

Sang, G. S., Gan, Z. M., & Yue, D. P. (2002). Landform Evolution and Soil Erosion on Zhouyuan During Human History Period. *Journal of Mountain Research*, 6.

Santos, E. B. D., Nogueira, F. M., & Talgatti, D. M. (2021). Plant Species Composition and the Perception of the Afforestation in Urban Public Green Spaces in a Municipality in Eastern Brazilian Amazon. *Sustainability, 13*(18), 1–15. doi:10.3390u131810332

Santra, P., Moharana, P. C., Kumar, M., Soni, M. L., Pandey, C. B., Chaudhari, S. K., & Sikka, A. K. (2017). Crop production and economic loss due to wind erosion in hot arid ecosystem of India. *Aeolian Research, 28*, 71–82. doi:10.1016/j.aeolia.2017.07.009

Sapkota, Y., & White, J. R. (2021). Long-term fate of rapidly eroding carbon stock soil profiles in coastal wetlands. *The Science of the Total Environment, 753*, 141913. doi:10.1016/j.scitotenv.2020.141913 PMID:32906042

Sayl, K. N., Muhammad, N. S., Yaseen, Z. M., & El-shafie, A. (2016). Estimation the physical variables of rainwater harvesting system using integrated GIS-based remote sensing approach. *Water Resources Management, 30*(9), 3299–3313.

Schaub, S., Buchmann, N., Lüscher, A., & Finger, R. (2020). Economic benefits from plant species diversity in intensively managed grasslands. *Ecological Economics, 168*, 106488. doi:10.1016/j.ecolecon.2019.106488

Schiavon, E., Taramelli, A., Tornato, A., & Pierangeli, F. (2021). Monitoring environmental and climate goals for European agriculture: User perspectives on the optimization of the Copernicus evolution offer. *Journal of Environmental Management, 296*, 113121. doi:10.1016/j.jenvman.2021.113121 PMID:34217939

Schmidt, S., Alewell, C., Panagos, P., & Meusburger, K. (2016). Regionalization of monthly rainfall erosivity patterns in Switzerland. *Hydrology and Earth System Sciences, 20*(10), 4359–4373.

Schmidt, S., Meusburger, K., de Figueiredo, T., & Alewell, C. (2017). Modelling Hot Spots of Soil Loss by Wind Erosion (SoLoWind) in Western Saxony, Germany. *Land Degradation & Development, 28*(3), 1100–1112. doi:10.1002/ldr.2652

Scholes, R. J., Montanarella, L., Brainich, E., Barger, N., Ten Brink, B., Cantele, M., Erasmus, B., Fisher, J., Gardner, T., Holland, T. G., Kohler, F., Kotiaho, J. S., Von Maltitz, G., Nangendo, G., Pandit, R., Parrotta, J., Potts, M. D., Prince, S., Sankaran, M., & Willemen, L. (2018). *IPBES (2018): Summary for policymakers of the assessment report on land degradation and restoration of the Intergovernmental Science-Policy Platform on Biodiversity and Ecosystem Services.* Academic Press.

Schumm, S. A. (1956). Evolution of Drainage Systems and Slopes in Badlands at Perth Amboy, New Jersey. *Geological Society of America Bulletin*, *6*(5), 597–646. doi:10.1130/0016-7606(1956)67[597:EODSAS]2.0.CO;2

Seferagić, D., & Popovski, V. (1989). Kvaliteta života kao cilj društvenog razvoja. *Revija za Sociologiju*, *20*(1-2), 73–87.

Sekularac, I. (2021). *More garbage than water: Serbia promises clean-up of hydro reservoir.* Reuters https://www.reuters.com/article/us-environment-serbia-dam-idUSKBN29B1D7

Serbian National Strategy for Protection and Rescue in Emergency Situations. Official Gazette of the Republic of Serbia, No 86/2011.

Shahanov, V. M., & Cirella, G. T. (2021). Shelterbelt Planning in Agriculture: Application from Bulgaria. In G. T. Cirella (Ed.), *Human Settlements. Advances in 21st Century Human Settlements.* Springer. doi:10.1007/978-981-16-4031-5_8

Sharpley, A. N., Smith, S. I., Iones, O. R., Berg, W. A., & Coleman, G. A. (1992). The transport of bioavailable phosphorus in agricultural runoff. *Journal of Environmental Quality*, *21*(1), 30–35. doi:10.2134/jeq1992.00472425002100010003x

Sharpley, A. N., Weld, J. L., Beegle, D. B., Kleinman, P. J. A., Gburek, W. J., Moore, P. A., & Mullins, G. (2003). Development of phosphorus indices for nutrient management planning strategies in the United States. *Journal of Soil and Water Conservation*, *58*(3), 137–152.

Sharratt, B. S., Tatarko, J., Abatzoglou, J. T., Fox, F. A., & Huggins, D. (2015). Implications of climate change on wind erosion of agricultural lands in the Columbia plateau. *Weather and Climate Extremes, 10*(A), 20-31. doi:10.1016/j.wace.2015.06.001

Sharratt, B., Vaddella, V. K., & Feng, G. (2013). Threshold friction velocity influenced by wetness of soils within the Columbia Plateau. *Aeolian Research*, *9*, 175–182. doi:10.1016/j.aeolia.2013.01.002

Sherwood, S., & Uphoff, N. (2000). Soil health: Research, practice and policy for a more regenerative agriculture. *Applied Soil Ecology*, *15*(1), 85–97. doi:10.1016/S0929-1393(00)00074-3

Shikangalah, R. N., Jeltsch, F., Blaum, N., & Mueller, E. N. (2016). A review on urban soil water erosion. *Journal for Studies in Humanities and Social Sciences*, 163-178.

Shi, P., Yan, P., Yuan, Y., & Nearing, M. A. (2004). Wind erosion research in China: Past, present and future. *Progress in Physical Geography*, *28*(3), 366–386. doi:10.1191/0309133304pp416ra

Shit, P. K., Pourghasemi, H. R., & Bhunia, G. S. (Eds.). (2020). *Gully erosion studies from India and surrounding regions.* Springer International Publishing.

Shucksmith, M., Cameron, S., Merridew, T., & Pichler, F. (2009). Urban–rural differences in quality of life across the European Union. *Regional Studies*, *43*(10), 1275–1289. doi:10.1080/00343400802378750

Shuster, W., Subler, S., & McCoy, E. (2001). Deep-burrowing earthworm additions changed the distribution of soil organic carbon in a chisel-tilled soil. *Soil Biology & Biochemistry*, *33*(7-8), 983–996. doi:10.1016/S0038-0717(01)00002-5

Simić, T. (2019). Uticaj klimatskih promena na ljudsku bezbednost u Republici Srbiji. In Ljudska bezbednost - Zbornik studentskih radova (pp. 33-43). Centar za istraživanje javnih politika.

Sirgy, M. J. (1986). A quality-of-life theory derived from Maslow's developmental perspective: 'Quality' is related to progressive satisfaction of a hierarchy of needs, lower order and higher. *American Journal of Economics and Sociology*, *45*(3), 329–342. doi:10.1111/j.1536-7150.1986.tb02394.x

Slavuj, L. (2012). Objective and subjective indicators in the research of the quality of life concept. *Geoadria*, *17*(1), 73–92. doi:10.15291/geoadria.238

Smith, K., & Petley, D. N. (2009). *Environmental hazards: Assessing risk and reducing disaster.* Routledge. doi:10.4324/9780203884805

Smith, M., Wallace, K., Lewis, L., & Wagner, C. (2015). A structured elicitation method to identify key direct risk factors for the management of natural resources. *Heliyon*, *1*(3), e00043. doi:10.1016/j.heliyon.2015.e00043 PMID:27441228

Smolka, A. (2006). Natural disasters and the challenge of extreme events: Risk management from an insurance perspective. *Philosophical Transactions of the Royal Society A*, *364*(1845), 2147–2165. doi:10.1098/rsta.2006.1818 PMID:16844653

Sokolović, D., Pičman, D., Lojo, A., Gurda, S., Bajrić, M., & Koljić, H. (2013). Određivanje optimalnog prostornog rasporeda mreže sekundarnih šumskih prometnica [Determining the optimal spatial distribution of the network of secondary forest roads]. *Sumarski List*, *137*(1-2), 7–22.

Sornette, D. (2004). *Why stock markets crash: Critical events in complex financial systems.* Princeton University Press.

Sposito, G. (2008). *The Chemistry of Soils* (2nd ed.). Oxford University Press Inc.

Stanković, S., Vasović, D., & Trajković, S. (2019). Model of sustainable water resources management in the conditions of extreme hydrological phenomena. *Journal of Environmental Protection and Ecology, 20*(3), 1393-1401. Available at: https://www.itu.int/en/ITU-D/Emergency-Telecommunications/Pages/Multi-hazard-early-warning-systems.aspx

Stanković, J. J., Janković-Milić, V., Marjanović, I., & Janjić, J. (2021a). An integrated approach of PCA and PROMETHEE in spatial assessment of circular economy indicators. *Waste Management*, *128*, 154–166. doi:10.1016/j.wasman.2021.04.057 PMID:33991754

Stanković, J. J., Marjanović, I., Papathanasiou, J., & Drezgić, S. (2021b). Social, Economic and Environmental Sustainability of Port Regions: MCDM Approach in Composite Index Creation. *Journal of Marine Science and Engineering*, *9*(1), 74. doi:10.3390/jmse9010074

Stanković, J. J., Marjanović, I., & Stojković, N. (2021c). DEA Assessment of Socio-economic Development of European Countries. *Management Journal of Sustainable Business & Management Solutions in Emerging Economies*, *26*(1).

Statute of the BD of B&H, Official Gazette of the Brčko District of B&H", No. 39/09, (2009), https://skupstinabd.ba/images/dokumenti/ba/statut-brcko-distrikta.pdf

Stefanidis, S., & Stathis, D. (2018). Effect of climate change on soil erosion in a mountainous mediterranean catchment (Central Pindus, Greece). *Water (Basel)*, *10*(10), 1469.

Stefanović, M., Gavrilović, Z., & Bajčetić, R. (2014). Local community and the issue of torrential floods. Organization for European Security and Co-operation, Mission to Serbia.

Steg, L., Abrahamse, W., Vlek, C., Wiersma, G., Noorman, K. J., Oostra, M., Van Kampen, P., Smeets, P., Graas, S., Grimberg, H., & Roos, J. (2002). *Energiebesparing door gedragsverandering. De invloed van informatie, feedback en commitment op energiegebruik, psychologische achtergronden en kwaliteit van leven (Reducing energy use via behavioural change).* University of Groningen, Department of Psychology.

Sterk, G. (2002). Causes, consequences and control of wind erosion in Sahelian Africa: A review. *Land Degradation & Development, 14*(1), 95–108. doi:10.1002/ldr.526

Stojanović, N., Ocokoljić, M., Anastasijević, V., Ralević, N., Veselinović, M., & Vasiljević, N. (2016). The Impact of Roadside Green Spaces on Ecological Conditions in the Urban Environment. *Poljoprivreda i Sumarstvo, 62*(3), 37. doi:10.17707/AgricultForest.62.3.02

Strahler, A. N. (1964). Quantitative Geomorphology of Drainage Basin and Channel Networks. McGraw Hill.

Strahler, A. N. (1957). Quantitative analysis of watershed geomorphology. *American Geophysical Union Transactions, 38*(6), 913–920. doi:10.1029/TR038i006p00913

Štrbac, N., Vuković, M., Voza, D., & Sokić, M. (2012). Održivi razvoj i zaštita životne sredine. *Zbornik radova: Reciklaža i održivi razvoj*, 18 – 29.

Stredova, H., Spacilova, B., Podhrazska, J., & Chuchma, F. (2015). A universal meterorologicak method to identify potential risk of wind erosion on heavy textured soils. Moravian Geographical Report 23(2), Czech Academy of Sciences, Institute of Geonics.

Stredova, H., Podhrazska, J., Litschmann, T., Streda, T., & Roznovsky, J. (2012). Aerodynamic parameters of windbreak based on its optical porosity. *Contributions to Geophysics and Geodesy, 42*(3), 213–226. doi:10.2478/v10126-012-0008-5

Štreimikienė, D. (2015). Environmental indicators for the assessment of quality of life. *Intelektinė Ekonomika, 9*(1), 67–79. doi:10.1016/j.intele.2015.10.001

Sun, R., Wang, D., Wang, Y., Zhang, L., & Gu, Y. (2020). Study on durability against dry-wet cycles and chloride ion erosion of concrete revetment materials at the water-level-fluctuations zone in yellow river delta wetlands. *Wetlands, 40*(6), 2713–2727. doi:10.100713157-020-01326-0

Swap, R., Garstang, M., Greco, S., Talbot, R., & Kållberg, P. (1992). Saharan dust in the Amazon Basin. *Tellus. Series B, Chemical and Physical Meteorology, 44*(2), 133–149. doi:10.3402/tellusb.v44i2.15434

Szigeti, N., Frank, N., & Vityi, A. (2020). The Multifunctional Role of Shelterbelts in Intensively Managed Agricultural Land – Silvoarable Agroforestry in Hungary. *Acta Silvatica et Lignaria Hungarica, 16*(1), 19-38.

Szlafsztein, C. F. (2015). Management of natural disasters in the Brazilian Amazon region. *Natural Hazards, 76*(3), 1745–1757. doi:10.100711069-014-1567-8

Tagorti, M. A. (1995). *Rapport d'étude bibliographique sur le secteur de la sebkha de Sidi El Hani*. Academic Press.

Talukder, B., Hipel, K. W., & vanLoon, G. W. (2018). Using multi-criteria decision analysis for assessing sustainability of agricultural systems. *Sustainable Development, 26*(6), 781–799. doi:10.1002d.1848

Tanaskovic, V. (2018). *Making the Sahara Desert Green - Sustainable transport of bottom lake sediments via pipelines to produce new agriculture land from desert* [PhD thesis]. Technical University of Vienna.

Tang, C. S., Yin, L., Jiang, N., Zhu, C., Zeng, H., Li, H., & Shi, B. (2020). Factors affecting the performance of microbial-induced carbonate precipitation (MICP) treated soil: A review. *Environmental Earth Sciences, 79*(5), 94. doi:10.100712665-020-8840-9

Tanović, M., Milanović, S., & Rađenović, Ž. (2020). Upravljanje preduzećem. In D. Živković & I. Simić (Eds.), *Inovativno preduzetništvo* (pp. 299–332). Innovation Centre Universty of Niš.

Telles, T. S., de Fátima Guimarães, M., & Dechen, S. C. F. (2011). The costs of soil erosion. *Revista Brasileira de Ciência do Solo, 35*(2), 287–298. Advance online publication. doi:10.1590/S0100-06832011000200001

Telles, T. S., Dechen, S. C. F., de Souza, L. G. A., & de Fátima Guimarães, M. (2013). Valuation and assessment of soil erosion costs. *Scientia Agrícola*, *70*(3), 209–216. doi:10.1590/S0103-90162013000300010

Thaler, T., Attems, M. S., Bonnefond, M., Clarke, D., Gatien-Tournat, A., Gralepois, M., Fournier, M., Murphy, C., Rauter, M., Papathoma-Köhle, M., Servain, S., & Fuchs, S. (2019). Drivers and barriers of adaptation initiatives–How societal transformation affects natural hazard management and risk mitigation in Europe. *The Science of the Total Environment*, *650*, 1073–1082. doi:10.1016/j.scitotenv.2018.08.306 PMID:30308796

The Intergovernmental Panel on Climate Change (IPCC). (n.d.). https://www.ipcc.ch/

Thevs, N., Strenge, E., Aliev, K., Eraaliev, M., Lang, P., Baibagysov, A., & Xu, J. (2017). Tree Shelterbelts as an Element to Improve Water Resource Management in Central Asia. *Water (Basel)*, *9*(11), 842. doi:10.3390/w9110842

Thogersen, J. (2006). Norms for environmentally responsible behaviour: An extended taxonomy. *Journal of Environmental Psychology*, *26*(4), 247–261. doi:10.1016/j.jenvp.2006.09.004

Thuyet, D. V., Van Do, T., Sato, T., & Hung, T. T. (2014). Effects of species and shelterbelt structure on wind speed reduction in shelter. *Agroforestry Systems*, *88*(2), 237–244. doi:10.100710457-013-9671-4

Tiranti, D., Cremonini, R., Marco, F., Gaeta, A. R., & Barbero, S. (2014). The DEFENSE (debris Flows triggEred by storms–nowcasting system): An early warning system for torrential processes by radar storm tracking using a Geographic Information System (GIS). *Computers & Geosciences*, *70*, 96–109.

Todorovski, L., & Džeroski, S. (2006). Integrating knowledge driven and data-driven approaches to modelling. *Ecological Modelling*, *194*(1), 3–13. doi:10.1016/j.ecolmodel.2005.10.001

Toimil, A., Camus, P., Losada, I. J., Le Cozannet, G., Nicholls, R. J., Idier, D., & Maspataud, A. (2020). Climate change-driven coastal erosion modelling in temperate sandy beaches: Methods and uncertainty treatment. *Earth-Science Reviews*, *202*, 103110. doi:10.1016/j.earscirev.2020.103110

Tóth, G., Hermann, T., Ravina da Silva, M., & Montanarella, L. (2018). Monitoring soil for sustainable development and land degradation neutrality. *Environmental Monitoring and Assessment*, *190*(57), 57. Advance online publication. doi:10.100710661-017-6415-3 PMID:29302746

Trivino-Tarradas, P., Gomez-Ariza, M. R., Basch, G., & Gonzalez-Sanchez, E. J. (2019). Sustainability assessment of annual and permanent crops: The Inspia model. *Sustainability*, *11*(3), 738. doi:10.3390u11030738

Tselios, V., & Tompkins, E. L. (2020). Can we prevent disasters using socioeconomic and political policy tools? *International Journal of Disaster Risk Reduction*, *51*, 101764. doi:10.1016/j.ijdrr.2020.101764

Turek, A. (2013). Sustainable Agriculture: Between Sustainable Development and Economic Competitiveness. In A. Jean-Vasile, T. Adrian, J. Subic, & D. Dusmanescu (Eds.), *Sustainable Technologies, Policies, and Constraints in the Green Economy* (pp. 219–235). IGI Global. doi:10.4018/978-1-4666-4098-6.ch012

Tzouramani, I., Mantziaris, S., & Karanikolas, P. (2020). Assessing sustainability performance at the farm level: Examples from greek agricultural systems. *Sustainability*, *12*(7), 2929. doi:10.3390u12072929

UN. (2013). *Draft Resolution Submitted by the Vice-Chair of the Committee, Ms. Farrah Brown (Jamaica), on the Basis of Informal Consultations on Draft Resolution A/C.2/68/L.21*. United Nations General Assembly. https://www.fao.org/fileadmin/user_upload/GSP/docs/iys/World_Soil_Day_and_International_Year_of_Soils__UNGA_Resolution_Dec._2013.pdf

UN. (2015). Transforming our world: the 2030 Agenda for Sustainable Development. Resolution adopted by the General Assembly on 25 September 2015. United Nations.

UNCCD. (1994). United Nations Convention to Combat Desertification. A/AC.214/27. Paris, October 1994.

UNCCD. (2016). *Scaling up Land Degradation Neutrality Target Setting. From Lessons to Actions: 14 Pilot Countries Experiences*. Available at: https://www.unccd.int/publications/scaling-land-degradation-neutrality-target-setting-lessons-actions-14-pilot-countries

UNDRR. (2020). *The Human Cost of Disasters 2000-2019*. Retrieved from https://www.preventionweb.net/files/74124_humancostofdisasters20002019reportu.pdf

UNISDR. (2004). *Living with risk - A global review of disaster reduction initiatives. Volume II*. United Nations Office for Disaster Risk Reduction. Retrieved from https://www.undrr.org/publication/living-risk-global-review-disaster-reduction-initiatives

UNISDR. (2015). *The Human Cost of Weather-Related Disasters 1995–2015*. CRED & UNISDR. Available online: https://reliefweb.int/sites/reliefweb.int/files/resources/COP21_WeatherDisastersReport_2015_FINAL.pdf

United Nations General Assembly. (2015). *The Sendai Framework for Disaster Risk Reduction 2015–2030* (Assembly Resolution A/Res/69/283). Author.

United Nations International Strategy for Disaster Reduction. (2009). *UNISDR terminology on disaster risk reduction*. https://www.unisdr.org/files/7817_UNISDRTerminologyEnglish.pdf

United Nations Treaty Collection. United Nations Framework Convention on Climate Change (1992). 1771, p. 107, https://treaties.un.org/pages/ViewDetailsIII.aspx?src=TREATY&mtdsg_no=XXVII-7&chapter=27&Temp=mtdsg3&clang=_en

United Nations, Climate Actions Division, Early Warning Systems. (n.d.). Available at: https://www.un.org/en/climatechange/climate-solutions/early-warning-systems

Upravljanje krizom – primjer poplava iz 2014. godine [Crisis management – the case of 2014. floods], Parlamentarna skupština Bosne I Hercegovine, Sekretarijat, Istraživački sektor, Sarajevo, 2015.

Vaezi, A. R., Abbasi, M., Keesstra, S., & Cerdà, A. (2017). Assessment of soil particle erodibility and sediment trapping using check dams in small semi-arid catchments. *Catena*, *157*, 227–240.

Van der Heuvel, H. M. (1992). The quality of the daily environment: Ways of cooperation. *Urban Ecology: Seventh Conference on Urban and Regional Research. Ankara: General Directorate of Technical Research and Implementation.*

Van Zonneveld, M., Turmel, M. S., & Hellin, J. (2020). Decision-making to diversify farm systems for climate change adaptation. *Frontiers in Sustainable Food Systems*, *4*, 32. doi:10.3389/fsufs.2020.00032

Vanwalleghem, T., Gómez, J. A., Infante Amate, J., González de Molina, M., Vanderlinden, K., Guzmán, G., Laguna, A., & Giráldez, J. V. (2017). Impact of historical land use and soil management change on soil erosion and agricultural sustainability during the Anthropocene. *Anthropocene*, *17*, 13–29. doi:10.1016/j.ancene.2017.01.002

Vasović, D., Janaćković, G., Malenović Nikolić, J., Milošević, L., & Mušicki, S. (2018b). Promoting reflective practice in resources protection area: a step to forecast outcomes in uncertainty. *Journal of Environmental Protection and Ecology, 19*(3), 1320-1329.

Vasović, D., Janaćković, G., Malenović Nikolić, J., Mušicki, S., & Marković, S. (2018a). Multimodality in the field of resources protection. *Journal of Environmental Protection and Ecology, 19*(4), 1519-1525.

Vasović, D., Malenovic Nikolic, J., & Janackovic, G. (2016). Evaluation and Assessment Model for Environmental Management under the Seveso III, IPPC/IED and Water Framework Directive. *Journal of Environmental Protection and Ecology, 17*(1), 356-365.

Vásquez-Méndez, R., Ventura-Ramos, E., Oleschko, K., Hernández-Sandoval, L., Parrot, J. F., & Nearing, M. A. (2010). Soil erosion and runoff in different vegetation patches from semiarid Central Mexico. *Catena, 80*(3), 162–169.

von Storch, H. (2005). Models of Global and Regional Climate. Encyclopedia of Hydrological Sciences. doi:10.1002/0470848944.hsa035

Wackernagel, M., Schulz, N. B., Deumling, D., Linares, A. C., Jenkins, M., Kapos, V., Monfreda, C., Loh, J., Myers, N., Norgaard, R., & Randers, J. (2002). Tracking the ecological overshoot of the human economy. *Proceedings of the National Academy of Sciences of the United States of America, 99*(14), 9266–9271. doi:10.1073/pnas.142033699 PMID:12089326

Wagreich, M., & Draganits, E. (2017). Mining-related geochemical anomalies from 3500 to 2000 BP as potential stratigraphic markers for the base of the Anthropocene. *EGUGA, 5934.*

Waldbusser, G. G., Voigt, E. P., Bergschneider, H., Green, M. A., & Newel, R. I. E. (2013). Biocalcification in the Eastern Oyster (*Crassostrea virginica*) in Relation to Long-term Trends in Chesapeake Bay pH. *Estuaries and Coasts, 34*(2), 221–231. doi:10.100712237-010-9307-0

Walega, A., Cupak, A., Amatya, D. M., & Drozdzal, E. (2017). Comparison of direct outflow calculated by modified SCS-CN methods for mountainous and highland catchments in upper Vistula Basin, Poland and lowland catchment in South Carolina, USA. *Acta Scientiarum Polonorum. Formatio Circumiectus, 16*(1), 187–207.

Wall, D. H., & Six, J. (2015). *Give soils their due.* Academic Press.

Wallace, R. B., Baumann, H., Grear, J. S., Aller, R. C., & Gobler, C. J. (2014). Coastal ocean acidification: The other eutrophication problem. *Estuarine, Coastal and Shelf Science, 148*, 1–13. doi:10.1016/j.ecss.2014.05.027

Walling, D. E. (2011). Human impact on the sediment loads of Asian rivers. *Sediment Problems and Sediment Management in Asian River Basins. IAHS Publ, 349*, 37–51.

Waltgenbach, S., Riechelmann, D. F., Spötl, C., Jochum, K. P., Fohlmeister, J., Schröder-Ritzrau, A., & Scholz, D. (2021). Climate Variability in Central Europe during the Last 2500 Years Reconstructed from Four High-Resolution Multi-Proxy Speleothem Records. *Geosciences, 11*(4), 166. doi:10.3390/geosciences11040166

Wang, F. (1901). *Grundriss der Wildbachverbauung, S.* Hirzel.

Wang, F., Xu, X., Zou, B., Guo, Z., Li, Z., & Zhu, W. (2013). Biomass Accumulation and Carbon Sequestration in Four Different Aged Casuarina equisetifolia Coastal Shelterbelt Plantations in South China. *PLoS One, 8*(10), e77449. doi:10.1371/journal.pone.0077449 PMID:24143236

Wang, Q., & Batkhishig, O. (2014). Impact of Overgrazing on Semiarid Ecosystem Soil Properties: A Case Study of the Eastern Hovsgol Lake Area, Mongolia. *Journal of Ecosystem & Ecography, 04*(01). Advance online publication. doi:10.4172/2157-7625.1000140

Wang, T., Jin, H., Fan, Y., Obembe, O., & Li, D. (2021). Farmers' adoption and perceived benefits of diversified crop rotations in the margins of US Corn Belt. *Journal of Environmental Management, 293*, 112903. doi:10.1016/j.jenvman.2021.112903 PMID:34102504

Wang, X., Quine, T. A., Zhang, H., Tian, G., & Yuan, W. (2019). Redistribution of soil organic carbon induced by soil erosion in the nine river basins of China. *Journal of Geophysical Research. Biogeosciences, 124*(4), 1018–1031. doi:10.1029/2018JG004781

Wang, Y., Dong, Y., Su, Z., Mudd, S., Zheng, Q., Hu, G., & Yan, D. (2020). Spatial distribution of water and wind erosion and their influence on the soil quality at the agropastoral ecotone of North China. *International Soil and Water Conservation Research*, *8*(3), 253–265. Advance online publication. doi:10.1016/j.iswcr.2020.05.001

Wardle, D. A., Bardgett, R. D., Klironomos, J. N., Setälä, H., Van Der Putten, W. H., & Wall, D. H. (2004). Ecological linkages between aboveground and belowground biota. *Science*, *304*(5677), 1629–1633.

Warr, P., & Suphannachart, W. (2021). Agricultural Productivity Growth and Poverty Reduction: Evidence from Thailand. *Journal of Agricultural Economics*, *72*(2), 525–546. doi:10.1111/1477-9552.12412

Wasli, A. (2016). *Holocene cyclostratigraphy within the sedimentary record of exoreic (Boujmel) and endorheic (Mhabeul) saline systems: eustatic and climatic controls* (Master Thesis). Higher Institute of Applied Sciences and Technology of Gabes.

Water Management Basis of the Republic of Serbia. (2002). Institute for Water Management "Jaroslav Cerni", AD, Belgrade.

Water management strategy on the territory of the Republic of Serbia until 2034, Official Gazette of RS No. 3/2017 (2017), https://www.paragraf.rs/propisi/strategija-upravljanja-vodama-u-srbiji-do-2034.html

Waters, C. N., Zalasiewicz, J., Summerhayes, C., Fairchild, I. J., & Rose, N. J. … Edgeworth, M. (2018). Global Boundary Stratotype Section and Point (GSSP) for the Anthropocene Series: Where and how to look for potential candidates. *Earth-Science Reviews*, *178*, 379–429. doi:10.1016/j.earscirev.2017.12.016

Weart, S. R. (1997). The discovery of the risk of global warming. *Physics Today*, *50*(1), 34–40. doi:10.1063/1.881664

Weaver, D., & Summers, R. (2001). Soil factors influencing eutrophication. In A handbook for understanding and managing agricultural soils. Department of Agriculture and Food.

Weinan, C., & Fryrear, D. W. (1996). Grain-size distributions of wind-eroded material above a flat bare soil. *Phys. Geography*, *17*, 554–58.

Wei, X., Cai, S., Ni, P., & Zhan, W. (2020). Impacts of climate change and human activities on the water discharge and sediment load of the Pearl River, southern China. *Scientific Reports*, *10*(1), 1–11.

Weninger, T., Scheper, S., Lackóová, L., Kitzler, B., Gartner, K., King, N. W., Cornelis, W., & Michel, K. (2021). Ecosystem services of tree windbreaks in rural landscapes: A systematic review. *Environmental Research Letters*, *16*(10), 103002. Advance online publication. doi:10.1088/1748-9326/ac1d0d

Wezel, A., Casagrande, M., Celette, F., Vian, J. F., Ferrer, A., & Peigné, J. (2014). Agroecological practices for sustainable agriculture. A review. *Agronomy for Sustainable Development*, *34*(1), 1–20. doi:10.100713593-013-0180-7

WHO. (1993). *Report of WHOQOL Focus Group Work*. World Health Organization.

WHO. (2017). *Flooding: Managing Health Risks in the WHO European Region*. Copenhagen, Denmark: World Health Organisation Regional Office for Europe. Retrieved from https://apps.who.int/iris/handle/10665/329518

Wiles, G. C., Barclay, D. J., Calkin, P. E., & Lowell, T. V. (2008). Century to millennial-scale temperature variations for the last two thousand years indicated from glacial geologic records of Southern Alaska. *Global and Planetary Change*, *60*(1-2), 115–125. doi:10.1016/j.gloplacha.2006.07.036

Wilson, M., & Maliszewska-Kordybach, B. (Eds.). (2000). *Soil quality, sustainable agriculture and environmental security in Central and Eastern Europe* (Vol. 69). Springer Science & Business Media. doi:10.1007/978-94-011-4181-9

Wisner, B., Blaikie, P., Cannon, T., & Davis, I. (2004). *At Risk: Natural Hazards, Peoples' Vulnerabilities and Disasters*. Routledge.

Woodruff, N. P., & Siddoway, F. H. (1965). A wind erosion equation. *Soil Science Society of America Proceedings, 29*, 602–608.

World Commission on Environment and Development. (1987). *Our common future*. Oxford University Press.

Wu, D., & Olson, D. L. (2010). Enterprise risk management: Coping with model risk in a large bank. *The Journal of the Operational Research Society, 61*(2), 179–190. doi:10.1057/jors.2008.144

Wuepper, D., Borrelli, P., & Finger, R. (2020). Countries and the global rate of soil erosion. *Nature Sustainability, 3*(1), 51–55.

Wu, T., Zhang, P., Zhang, L., Wang, J., Yu, M., Zhou, X., & Wang, G. (2018). Relationships between shelter effects and optical porosity: A meta-analysis for tree windbreaks. *Agricultural and Forest Meteorology, 259*, 75–81. doi:10.1016/j.agrformet.2018.04.013

Wu, X., Wang, Z., Guo, S., Liao, W., Zeng, Z., & Chen, X. (2017). Scenario-based projections of future urban inundation within a coupled hydrodynamic model framework: A case study in Dongguan City, China. *Journal of Hydrology (Amsterdam), 547*, 428–442.

Xie, H., Wang, G. G., & Yu, M. (2018). Ecosystem multifunctionality is highly related to the shelterbelt structure and plant species diversity in mixed shelterbelts of eastern China. *Global Ecology and Conservation, 470*. Advance online publication. doi:10.1016/j.gecco.2018.e00470

Xie, J., Chen, H., Liao, Z., Gu, X., Zhu, D., & Zhang, J. (2017). An integrated assessment of urban flooding mitigation strategies for robust decision making. *Environmental Modelling & Software, 95*, 143–155.

Yang, F., & Lu, C. (2016). Assessing changes in wind erosion climatic erosivity in China's dryland region during 1961–2012. *Journal of Geographical Sciences, 26*, 1263-1276.

Yang, R., Wu, D., Li, Z., Yuan, Z., Niu, L., Zhang, H., Chen, J., & Zhou, A. (2021). Holocene-Anthropocene transition in northwestern Yunnan revealed by records of soil erosion and trace metal pollution from the sediments of Lake Jian, southwestern China. *Journal of Paleolimnology*, 1–12. doi:10.100710933-021-00193-x

Yavari, S., McIntyre, N., & Baumgartl, T. (2021). An Empirical Analysis of Sediment Export Dynamics from a Constructed Landform in the Wet Tropics. *Water (Basel), 13*(8), 1087.

Young, J. A., & Evans, R. A. (1986). Erosion and deposition of fine sediments from playas. *Journal of Arid Environments, 10*(2), 103–115. doi:10.1016/S0140-1963(18)31251-5

Zambrano, A. M., Calderón, X., Jaramillo, S., Zambrano, O. M., Esteve, M., & Palau, C. (2017). Public Warning Applications: Requirements and Examples. Wireless Public Safety Networks, 3, 39-66. doi:10.1016/B978-1-78548-053-9.50003-2

Zeljko, D., Sarač, M., & Mičivoda, H. (2019). Sistem prognoze i upozorenja na poplave u slivu reke Save [Flood Forecasting and Warning System in the Sava River Basin]. In *Proceedings, Second Water Congress*. Udruženje konsultanata inženjera BiH.

Zhang, H., Fan, J., Cao, W., Harris, W., Li, Y., Chi., W., & Wang, S. (2018). Response of wind erosion dynamics to climate change and human activity in Inner Mongolia, China during 1990 to 2015. *Sci Total Environ, 639*, 1038-1050.

Zhang, J. Q., Okada, N., & Tatano, H. (2006). Integrated natural disaster risk management: Comprehensive and integrated model and Chinese strategy choice. *Ziran Zaihai Xuebao, 15*(1), 115–128.

Zhang, Y., Li, Y., Jiang, L., Tian, C., Li, J., & Xiao, Z. (2011). Potential of perennial crop on environmental sustainability of agriculture. *Procedia Environmental Sciences*, *10*, 1141–1147. doi:10.1016/j.proenv.2011.09.182

Zhao, G., Mu, X., Wen, Z., Wang, F., & Gao, P. (2013). Soil erosion, conservation, and eco-environment changes in the Loess Plateau of China. *Land Degradation & Development*, *24*(5), 499–510.

Zheng, X., Zhu, J., & Xing, Z. (2016). Assessment of the effects of shelterbelts on crop yields at the regional scale in Northeast China. *Agricultural Systems*, *143*, 49–60. Advance online publication. doi:10.1016/j.agsy.2015.12.008

Zheng, Y. (2010). Association analysis on pro-environmental behaviors and environmental consciousness in main cities of East Asia. *Behaviormetrika*, *37*(1), 55–69. doi:10.2333/bhmk.37.55

Zhibao, D., Xunming, W., & Lianyou, L. (2000). Wind Erosion in Arind and Semiarid Chin: An Overview. *Journal of Soil and Water Conservation*, *55*(4), 439–444.

Zhou, Y., Zhang, B., Qin, W., Deng, Q., Luo, J., Liu, H., & Zhao, Y. (2021). Primary environmental factors controlling gully distribution at the local and regional scale: An example from Northeastern China. *International Soil and Water Conservation Research*, *9*(1), 58–68.

Zittis, G., Hadjinicolaou, P., Klangidou, M., Proestos, Y., & Lelieveld, J. (2019). A multi-model, multi-scenario, and multi-domain analysis of regional climate projections for the Mediterranean. *Regional Environmental Change*, *19*(8), 2621–2635.

Zobeck, T.M., Baddock, M., Van Pelt, R.S., Tatarko, J., & Acosta-Martinez, V. (2013). Soil property effects on wind erosion of organic soils. *Aeolian Research, 10*, 43–51.

Zschau, J., & Küppers, A. N. (2003). *Early warning systems for natural disaster reduction*. Springer-Verlag. doi:10.1007/978-3-642-55903-7

Zuccaro, G., Leone, M. F., & Martucci, C. (2020). Future research and innovation priorities in the field of natural hazards, disaster risk reduction, disaster risk management and climate change adaptation: A shared vision from the ESPREssO project. *International Journal of Disaster Risk Reduction*, *51*, 101783. doi:10.1016/j.ijdrr.2020.101783

About the Contributors

Slobodan Milutinović, PhD, is a Full Professor at Faculty of Occupational Safety, University of Niš. Principally trained as an architect, he earned his PhD in Environmental Engineering at Faculty of Occupational Safety, University of Niš. He is an internationally recognized expert in environment and climate change with 25+ years of professional experience. Since 1994 he has been lecturing at University of Niš (from 2004 as a tenured professor). He was also lecturing at the University of Pennsylvania and University of North Carolina (USA), University of Padua (Italy), Central European University (Hungary), and Technical University of Darmstadt (Germany) as a Fulbright and Erasmus visiting professor. His scientific interest covers Sustainable Development, Environmental Politics and Policies, Urbanization, and Urban Sustainability. His publication record covers 14 peer-reviewed books and book chapters, as well as more than 100 scientific papers in leading scientific journals.

Snežana Živković is a Full Professor at Faculty of Occupational Safety, University of Niš. She received a PhD degree in Occupational Safety from Faculty of Occupational Safety, University of Niš. She has published over 250 papers in international monographs and leading scientific journals, including Renewable and Sustainable Energy Reviews, Environmental Science and Pollution Research and International Journal of Injury Control and Safety Promotion. Professor Živković has been the author of 4 international monographs, 3 national monographs and 5 handbooks. Her research interests include occupational safety, organizational behavior, management and human resources, emergency management, protection of living and working environment. She is an executive editor of the scientific journal "Economics of Sustainable Development". She has been the member of the organizing committee of numerous scientific conferences. In her rich scientific career, she worked as a reviewer for many scientific journals, including: Themes, Economics, International Journal of Injury Control and Safety Promotion, Safety Engineering, Economics of Sustainable Development, etc.

* * *

Muhamed Bajric defended his doctorate at the University of Sarajevo - Faculty of Forestry on "Development of erosion processes on skid trails of different longitudinal slope". The narrow field of interest is the development of erosion processes and rehabilitation of secondary networks of the forest traffic infrastructure. He has participated in several projects of solving that very problem. In the teaching process, he is involved in issues of arranging torrents and rehabilitation of the damaged forest terrain.

Atila Bezdan is an associate professor and Director of the Department of Water Management at the Faculty of Agriculture, University of Novi Sad, Serbia. His research focuses on hydrologic and hydraulic modelling of drainage and irrigation systems, application of GIS and Remote Sensing in Agricultural Water Management, especially on risk assessment of occurrence of extreme hydro-meteorological events like droughts and floods.

Boško Blagojević is Associate Professor at Department of Water Management, Faculty of Agriculture, University of Novi Sad. His research interests includes various aspects of multi-criteria decision analysis (MCDA) in individual and group settings with a main focus on application of MCDA methods in agriculture, water resources management and forest management.

Ivan Blinkov has more than 34 years of experience in university education as well as in scientific and applied projects, national strategies, action plans in the fields of: land, water and risk/hazard management i.e. erosion and torrent control, watershed management, torrent hazard/risk management, land degradation and desertification… Participant in > 80 project, Author/coauthor of 100 published papers, 4 books, 6 scripts, chapters in 5 monographs, 5 printed projects, 4 brochures, more than 300 presentations on various events. Member of various international and national professional bodies. Reviewer of scientific papers and projects, textbooks; Reviewer for academic assignment in the country and Serbia; Member of various national and international working groups; Member of scientific committee of international scientific conferences; Mentor of 3 PhD, 4 MSc; Member of commission for reviewing PhD Thesis in the country and Serbia; Member of the expert council of the Ministry of environment and physical planning; Member of the Scientific council of Centre for Environment and materials, within MANU (Macedonian Academy of Science and Art) (2014 -), Member of the Council of WASWAC (World Association for Soil and Water Conservation) Awarded as the BEST SCIENTIST in the field of Bio-technical Sciences – Ss. Cyril and Methodius University, 9.11.2018.

Milica Caković is a junior researcher at the Faculty of Forestry University of Belgrade (Department of Ecological Engineering in Soil and Water Resources Protection). She was finishing her Master's degree in the Faculty of Forestry in 2018; she enrolled in Ph.D. studies in the same year. She is working as a researcher within the project "Studying climate change and its influence on the environment: impacts, adaptation and mitigation" financed by the Ministry of Education and Science of the Republic of Serbia as well as a non-key expert on the Second Institutional Development and Agriculture Strengthening Project (MIDAS 2) (title of consulting services: Development of the Atlas of Honey Plant with the Apiary Cadastre). Her Ph.D. thesis theme is to analyze diffuse water pollution, phytoremediation, heavy metals in soil, and ecological revitalization. She got award for the best student in generation on bachelor studies.

Swathi Dash is pursuing her doctoral degree focusing on problems associated with surface water through sustainable solutions.

Slađan Đikić is a Director of Environmental Protection Department at Leskovac City Administration. He graduated from the Faculty of Occupational Safety, University of Niš on the module Environmental Protection. During his professional carrier, he has successfully finished numerous courses regarding Environmental and Fire Protection. Mr. Đikić has been a project coordinator of numerous projects at the local level in the City of Leskovac. He is a member of the Headquarter for the Emergency Situations in the City of Leskovac and the Eco-commissioner Network within the Standing Conference of Towns and Municipalities (National Association of Local Authorities in Serbia).

Nada Dragović is a full professor and head of the Department of Ecological Engineering in Soil and Water Resources Protection. She was educated at the Faculty of Forestry in the field of erosion and torrent control. Prof. Dragović has published more than 170 bibliographic units so far, alone or in cooperation with other authors, in the following research fields: soil erosion and torrent control, the application of modern planning methods in the organization of works on erosion and torrent control, project management in the field of torrent control, watershed management, research in erosion processes and sediment transport.

Bhavya Dwarapureddi specializes in addressing problems associated with water and wastewater providing sustainable solutions. With nearly 3 years of teaching and 6 years of research experience the author is specialized in environmental science and allied areas.

Elhoucine Essefi obtained his PhD form National engineering School of Sfax titled 'Wet Aeolian Sedimentology and Sequence Stratigraphy within the Terrestrial Analogues in Eastern Tunisia: Implications for Wet Aeolian Sedimentology and Sequence Stratigraphy on Mars', in 2013. In 2011, he obtained License in Linguistics and Anglophone Civilization from the Faculty of Arts and Humanities of Sfax. In 2009, he obtained the Master degree from the Faculty of Sciences of Sfax titled Multidisciplinary study of Sidi El Hani Saline Environment: the Geological History and the Climatic Variability. In 2003, he obtained the 'Maitrise' degree in Earth Sciences from the Faculty of Sciences of Sfax.

Tatjana D. Golubović was born in Niš, Serbia. She received the diploma and the Ph.D. degree in chemistry from the University of Niš, Faculty of Science and Mathematics. Her main areas of research include chemistry, environmental chemistry, soil protection, eco-toxicology, etc. She is currently working as a full professor at the University of Niš, Faculty of Occupational Safety in Niš.

Soumaya Hajji is an Engineer in Geology from National Engineering School in Sfax (ENIS), Tunisia in 2000. She obtained her PhD in Modeling in Hydrogeology in 2015. She obtained her HDR in Environnemntal Hydrogeology and geo-Informatic in 2021. She is a researcher at geochemistry and hydrogeology, modelling and geostatistics at 'Water-Energy-Environment' Research Laboratory (LR3E) in Engineering School Sfax, Tunisia. Her research activities deal with geochemistry, modeling, analysis, optimisation, simulation of groundwater resources, groundwater quality and. She is an Associate Professor in Geology Departement in Faculty of Sciences of Sfax, Tunisia.

Johannes Hübl is appointed to the field of Natural Hazards and Risk Management and heads the Institute of Mountain Risk Engineering (BOKU-IAN). He holds a PhD from BOKU-University (1995) and a Habilitation in the field of Torrent Control (2005). His key qualifications encompass monitoring of torrential activities, documentation of natural hazards, hazard analysis and design of torrent mitigation works.

Ivana Janjić works as a junior research assistant at Innovation Centre, University of Niš and she is also a PhD student, module Business management at Faculty of Economics, University of Niš, Serbia.

She was a scholarship holder at basic academic studies, master academic studies and doctoral academic studies, supported by Ministry of Education, Science and Technological Development of the Republic of Serbia. Ivana Janjić, at the present, is in the process of research and writing a doctoral dissertation in the area of intellectual capital management. Her fields of interest are focused on: R&D management, R&D investments, intellectual capital management, intangible asset management, business performance management. She has authored and published numerous scientific and research articles in scientific journals and book collections of papers, and she has participated at conferences in the country and abroad.

Milica Jovanović Vujatović is a research assistant at Innovation Centre, University of Niš, Serbia and a PhD student at Faculty of Economics, University of Niš, Serbia. She was a scholarship holder of the Ministry of Education, Science and Technological Development of the Republic of Serbia at all levels of study. Milica Jovanović is currently in the process of writing a doctoral dissertation in the field of intellectual capital management. Her areas of interest are based on the following: intellectual capital management, business performance management, competitiveness, intangible asset valuation, intangible asset management, brand valuation, brand management. She has published over 20 articles and has participated in a number of domestic and international conferences. Milica Jovanović is a member of the Society of Economists "Ekonomika" and Serbian Marketing Association "SeMA". https://orcid.org/0000-0002-6410-0938.

Sara Lukić is Associate Professor at University of Belgrade Faculty of Forestry - Department of Ecological engineering for soil and water resources protection - Chair of Amelioration. Her research interests are focused on the forest restoration in various environments and emerging ecosystem services such as control of land degradation and desertification, carbon sequestration and climate change mitigation, maintenance of biodiversity. She deals with the study of the provisioning and regulating ecosystem services of the forests established on the land affected by various forms of degradation, including extremely degraded lands such as bare land and sites of former mining activities. She also deals with the study of protective forest belts (shelterbelts and windbreaks) and the ecosystem services they provide in agroforestry systems. The aim of these studies is to identify the potentials of water and wind erosion control measures and techniques for sustainable land management. She has participated as a researcher in ten scientific projects and a number of professional projects. She published more than fifty scientific articles in the field of forest restoration, afforestation, protective forest belts/shelterbelts, land reclamation and carbon sequestration. Also, she is author of the textbook "Protective Forest Belts".

Ivana Marjanović is a PhD candidate and a teaching assistant at the Faculty of Economics, University of Niš. Her area of interest is the application of quantitative methods in economics, with the main focus on the application of operational research and decision theory methods and models. She has noteworthy experience in conducting quantitative and qualitative research, in the area of urban studies, local development, and public procurements. In addition, she has a high level of theoretical knowledge about statistics as well as practical knowledge and skills related to the methodology of survey implementation, data collection, data processing, and data analyses.

Milan Marković, Ph.D. in Economics, is a research associate at the Innovation Center of the University of Niš. During studies, he received recognition by the Faculty of Economics, the University of Niš and the City of Niš. He was the recipient of scholarship from the Young Talent Fund of the Republic of

Serbia. From 2014 to 2019 he was engaged in the project "Improving the Competitiveness of the Public and Private Sector by Networking Competences in the Process of European Integration of Serbia" funded by the Ministry of Education, Science and Technological Development of Serbia. Milan Marković is a member of the editorial board of the Journal "Economics of Sustainable Development", a member of the Association of Economists and Managers of the Balkans and the Balkan Scientific Association of Agricultural Economists, and a reviewer of many scientific publications. His key areas of interest are national economics, economics of agriculture and international finance.

Sandra Milanović is a junior research assistant at the Innovation Center of the University of Niš and a doctoral student at the Faculty of Economics of the University of Niš. She has volunteered as a junior teaching assistant for as many as nine subjects at the Faculty of Economics in Niš, the Department of Business Management. During spring semester in 2018 she was enrolled in Erasmus+ program for student mobility at Dimitar A. Tsenov Academy of Economics in Svishtov, Bulgaria. Starting from the 2014 until 2018, she was a scholar of the Ministry of Education, Science and Technological Development. She is the author of many scientific papers published in domestic and international journals and presented at conferences. Key area of her interest is business management and human resource management.

Raman Nautiyal is a Statistician, Retd. Scientist, Indian Council of Forestry Research and Education, India.

Vijender Panwar is a Scientist, Forest Ecology and Climate Change, Forest Research Institute, Dehradun.

Paolo Porto is currently Professor in Agricultural Hydraulics and Torrent control, at the University Mediterranea of Reggio Calabria (Italy). He obtained the National Scientific Qualification of full professor of Agricultural, Forest, and Biosystems Engineering in January 2014. He completed his PhD at the University of Bari (southern Italy) in 1994 where he worked in the area of catchment hydrology and torrent control. While in Exeter Paolo Porto was awarded a 2-year European Marie Curie fellowship by which he continued his research activity in Exeter in the field of soil erosion and radionuclide tracers supervised by Prof. Walling. During his activity in Reggio Calabria, Paolo Porto was nominated Erasmus Coordinator of the Agricultural Faculty promoting new teaching and research links with other European countries. Since 2015 to date, Paolo Porto is the Coordinator of the Bachelor Degree in Environmental and Forest Sciences at the University Mediterranea of Reggio Calabria. His research activity (H-index=22) in soil erosion, radionuclides, hydrology, torrent control is well documented in many papers in international and national journals (for the ISI papers see https://www.scopus.com/authid/detail.uri?origin=resultslist&authorId=7003774123&zone=). He has served as a Referee and Associated editor for many international journals and he is a member of the editorial board of Pedosphere (Elsevier). Paolo Porto has been involved with the activities of the International Commission on Continental Erosion (ICCE) since 2004 and was elected Secretary for the terms 2011-2015 and 2015-2019.

Ratko Ristić, a corresponding member of the Academy of Engineering Sciences of Serbia (AINS), is a full professor and Dean of the University of Belgrade- Faculty of Forestry. He was born on 26 July 1961. In 1985, he graduated from the Department of Water Resources Management in Erosive Regions of the Faculty of Forestry, and defended his Master of Science Thesis in 1993. He defended his doctoral dissertation in 2000. Since March 1st 1990, he has been employed at the Faculty of Forestry as an assistant trainee. Professor in 2001, Associate Professor in 2006 and a full-time professor in 2011. He worked in the Water Management Organization of Belgrade, as a designer-collaborator on several conceptual and main projects for the regulation of riverbeds of torrential flows (1986-1987). During 1989, he worked at the Belgrade Research Station. He is a member of IAHS (International Association of Hydrological Sciences) and WASWC (World Association for Soil and Water Conservation), IKS (Serbian Chamber of Engineers, Licenses No. 375791204 and 373791104) and APPS (Association of Spatial Planners of Serbia). He is engaged at all three study levels at the University of Belgrade - Faculty of Forestry. He introduced new subjects: "Spatial arrangement of erosive regions" and "Application of GIS in the protection of soil and water resources". His textbook "Hydrology of torrential flows" was published in 2011. He is the mentor of 7 defended doctoral dissertations, 15 Master of Science theses. He is currently the mentor of 5 doctoral dissertations being prepared. So far, he has published over 130 bibliographic units dealing with erosion and torrent control, hydrology, sustainable use of watersheds and natural disasters prevention. The Ratko Ristić participated in the implementation of 61 main projects for economic purposes, both as a responsible designer or associate designer, and 70 other projects (conceptual projects, feasibility studies, environmental impact assessment studies, elaborates). These projects resulted in the construction of a total of 56 facilities, of which the candidate managed 17. In 2017, he was appointed representative of Serbia in the UNCCD (United Nation Convention Combat Desertification) Committee on Science and Technology. He initiated and realized cooperation between the University of Belgrade and the Universities of Waterloo (Canada, 2011) and Volgateh (Russia, 2017). He is married and father of two children.

Mohamed Ali Tagorti is a Professor in the Higher Institute of Applied Sciences and Technology of Sousse ISSAT Sousse (Tunisia). He is a Director of the Department of Biological Sciences at the Higher Institute of Biotechnology of Monastir. Professor Tagorti is a Member of the foundation committee and treasurer of the Tunisian Association for the Valorization of Bioressources at the Higher Institute of Biotechnology of Monastir; Member of the founding committee of the Promotion and Culture Association in Sousse Riadh Town, Member of the Foundation Committee and Vice-President of the "ELFERDAWS" Charity Association in Sousse Riadh Town. He is also a member of numerous Scientific and Organizing Committees of International Conferences.

Vesela Tanaskovic Gassner is a speaker, delegate and panelist at many national and international scientific events, conferences, and debates. Recognized as a thought leader, researcher and lecturer in a multidisciplinary framework of her research field, using bottom lake sediments from dams to rebuild soils in afforestation and reforestation projects.

Saritha Vara has 17 years of teaching and research experience. Specializations of the author include environmental planning and sustainable development, addressing challenges of potable water, and remote sensing technology. The author has published 90 research papers and is also active in addressing issues of public and occupational health and corrosion studies.

Dejan Vasović received the B.S. and M.S. degrees in Environmental Protection Engineering from University of Niš, Faculty of Occupational Safety, Niš, Serbia, in 2006 and 2011, respectively, and the Ph.D. degree in the same field from University of Niš, Faculty of Occupational Safety, Niš, Serbia, in

2016. Since 2017, he works as an Assistant Professor at the Faculty of Occupational Safety, Environmental Protection Department. Currently, he is the Vice - Chair of the Department of Environmental and Occupational Quality Management. He has authored three international book chapters and more than twenty papers in journals with impact factor. His research interests include environmental protection, water resources management, system standards, emergency management, natural hazards.

Tijana Vulevic is an Assistant Professor at the Faculty of Forestry, University of Belgrade. She is teaching in the group of subjects related to the organisation of soil erosion and torrent control works, project management and decision support in soil and water resource management. She has published a lot of papers related to soil erosion and torrent control concerning organisation, legislation, financing, as well as decision making and risk analysis.

Index

E

F

G

H

I

L

M

N

O

P

Q

R

S

T

V

W

www.ingramcontent.com/pod-product-compliance
Lightning Source LLC
LaVergne TN
LVHW062138300926
849161LV00010B/2568